Polysaccharides

Polysaccharides offer unique and valuable functional properties, persisting in technological importance and poised to grow more critical due to sustainability demands and emerging applications in medical and life sciences. This book presents comprehensive information about carbohydrate polymers, providing readers with an enhanced appreciation of carbohydrate structure and function, a new enzyme library, and extraction strategies that will help to advance a number of exciting domains of research, including genomics, proteomics, chemical synthesis, materials science, and engineering.

Key Features

- Details the source, production, structures, properties, and current and potential applications of polysaccharides.
- Discusses general strategies of isolation, separation, and characterization of polysaccharides.
- Describes botanical, algal, animal, and microbial sources of polysaccharides.
- Demonstrates the importance of carbohydrates in new lead generation.
- Highlights the range of possibilities for polysaccharides to make real-world impact.

Bhasha Sharma, Assistant Professor, Department of Chemistry, Shivaji College, University of Delhi, India.

Md Enamul Hoque, Professor, Department of Biomedical Engineering, Military Institute of Science and Technology (MIST), Dhaka, Bangladesh.

Polysaccharides
Advanced Polymeric Materials

Edited by
Bhasha Sharma and Md Enamul Hoque

CRC Press
Taylor & Francis Group
Boca Raton London New York

CRC Press is an imprint of the
Taylor & Francis Group, an **informa** business

Designed cover image: © Shutterstock

First edition published 2023
by CRC Press
2385 NW Executive Center Drive, Suite 320, Boca Raton FL 33431

and by CRC Press
4 Park Square, Milton Park, Abingdon, Oxon, OX14 4RN

CRC Press is an imprint of Taylor & Francis Group, LLC

ISBN: 9781032207506 (hbk)
ISBN: 9781032207513 (pbk)
ISBN: 9781003265054 (ebk)

DOI: 10.1201/9781003265054

Typeset in Times
by codeMantra

Contents

Chapter 3   Synthetic Polysaccharides: Adored, Deplored and Ubiquitous...............48

*Rois Uddin Mahmud, Md. Raijul Islam, Md. Abdur Rouf, Md. Rubel Alam,
Asif Mahmud Rayhan, and Md Enamul Hoque*

Chapter 16 Polysaccharide-Based Fluorescent Materials for Sensing and
Security Applications 287

Akhil Padmakumar, Drishya Elizebath, Jith C. Janardhanan,
Rakesh K. Mishra, and Vakayil K. Praveen

Preface

The smart properties of polysaccharides were used by mankind long before the term "smart" was ever applied to materials. During the last half of the century, the allure of polysaccharides was eclipsed by large investments in chemical synthesis and molecular biology that yielded remarkable advances in the ability to engineer functional materials from synthetic polymers, proteins, and nucleic acids. Yet polysaccharides persist in technological importance, and they promise to become more important due to the recent trend toward sustainability and the emergence of applications in medical and life sciences. Importantly, polysaccharides offer unique and valuable functional properties. The work presents comprehensive information about carbohydrate polymers. The term "carbohydrate polymers" describe ubiquity in the nature of molecular systems containing carbohydrates. The property of conformational restriction in polysaccharides makes them candidates for being the initial self-ordering molecules of prebiotic evolution. This book details the novel synthetic methods involving carbohydrates, oligosaccharides, and glycoconjugates; the use of chemical methods to address aspects of glycobiology; physicochemical studies involving carbohydrates; and the chemistry and biochemistry of carbohydrate polymers. Several chapters are devoted to the study and exploitation of polysaccharides, which have current or potential applications in areas such as bioenergy, bioplastics, biomaterials, biorefining, chemistry, drug delivery, food, health, nanotechnology, packaging, paper, pharmaceuticals, medicine, oil recovery, textiles, tissue engineering and wood, and other aspects of glycoscience. This book details polysaccharides covering their source, production, structures, properties, and current and potential application. It includes a systematic discussion of the general strategies of isolation, separation, and characterization of polysaccharides. Subsequent chapters are devoted to polysaccharides obtained from various sources, including botanical, algal, animal and microbial, and their applications in respective fields. This book begins with an overview of the almost universal occurrence of natural macromolecules in living organisms where they form a variety of functions. The text then examines the synthesis of polysaccharides. Importantly, polysaccharides offer unique and valuable functional properties. We anticipate that new experimental and theoretical tools will emerge to provide the necessary understanding of the structure–property–function relations that will enable polysaccharide-smartness to be understood and controlled. This enhanced understanding will complement the intrinsic characteristics of polysaccharides as sustainable, environmentally friendly, and biologically compatible materials. Thus, we envision polysaccharides will continue to be well-positioned for applications in foods, cosmetics, and medicine.

About the Editors

Dr Md Enamul Hoque is a Professor in the Department of Biomedical Engineering at the Military Institute of Science and Technology (MIST), Dhaka, Bangladesh. Before joining MIST, he served in several leading positions in some other global universities such as Head of the Department of Biomedical Engineering at King Faisal University (KFU), Saudi Arabia; Founding Head of Bioengineering Division, University of Nottingham Malaysia Campus (UNMC), and so on. He received his PhD in 2007 from the National University of Singapore (NUS), Singapore (1st in Asia and 8th in the World in QS World University Rankings 2024) with a globally prestigious scholarship from the Singapore Government. He also obtained his PGCHE (Post Graduate Certificate in Higher Education) from the University of Nottingham, UK (18th in UK and 100th in QS World University Rankings 2024) in 2015. He is a Chartered Engineer (CEng) certified by the Engineering Council, UK; Fellow of the Institute of Mechanical Engineering (FIMechE), UK; Fellow of Higher Education Academy (FHEA), UK and Member, World Academy of Science, Engineering and Technology. He received the Highest Publishing Scientist Award at MIST on 14 March 2023; Outstanding Nano-scientist Award in the International Workshop on Recent Advances in Nanotechnology and Applications (RANA-2018) held on 7–8 September 2018, AMET University, Chennai, India. His major areas of research interests include (but are not limited to) Biomedical Engineering, Biomaterials, Biocomposites, Biopolymers, Nanomaterials, Nanotechnology, Biomedical Implants, Rehabilitation Engineering, Rapid Prototyping Technology, 3D Printing, Stem Cells and Tissue Engineering. Recently, he has been ranked the Best Biomedical Engineering Scientist in Bangladesh; https://www.adscientificindex.com/scientist/md-enamul-hoque/424943, and also, Top 2% Scientists in the World; https://elsevier.digitalcommonsdata.com/datasets/btchxktzyw/6.

Dr. Bhasha Sharma is currently working as Assistant Professor in Department of Chemistry, Shivaji College, University of Delhi, India. She received her BSc (2011) in Polymer Sciences from NSIT. Dr. Sharma completed her Ph.D. in Chemistry in 2019 under the guidance of Prof Purnima Jain from the University of Delhi. She has more than 7 years of teaching experience. She has published more than 40 research publications in reputed international journals. Her recently edited book titled "Graphene-based biopolymer nanocomposites" has been published in Springer Nature. Her authored book "3D Printing Technology for Sustainable Polymers" and edited books "Biodegradability of Conventional Plastics: Opportunities, Challenges, and Misconceptions", and "Sustainable Packaging: Gaps, Challenges, and Opportunities," has been accepted in Wiley, Elsevier, and Taylor Francis, respectively. She has received **Commendable Research Award** from **Netaji Subhas University of Technology**, Delhi, India in 2023. Her research interests revolve around sustainable polymers for packaging applications, environmentally benign approaches for biodegradation of plastic wastes, fabrication of bionanocomposites, and finding strategies to ameliorate the electrochemical activity of biopolymers.

Contributors

Qazi Adfar
Materials Chemistry & Engineering Research
  Laboratory, Department of Chemistry,
  National Institute of Technology, Srinagar,
  India
and
Vishwa Bharti Degree College, Srinagar, India

Aiswarya P. R.
School of Chemical Science
Mahatma Gandhi University
Kottayam, India

Raheela Akhter
Materials Chemistry & Engineering
Laboratory, Department of Chemistry
National Institute of Technology
Srinagar, India

Md. Rubel Alam
Department of Knitwear Engineering
BGMEA University of Fashion & Technology
  (BUFT)
Dhaka, Bangladesh

Anila Antony
Central Institute of Petrochemicals Engineering
  & Technology, CIPET KOCHI
Kochi, India

Mohammad Aslam
Biofuels Research Laboratory, Department of
  Chemistry
National Institute of Technology
Srinagar, India

Humira Assad
Department of Chemistry, Faculty of
  Technology and Science
Lovely Professional University
Phagwara, India

Biplob Kumar Biswas
Department of Chemical Engineering
Jashore University of Science and Technology
Jashore, Bangladesh

Rachid Bouhfid
Composites and Nanocomposites Center
Moroccan Foundation for Advanced Science,
  Innovation and Research
Rabat, Morocco
and
Mohammed VI Polytechnic University
Ben Guerir, Morocco

Tanvir Mahady Dip
Bangladesh University of Textiles
Dhaka, Bangladesh

Drishya Elizebath
Photosciences and Photonics Section, Chemical
  Sciences and Technology Division
CSIR-National Institute for Interdisciplinary
  Science and Technology (CSIR-NIIST)
Thiruvananthapuram, India
and
Academy of Scientific and Innovative Research
  (AcSIR)
Ghaziabad, India

Hamid Essabir
Composites and Nanocomposites Center
Moroccan Foundation for Advanced Science,
  Innovation and Research
Rabat, Morocco
and
Mohammed VI Polytechnic University
Ben Guerir, Morocco

Richika Ganjoo
Department of Chemistry, Faculty of
  Technology and Science
Lovely Professional University
Phagwara, India

Seema Garg
Department of Chemistry, Amity Institute of
  Applied Sciences
Amity University
Noida, India

Surendra G. Gattani
School of Pharmacy
Swami Ramanand Teerth Marathwada
    University
Nanded, India

Abdellah Halloub
Composites and Nanocomposites Center
Moroccan Foundation for Advanced Science,
    Innovation and Research
Rabat, Morocco
and
Mohammed VI Polytechnic University
Ben Guerir, Morocco

Liqaa Hamid
Graz University of Technology (TU Graz),
    Austria

Md Enamul Hoque
Department of Biomedical Engineering
Military Institute of Science and Technology
    (MIST)
Dhaka, Bangladesh

Shokat Hussain
Materials Chemistry & Engineering Research
    Laboratory, Department of Chemistry,
    National Institute of Technology
Srinagar, India

Md. Raijul Islam
Department of Textile Engineering
BGMEA University of Fashion & Technology
    (BUFT)
Dhaka, Bangladesh

Jith C. Janardhanan
Photosciences and Photonics Section, Chemical
    Sciences and Technology Division
CSIR-National Institute for Interdisciplinary
    Science and Technology (CSIR-NIIST)
Thiruvananthapuram, India

Neetha John
Central Institute of Petrochemicals Engineering
    & Technology, CIPET Kochi
Kochi, India

Jeffy Joji
Central Institute of Petrochemicals Engineering
    & Technology, CIPET KOCHI
Kochi, India

Swetha K. S.
Central Institute of Petrochemicals Engineering
    & Technology, CIPET KOCHI
Kochi, India

Md Humayun Kabir
Institute of Pure and Applied Sciences
Marmara University
Istanbul, Turkey

M. Azizur R. Khan
Department of Chemistry
Jashore University of Science and Technology
Jashore, Bangladesh

Nishat Khan
Department of Chemistry, Amity Institute of
    Applied Sciences
Amity University
Noida, India

Tabassum Khan
SVKM's Dr. Bhanuben Nanavati College of
    Pharmacy
Mithibai College Campus
Mumbai, India

Most. Afroza Khatun
Department of Chemical Engineering
Jashore University of Science and Technology
Jashore, Bangladesh

Ashish Kumar
Nalanda College of Engineering, Department
    of Science and Technology and Technical
    Education
Bihar Engineering University
Government of Bihar
Bihar, India

Rois Uddin Mahmud
Department of Textile Engineering
BGMEA University of Fashion & Technology
    (BUFT)
Dhaka, Bangladesh

Shrikant S. Maktedar
Materials Chemistry & Engineering
   Laboratory, Department of Chemistry
National Institute of Technology
Srinagar, India

Raji Marya
Composites and Nanocomposites Center
   Moroccan Foundation for Advanced
   Science,
Innovation and Research
Rabat, Morocco
and
Mohammed VI Polytechnic University
Ben Guerir, Morocco

Rakesh K. Mishra
Department of Chemistry
National Institute of Technology, Uttarakhand
   (NITUK)
Pauri (Garhwal), India

Subrata Mondal
Department of Mechanical Engineering
National Institute of Technical Teachers'
   Training and Research (NITTTR) Kolkata
Kolkata, India

Nizam P.A.
School of Chemical Science
Mahatma Gandhi University
Kottayam, India

Akhil Padmakumar
Photosciences and Photonics Section, Chemical
   Sciences and Technology Division
CSIR-National Institute for Interdisciplinary
   Science and Technology (CSIR-NIIST)
Thiruvananthapuram, India
and
Academy of Scientific and Innovative Research
   (AcSIR)
Ghaziabad, India

Shahanaz Parvin
Department of Chemical Engineering
Jashore University of Science and Technology
Jashore, Bangladesh

Vakayil K. Praveen
Photosciences and Photonics Section, Chemical
   Sciences and Technology Division
CSIR-National Institute for Interdisciplinary
   Science and Technology (CSIR-NIIST)
Thiruvananthapuram, India
and
Academy of Scientific and Innovative Research
   (AcSIR)
Ghaziabad, India

Abou el kacem Qaiss
Composites and Nanocomposites Center
Moroccan Foundation for Advanced Science,
   Innovation and Research
Rabat, Morocco
and
Mohammed VI Polytechnic University
Ben Guerir, Morocco

Md. Wasikur Rahman
Department of Chemical Engineering
Jashore University of Science and Technology
Jashore, Bangladesh

Md. Sohel Rana
Department of Chemical Engineering
Jashore University of Science and Technology
Jashore, Bangladesh

Asif Mahmud Rayhan
Department of Mechanical and Aerospace
   Engineering
Western Michigan University
Michigan, USA

Md. Abdur Rouf
Department of Knitwear Engineering
BGMEA University of Fashion & Technology
   (BUFT)
Dhaka, Bangladesh

Irene Samy
Smart Engineering Systems Research Center
   (SESC)
Nile University
Sheikh Zayed City, Egypt

Nikita Sanap
SVKM's Dr. Bhanuben Nanavati College of
    Pharmacy
Mithibai College Campus
Mumbai, India

Bhasha Sharma
Department of Chemistry, Shivaji College
University of Delhi
Delhi, India

Shveta Sharma
Department of Chemistry, Faculty of
    Technology and Science
Lovely Professional University
Phagwara, India

Sk Md Ali Zaker Shawon
Department of Chemical and Biomolecular
    Engineering
Vanderbilt University
Nashville, Tennessee, United States

Nawrin Rahman Shefa
Department of Chemical Engineering
Jashore University of Science and Technology
Jashore, Bangladesh

Enock Siankwilimba
Graduate School of Business
University of Zambia
Lusaka, Zambia

Abhinay Thakur
Department of Chemistry, Faculty of
    Technology and Science
Lovely Professional University
Phagwara, India

Sabu Thomas
International and Inter-University Centre for
    Nanoscience and Nanotechnology
Mahatma Gandhi University
Kottayam, India

Deepika Tiwari
SVKM's Dr. Bhanuben Nanavati College of
    Pharmacy
Mithibai College Campus
Mumbai, India

Shradha S. Tiwari
Department of Pharmaceutics
Annasaheb Dange College of B. Pharmacy
Ashta, India

Muhammet Uzun
Textile Engineering Department, Faculty of
    Technology, Marmara University Istanbul,
    Türkiye
and
Center of Nanotechnology and Biomaterials
    Applied and Research Marmara University
    Istanbul, Türkiye

1 Polysaccharide-Based Polymeric Gels

Structure, Properties, and Applications

Qazi Adfar
National Institute of Technology
Vishwa Bharti Degree College

Raheela Akhter, Shokat Hussain,
Mohammad Aslam, and Shrikant S Maktedar
National Institute of Technology

1.1 INTRODUCTION

Polymers are a class of naturally occurring or synthetically made substances, which are composed of materials made of repeating chains of molecules or chemical units called monomers. These substances, which contain a large number of structural units, are joined by the same type of linkage, giving rise to macromolecules. The macromolecules are very large molecules with molecular weights ranging from a few thousand to as high as millions, grams/mole. These materials have unique properties depending on the type of molecules being bonded and how they are bonded. Naturally occurring polymers have been used by man since prehistoric times. They have been modified and processed empirically over many centuries for various applications such as textiles for clothing, industrial fibers, adherences, food products, and medicines. Natural polymers may be of inorganic and organic origin. Inorganic polymers include diamond and graphite. Both are composed of carbon [1]. Furthermore, an artificial carbon allotrope notably known as graphene have confirmed multifunctional applications due to the surface functionalization of graphene oxide [2]. In line with multifaceted applications, P. Malik et. al. 2020 has successfully demonstrated the modulation of graphene oxide towards antioxidant behavior [3]. Organic polymers also make up many materials in living organisms. They play a crucial role in living things, providing structural materials and participating in vital life processes; e.g., solid parts of all plants are made up of natural polymers. These include cellulose, lignin, and various resins. More often, the term polymer is used to describe plastics, which are synthetic polymers. Vulcanized rubber is also a type of synthetic polymer. Polymers are the basis of many minerals and man-made materials. These materials now constitute one of the most successful and useful classes of materials that possesses a broad range of physical properties. Besides, synthetic polymers play a crucial role in most of the consumer products in the present-day world. They are present everywhere. Just look around, your plastic bottles, the silicon rubber tips on your phones earbuds. The nylon and polyester in your jackets and sneakers. The rubber in the types of your family car.

Polymer chemists have designed and synthesized polymers that vary in hardness, flexibility, softening temperature, solubility in water, biodegradability, and biocompatibility. They have produced polymeric materials that are as strong as steel yet lighter and more resistant to corrosion. Oil, natural gas, and water pipelines are now routinely constructed of plastic pipes. In recent years, automobiles have increased their use of plastic components to build lighter vehicles that consume less fuel. Other industries such as those involved in the manufacture of textiles, rubber, and paper packaging materials are built upon the polymer chemistry [4–6]. Polymeric materials from natural resources can also be very useful for synthesis of hydrogels, coatings and other useful materials [7,8].

1.1.1 POLYMERIC GELS

To meet new challenges, the demand of natural polymers is expected to grow by 7.1% every year. Their low toxicity and an excellent biodegradability have attracted researchers to pay attention toward their widespread applications, which range from health sciences such as agents for various kinds of drug delivery systems to water purification. While different types of polymers give rise to various kinds of materials, among them one of the prominent materials are those of polymeric gels which have gained importance over the years due to their several unique characteristics such as their rheological and mechanical properties [9–12]. They have become indispensable part of new advanced and smart materials, which are used in biological, biomedical, electronic, and environmental fields [13]. **Gels:** Gels are soft, solid, or liquid-like unique three-dimensional polymeric networks where the polymer matrix is filled with liquid or gas. The three-dimensional networks are composed of several components such as long polymeric chains, species of small molecules, and a large amount of solvent. The network formation occurs due to physical, chemical, or supramolecular crosslinking [14,15]. In case of physical gels, the network is due to various weak interactions, like the entanglement of the polymer chains, hydrogen bonds, or Van der Waals interactions. Such structures are usually not permanent, and they dissolve over the time when immersed in their solvents. However, the polymer chains can also be cross linked through chemical reactions leading to strong covalent or electrostatic bonds. The chemically crosslinked network is much more stable and cannot be dissolved with the degradation of the polymer. Therefore, chemical gels are usually preferable in the majority of the application fields.

The weight and size of gels are more like a liquid, but they are treated like a solid. Under certain conditions, the liquid in gel is rendered solid by more or less rigid network of microscopic particles dispersed throughout its volume that results in a honeycomb-like structure, enclosing within it all the liquid (Figure 1.1).

A gel is overall a semisolid that can have properties ranging from soft and weak to hard and tough [16–18]. The word "gel" was coined in the 19th century by Scottish Chemist Thomas Graham, who derived this name from the characteristics of gelatin. The two important characteristics of gels are phase state and their rheological properties. The gel phase is accompanied by a fundamental molecular reorganization that leads to increase in order and a loss of lateral mobility and exhibits no flow when in steady state. They do not show the Tyndall effect, Brownian movement, and electrophoresis. Gels are classified into elastic gels and non-elastic gels. Elastic gels possess the property of elasticity, i.e., they change to solid mass on dehydration which can again be converted into gel

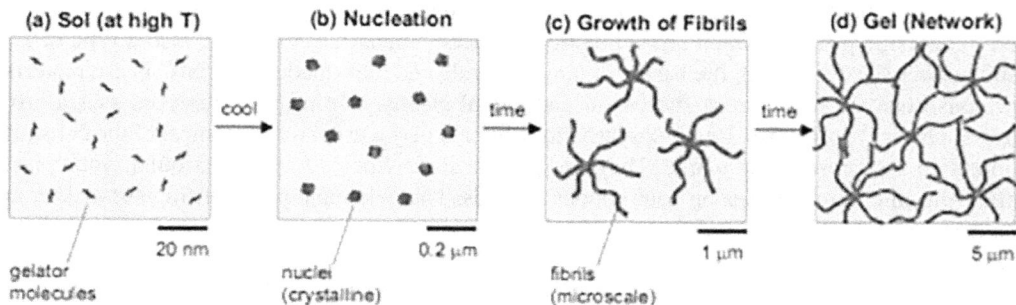

FIGURE 1.1 Gel formation [15]

by addition of water followed by heating and cooling. When they are placed in contact with water, they absorb water and swell. Elastic gels (or reversible gels) are lyophilic and are mostly made up of organic substances. Some examples of elastic gels are gelatin, agar, starch, etc. [19].

Non-elastic gels (or irreversible gels) do not possess the property of elasticity, i.e., they change to solid mass by dehydration which becomes rigid and cannot be converted into original form by heating with water. They are lyophobic in nature. Some examples of non-elastic gels are silica acid, hydroxides of Fe, Al, Cr, solid alcohol, etc. When contraction of gel causes separation of liquid from it, the process is called syneresis. Many gels have tendency to absorb liquid and swell. This property is known as imbibition. These hydrophilic gels can swell in water and hold large amount of water while maintaining the structure due to physical and/or chemical crosslinking of individual polymer chains. These are called hydrogel.

In certain gels when the liquid component of the gel has been replaced with an air or gas without significant collapse of the gel structure, the result is mesophorous solid foam with extremely low density and extremely low conductivity. These are called aerogels. The term mesoporous here refers to a material that contains pores ranging from 2 to 50 nm in diameter. The shrinkage in aerogels is less than 15%. We may also have nanogels. A nanogel is a particle of nano dimension composed of a hydrogel. It is a crosslinked hydrophilic polymer network where polymerization and crosslinking are being held concomitantly. A gel is named as a xerogel, when the liquid phase of a gel is removed by evaporation at room temperature. It may retain its original shape but often cracks due to the extreme shrinkage that is experienced while being dried. The shrinkage is more than 90%. Xerogel usually retains high porosity (15%–50%) and enormous surface area, along with small pore size (1–10 nm).

Solvent removal under supercritical conditions results in aerogels. Therefore, the method of drying will dictate whether an aerogel or a xerogel will be formed. Cryogel is yet another form of gel that is formed by freeze-drying. This enables the solvent to convert to gas phase directly from solid phase. It results in a flexible composite blanket designed for insulating cold temperature environments ranging from cryogenic to ambient.

From the aforementioned discussion, it is clear that all other types of gels originate from hydrogels. So hydrogel can be called the fundamental type of gel system on which the formation of other types of gel systems such as aerogels, xerogels, and cryogenic gels depend.

- Hydrogels have good cell attachment, molecular response, structural integrity, biodegradability, biocompatibility, and good mechanism for solute transport. In addition to these, hydrogels have some basic characteristics, which make them a preferred choice of gel systems [20].
- Hydrogels are inherently heterogeneous. They have solid-rich regions distributed within liquid environment.
- Water can be freely diffused or be loosely bound or tightly associated with the network.
- Hydrogels have solid-like characterstics with infinite viscosity, defined shape, and modulus. At the same time, they exhibit liquid-like characterstics, allowing solutes to diffuse freely as long as they are not larger than the average mesh size.

The basic criterion for a material or polymer to be called hydrogel is that water must constitute at least 10% of the total weight (or volume). When the content of water exceeds 95% of the total weight (or volume), the hydrogel is said to be superabsorbent. Apart from the swelling factor in hydrogels or for that matter any type of gel system, the factor that matters the most is the type of crosslinking. Based on the nature of crosslinking, we have the following types of hydrogels:

- **Physical hydrogels:** involve physical interactions such as molecular entanglement, ionic interactions, hydrogen bonding, and Van der Waals forces. E.g. Gelatin, agar.
- **Chemical hydrogels:** involve the formation of covalent bonds, e.g., poly(methyl methacrylate) (PMMA). Based on the charge of building blocks, these are neutral (e.g., dextran); anionic (e.g., carrageenam, alginic acid, etc.); and cationic (e.g., chitosan); and ampholytic

(that can behave as both positive and negative) (e.g., collagen carboxymethyl chitin, fibrin, etc.). The crosslinkages in hydrogels are based on the following methods of preparation:

- **Homopolymer:** involves the crosslinking network of a single hydrophilic monomer type.
- **Copolymer:** involves crosslinking of two (or more) monomer units, at least one of which is hydrophilic to render them swellable.
- **Multipolymer:** involves three or more comonomers reacting together.
- **Interpenetrating polymer networks:** comprises two or more networks which are at least partially interlaced on a polymer scale but are not covalently bonded to each other.

1.1.2 CROSSLINKING METHODS

Preparing hydrogels with certain aspects like mesh size, swelling behavior, and permeability characteristics depend on the method of preparation. The method may be of a physical and/or chemical crosslinking, graft polymerization, or radiation crosslinking. The last one may cause either physical or chemical crosslinking depending on the kind of polymer which is being used.

1.1.2.1 Physical Methods

The main advantage of physical crosslinking is that no other crosslinker is needed at all, and this makes it very safe. The crosslinker may be toxic, or even if it is not toxic, its effects may be undesired. Physical crosslinking is a relatively easy process, but it often results in low mechanical strength for most physical crosslinkings due to the weak nature of interactions. In physical crosslinkings, besides ionic interactions and hydrogen bonds, we also have alternate heating or cooling cycles and crystallization.

When heating/cooling cycles are repeated multiple times, it ends up in the formation of coils (as shown in Figure 1.2). This heating/cooling is done along with sodium ions or potassium ions. These established physical crosslinks contribute to the stabilization of hydrogels to form a hydrogels, e.g., PEO, PEG, and PLA.

Crystallization is another common method used for crosslinking. In this, freezing/thawing is repeated multiple times which again results in the formation of coils (as shown in Figure 1.2). Finally, on freezing, these coils interact with each other and form a very strong hydrogel. Crystallization is actually a technique that provides reasonably a good mechanical strength. In many cases, it gives mechanical strength which is better than chemical crosslinking, e.g., PVA and Xanthan.

1.1.2.2 Chemical Methods

In many cases, chemical crosslinking is an irreversible crosslinked network due to the formation of covalent bonds. It is basically a direct reaction of a polymer or its branches with the bifunctional component of small molecular weight, which is a crosslinking agent or a crosslinker. In this, one group of the bifunctional crosslinker reacts with one polymer chain on one end, and with its another functional group, it reacts with the another polymer chain on the other end, thus creating crosslinks

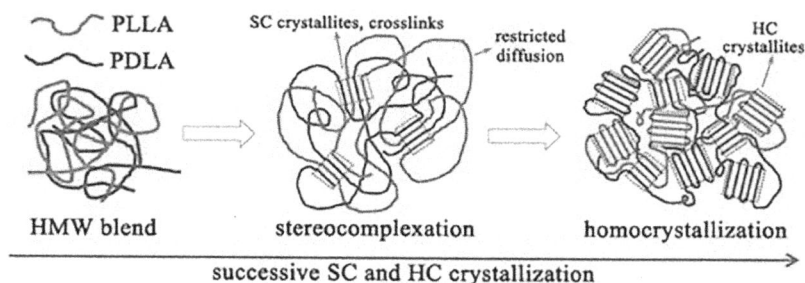

FIGURE 1.2 Effect of heating/cooling cycles and crystallization on respective polymer coils [21].

between polymer chains. One such common example is that of gluteraldehyde, where you have two aldehyde groups, which easily react with amines, etc. Another type of chemical crosslinking can be done by grafting, where polymerization of a monomer is performed on the backbone of a previously formed polymer or its branches. Apart from the types of crosslinkings, it is the nature of the polymer chains that defines the particular type of gel they form. Polymeric gels comprise a great variety of different polymeric components, which are either derived from natural resources such as peptides, proteins, polysaccharides, and nucleic acids or from synthetically prepared polymers such as polyesters, which include PEG-PLA-PEG, PEG-PLGA-PEG, PHB, etc., and from other synthetic polymers such as PEG-bis(PLA-acrylate), PEG-g-p(AAm-co-vamine), PAAm, etc. [21].

1.2 POLYSACCHARIDE-BASED POLYMERIC GELS

While most commercially available gels are synthetic polymer-based, there is increasing interest in the use of natural polymers that have renewable advantageous properties like biodegradability and good biocompatibility [22]. Polysaccharides are one such important group of renewable materials due to their abundance and low cost. The non-polysaccharide component can range from other renewable resources like polypeptides, synthetic polymers to inorganic additives. In this chapter, we will more specifically elucidate the structural aspects, properties, and applications of polysaccharide-based polymeric gels.

1.3 STRUCTURAL ASPECTS

1.3.1 CELLULOSE

Cellulose is the most available polysaccharide biomolecule in nature. It is estimated to account for 50% of the world's organic carbon. It is present mostly in plant cells. It plays an integral role in keeping the structure of plant cell walls stable. Cellulose can be also synthesized via certain types of bacteria (e.g., *Acetobacter xylinum*) with slightly different physical and macromolecular properties.

Structure: The primary structure of cellulose is comprised of a long chain of anhydrous d-glucose units attached together by β-1,4-glucosidic linkage (Figure 1.3). It shows high degree of crystallinity usually 40%–60% in plant cellulose and 61%–70% in bacterial cellulose. Such high crystallinity is due to the extended hydrogen bonding along the cellulose chains resulting in very tight and well-packed chains, which hinder cellulose dissolution in common solvents [24] (Figure 1.4).

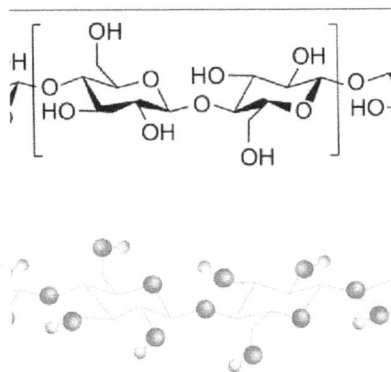

FIGURE 1.3 Structure of cellulose [23].

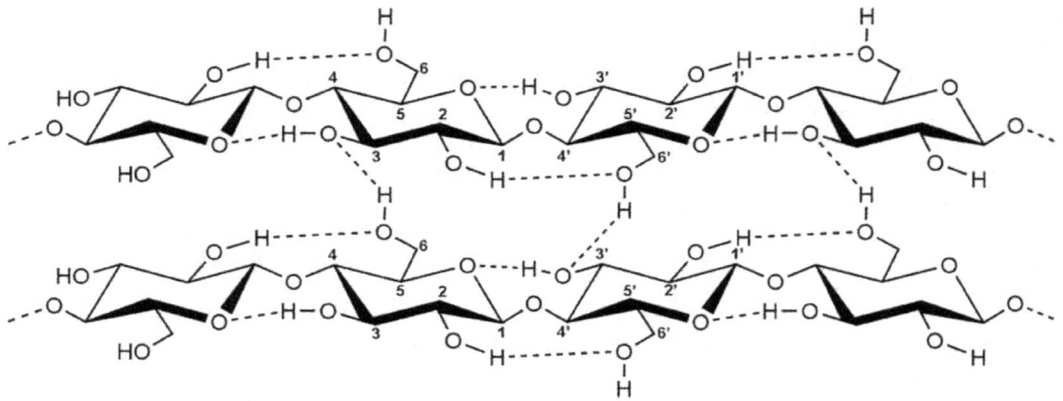

FIGURE 1.4 Cellulose structure showing β-1,4-glycosidic linkages and hydrogen bonding between cellulose chains [25].

Glucose unit in cellulose has three hydroxyl functional groups in carbon atoms six (primary) and two and three (secondary), which open wide possibilities of chemical modifications in order to produce cellulose derivatives with distinguished characteristics.

1.3.1.1 Cellulose-Based Hydrogels

As mentioned earlier, hydrogel forms when hydrophilic polymer chains crosslink either by physical or chemical bonds to create a 3D network [26–30]. Such mechanism requires water or solvent-soluble polymers in order to facilitate the physical or chemical interactions with the crosslinker. Special solvent systems have been used for such purposes such as N-Methylmorpholine-N-oxide (NMMO) ionic liquids (ILs) and alkali/urea [31,32]. These new solvent systems have opened new scope for native cellulose for chemical modifications such as reversible and irreversible cellulose-based hydrogels [33]. Due to strong intermolecular interactions, the cellulose is insoluble in water. However, the partial substitution of hydrogel groups results in weakening of these bonds, leading to improved solubility. Water solubility increases with the degree of substitution. For the hydrogel synthesis, mostly cellulose ethers are utilized (Figure 1.5).

Besides, cellulose derivatives, such as carboxymethyl cellulose and hydoxyethyl cellulose [34,35], have been widely utilized in hydrogel preparation (Figure 1.6).

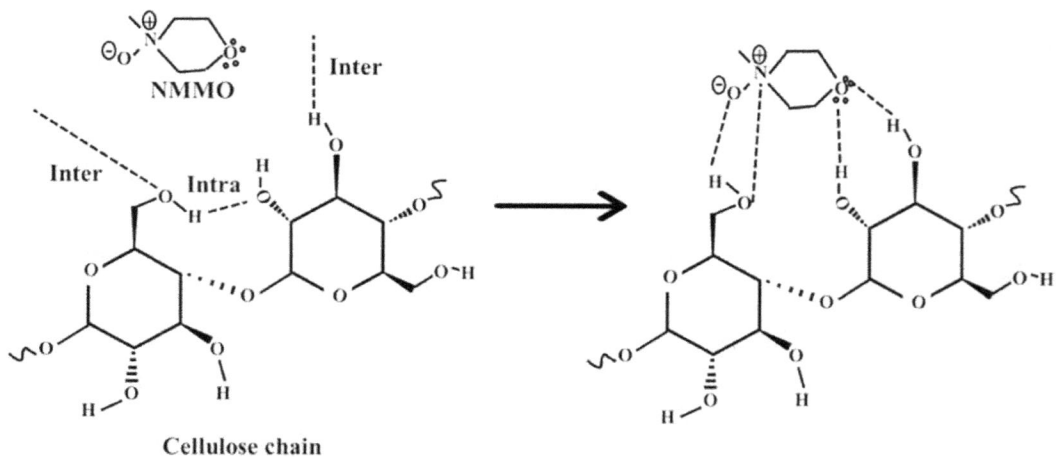

FIGURE 1.5 NMMO-Cellulose dissolution mechanisms showing physical crosslinking [31].

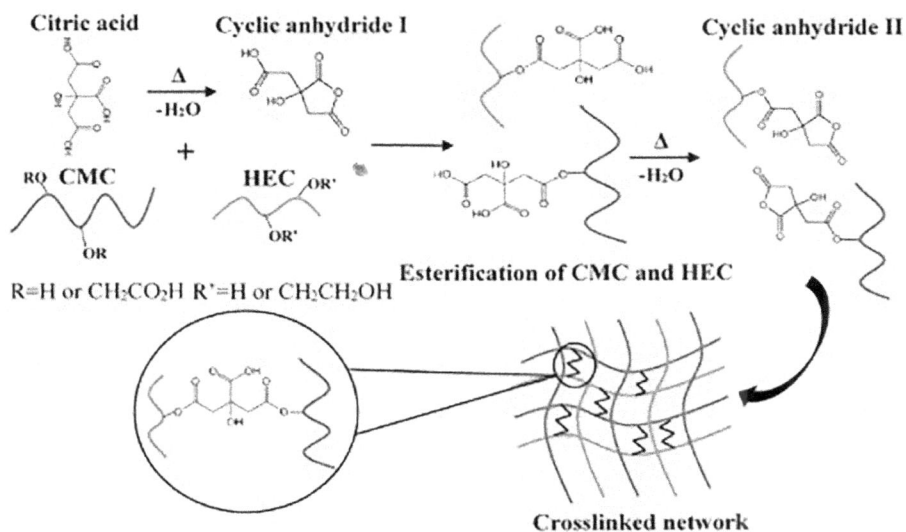

FIGURE 1.6 Crosslinked carboxymethyl cellulose/hydroxyethyl cellulose hydrogel [36].

Native cellulose solution which is dissolved in lithium chloride/dimethyl acetamide (LiCl / DMAC) system leads to 3D hydrogel upon dropping into non-solvent systems, such as azeotropic methanol and isopropanol. The same technique of hydrogel formation has been reported by using different dissolution systems, such as paraformadehyde/dimethyl sulphoxide (DMSO), triethyl ammonium chloride /DMSO, and tetrabutylammonium chloride/DMSO [37] (Figure 1.7).

Chemical crosslinking of cellulose and mainly cellulose derivatives depend on few small molecules that react with the free hydroxyl functional groups of the glucose units, such as citric acid, divinyl sulphone [34,35], diglycidyl ether [39], epichlorohydrin [35,40], and carbodiimide [41] (Figure 1.8).

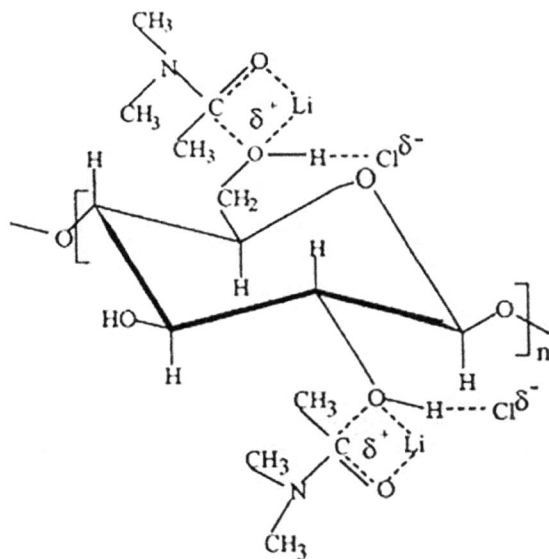

FIGURE 1.7 Homogeneous etherification of cellulose in LiCl/DMAC [38].

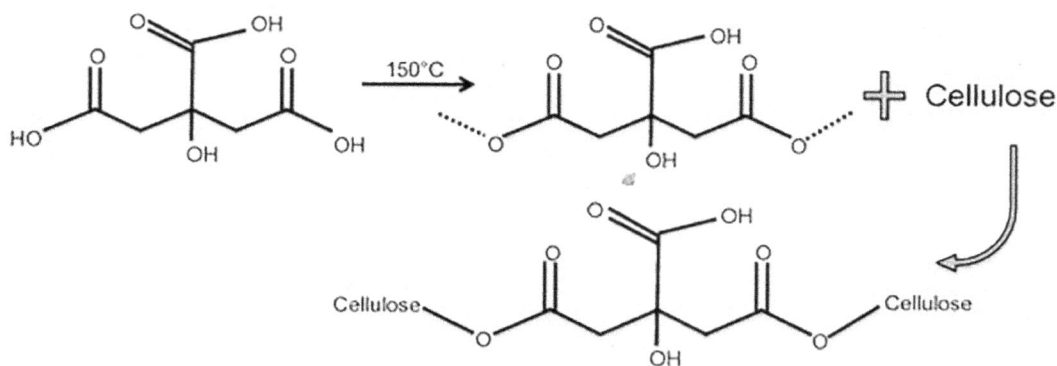

FIGURE 1.8 Cellulose-based hydrogel crosslinked with citric acid [42].

1.3.2 STARCH

Starch is mostly present in plant cells as granules of energy storage [43]. It is another polysaccharide having its monomers as α-D-glucose units.

Structure: Starch is constituted of two different macromolecular structures: amylose and amylopectin. The former shows α-1,4-linkage and the later has α-1,6-linkage. Amylose is a comparatively smaller macromolecule. The chains form double helices with crystalline structures. Starch exists in two allomorphic forms: type "A" and type" B" [44] "A" type helices are close packed with water molecules positioned between the helices [45–47]. In type "B", the helices form a hexagonal lattice, and water molecules lie inside the hexagonal structure (Figure 1.9).

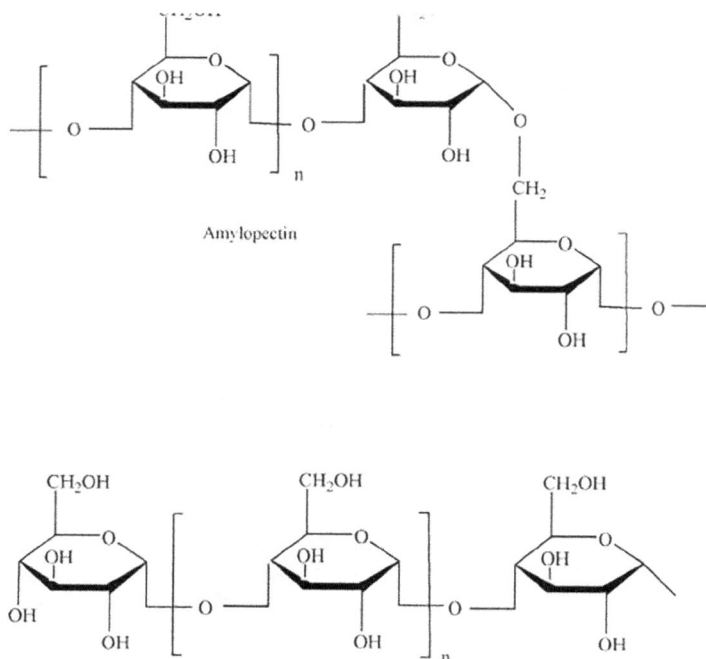

FIGURE 1.9   Structure of amylopectin and amylose [48].

1.3.2.1   Starch-Based Hydrogels

The granular structure of starch gets destroyed while hydrogel synthesis. Bioactive carboxymethylated starch is considered to be relevant for its gel synthesis (Figure 1.10 [49]). Ecofriendly hydrogel synthesis of starch is carried out with citric acid and itaconic acid (Figure 1.11, [50]).

In situ hydrogel is constructed by using starch-based nanoparticles via Schiff base (Figure 1.12).

1.3.3   CHITIN

Chitin is the second abundant material, next to cellulose. It is a very important structural material in nature. It is found in the exoskeletons of insects, the cell walls of fungi, and certain hard structures in invertebrates and fish.

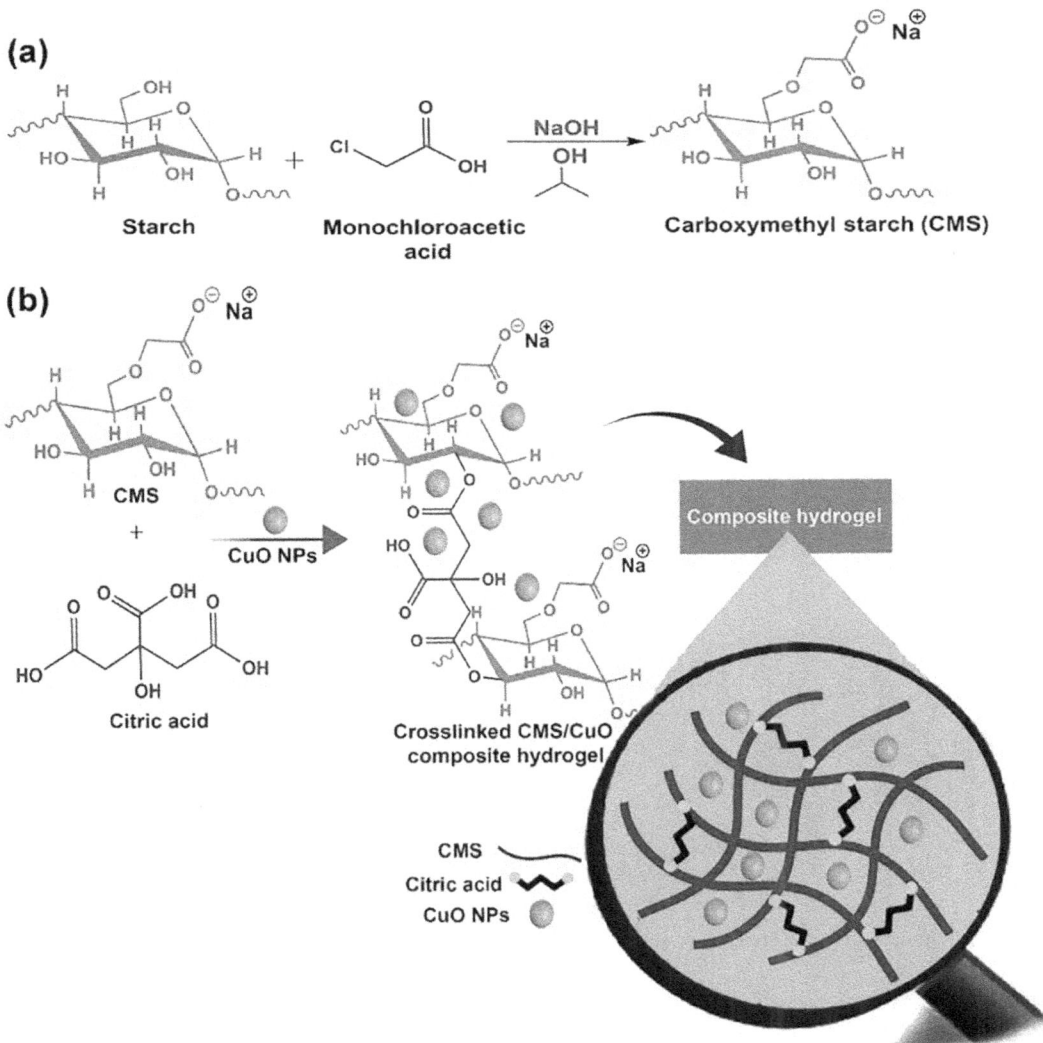

FIGURE 1.10   Crosslinking of CMS and citric acid using CuONPs to form hydrogel [49].

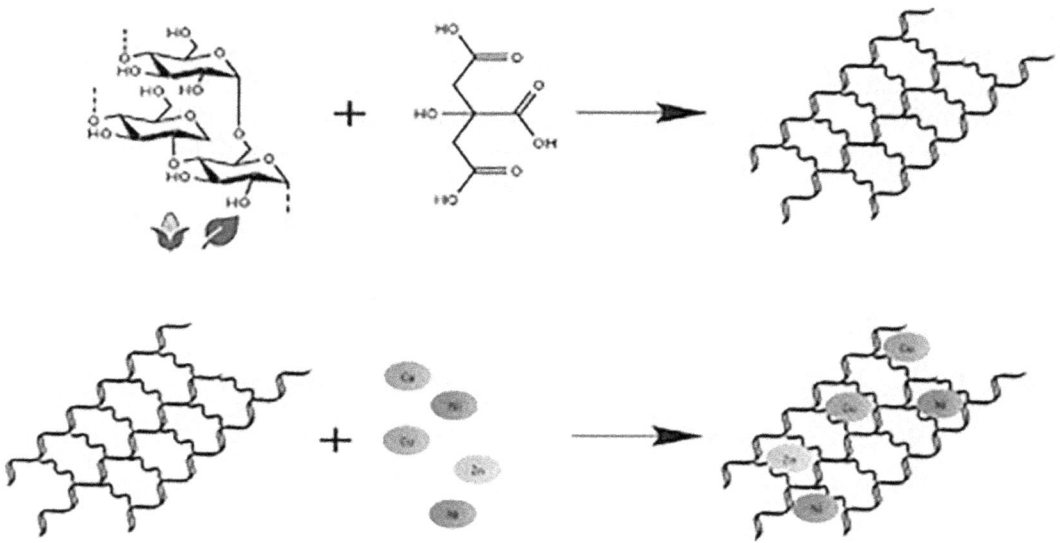

FIGURE 1.11 Starch-based hydrogel from citric acid and itaconic acid [50].

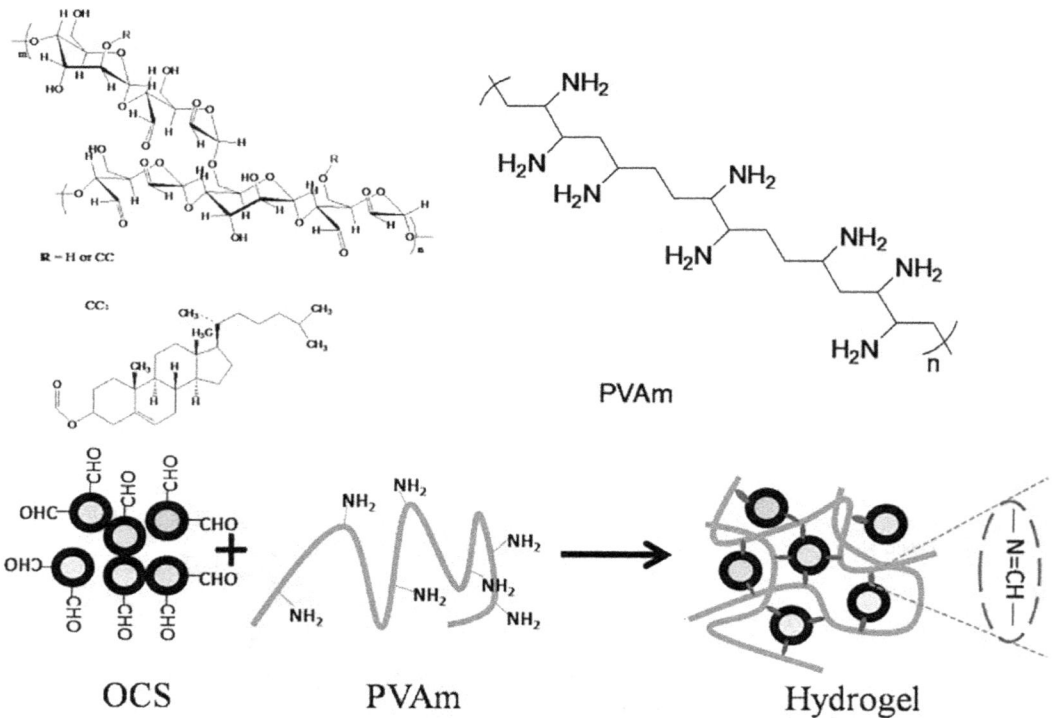

FIGURE 1.12 Hydrogel formation of starch-based nanoparticles via a Schiff base [51].

Structure: Chitin is comprised of a long chain polymer of *N*-acetylglucosamine units, which is an amide derivative of glucose. It has glycosidic bonds joining the series of substituted glucose molecules possessing acetylated amine group instead of OH group in glucose (Figure 1.13).

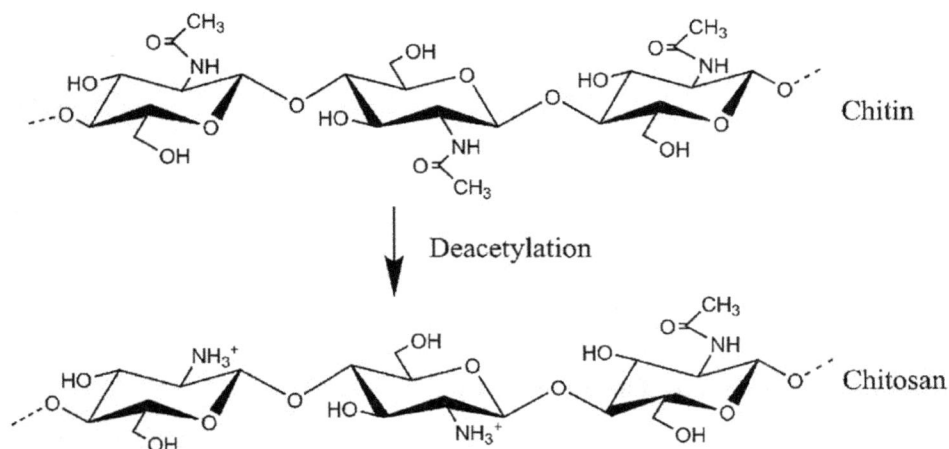

FIGURE 1.13 Structures of chitin and chitosan [52].

1.3.3.1 Chitin-Based Hydrogels

Biocompatible chitin/CNTs (carbon nanotubes) composite hydrogels were prepared with chitin solution in 11 wt% NaoH/4 wt% urea aqueous solution and subsequently regenerating in ethanol [53]. In these, CNTs were dispersed homogeneously in chitin matrix and combined with chitin nanofibers to form a compact and neat chitin/CNT nanofibrous network (as shown in Figure 1.14) through intermolecular interactions, such as electrostatic interactions, hydrogen bonding, and amphiphilic interactions [54].

In another latest development, [55] a novel, chitin-based composite hydrogel reinforced by tannic acid and modified reduced graphene oxide was prepared via a facile freezing-thawing approach. Epicholrohydrin and TRGO sheets were employed as efficient crosslinkers to fabricate dually crosslinked TRGO/chitin composite hydrogels (Figure 1.15).

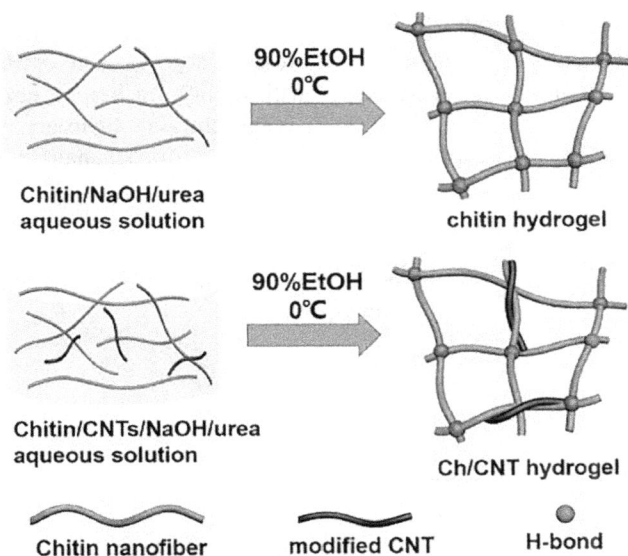

FIGURE 1.14 Chitin/CNTs composite hydrogel [54].

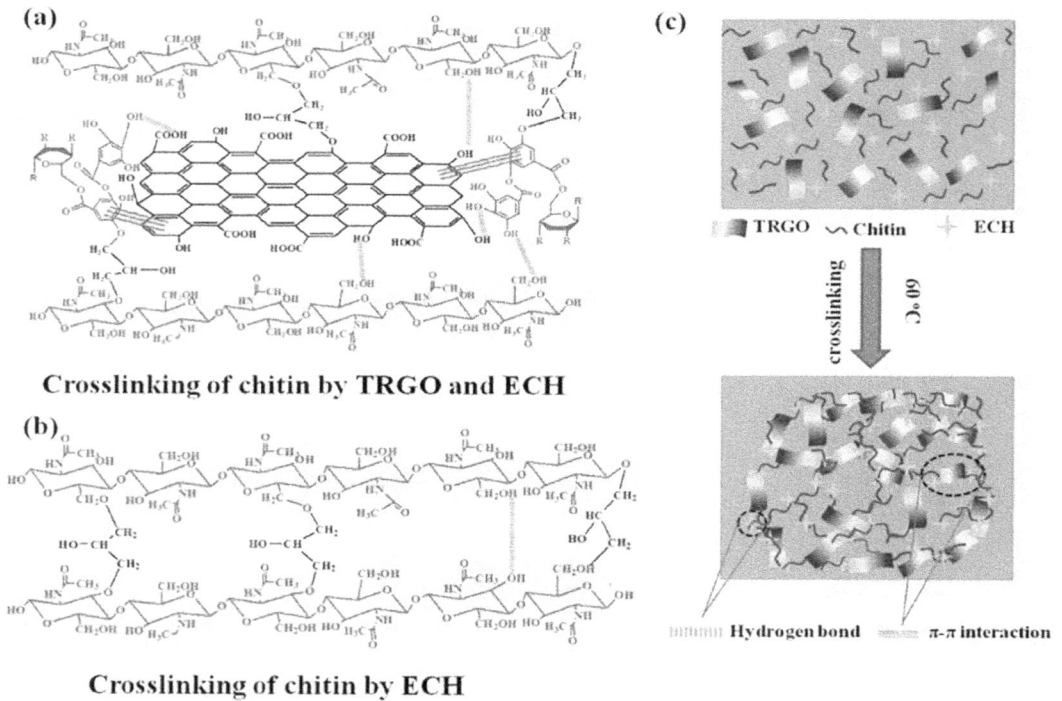

FIGURE 1.15 TRGO/chitin composite hydrogel [55].

1.3.4 CHITOSAN

The most important derivative of chitin is its deacetylated form, called chitosan, which contains amino groups in C(2) position. Chitosan is a polymeric structure that is composed of linear sequence of D-glucosamine and N-acetyl-D-glucosamine (Figure 1.13).

1.3.4.1 Chitosan-Based Hydrogels

Chitosan is a natural cationic copolymer that presents a great deal of interest for hydrogel structures. The polymer has hydrophilic nature with ability via human enzymes, which result in its biocompatibility and biodegradability. To make chitosan hydrogel, you can use cross-linker such as glutarudehyde, genipin (Figure 1.16, below), diisocyanate, and tripolyphosphate.

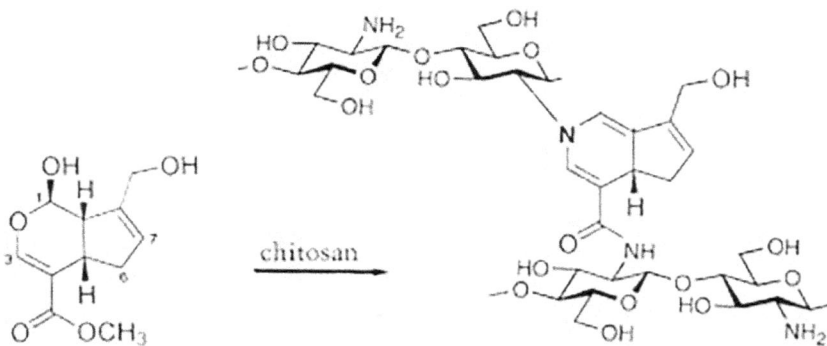

FIGURE 1.16 Crosslinking of chitosan with genipin [57].

Physical crosslinking method such as thawing and freezing is also made use of for the purpose. Apart from traditional methods, a novel chitosan hydrogels reinforced by silver nanoparticles can be constructed (Figure 1.17, [56]). In this method, LiOH/urea solvent system is used and then Ag nanoparticles are integrated into the chitosan network, which results in the formation of 3D network due to intermolecular and intramolecular interactions with ultra-high mechanical properties and improved antibacterial property.

1.3.5 ALGINATES

Alginates are naturally occurring linear anionic polysaccharides extracts found in brown marine algae and several bacteria strains. It is biocompatible, biodegradable, nontoxic, and easy to gel. Alginates can be processed into various forms such as hydrogels, microspheres, fibers, and sponges [58]. They have been widely applied in medical field.

Structure: Chemically alginates are copolymers mainly composed of β-D-mannuronic acid (M) and its C-5 epimer α-L-guluronic acid (G) residues linked via 1,4-glycosidic bond in an irregular blockwise manner (as shown in Figure 1.18).

The M and G residues are organized in homopolymeric blocks of G units (GG) or M units as (MM blocks) and heteropolymeric sequences of randomly coupled G and M units (GM or MG blocks). There occurrence, proportion, and distribution may differ significantly depending on their natural source.

1.3.5.1 Alginate-Based Gels

The most important property of alginates is their ability to form ionic gel in the presence of the polyvalent cations. The gelling is the result of ion exchange between monovalent ions of alginate solution (most often, Na^+ ions) and polyvalent cations, followed by subsequent coordination of polyvalent metal

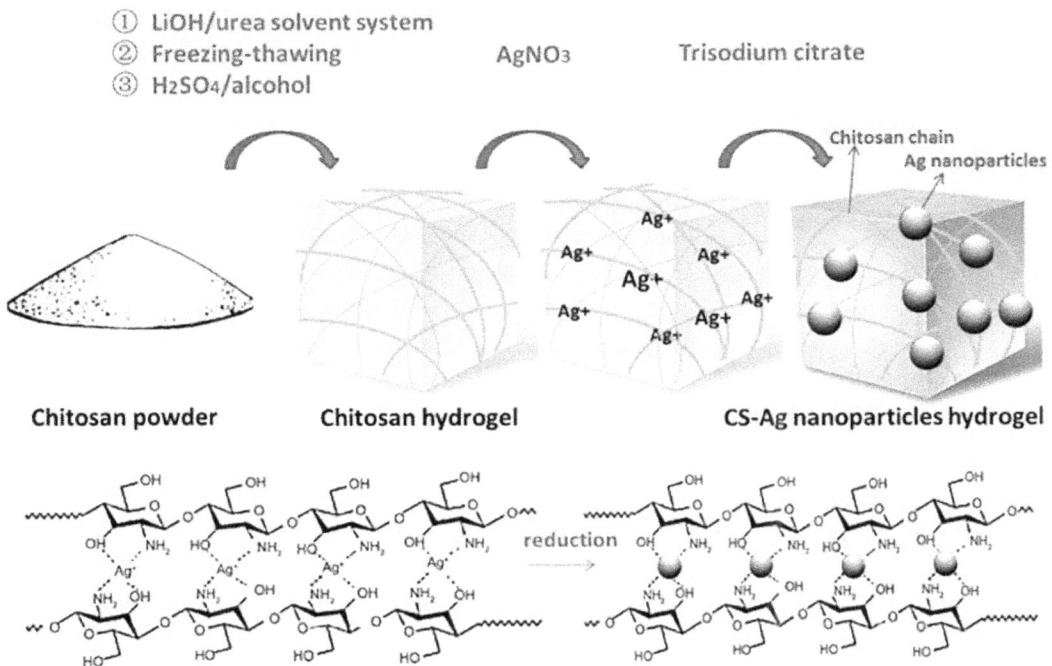

FIGURE 1.17 Chitosan hydrogel reinforced by silver nanoparticles [56].

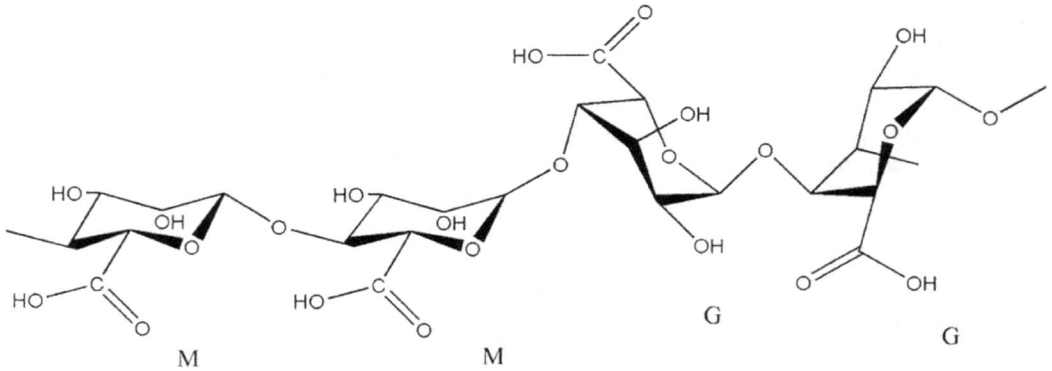

FIGURE 1.18   Structure of alginate [59].

ions with alginate macromolecule. Calcium is the most common cation applied to form alginate gels. The typical gelation mechanism involves coordination and chelating structures in the model of egg box, in which G units selectively form higher order junction zones (Figure 1.19), which is composed of two or more chains and crosslink via hydrogen bonding, with oxygen atoms in the G blocks of the two adjacent polymer chains. In 3D network of egg box, each cation is bound with four G residues, thus there should be 8–12 adjacent G residues in order to form a stable junction for Ca^{2+} alginate gels.

Several other divalent cations can bind to alginates, but with different affinity which is in the order of:

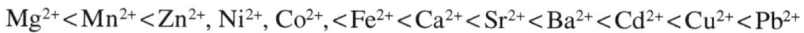

$$Mg^{2+} < Mn^{2+} < Zn^{2+}, Ni^{2+}, Co^{2+}, < Fe^{2+} < Ca^{2+} < Sr^{2+} < Ba^{2+} < Cd^{2+} < Cu^{2+} < Pb^{2+}$$

Alginate can also form irreversible hydrogel by creating permanent bonds between alginate chains either by chemical crosslinking agents or by photo crosslinking using photoinitiators [61].

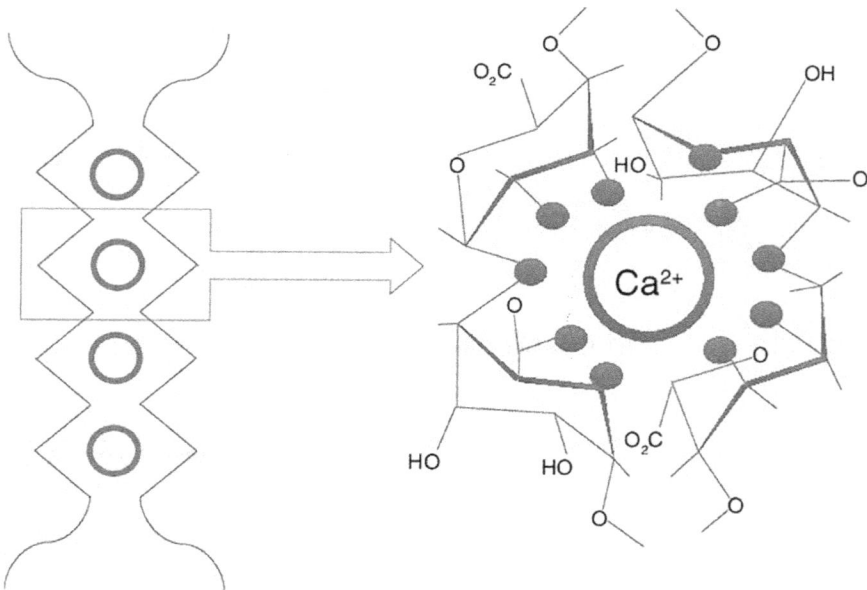

FIGURE 1.19   The junction zone in the egg box model of calcium alginate gel [60].

This is done with the help of bifunctional or multifunctional crosslinkers. For instance, gluteraldehyde, a bifunctional chemical crosslinker, creates an acetal link with alginate hydroxyl group [62] in the presence of poly(ethylene glycol) diamine. Reductive amination takes place between amino groups of the crosslinker and the aldehyde group of oxidized alginate [63] (Figure 1.20). Photo crosslinking of alginate hydrogel has also been reported for medical purposes [64] (Figure 1.21). It requires modifying alginate with photo-active groups such as methyl acrylate, etc. [65,66].

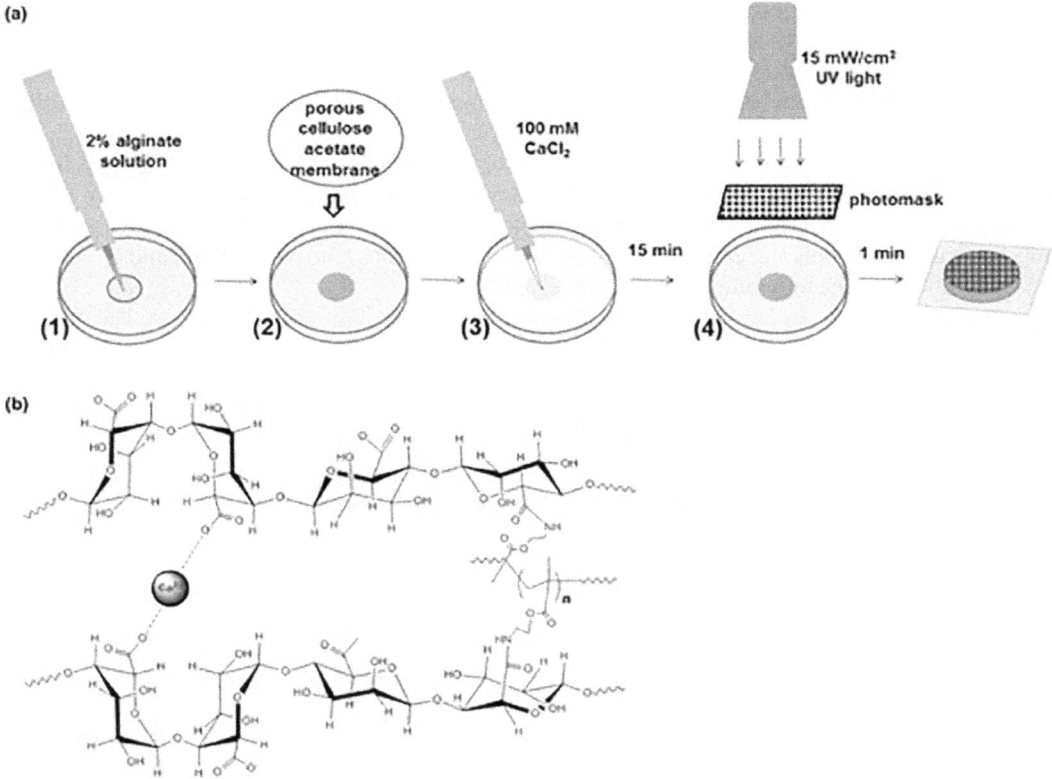

FIGURE 1.20   Photo crosslinking of alginate hydrogels [67].

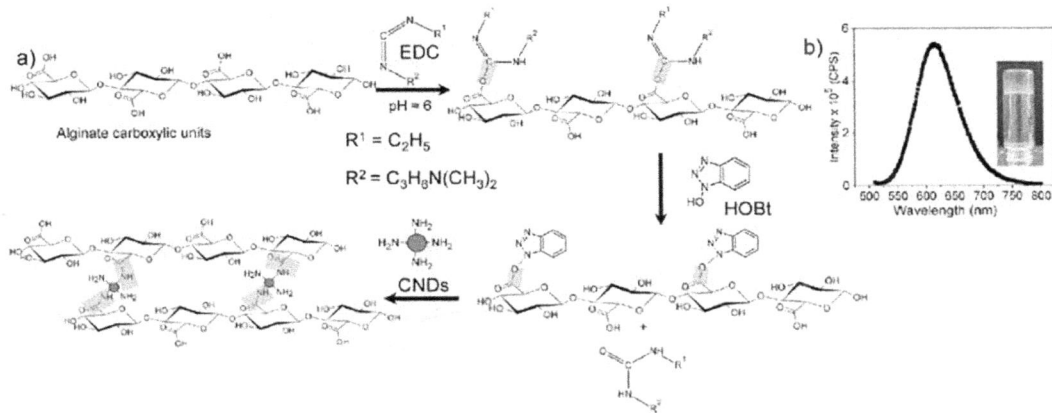

FIGURE 1.21   Chemical crosslinkers in alginate gels [68].

1.3.6 CARRAGEENANS

Carrageenans are a family of natural linear sulfated polysacchasides that are extracted from red edible seaweeds. The most prominent among them is *Chondrus crispus* which is a dark-red parsley like plant that grows attached to rocks. Gelatinons extracts of the *Chondrus crispus* seaweeds have been used as food additives since approximately the 15th century. Carrageenan is a vegetarian and vegan alternative to gelatin in diary, confectionery, and other foods. Besides, it is also used in meat products due to their strong bonding to food proteins.

Structure of Carrageenans: Carrageenans contain 15%–40% ester-sulfate content, which makes them anionic polysaccharides. Based on their sulfate content, carrageenans are categorized into three different classes: (kapa) κ-carrageenan, which has one sulfate group per disaccharide; (iota) ι-carrageenan which has two; and (lambda) λ-carrageenan, which has three sulfate groups (Figure 1.22).

κ-Carrageenan is a polysaccharide that consists of alternating disaccharide units of O-3 linked β-D-galactopyranosyle-4-sulfate and O-4-linked 3,6-anhydro-D-galactopyranosyl residues (below in Figure 1.22), and ι-carrageenan is more highly sulfated galactan and is also most characterized polysaccharide. The polymer consists of O-3-linked β-D-galactopyranosyl-4-sulfate and O-4-linked 3,6-anhydro-α-D-galactopyranosyl-2-sulfate residues.

1.3.6.1 Gelation Mechanism of κ-Carrageenan

κ-Carrageenan is well known for its gel-forming property. It is a cation-selective bonding polymer, which gels in the presence of large size univalent cations such as K^+ (Van der Waals rad 275 pm), Rb^+ (303 pm), and Cs^+ (343 pm) but does not do so in the presence of small cations, i.e., Na^+ (227 pm) and Li^+ (182 pm) (Masakuni Tako 295). Even at a concentration range of 0.1%–10% at room temperature, it changes into ice-like structure with hydrogen bonding between polymer

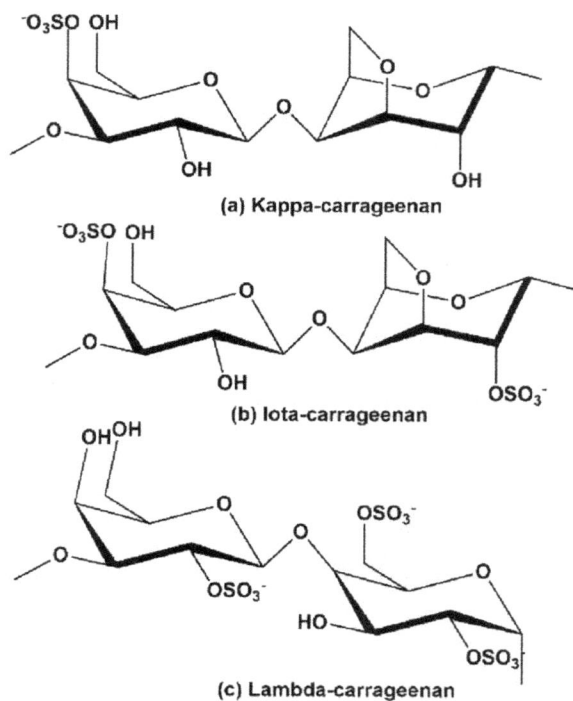

(a) Kappa-carrageenan

(b) Iota-carrageenan

(c) Lambda-carrageenan

FIGURE 1.22 Chemical structures for three basic types of carrageenans [69].

and water molecules and between water–water molecules, resulting in gelling. In κ-carrageenan, an intramolecular K^+ bridge is developed between sulfate oxygen C-4, which is oriented at axial configuration of D-galactopyranosyl residues and 3,6 oxygen of 3,6-anhydro-α-D-galactopyranosyl residue (Figure 1.23). This type of bridge does not exist with Na^+ ions because the radius of Na^+ ions is too small for an association with ring oxygen group and also due to the presence of too much hydration, which prevents electrostatic forces of attraction being involved in the bridge. Many intramolecular K^+ cation bridges serve to keep the polymer chains rigid, which further lead to intermolecular association. However, the transition temperature is observed at 25°C where inter and intramolecular associations dissociate above the given temperature.

1.3.6.2 Gelation Mechanism of ι-Carrageenan

Gelation in ι-carrageenan occurs with Ca-salt on cooling but does not do so with K-salt or Na-salt. Ca-salt of ι-carrageenan involves intramolecular association through Ca cation, with the ionic bonding between sulfate oxygen (as shown in Figure 1.24). Each Ca cation is coordinated to two sulfate

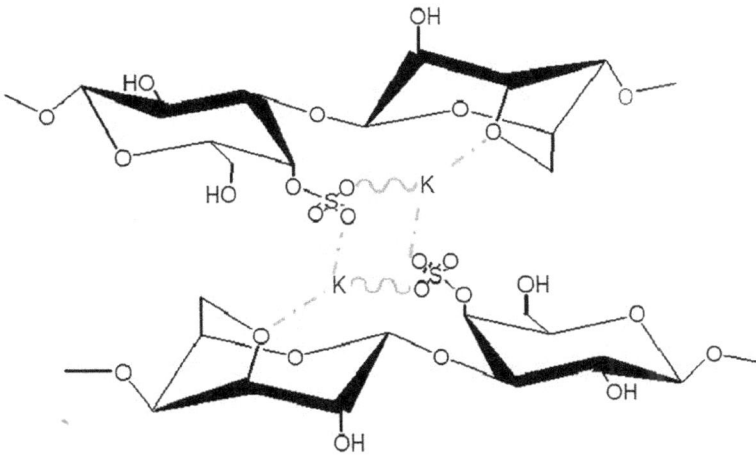

FIGURE 1.23 Gelation mechanism of κ-carrageenan ionic bonding (curved lines) and electrostatic forces of attraction (dotted lines) [45].

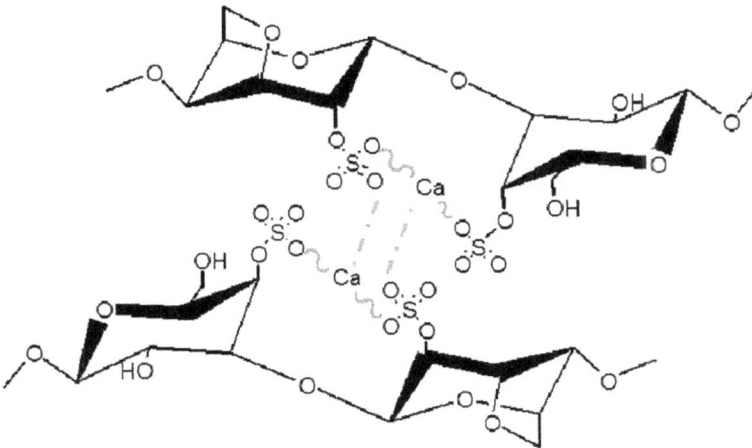

FIGURE 1.24 Gelation mechanism of ι-carrageenan, ionic bonding (curved lines), and electrostatic forces of attraction (dotted lines) [45].

groups by ionic bonding, which is stronger than electrostatic forces of attraction. This makes the molecular chain rigid even at room temperature, and therefore, intermolecular associations take place. Thus intramolecular Ca-bridge is developed, which further leads to intermolecular associations of Ca-salt of ι-carrageenan. These two types of bridges differ essentially from that of the K-salt bridges of κ-carrageenan. The former consists of double ionic forces and an electrostatic force of attraction, whereas the later consists of single bond and triple electrostatic forces of attraction. These differences give rise to conformational transition of ι-carrageenan and κ-carrageenan in aqueous solution. Since the ionic bonding is stronger than electrostatic forces of attraction, the molecular chain can keep rigid even up to 45°C, which is the transition temperature for ι-carrageenan, beyond 45°C intra and intermolecular Ca-bridges dissociate.

1.3.7   AGAROSE

Agarose is the major gelling constituent of agar extracted from red sea weeds and is also found as a support structure of cell wall for marine algae. It is a polysaccharide which together with agropectin creates agar. It is used especially as a supporting medium in gel electrophoresis.

Structure of Agarose: Agarose consists of a copolymer alternating O-3 linked beta-D-galacto-pyranosyl and O-4 linked 3,6-anhydro-alpha-L-galactopyranosyl residues. Its structure is similar to κ-carrageenan and ι-carrageenan, except for the sulfate content and L-configuration (as shown in Figure 1.25). In fact, agarose is a diastereomeric derivative of κ-carrageenan and ι-carrageenan.

1.3.7.1   Gelation Mechanism of Agarose

Gelation occurs at a concenteration of 0.08% at low temperature (0°C), stays constant during increase in temperature upto 60°C in the presence of 0.1% of salt (NaCl, KCl, $CaCl_2$ or $MgCl_2$), and then decreases rapidly with further increase in temperature [45,71]. This can be caused by "salting out" effect which might take place because of the rigidity of the molecular chains of agarose through the formation of intra and intermolecular association, resulting from the tetrahedral distribution of anhydro α-L-galactosyl residues and water molecules. However, the least elastic modulus was observed in the presence of area (4.0 M). Thus, it was proposed that there exists an intramolecular hydrogen bonding between OH-4, which is oriented at axial configuration of the β-D-galactopyranosyl and the adjacent hemiacetal oxygen atom of 3,6-anhydro-α-L-galactopyranosyl residues (Figure 1.26). The 3,6-anhydro-L-galactopyranosyl residues is a cage-like sugar that contributes by stabilizing the proposed intramolecular hydrogen bonding, even at high temperature >60°C. Intermolecular hydrogen bonding was also proposed between the ring O-3, 6-atom and OH-2 which is oriented at axial configuration of 3,6-anhydro-α-L-galactopyranosyl residues on different molecules (Figure 1.26). This intermolecular hydrogen bonding results from the cage effect of the 3,6-anhydro-α-L-galactopyranosyl residues, which adopt a tetrahedral distribution and therefore also attract water molecules with hydrogen bonding (Figure 1.26). The mode of intra and intermolecular hydrogen bonding of agarose molecules has been supported by 1H and 13C NMR spectroscopy.

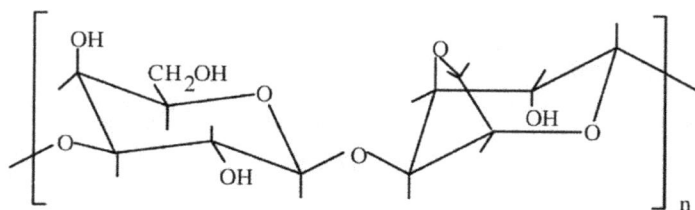

FIGURE 1.25   Structure of agarose [70].

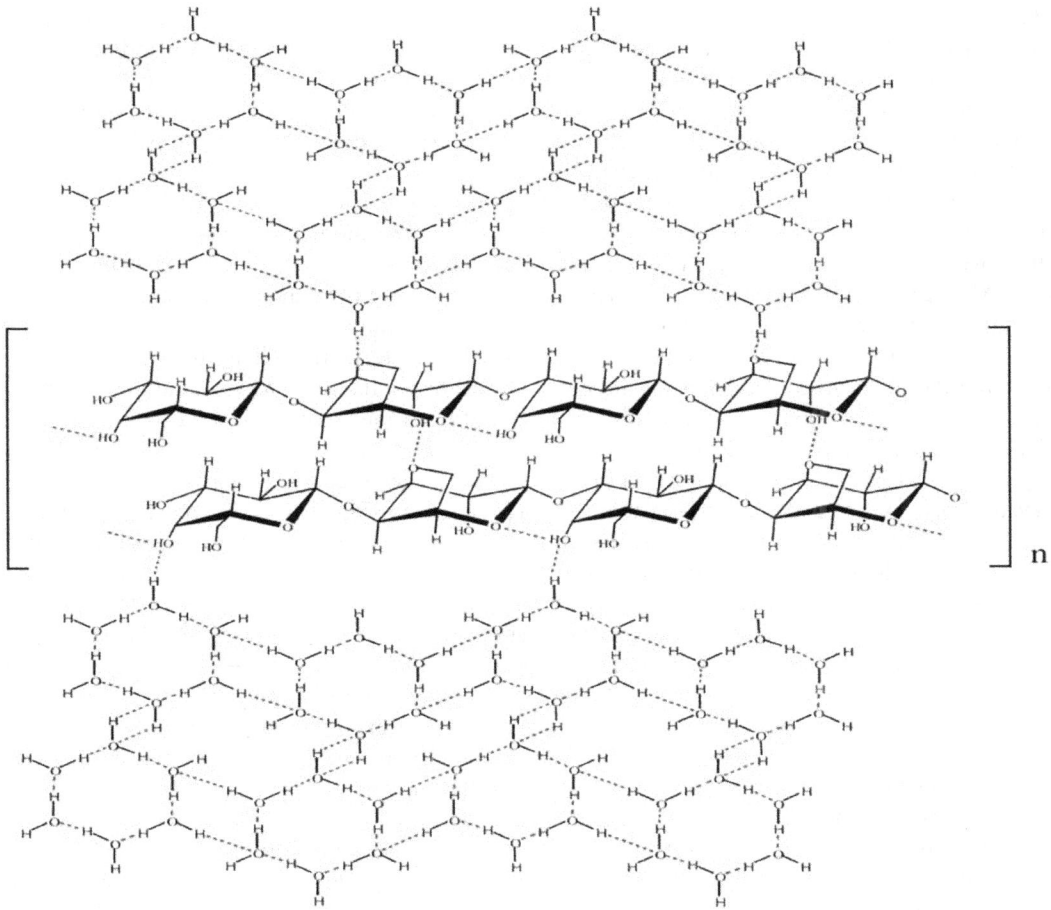

FIGURE 1.26 Gelation mechanism of agarose showing intra and intermolecular (---) hydrogen bonding and with water molecules also [45].

1.4   PROPERTIES

Properties of hydrogels include their mechanical, rheological, and swelling properties. Apart from these, they have their respective thermal, electrical, and physicochemical properties. In this section, we will focus on the properties which pertain to some polysaccharide-based polymeric gels, in particular, and also on some principle properties which pertain to hydrogels in general.

1.4.1   MECHANICAL PROPERTIES

To explain mechanical behaviors of hydrogels, rubber elastic theory is regarded as the most relevant and efficient approach because it associates important bulk property of hydrogels: rubbery modulus scales, temperature, and crosslinking density [72, 73]. Hydrogels show elastic property like rubber. Peppas et al. [74] developed an equation (Eq. 1.1). It is known as rubber elastic theory, which gives the structure of hydrogels in solvent.

$$\tau = pRT/Mc \ (1\_2Mc/Mn) \ (a\_1/a2) \ 3\sqrt{} \ (V2,s/V2, r) \tag{1.1}$$

Here, "τ" (tau) is the applied stress to the polymer as a function of elongation, "ρ" (rho) is the polymer density, "R" is the universal gas constant, "T" is the absolute temperature, Mc and Mn are the average molecular weights between and without crosslink, respectively, "a" is known as the extension ratio, and V2,s and V2,r are the polymer volume fraction in the swollen and relaxed forms.

However, the original rubber theory cannot properly explain the nature of hydrogels in aqueous medium and demands some modifications [75].

Yet, other parameters of mechanical properties and hydrogel stiffness can be demonstrated by using oscillatory shear measurements of elastic (G′) and viscous modulus (G″) of hydrogel matrices that have been measured at varied temperature using constant stress rheometer. Given below is typical DMA scan (dynamic mechanical analysis) of thermoelastic polymers which is used to describe the viscoelastic behavior of a polymer in the test temperature range.

DMA is a frequently used technique in material characterization. It is the most useful for studying the viscoelastic behavior of polymers. DMA mechanically deforms a sample and measures the sample response as a function of temperature, time, frequency, and amplitude. The DMA measurements are used mainly to determine glass transition temperature, moduli, damping behavior, relaxation behavior, molecular interactions, creep recovery, degree of crosslinking, crystallinity, and effect of fillers on material properties. In Figure 1.27, G′ represents storage modulus, G″ represents loss modulus, and "tan delta" represents function of temperature. It is also called the loss factor and is the ratio of loss modulus to storage modulus. These three parameters describe viscoelastic behavior of polymers in the test temperature range.

In the "glassy state," polymers are ideally elastic and the value of storage modulus G′ is very high. As the temperature goes up, short chain mobility starts in the sample, and we enter the glass transition range. In this range, the sample becomes leathery and soft, and therefore, the storage modulus G′ decreases sharply. Also in this range, the value of loss modulus G″ and "tan delta" attain their peaks. After the glass transition range, we enter the so-called rubbery plateau in which loss modulus G″ and "tan delta" are small. Here the thermoplastic polymer exhibits rubbery-like properties and can be permanently deformed. The width of the rubbery plateau increases with the increase in molecular weight of the polymer. After this, at higher temperature, the material starts to flow, and this is called the flow region. In this state, the storage modulus G′ decreases while the loss modulus G″ and "tan delta" increase. The material behaves like a liquid because in melted state the polymer behaves like amorphous polymer.

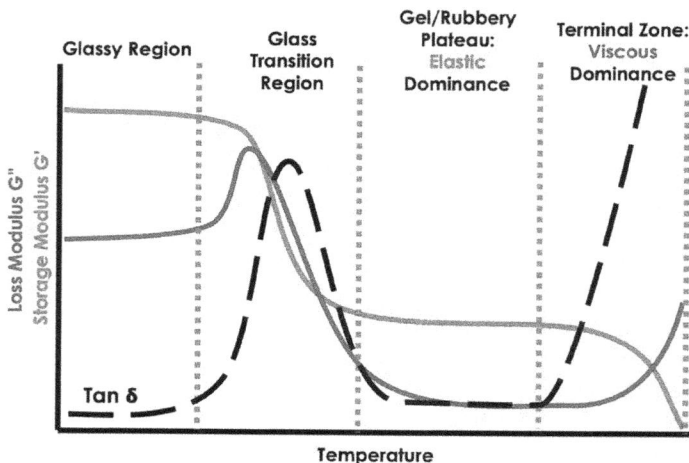

FIGURE 1.27   Typical DMA scan (Center for Industrial Rheology).

G' relates to the energy stored in the material, G" (is the loss factor), which relates to the energy dissipated in the form of heat, and the "tan delta" is the measure of how well the energy is dissipated.

Examples of mechanical characteristics: Storage modulus (G') and loss modulus (G") of polysaccharide hydrogels in native and squeezed situations. Data are reported as mean of three value ± standard error [22].

For low values of oscillation stress, the storage moduls (G') is higher than the viscous one (G") indicating the gel like nature of the material. Above the cross point, i.e., the oscillation stress at which G' and G" assume the same value, the viscous modulus is higher than the storage modulus, which is characteristic of a liquid. This implies a sol-gel transition of a material. However, when the oscillation stress is decreased (step2), G' and G" show always lower values than those found during the increase in oscillation stress (Figure 1.28). By repeating the same sequence of steps in step 3 and 4 (Figure 1.29), G' and G" values follow the same trend and maintain the same values as those determined in step 2. This result indicates a sort of "recovery effect" of the G' and G" values once the material is subjected to further cycles of stress.

In case of squeezed hydrogels (Figure 1.29), both the storage modulus G' and loss modulus G" values are reduced due to the passage from the syringe with respect to those of the native hydrogels (Table 1.1). However, by repeating the same cycles of stress several times on the squeezed hydrogels as done on native hydrogels (Figure 1.29), G' and G" values remain same. This does not occur for a native hydrogel (step 1 and 2- Figure 1.28). The squeezing process induces the same effect on the hydrogel as the stress applied by the rheometer. The lower values of G' and G" of a squeezed hydrogel with respect to its native sample indicate that the hydrogel network is softened by the application of shear stress such as squeezing through a syringe or by the stress applied by a rheometer. However, additional stress does not lead to further alterations of the hydrogels mechanical properties [22].

- In another study, four different gel-forming polysaccharides, ι-carrageenan, κ-carrageenan, xantham gum, and Konjac gum were used for making hydrogels (for wound healing) [76]. These hydrogels are biocompatible, and they form helical structures and thus are stronger materials. DMA data shows some unusually high modulus for these hydrogels. In some cases, it is higher than those of contact lenses, which are considered to be some of the strongest known biodegradable hydrogels. DMA reports show that in these polysaccharide hydrogels the strength increases considerably after drying.

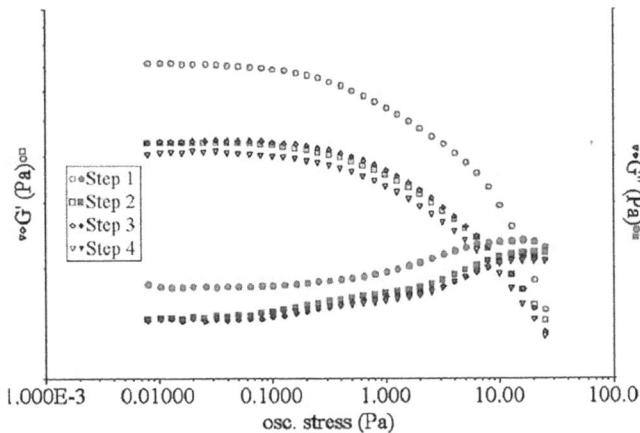

FIGURE 1.28 Typical trend of G' and G" values vs oscillation stress during sweep test performed on a polysaccharide-based hydrogel. The figure shows the G' and G" value vs oscillation stress for each step of test [22].

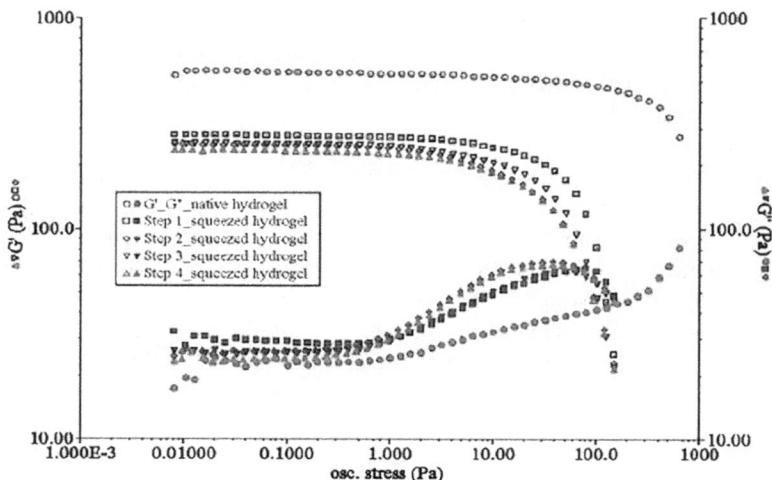

FIGURE 1.29 Stress sweep test performed on (a) CMC hydrogel, (b) hyal hydrogel, (c) CHT hydrogels, native and after being squeezed through a syringe. The figure shows the G' and G" value vs oscillation stress for each step of the test [22].

TABLE 1.1
Mechanical Properties of Different Types of Hydrogels

S. No.	Hydrogel Type	Status	G' (Pa)	G" (Pa)
1	CMC	Native	550 ± 30	25 ± 1
2	Carboxyl methyl cellulose	Squeezed	240 ± 20	20 ± 2
3	Hydal (hyaluronan)	Native squeezed	970 ± 25	65 ± 10
			340 ± 20	45 ± 5
4	CHT chitosan	Native squeezed	4,350 ± 65	165 ± 45
			2,460 ± 150	220 ± 15

- In yet another study, cellulose gel membrane was used as GPE (gel polymer electrolyte) in lithium-ion battery. This showed mechanical properties of GPE [17]. In this GPE based on cellulose membrane with 5% crosslinker, possessed good tensile fracture strength robust which increased along with increase of ECH from 5% to 9%. The tensile strength significantly increased from 14.61 to 46.21 MPa.

1.4.2 RHEOLOGICAL PROPERTIES

Rheology is a branch of mechanics, which deals with the study of deformation and flow of objects under an external force. The study of the rheological properties of polysaccharides is useful for applications in food industry, e.g., prediction of their gelation or thickening properties can aid the manufacture, distribution, storage, and consumption of food products [77]. Many naturally occurring polysaccharides are used as gelling agents, thickeners, and emulsifiers due to their unique rheological properties. Xanthan gum and gellan gum are most commonly used for such purposes. Examples:

- A polysaccharide fraction (FMPS) was isolated from floral mushrooms cultivated in Huangshan Mountain in Anhui Province of China. Their rheological properties in aqueous solutions were investigated [78]. The FMPS solution showed shear thinning behavior at 25°C. Dynamic viscoelastic tests revealed that G′ and G″ exhibited strong dependences on the concentration and temperature. The exponent n of G′ ~ Wn and "tan delta" also exhibited strong dependence on concenteration and temperature. The gel point (C gel) of FMPS solution was 1.16×10^{-2} gel mL^{-1} at 15°C, and the "T gel" of 1.14×10^{-2} g mL^{-1} FMPS solution was 20.6°C. Dynamic frequency sweep measurements indicated that the FMPS gel system was stable in the selected range of frequency. The heating cooling process proved that the sol-gel transition of FMPS in aqueous solutions was thermally reversibly.

- In another study, rheological properties of starch isolated from Avocado seeds were evaluated [79]. Avocado seed powder was suspended in solution containing 2 mM Tris, 7.5 mM NaCl, and 80 mM NaHSO$_3$ (solvent A) or sodium bisulfite solution (1,500 ppm SO$_2$, solvent B). Solvent type had no influence (p>0.05) on starch properties. Amylose content was 15%–16%. Gelatinization temperature range was 56°C–74°C, and peak temperature was 65.7°C. Maximum viscosity 380–390 Bu, breakdown -2 Bu, consistency 200 Bu, and setback 198 Bu. Avocado seed starch dispersions (5% W/V) were characterized as viscoelastic systems with G′>G″.

Avocado seed starch has potential applications as thickening agent and gelling agent in food systems, as a vehicle in pharmaceutical systems, and as an ingredient in biodegradable polymers for food packaging.

1.4.3 SWELLING PROPERTIES

The factors on which swelling property and the degree of swelling depend are the network density, solvent nature, polymer solvent interactions, crosslinking density, temperature, and surface-to-volume ratio. Besides all these factors, water or solvent appears as a plasticizer in a hydrophilic polymer network system [80,81]. Swelling theory of hydrogels was pioneered by Flory and Rehner 1943. This provided a simple conceptual and mathematical approach of swelling which further attracted a lot of interest in research perusals that have helped in its better understanding. The theory is related to the molecular weight of the chains which is said to be proportional to the extent of swelling (Figure 1.30) with larger values allowing more swelling.

When a solvent is added to a crosslinked polymer, it swells. The amount of swelling depends on two factors:

1. How soluble (uncrosslinked) the polymer is in the solvent and
2. The length (or in this theory the molecular weight) of the chains between crosslinks. Obviously, the shorter the length (the higher the crosslink density), the lesser the swelling.

Moreover, the quantity of water absorption by a hydrogel is an important measure, which is referred to as degree of swelling.

$$\text{Degree of swelling} = (\text{wet wt.} - \text{dry wt.})/\text{dry wt.} \times 100\%$$

The degree of swelling helps to find:

- The solute diffusion coefficient through the hydrogel
- The surface properties and surface mobility and also the mechanical properties.

Examples of highly swollen hydrogels are those of cellulose, poly(vinyl alcohol), and poly(ethylene glycol), since all these instances have lot of OH groups. Besides this concept, other factors which

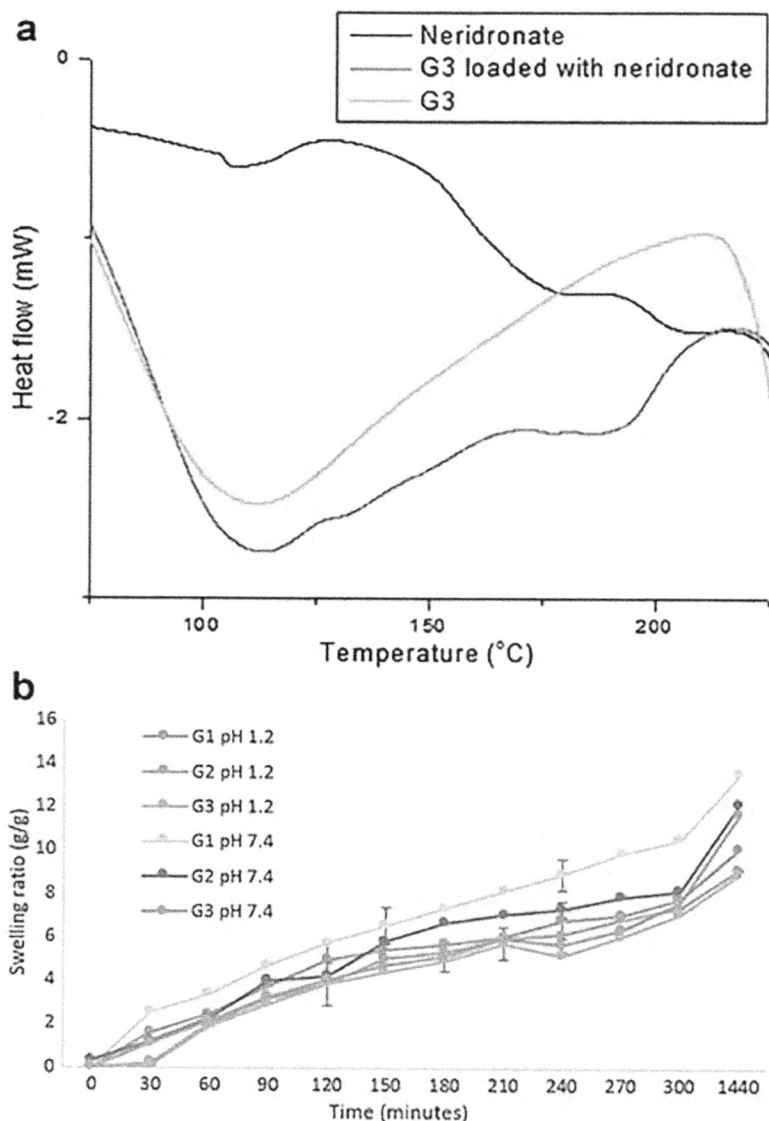

FIGURE 1.30 Swelling shown as ratio of (a) heat flow vs temperature. (b) Swelling vs time [82].

influence the swelling capacity are such as osmotic pressure, rise in temperature of water, and surface-to-volume ratio.

When content of water exceeds 95% of total weight (or volume), the hydrogels are categorized as SAPs (superabsorbent polymers). Some SAPS undergo distinct physical changes upon small environmental variations. These so-called "smart polymers" have the ability to sense environmental stimuli including changes in pH, temperature, light, and pressure [83]. Nowadays, these SAPs are usually made from synthetic polymers and are aimed to be replaced by naturally occurring polysaccharides and proteins, since these are biocompatible, biodegradable, renewable, and have less environmental impacts. Currently, the polysaccharides that are used in the manufacture of SAPs are based on alginate, chitosan, agrose, carrageenans, dextrin, cellulose, starch, gellan gum, xanthan, etc. These polysaccharides on their own or along with some suitable components show a great deal of swelling properties.

1.4.4 Miscellaneous Properties

Apart from the properties discussed above, polysaccharide-based polymeric gels have exhibited (and continue to do so) various physicochemical and some other properties. Several instances are highlighted as follows:

- Carbon nanotubes (CNTs) were incorporated into cellulose nanofibers (CNF) gel. The gelation process caused the shrinking of CNT/CNF gel film. These CNF/CNT gel films showed a conductivity of 5.20 s/cm after treatment with alkali, which is almost threefold higher than the CNF/CNT film (without alkali treatment), at 20 wt% CNTs, Conductivity further improved to, as high as 17.4 s/cm, when adding 50 wt% CNTs. This high-density 3D network provides for adequate electron transport pathways and gives the gel film remarkable electrical conductive properties [84].
- In the last decades, there has been a notable increasing interest in the development of plastics from polysaccharides as an alternative to replace synthetic plastics films. Therefore, a plasticized cassava starch matrix composite reinforced by a multiwalled carbon nanotube (MWCNT) and hercynite ($FeAl_2O_4$) nanomaterial was developed. This nanofiller hybrid shows strong mechanical interlocking with the matrix and provides excellent stability in water, ensuring a good dispersion in the starch matrix. The composite containing 0.04 wt% of the nanohybrid filler displays tremendous increments of 370% in Young's modulus, 350% in tensile toughness, and 70% decrease in water vapor permeability. The introduction of plasticizers such as glycerol and water induces crystallization without deterioration of the tensile strength [85].
- The effects on physicochemical properties of "in vitro" starch digestibility of composite potato starch/protein blends have been investigated. The proteins in blends were taken as 0%, 5%, 10%, or 15% during processing (i.e., cooking, cooling, and reheating). The results showed that the protein in the blend proportionally restricted starch granule swelling during cooking and facilitated amylopectin recrystallization during cold storage. It was seen that starch protein interaction reduced starch digestibility of the processed blends [86].

1.5 APPLICATIONS

The fascinating physicochemical, mechanical, rheological, structural and other unique properties of Polysaccharide-based polymeric gels made them to find applications in various fields of technology and are used in innumerable systems. It is not possible to discuss every aspect in an elaborated manner, yet we have tried to list some prominent applications and presented several instances to elucidate their wide range of scope and prospects in field of biomedics, industry, and agriculture.

1.5.1 Biomedical Applications

1.5.1.1 Drug Delivery

- **Controlled release:** Smart hydrogels, with high water content, soft and rubbery consistency, and low interfacial tension, and biological fluids are considered as excellent drug delivery systems for various bioactive molecules [87]. Characteristics like the swelling behavior in an aqueous medium, pH and/or temperature sensitivity, and zero-order kinetics play important roles in the development of hydrogel-based drug delivery systems [88]. The pH-responsive bacterial cellulose (BC)-g-PAA hydrogels prepared by electron beam (EB) irradiation were explored as an oral delivery system for proteins by using bovine serum albumin (BSA) as a model protein drug. BSA is loaded in the hydrogel via swelling diffusion method, and the entrapment efficiency (EE) of BSA was 4,055%. Hydrogels with greater swelling exhibited great EE. It was found that less than 10% of BSA was released in simulated gastric fluids (SGF) due to lower swelling of hydrogels at acidic pH 1.6, and a maximum release of 90% was observed at simulated intestinal fluids (SIF), because of higher swelling of hydrogel at higher pH 6.8. Subsequent conformational stability analyses indicated that the structural integrity and bioactivity of released BSA were maintained [89].

- **Targeted release:** Some dangerous pathogenic strains such as Methicillium-resistant *Staphylococcus aureus* are resistant to the wide range of conventional antibiotics. Antimicrobial peptides (AmPs) are remarkable antibacterial drugs, which can be used instead of conventional antibiotics due to their membrane targeting activity, especially against both gram-negative and gram-positive multidrug-resistant strains of pathogenic bacteria. Among these novel carriers, polysaccharide-based nanogels have incredible properties such as high-loading capacity, stimuli-responsive release behavior, simple large-scale production, and thermodynamic stability.

1.5.1.2 Tissue Engineering

Recently, some successful strategies have been constructed to prepare hydrogels with enhanced mechanical performance and other attractive properties, such as double network hydrogels and nanocomposite hydrogel, which are promising for tissue engineering applications [90,91].

CNCs (cellulose nanocrystals), which are typically isolated from cellulose, have been widely used as fillers to reinforce polymeric hydrogels due to their favorable mechanical properties and intrinsic biocompatibility. The unique aspect ratio and plenty of active –OH groups on their surface make CNCs a remarkable precursor to prepare a solid stabilizer. Yang et al. reported injectable hydrogels based on adipic acid dihydrazine-modified CMC, and aldehyde-modified dextran reinforced with CNCs, and aldehyde-functionalized CNCs (CHO-CNCs). Gelation occurred within seconds as the hydrogel components were extruded from double-barrel syringe, and the CNCs were observed to be evenly distributed throughout the composites. Swelling tests showed that all the CNCs reinforced hydrogels could maintain their structure for more than 60 days both in water and in 10 mM PBS, suggesting the possibility of long-term applications. The CHO-CNCs could act both as filler and a chemical crosslinker, making the hydrogels more elastic, more stable, and capable of facilitating higher NP loadings without compromising the mechanical strength compared to hydrogels with unmodified CNCs. Cytotoxicity tests found that both the starting components and the hydrogels showed good cytocompatiblity against NIH 3T3 fibroblast cells. Combined with their syringeability, the prepared CNC-reinforced injectable hydrogels could fill in irregular cavities and shapes without the need for pre-shape forming processing, thus they were found to be promising for bone tissue engineering.

The introduction of inorganic NPs (nanoparticles), such as LAPONITE, hydroxyapatite (HAp), and titanium doxide, can also reinforce cellulose-based hydrogels for bone tissue engineering applications. Boyer and colleagues developed a LAPONITE NP reinforce silated HPMC (Si-HPMC) hydrogel, in which the LAPONITE NPs were self-setting within the gel structure of Si_HPMS, resulting in a hybrid interpenetrating network (IPN). This IPN structure increased the mechanical properties of the hydrogel without interfering with the O_2 diffusion and cell viability after gelification. The ability of the hybrid scaffold containing the Si-HPMC/LAPOTINE hydrogel and chondrogenic cells to form cartilaginous tissue in vivo was investigated during 6 week implantation in subcutaneous pockets of nude mice. Histological analysis of the composite constructs revealed the formation of cartilage-like tissue with the extracellular matrix containing glycosaminoglycans and collagens. These results indicated the prepared Si-HPMC/LAPOTINE hydrogels have great potential for cartilage defect repairs.

Moreover, the morphology, inner pore structure, and uniformity of hydrogels play an important role in determining their biomedical applications. 3D printing technology provides a unique opportunity to develop biomedical scaffolds with uniform or gradient porosity required for tissue rapair and regeneration using biocompatible cellular-based nanocomposites. By 3D printing technology, Sultan et al. fabricated CNC-based hydrogel scaffolds with uniform and/or gradient porosity from a hydrogel ink of sodium alginate (SA) and gelatin reinforced with CNCs. The CNCs provided favorable rheological properties required for 3D printing. The printed scaffolds were crosslinked sequentially via covalent and ionic interactions, resulting in dimensionally stable hydrogel scaffolds with a pore size of 80–2,125 run and nanoscaled pore wall roughness which were favorable for cell interaction [89].

1.5.1.3 Wound Healing

Cellulose-based hydrogels are promising wound dressing materials since they are biocompatible, biodegradable, and environmentally friendly. Anjum et al. [92] developed a composite material containing nano-silver nanohydrogels (nSnH) along with Aloe vera and curcumin that promote antimicrobial activity, wound healing, and injection control. A PVA/polyethylene oxide/CMC matrix was used as a gel system to blend with nSnH, Aloe vera, and curcumin and was coated onto hydrogel polyester fabric to develop antimicrobial dressings. The cumulative release of Ag from the dressing was found to be about 42% of the total loading after 48 hours. The antimicrobial activity of the dressings was demonstrated against both *S. aureus* and *E. coli*. All of the dressings showed bacteria colony reduction above 70%, and the Gel/nSnH/Aloe was found to lead to complete (100%) reduction of *S. aureus*. In vivo wound healing was carried out on full thickness skin wounds created on Swiss albino mice. After 16 days, the Gel/nSnH/Aloe treated wounds were healed 100% with minimum scarring. By histological studies, Gel/nSnH/Aloe treated wound tissues were found to exhibit a thinner epidermal layer and organized collagen deposition without inflammation of healed tissues [93].

These results indicated that nSnH and Aloe vera–based dressing materials could be promising candidates for wound dressing [89].

1.5.1.4 Ophthalmic Uses

Cellulose derivatives such as methylcellulose (MC), carboxymethyl cellulose (CMC), and hydroxyl propyl methyl cellulose (HPMC) are naturally occurring polymers with thermo-responsive properties. They are commercially available in various products for the treatment of dry eyes, including Murocel (MC), Celluvisc (CMC sodium), and Ultra Tears (HPMC). Combining cellulose derivatives with poloxamers generates gelling systems with enhanced properties to favor ophthalmic drug delivery, e.g., huronic F-127 in combination with HPMC as a viscosity enhancing agent, and prolongs the ocular residence time of the drug. Other polymers which are investigated for enhancing ocular drug delivery include Xyloglucan.

A novel stimuli-responsive gel was designed by Yu et al. [94] for ophthalmic drug delivery. The hydrogel was made by chemical crosslinking of the biocompatible and pH-sensitive carboxymethyl chitosan (CMC) with the temperature-sensitive poloxamer 407 (F127) using gluteraldehyde as a crosslinking agent to form a pH-induced thermo-sensitive sol-gel transition at a very low concentration. The gel maintained controlled release of a model drug and showed no toxicity to human corneal epithelial cells [95].

Moreover, preparation of polysaccharide-based contact lenses, made from chitosan/gelatin composites were successfully synthesized and the products showed good properties [96]. The incorporation of non-crosslinked hydrophilic polysaccharides such as hyaluronic acid [97] and hydroxyl propylmethyl cellulose [98] in contact lenses provides a reduction in ocular dryness caused by the lenses and minimizes protein sorption through their slow release and effectively function as a wetting agent on the lens surface.

1.5.1.5 Bio-imaging

Bio-imaging such as fluorescent imaging (FLI), MRI, USI, photoacoustic imaging and tomographic scanning, has been recognized as an important technology in life science and medical fields. Fluorescent materials can be embedded in cellulose-based hydrogels, these materials include CdSe/ZnS QDs carbon dots (CDs), rare-earth doped phosphor ($SrAl_2O_4$;Eu^{2+}, Dy^{3+}), and S,N-codoped graphene QDs, For instance, fluorescent cellulose QDs hydrogels were obtained by embedding CdSe/

ZnS QDS in cellulose materials in NaoH/urea aqueous system via a mild chemical crosslinking process. These cellulose QDs hydrogels were found to exhibit a transparent appearance under visible light, while under a UV lamp, a single color emission (yellowish-green) was observed with QDs (0.06 wt%) indicating a high efficiency for the fluorescent emission (Figure 1.31). QDs with sizes in the range of 2.8–3.6 nm emitted different colors from green to red, depending on the size of the QDs. This suggested that the optical properties of QDs could be well maintained in the hydrogels, which are promising for applications in the fields of fluoroimmunoassay and biological labeling.

Finger print FLI is one of the most prominent technologies in the field of forensic medicine. Recently, Hai et al. [100] developed a C10-SCN stimulated reversible response lanthanide luminescent Tb (*m*)-CMC complex hydrogel, which consisted of 2,6-dimethyl pyridin-4-amine, functionalized with a CMC-binding aptamer and a terburin (*m*)-2-(2-aminobenzamide) benzoic acid complex modified with CMC, upon irradiation with 270 nm UV light, the imaging information of the fingerprint could be quenched and recovered by ClO-/SCN-regulation, respectively. This results in reversible on/off conversion of the luminescence signals for the encryption and decryption of multilevels of information as shown in Figure 1.32. These complex hydrogels can ensure the confidentiality of fingerprint information and thus opens the possibility of using tunable luminescent hydrogels for fingerprint information detection, security protection, and storage.

1.5.1.6 Self-Healing Sensors

The development of smart wearable devices has become a research hotspot due to their potential applications in health monitoring. Self-healing devices can restore their structure and functionality after damaging, endowing them with enhanced durability, reliability, and safety. Self-healing hydrogels have attracted significant attention for the development of self-healing wearable devices (pressure sensors) for human motion detection, due to their viscoelasticity, conductivity, and biocompatibility. Self-healing hydrogels should simultaneously meet the required mechanical toughness and excellent mobility properties, which are two contradictory characteristics. These criteria are met by CNCs which are commonly used as fillers to reinforce hydrogels due to their unique properties. Recently, CNCs have been successfully used to create integrated conductive hydrogels with both excellent self-healing and mechanical performance. Liu and coworkers developed a conductive, elastic, self-healing, and strain-sensitive polymer network hydrogel called "F-hydrogel". This was done via covalent crosslinking of PVA and PVP, in which a "hard" Fe^{3+} crosslinked CNC network with dynamic CNC-Fe^{3+} coordination bonds was interconnected as reinforcing domains. These "F-hydrogels" displayed

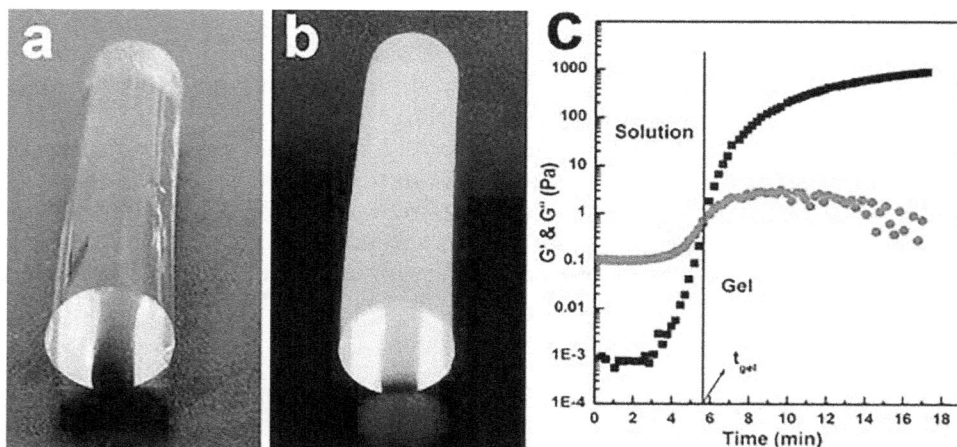

FIGURE 1.31 A cellulose QDs hydrogel with 0.06 wt% QDs (3.0 nm) under visible and UV lamp, respectively [99].

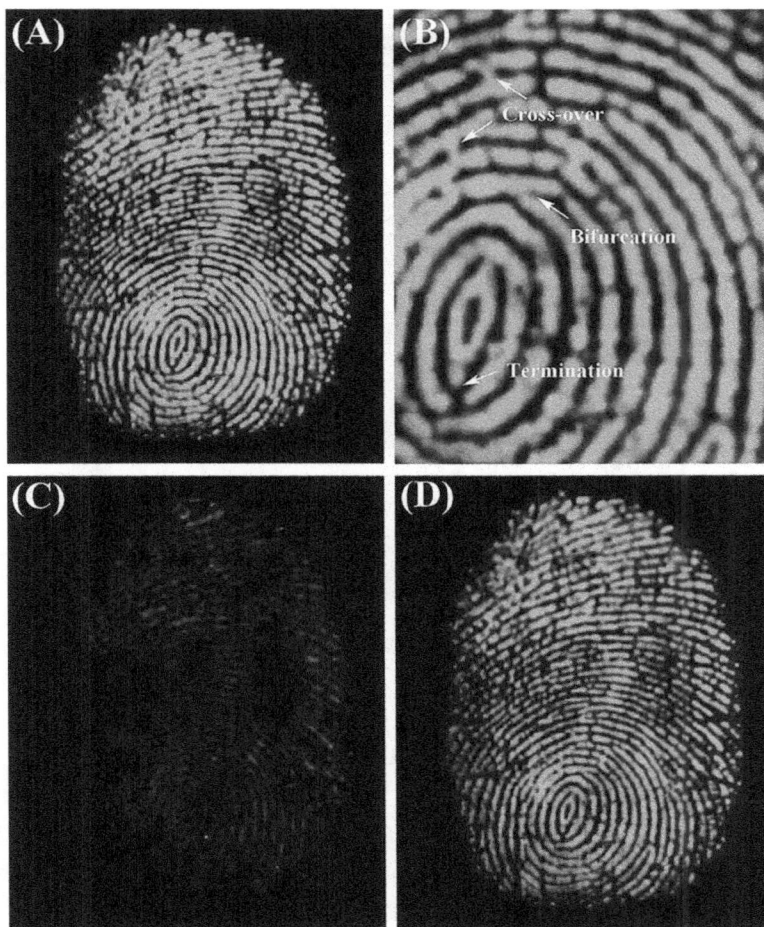

FIGURE 1.32 (a) Luminescence image of a fingerprint on a microscope slide. (b) Corresponding magnified image. (c and d) Luminescence images of the fingerprint in response to ClO- (c) and SCN- (d) [100].

ultrasensitive, stable, and repeatable resistance variations upon mechanical deformations and could be applied tightly on the skin surface, which were promising in the application of wearable devices. Based on these characteristics of the "F-hydrogels", a wearable soft strain sensor was assembled to monitor finger-point motions, breathing, and even a slight blood pulse. When the "F-hydrogel" sensor was put on the wrist of a volunteer, the blood pulse changes before and after the exercise.

Progress has been made in recent past in self-healing hydrogels, which are conductive and mechanically durable. Different methods are adopted for designing self-healing sensors based on hydrogels [101], which are biocompatible and respond to various stimuli such as strain, concenteration, temperature, pH, etc., and these changes are detected via electrical signals (Figure 1.33).

1.5.2 Industrial Applications

1.5.2.1 Food Packaging

Polysaccharides have played a tremendous role in replacing synthetic polymers for being biodegradable and biocompatible. They are used in food packaging in the form of edible films, membranes, coatings, and even modified to be used as thermoplastics for making packaging materials [102].

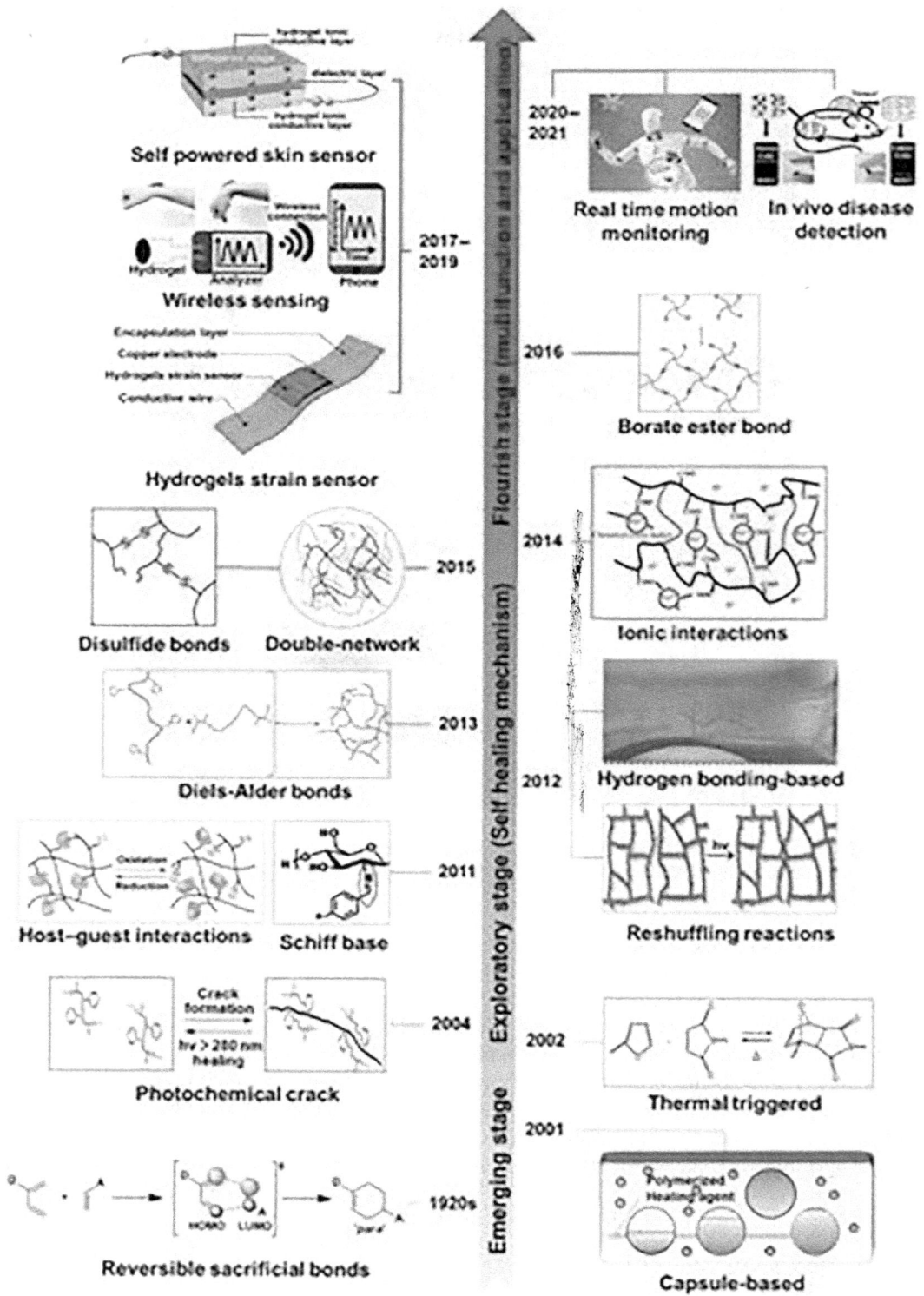

FIGURE 1.33 The figure above shows various types of bonding in hydrogels and their response as sensors via electrical signals [101].

1.5.2.2 Food Additives

They are also used as approved food additives; as thickeners, emulsifiers, stabilizers, and gelling agents in various food items. They are also used as edible films for preserving fish, fruits, vegetables, and dairy products. Polysaccharide gels are used in pharmaceuticals, cosmetics, fibers, textile dying, and textile printing, and also in the manufacture of various petroleum products. Polysaccharide nano-composites have become increasingly important in offering a green alternative to synthetic polymers for preparing soft nanomaterials. They have been used in composites with hard nanomaterials, such as metal nanoparticles and carbon-based nanomaterials [103]. Polysaccharide-based hydrogels due to their swelling property are made use of for making SAPs which in turn find wide range of applications in field of biomedics, agriculture, wastewater treatment, and in the manufacture of different types of diapers. They are also used in designing smart polymer. These so-called "smart polymers" have ability to sense environmental stimuli including very small changes in pH, temperature, light, pressure, etc. In batteries and energy storages devices, polysaccharide gels such as dense cellulose-based polymer gels are used as electrolytes in lithium-ion batteries. These are renewable low cost and have high discharge capacity. Meanwhile, researches are on for designing the next-generation, all-cellulose energy storage devices, having improved cyclic stability and specific capacitance [104].

1.5.3 Agricultural Applications

1.5.3.1 Fertilizer Release

Polysaccharide-based gels are used to release compounds like fertilizers into the soil in a controlled manner, by which their benefit remains to be there, for the longer period of time. The desired bioactive agents are incorporated into polymer matrix, and the gel expansion is controlled by the presence of moisture, which leads to their swelling. The controlled release effect is optimum usually when the soil water content is lower than 45% (w/w), temperature is below 35°C, and the soil pH is in the range from weak acid to neutral [105].

1.5.3.2 Pest Control

Hydrogels are also used for the controlled allowance or release of other chemicals which protect the yield against pests (pesticides), weeds (herbicides), and fungus (fungicides).

1.5.3.3 Soil Conditioning

In agricultural field, hydrogels are used for increasing the amount of available moisture in the root zones, thus implying longer intervals between irrigation and for improving the physical properties of soil which include water-holding pores, bulk density, porosity, structural stability, infiltration, and hydraulic conductivity. With controlled release mechanism, the functionalized polymers can be increased for enhancing the absorption of nutrients by plants.

1.6 CONCLUSION

Most of the polysaccharides are cheap, easily available, and important ones like cellulose and chitin are present in nature in abundance. Due to their biodegradable, biocompatible, and renewable nature, as well as their nontoxic and minimal environmental impact, polysaccharides naturally become the most preferred choice for the application of hydrogels and all other types of gels. For hydrogels preparations, they have the ability to be used alone or with various kinds of copolymers or/and nanomaterials to form unique composites which show extraordinary performances. Polysaccharide-based polymeric gels have usually a lot of hydroxyl groups present in their structures, which is a pathway for their various types of derivatizations. This makes them useful in diverse applications. With desirable and modifiable properties such as mechanical, rheological, swelling, physicochemical and being stimuli-responsive, etc.; polysaccharide-based polymeric gels presently have numerous applications and a great potential for interesting research with bright future prospects.

ACKNOWLEDGMENTS

QA and SSM are highly thankful to Principal, Vishwa Bharti Degree College, Rainawari, Srinagar, and Director, National Institute of Technology, Srinagar, for providing basic infrastructural and laboratory facilities, respectively. SH is highly thankful for the fellowship under CSIR-JRF Scheme [File No: 09/0984(15809)2022-EMR-I]. RA is highly thankful to Ministry of Education (MoE), for financial assistance through Institute fellowship.

REFERENCES

1. Beatrice CAG, Rosa-Sibakov N, Lille M, Sözer N, Poutanen K, Ketoja JA. Structural properties and foaming of plant cell wall polysaccharide dispersions. *Carbohydrate Polymers* 2017;173:508–18. https://doi.org/10.1016/j.carbpol.2017.06.028.
2. Avashthi G, Maktedar SS, Singh M. Surface-induced in situ sonothermodynamically controlled functionalized graphene oxide for in vitro cytotoxicity and antioxidant evaluations. *ACS Omega* 2019;4:16385–401. https://doi.org/10.1021/acsomega.9b01939.
3. Malik P, Maktedar SS, Avashthi G, Mukherjee TK, Singh M. Robust curcumin-mustard oil emulsions for pro to anti-oxidant modulation of graphene oxide. *Arabian Journal of Chemistry* 2020;13:4606–28. https://doi.org/10.1016/j.arabjc.2019.10.011.
4. Chanda M. *Introduction to Polymer Science and Chemistry: A Problem-Solving Approach*, 2nd ed., CRC Press; 2013. https://doi.org/10.1201/b14577.
5. Nayak AK, Das B. Introduction to polymeric gels. *Polymeric Gels* 2018:3–27. https://doi.org/10.1016/b978-0-08-102179-8.00001-6.
6. Nayak AK, Bera H. In situ polysaccharide-based gels for topical drug delivery applications. *Polysaccharide Carriers for Drug Delivery* 2019:615–38. https://doi.org/10.1016/b978-0-08-102553-6.00021-0.
7. Yousuf S, Maktedar SS. Influence of quince seed mucilage-alginate composite hydrogel coatings on quality of fresh walnut kernels during refrigerated storage. *Journal of Food Science and Technology* 2022;59:4801–11. https://doi.org/10.1007/s13197-022-05566-2.
8. Yousuf S, Maktedar SS. Utilization of quince (Cydonia oblonga) seeds for production of mucilage: functional, thermal and rheological characterization. *Sustainable Food Technology* 2023;1:107–15. https://doi.org/10.1039/d2fb00010e.
9. Ibrahim NA, Nada AA, Eid BM. Polysaccharide-based polymer gels and their potential applications. *Polymer Gels* 2018:97–126. https://doi.org/10.1007/978-981-10-6083-0_4.
10. Thakur VK, Thakur MK, Voicu SI (Eds.). *Polymer Gels: Perspectives and Applications*. Berlin, Springer; 2018. https://doi.org/10.1007/978-981-10-6080-9.
11. Fekete T, Borsa J. Polysaccharide-based polymer gels. *Polymer Gels* 2018:147–229. https://doi.org/10.1007/978-981-10-6086-1_5.
12. Buruiana LI, Ioan S. Polymer gel composites for bio-applications. *Polymer Gels* 2018:111–23. https://doi.org/10.1007/978-981-10-6080-9_5.
13. Li MX, Wang XW, Yang YQ, Chang Z, Wu YP, Holze R. A dense cellulose-based membrane as a renewable host for gel polymer electrolyte of lithium ion batteries. *Journal of Membrane Science* 2015;476:112–8. https://doi.org/10.1016/j.memsci.2014.10.056.
14. Tokita M, Nishinari K. Gels: structures, properties, and functions. *Surface and Colloid Science* 2004;129:95-104. https://doi.org/10.1007/978-3-642-00865-8.
15. Raghavan SR. Distinct character of surfactant gels: a smooth progression from micelles to fibrillar networks. *Langmuir* 2009;25:8382–5. https://doi.org/10.1021/la901513w.
16. Echeverria C, Fernandes S, Godinho M, Borges J, Soares P. Functional stimuli-responsive gels: hydrogels and microgels. *Gels* 2018;4:54. https://doi.org/10.3390/gels4020054.
17. Du Z, Su Y, Qu Y, Zhao L, Jia X, Mo Y, et al. A mechanically robust, biodegradable and high performance cellulose gel membrane as gel polymer electrolyte of lithium-ion battery. *Electrochimica Acta* 2019;299:19–26. https://doi.org/10.1016/j.electacta.2018.12.173.
18. Al-Kinani AA, Zidan G, Elsaid N, Seyfoddin A, Alani AWG, Alany RG. Ophthalmic gels: Past, present and future. *Advanced Drug Delivery Reviews* 2018;126:113–26. https://doi.org/10.1016/j.addr.2017.12.017.
19. Serrero A, Trombotto S, Cassagnau P, Bayon Y, Gravagna P, Montanari S, et al. Polysaccharide gels based on chitosan and modified starch: Structural characterization and linear viscoelastic behavior. *Biomacromolecules* 2010;11:1534–43. https://doi.org/10.1021/bm1001813.

20. Hussain S, Maktedar SS. Structural, functional and mechanical performance of advanced graphene-based composite hydrogels. *Results in Chemistry* 2023;6:101029. https://doi.org/10.1016/j.rechem.2023.101029.
21. Bai H, Deng S, Bai D, Zhang Q, Fu Q. Recent advances in processing of stereocomplex-type polylactide. *Macromolecular Rapid Communications* 2017;38:1700454. https://doi.org/10.1002/marc.201700454.
22. Pasqui D, De Cagna M, Barbucci R. Polysaccharide-based hydrogels: The key role of water in affecting mechanical properties. *Polymers* 2012;4:1517–34. https://doi.org/10.3390/polym4031517.
23. Eo MY, Fan H, Cho YJ, Kim SM, Lee SK. Cellulose membrane as a biomaterial: from hydrolysis to depolymerization with electron beam. *Biomaterials Research* 2016;20. https://doi.org/10.1186/s40824-016-0065-3.
24. Sannino A, Demitri C, Madaghiele M. Biodegradable cellulose-based hydrogels: Design and applications. *Materials* 2009;2:353–73. https://doi.org/10.3390/ma2020353.
25. Chen Y-L, Zhang X, You T-T, Xu F. Deep eutectic solvents (DESs) for cellulose dissolution: a mini-review. *Cellulose* 2018;26:205–13. https://doi.org/10.1007/s10570-018-2130-7.
26. Chang C, Duan B, Cai J, Zhang L. Superabsorbent hydrogels based on cellulose for smart swelling and controllable delivery. *European Polymer Journal* 2010;46:92–100. https://doi.org/10.1016/j.eurpolymj.2009.04.033.
27. Chang C, He M, Zhou J, Zhang L. Swelling behaviors of pH- and salt-responsive cellulose-based hydrogels. *Macromolecules* 2011;44:1642–8. https://doi.org/10.1021/ma102801f.
28. Pérez-Madrigal MM, Edo MG, Alemán C. Powering the future: Application of cellulose-based materials for supercapacitors. *Green Chemistry* 2016;18:5930–56. https://doi.org/10.1039/c6gc02086k.
29. Ding F, Wu S, Wang S, Xiong Y, Li Y, Li B, et al. A dynamic and self-crosslinked polysaccharide hydrogel with autonomous self-healing ability. *Soft Matter* 2015;11:3971–6. https://doi.org/10.1039/c5sm00587f.
30. Kabir SMF, Sikdar PP, Haque B, Bhuiyan MAR, Ali A, Islam MN. Cellulose-based hydrogel materials: chemistry, properties and their prospective applications. *Progress in Biomaterials* 2018;7:153–74. https://doi.org/10.1007/s40204-018-0095-0.
31. Sayyed AJ, Deshmukh NA, Pinjari DV. A critical review of manufacturing processes used in regenerated cellulosic fibres: viscose, cellulose acetate, cuprammonium, LiCl/DMAc, ionic liquids, and NMMO based lyocell. *Cellulose* 2019;26:2913–40. https://doi.org/10.1007/s10570-019-02318-y.
32. Zhou J, Zhang L, Deng Q, Wu X. Synthesis and characterization of cellulose derivatives prepared in NaOH/urea aqueous solutions. *Journal of Polymer Science Part A: Polymer Chemistry* 2004;42:5911–20. https://doi.org/10.1002/pola.20431.
33. Chang C, Zhang L. Cellulose-based hydrogels: Present status and application prospects. *Carbohydrate Polymers* 2011;84:40–53. https://doi.org/10.1016/j.carbpol.2010.12.023.
34. Marc? G, Mele G, Palmisano L, Pulito P, Sannino A. Environmentally sustainable production of cellulose-based superabsorbent hydrogels. *Green Chemistry* 2006;8:439. https://doi.org/10.1039/b515247j.
35. Dong W, Han B, Feng Y, Song F, Chang J, Jiang H, et al. Pharmacokinetics and biodegradation mechanisms of a versatile carboxymethyl derivative of chitosan in rats: In vivo and in vitro evaluation. *Biomacromolecules* 2010;11:1527–33. https://doi.org/10.1021/bm100158p.
36. Ayouch I, Kassem I, Kassab Z, Barrak I, Barhoun A, Jacquemin J, et al. Crosslinked carboxymethyl cellulose-hydroxyethyl cellulose hydrogel films for adsorption of cadmium and methylene blue from aqueous solutions. *Surfaces and Interfaces* 2021;24:101124. https://doi.org/10.1016/j.surfin.2021.101124.
37. Voets M, Antes I, Scherer C, Müller-Vieira U, Biemel K, Marchais-Oberwinkler S, et al. Synthesis and evaluation of heteroaryl-substituted dihydronaphthalenes and indenes: Potent and selective inhibitors of aldosterone synthase (CYP11B2) for the treatment of congestive heart failure and myocardial fibrosis. *Journal of Medicinal Chemistry* 2006;49:2222–31. https://doi.org/10.1021/jm060055x.
38. Tosh B, Saikia CN, Dass NN. Homogeneous esterification of cellulose in the lithium chloride–N,N-dimethylacetamide solvent system: Effect of temperature and catalyst. *Carbohydrate Research* 2000;327:345–52. https://doi.org/10.1016/s0008-6215(00)00033-1.
39. Mathur AM, Moorjani SK, Scranton AB. Methods for synthesis of hydrogel networks: A review. *Journal of Macromolecular Science, Part C: Polymer Reviews* 1996;36:405–30. https://doi.org/10.1080/15321799608015226.
40. Zhou J, Chang C, Zhang R, Zhang L. Hydrogels prepared from unsubstituted cellulose in NaOH/urea aqueous solution. *Macromolecular Bioscience* 2007;7:804–9. https://doi.org/10.1002/mabi.200700007.

41. Sannino A, Madaghiele M, Lionetto MG, Schettino T, Maffezzoli A. A cellulose-based hydrogel as a potential bulking agent for hypocaloric diets: An in vitro biocompatibility study on rat intestine. *Journal of Applied Polymer Science* 2006;102:1524–30. https://doi.org/10.1002/app.24468.

42. Herrera MA, Mathew AP, Oksman K. Barrier and mechanical properties of plasticized and cross-linked nanocellulose coatings for paper packaging applications. *Cellulose* 2017;24:3969–80. https://doi.org/10.1007/s10570-017-1405-8.

43. Lu Z-H, Donner E, Yada RY, Liu Q. Physicochemical properties and in vitro starch digestibility of potato starch/protein blends. *Carbohydrate Polymers* 2016;154:214–22. https://doi.org/10.1016/j.carbpol.2016.08.055.

44. Buléon A, Colonna P, Planchot V, Ball S. Starch granules: structure and biosynthesis. *International Journal of Biological Macromolecules* 1998;23:85–112. https://doi.org/10.1016/s0141-8130(98)00040-3.

45. Tako M. The principle of polysaccharide gels. *Advances in Bioscience and Biotechnology* 2015;06:22–36. https://doi.org/10.4236/abb.2015.61004.

46. Abd El-Mohdy HL, Hegazy EA, El-Nesr EM, El-Wahab MA. Synthesis, characterization and properties of radiation-induced Starch/(EG-co-MAA) hydrogels. *Arabian Journal of Chemistry* 2016;9:S1627–35. https://doi.org/10.1016/j.arabjc.2012.04.022.

47. Aegerter MA. Welcoming new coeditor Andrei Jitianu. *Journal of Sol-Gel Science and Technology* 2020;93:471–2. https://doi.org/10.1007/s10971-020-05246-7.

48. Ismail H, Irani M, Ahmad Z. Starch-based hydrogels: Present status and applications. *International Journal of Polymeric Materials* 2013;62:411–20. https://doi.org/10.1080/00914037.2012.719141.

49. Abdollahi Z, Zare EN, Salimi F, Goudarzi I, Tay FR, Makvandi P. Bioactive carboxymethyl starch-based hydrogels decorated with CuO nanoparticles: Antioxidant and antimicrobial properties and accelerated wound healing in vivo. *International Journal of Molecular Sciences* 2021;22:2531. https://doi.org/10.3390/ijms22052531.

50. Duquette D, Nzediegwu C, Portillo-Perez G, Dumont M, Prasher S. Eco-friendly synthesis of hydrogels from starch, citric acid, and itaconic acid: Swelling capacity and metal chelation properties. *Starch - Stärke* 2020;72:1900008. https://doi.org/10.1002/star.201900008.

51. Li Y, Liu C, Tan Y, Xu K, Lu C, Wang P. In situ hydrogel constructed by starch-based nanoparticles via a Schiff base reaction. *Carbohydrate Polymers* 2014;110:87–94. https://doi.org/10.1016/j.carbpol.2014.03.058.

52. Nilsen-Nygaard J, Strand S, Vårum K, Draget K, Nordgård C. Chitosan: Gels and interfacial properties. *Polymers* 2015;7:552–79. https://doi.org/10.3390/polym7030552.

53. Shen X, Shamshina JL, Berton P, Gurau G, Rogers RD. Hydrogels based on cellulose and chitin: fabrication, properties, and applications. *Green Chemistry* 2016;18:53–75. https://doi.org/10.1039/c5gc02396c.

54. Wu S, Duan B, Lu A, Wang Y, Ye Q, Zhang L. Biocompatible chitin/carbon nanotubes composite hydrogels as neuronal growth substrates. *Carbohydrate Polymers* 2017;174:830–40. https://doi.org/10.1016/j.carbpol.2017.06.101.

55. Liu C, Liu H, Tang K, Zhang K, Zou Z, Gao X. High-strength chitin based hydrogels reinforced by tannic acid functionalized graphene for congo red adsorption. *Journal of Polymers and the Environment* 2020;28:984–94. https://doi.org/10.1007/s10924-020-01663-5.

56. Xie Y, Liao X, Zhang J, Yang F, Fan Z. Novel chitosan hydrogels reinforced by silver nanoparticles with ultrahigh mechanical and high antibacterial properties for accelerating wound healing. *International Journal of Biological Macromolecules* 2018;119:402–12. https://doi.org/10.1016/j.ijbiomac.2018.07.060.

57. Muzzarelli RAA. Genipin-crosslinked chitosan hydrogels as biomedical and pharmaceutical aids. *Carbohydrate Polymers* 2009;77:1–9. https://doi.org/10.1016/j.carbpol.2009.01.016.

58. Qureshi D, Nayak AK, Kim D, Maji S, Anis A, Mohanty B, Pal K. Polysaccharide-based polymeric gels as drug delivery vehicles. In *Advances and Challenges in Pharmaceutical Technology* (pp. 283-325). Academic Press; 2021. https://doi.org/10.1016/B978-0-12-820043-8.00013-X.

59. Gomathi T, Susi, S, Abirami, D, Sudha, PN. Size optimization and thermal studies on calcium alginate nanoparticles. *IOSR Journal of Pharmacy* n.d.:01–7.

60. Templeman JR, Rogers MA, Cant JP, McBride BW, Osborne VR. Effects of a wax organogel and alginate gel complex on holy basil (Ocimum sanctum) in vitro ruminal dry matter disappearance and gas production. *Journal of the Science of Food and Agriculture* 2018;98:4488–94. https://doi.org/10.1002/jsfa.8973.

61. Chan AW, Whitney RA, Neufeld RJ. Semisynthesis of a controlled stimuli-responsive alginate hydrogel. *Biomacromolecules* 2009;10:609–16. https://doi.org/10.1021/bm801316z.

62. Li S, Yan Y, Xiong Z, Zhang CWR, Wang X. Gradient hydrogel construct based on an improved cell assembling system. *Journal of Bioactive and Compatible Polymers* 2009;24:84–99. https://doi.org/10.1177/0883911509103357.

63. Yang J-S, Xie Y-J, He W. Research progress on chemical modification of alginate: A review. *Carbohydrate Polymers* 2011;84:33–9. https://doi.org/10.1016/j.carbpol.2010.11.048.

64. Jeon O, Bouhadir KH, Mansour JM, Alsberg E. Photocrosslinked alginate hydrogels with tunable bio-degradation rates and mechanical properties. *Biomaterials* 2009;30:2724–34. https://doi.org/10.1016/j.biomaterials.2009.01.034.

65. Giri TK, Kumarasamy D, Mukherjee S, Das M. Prospect of plant and algal polysaccharides-based hydrogels. *Plant and Algal Hydrogels for Drug Delivery and Regenerative Medicine* 2021:37–73. https://doi.org/10.1016/b978-0-12-821649-1.00009-x.

66. Rinaudo M. Gelation of polysaccharides. *Journal of Intelligent Material Systems and Structures* 1993;4:210–5. https://doi.org/10.1177/1045389x9300400210.

67. JE, Morlock CM, Alsberg E. Dual ionic and photo-crosslinked alginate hydrogels for micropatterned spatial control of material properties and cell behavior. *Bioconjugate Chemistry* 2015;26:1339–47. https://doi.org/10.1021/acs.bioconjchem.5b00117.

68. Wijayapala R, Hashemnejad SM, Kundu S. Carbon nanodots crosslinked photoluminescent alginate hydrogels. *RSC Advances* 2017;7:50389–95. https://doi.org/10.1039/c7ra09805g.

69. Yashaswini DGV, Venkatesan J, Anil S. *Hydrocolloids from Marine Macroalgae: Isolation and Applications. Algae for Food.* 1st ed., CRC Press; 2021, p. 16.

70. Garcia RB, Vidal RRL, Rinaudo M. Preparation and structural characterization of O-acetyl agarose with low degree of substitution. *Polímeros* 2000;10:155–61. https://doi.org/10.1590/s0104-14282000000300012.

71. Tako M, Tamaki Y, Teruya T, Takeda Y. The principles of starch gelatinization and retrogradation. *Food And Nutrition Sciences* 2014;05:280–91. https://doi.org/10.4236/fns.2014.53035.

72. de Gennes PG, Witten TA. Scaling concepts in polymer physics. *Physics Today* 1980;33:51–4. https://doi.org/10.1063/1.2914118.

73. Flory PJ, Rehner J. Statistical mechanics of cross-linked polymer networks I. Rubberlike elasticity. *The Journal of Chemical Physics* 1943;11:512–20. https://doi.org/10.1063/1.1723791.

74. Peppas N. Hydrogels in pharmaceutical formulations. *European Journal of Pharmaceutics and Biopharmaceutics* 2000;50:27–46. https://doi.org/10.1016/s0939-6411(00)00090-4.

75. Wee JS-H, Chai AB, Ho J-H. Fabrication of shape memory natural rubber using palmitic acid. *Journal of King Saud University - Science* 2017;29:494–501. https://doi.org/10.1016/j.jksus.2017.09.003.

76. Juris S, Mueller A, Smith B, Johnston S, Walker R, Kross R. Biodegradable polysaccharide gels for skin scaffolds. *Journal of Biomaterials and Nanobiotechnology* 2011;02:216–25. https://doi.org/10.4236/jbnb.2011.23027.

77. Li Z, Lin Z. Recent advances in polysaccharide-based hydrogels for synthesis and applications. *Aggregate* 2021;2. https://doi.org/10.1002/agt2.21.

78. Xu J-L, Zhang J-C, Liu Y, Sun H-J, Wang J-H. Rheological properties of a polysaccharide from floral mushrooms cultivated in Huangshan Mountain. *Carbohydrate Polymers* 2016;139:43–9. https://doi.org/10.1016/j.carbpol.2015.12.011.

79. Chel-Guerrero L, Barbosa-Martín E, Martínez-Antonio A, González-Mondragón E, Betancur-Ancona D. Some physicochemical and rheological properties of starch isolated from avocado seeds. *International Journal of Biological Macromolecules* 2016;86:302–8. https://doi.org/10.1016/j.ijbiomac.2016.01.052.

80. Lima-Tenório MK, Tenório-Neto ET, Guilherme MR, Garcia FP, Nakamura CV, Pineda EAG, et al. Water transport properties through starch-based hydrogel nanocomposites responding to both pH and a remote magnetic field. *Chemical Engineering Journal* 2015;259:620–9. https://doi.org/10.1016/j.cej.2014.08.045.

81. Otsuka E, Suzuki A. Swelling properties of physically cross-linked PVA gels prepared by a cast-drying method. *Gels: Structures, Properties, and Functions* 2009:121–6. https://doi.org/10.1007/978-3-642-00865-8_17.

82. Aderibigbe BA, Ray SS. Gum acacia polysaccharide-based pH sensitive gels for targeted delivery of neridronate. *Polymer Bulletin* 2016;74:2641–55. https://doi.org/10.1007/s00289-016-1857-2.

83. Schierbaum. Elias, Hans Georg. *An Introduction to Polymer Science.* XXII, 470 pages, 167 Figures, 65 Tables, Hardcover, VCH Verlagsgesellschaft mbH, Weinheim, New York, Basel, Cambridge, Tokyo, 1997. DM 88,–/öS 642,–/sF 80. ISBN 3-527-28290-6. Starch - Stärke 1997;49:329–329. https://doi.org/10.1002/star.19970490716.

84. Chen C, Mo M, Chen W, Pan M, Xu Z, Wang H, et al. Highly conductive nanocomposites based on cellulose nanofiber networks via NaOH treatments. *Composites Science and Technology* 2018;156:103–8. https://doi.org/10.1016/j.compscitech.2017.12.029.

85. Morales NJ, Candal R, Famá L, Goyanes S, Rubiolo GH. Improving the physical properties of starch using a new kind of water dispersible nano-hybrid reinforcement. *Carbohydrate Polymers* 2015;127:291–9. https://doi.org/10.1016/j.carbpol.2015.03.071.

86. Lu Z-H, Donner E, Yada RY, Liu Q. Physicochemical properties and in vitro starch digestibility of potato starch/protein blends. *Carbohydrate Polymers* 2016;154:214–22. https://doi.org/10.1016/j.carbpol.2016.08.055.

87. Nayak AK, Bera H. In situ polysaccharide-based gels for topical drug delivery applications. *Polysaccharide Carriers for Drug Delivery* 2019;615–38. https://doi.org/10.1016/b978-0-08-102553-6.00021-0.

88. Lodhi BA, Hussain MA, Sher M, Haseeb MT, Ashraf MU, Hussain SZ, et al. Polysaccharide-based super-porous, superabsorbent, and stimuli responsive hydrogel from sweet basil: A novel material for sustained drug release. *Advances in Polymer Technology* 2019;2019:1–11. https://doi.org/10.1155/2019/9583516.

89. Fu L-H, Qi C, Ma M-G, Wan P. Multifunctional cellulose-based hydrogels for biomedical applications. *Journal of Materials Chemistry B* 2019;7:1541–62. https://doi.org/10.1039/c8tb02331j.

90. Mohtashamian S, Boddohi S. Nanostructured polysaccharide-based carriers for antimicrobial peptide delivery. *Journal of Pharmaceutical Investigation* 2016;47:85–94. https://doi.org/10.1007/s40005-016-0289-1.

91. Diekjürgen D, Grainger DW. Polysaccharide matrices used in 3D in vitro cell culture systems. *Biomaterials* 2017;141:96–115. https://doi.org/10.1016/j.biomaterials.2017.06.020.

92. Nazir F, Ashraf I, Iqbal M, Ahmad T, Anjum S. 6-deoxy-aminocellulose derivatives embedded soft gelatin methacryloyl (GelMA) hydrogels for improved wound healing applications: In vitro and in vivo studies. *International Journal of Biological Macromolecules* 2021;185:419–33. https://doi.org/10.1016/j.ijbiomac.2021.06.112.

93. Zhu T, Mao J, Cheng Y, Liu H, Lv L, Ge M, et al. Recent progress of polysaccharide-based hydrogel interfaces for wound healing and tissue engineering. *Advanced Materials Interfaces* 2019;6:1900761. https://doi.org/10.1002/admi.201900761.

94. Xu Y, Fu S, Liu F, Yu H, Gao J. Multi-stimuli-responsiveness of a novel polydiacetylene-based supramolecular gel. *Soft Matter* 2018;14:8044–50. https://doi.org/10.1039/c8sm01515e.

95. Al-Kinani AA, Zidan G, Elsaid N, Seyfoddin A, Alani AWG, Alany RG. Ophthalmic gels: Past, present and future. *Advanced Drug Delivery Reviews* 2018;126:113–26. https://doi.org/10.1016/j.addr.2017.12.017.

96. Xin-Yuan S, Tian-Wei T. New contact lens based on chitosan/gelatin composites. *Journal of Bioactive and Compatible Polymers* 2004;19:467–79. https://doi.org/10.1177/0883911504048410.

97. Ali M, Byrne ME. Controlled release of high molecular weight hyaluronic acid from molecularly imprinted hydrogel contact lenses. *Pharmaceutical Research* 2009;26:714–26. https://doi.org/10.1007/s11095-008-9818-6.

98. White CJ, McBride MK, Pate KM, Tieppo A, Byrne ME. Extended release of high molecular weight hydroxypropyl methylcellulose from molecularly imprinted, extended wear silicone hydrogel contact lenses. *Biomaterials* 2011;32:5698–705. https://doi.org/10.1016/j.biomaterials.2011.04.044.

99. Chang C, Peng J, Zhang L, Pang D-W. Strongly fluorescent hydrogels with quantum dots embedded in cellulose matrices. *Journal of Materials Chemistry* 2009;19:7771. https://doi.org/10.1039/b908835k.

100. Hai J, Li T, Su J, Liu W, Ju Y, Wang B, et al. Reversible response of luminescent terbium(III)-nanocellulose hydrogels to anions for latent fingerprint detection and encryption. *Angewandte Chemie* 2018;130:6902–6. https://doi.org/10.1002/ange.201800119.

101. Qin T, Liao W, Yu L, Zhu J, Wu M, Peng Q, et al. Recent progress in conductive self-healing hydrogels for flexible sensors. *Journal of Polymer Science* 2022;60:2607–34. https://doi.org/10.1002/pol.20210899.

102. Ferreira A, Alves V, Coelhoso I. Polysaccharide-based membranes in food packaging applications. *Membranes* 2016;6:22. https://doi.org/10.3390/membranes6020022.

103. Zheng Y, Monty J, Linhardt RJ. Polysaccharide-based nanocomposites and their applications. *Carbohydrate Research* 2015;405:23–32. https://doi.org/10.1016/j.carres.2014.07.016.

104. Pérez-Madrigal MM, Edo MG, Alemán C. Powering the future: application of cellulose-based materials for supercapacitors. *Green Chemistry* 2016;18:5930–56. https://doi.org/10.1039/c6gc02086k.

105. Yang X, Geng J, Li C, Zhang M, Tian X. Cumulative release characteristics of controlled-release nitrogen and potassium fertilizers and their effects on soil fertility, and cotton growth. *Scientific Reports* 2016;6. https://doi.org/10.1038/srep39030.

2 Environmental Aspects, Recycling, and Sustainability of Polysaccharides

Liqaa Hamid
Graz University of Technology (TU Graz)

Irene Samy
Nile University

2.1 PREFACE

As the world's ability to cope with the fast-rising manufacturing of disposable plastic products overwhelms the world's ability to deal with them, plastic pollution has become one of the most important environmental challenges. The low cost of producing primary plastic packaging materials from petroleum resources, when compared to recycled plastics, has resulted in the largest virgin plastic market, resulting in significant amounts of plastic in landfills and the natural environment [1]. Plastic pollution is especially noticeable in impoverished Asian and African countries, where rubbish collection systems are sometimes ineffective or non-existent [2]. However, the industrialized world, particularly in nations with poor recycling rates, has difficulty collecting discarded plastics properly. Plastic waste has become so pervasive that efforts are underway to draft a global treaty. Plastics are synthetic polymers that are qualified for being utilized in plenty of industries, whether as tools or as products. The manufacturing and development of thousands of new plastic products have surged in recent years, altering the modern age to the point that life would be unrecognizable without them [2]. However, the conveniences that plastics give have resulted in a throw-away culture that exposes the material's dark side: single-use plastics (SUPs) now account for a large percentage of all plastic manufactured every day. Many of these products, such as plastic bags and food wrappers, have a short lifespan but because of their durability, they often stay intact for a long time in the environment resulting in fragmentation that produces microplastics. Over time, more discoveries of SUP pollution have been recorded in different environmental media globally, including soil and oceans. There is a growing push to reduce SUPs because of the accumulating research and results indicating that they pose a threat to plant growth, land animals, and marine ecosystems. As a result, there have been proposals of guidelines and plans to reduce SUP usage, as well as some suggestions for reducing SUP waste [3].

Plastic pollution is the accumulation of synthetic plastic items in the environment to the point where they pose a threat to wildlife habitats, along with human populations. Bakelite's development in 1907 ushered in a material revolution by introducing genuinely synthetic plastic resins into global commerce. Plastics had been discovered to be persistent pollutants in numerous environmental niches, from Mount Everest to the bottom of the sea, by the end of the 20th century. Plastics have gained increased attention as large-scale pollution, whether they are mistaken for food by animals or just cause substantial visual blight.

Can plastic pollution major current environmental issue be addressed by biomaterials? This type of material, whether alone or in combination, can achieve commercially functional properties of a matter known as bioplastics and biocomposites. Made from renewable resources, they have the tendency of being naturally recycled [4]. Waste management is the main key player. The negative environmental implications of the extensive use of synthetic plastics in the production process

DOI: 10.1201/9781003265054-2

37

have given the right impetus to biopolymers produced by biologically derived plastics as a likely viable replacement in a variety of fields. The true edge, though, is creating novel substitutes exploiting food waste whether it is industrially or agriculturally sourced. In the past few years, there has been a considerable rise in the test results of materials produced using biowaste. Research institutions and project managers have been drawn to the great number of raw materials which is abundant and may be utilized to produce new alternatives, in addition to the many young researchers that have embraced these different approaches in the time now being [5]. They have emerged as a suitable choice for consumers and producers looking for an alternative to ubiquitous plastics. Furthermore, the environmental fate of such materials solos, or blends, is remarkably friendly [6]. The name appears promising at first glance, suggesting a sustainable product, yet, is bioplastic, however, the answer to our environmental woes? A single-use item with the feel of plastic but none of the guilt? Such an effective alternative requires a thorough knowledge of both the source of the widespread use of plastics and the ecological consequences, therefore, sound multidisciplinary research to outreach reliable solutions [2].

2.2 LIMITATIONS, CHALLENGES, AND OPPORTUNITIES

Despite having many advantages, plastics are denounced because of their key features; durability and chemical stability which elongate their degradation process [7]. The versatility of plastic utilization has led to intensifying our consumption of such lightweight, cheap, and durable material. Flexible packaging, often known as film-based packaging, is any piece of material that might be easily reshaped, bags, labels, and wraps for instance. Flexible packaging is made using a specialized film that has a specific set of properties for a given use. In another sense, film-based packaging is used because of its large spectrum of protective functions which gives the greatest properties of materials like plastic, paper, and aluminum while using a fewer amount of them [8]. Low-density polyethylene (LDPE) films, for example, offer a high degree of transparency and moderate stretchability, making them ideal for use as shopping bags. High-density polyethylene (HDPE) films, on the other hand, have a certain degree of opacity and limited stretchability, making them suitable for use as drinks packages [9]. Building on the industry of consumer-level products, it is critical to investigate the challenges of regulations that may be in the favor of the environment, especially the techniques of handling the long-lasting stage in the life of film-based packaging, to be able to design and set up sustainable solutions [10].

Plastic waste can be handled in a variety of ways, including dumping in the landfills, burning, and reincorporating into new products. However, because each of these activities has different effects on the environment, determining the optimum option can be challenging. Discussing the burden of plastics and their consequences on behalf; treatment strategies and patterns, recycling technologies, and policies are a must to assess the possibilities of reducing the negative consequences on the environment [11]. A Life Cycle Assessment (LCA), perhaps, is a tool for evaluating all the different impacts of how a process could interact with the environment and comparing it with other processes which may be one of many needed solutions to take a step forward into reducing such negative effect.

Although plastics are easily pliable, which means they could be molded to form various shapes. Moreover, when adding the right additive, stabilizer for example, they can have several transparency, thinness, elasticity, and thermal qualities, regardless, plastic production, consumption, and waste generation are on the rise, posing a growing threat to the marine and terrestrial ecosystems and turning their most appealing aspects into a curse. Such pollution has gotten a lot of attention in the last years, when it polluted ecosystems, ruined the aesthetics of seas, and had severe effects on marine creatures. Pollution levels in the water are so high that it is impacting and threatening all ocean life. Plastic is one of the most prevalent causes of marine pollution; bags are frequently swallowed by marine species, resulting in the deaths of many. It dissolves into microplastic, which is a major hazard of untreated plastic waste. Micro and nano-plastics have been shown to penetrate the tissues of animals, marine organisms, and humans, posing serious risks. Furthermore, to extend the life of plastic, manufacturers are adding stabilizers, causing more problems. Plastics can adsorb and release pollutants to and from the environment due to their production process which includes dangerous chemicals. These are all found in the bodies of animals, according to research published in 2018, by The State of Plastics, and eventually, contaminate the human food chain [7].

2.2.1 THE THREE MAJOR PHENOMENA

The pollution is rising, three separate phenomena have resulted which are the origin of causing such this devastating mess… First is shopping bags and film materials thrown all over the places, being single-use non-biodegradable plastics. Second, the non-recyclability of plastics: the necessity for high-quality recycling must be valued by society and the market. The third is human behavior: our lack of knowledge about pollution and apathy toward it. Comparatively, little has been done to persuade society to discourage excessive plastic use, implement recycling practices, and raise awareness of the various types of plastic and their potential to harm as well as benefit human health and other species. This contrasts with the speed at which plastics have permeated millions of uses across all cultures [12].

2.3 SINGLE-USE, NON-BIODEGRADABLE PLASTICS

The majority of increasement in plastic production has been attributed to plastic packaging, which makes up more than 40% of non-fiber materials. Plastic packaging makes about half of all world-wide plastic waste because most plastic components are only meant to be used once [7]. SUPs are gaining popularity around the world, especially with regard to packaging and consumables like shopping bags and disposable tableware that are intended to be used only once [13]. A considerable portion of the material generated has a transient purpose and is quickly discarded. SUPs that are discarded, as well as various shapes and sizes of non-recycled plastic materials, pose the greatest threat. All of them eventually make their way to the seas via landfills, water bodies, and rivers. Plastic has the ability to endure in sea surface waters, where it will eventually assemble in remote areas of the world's oceans. In subtropical areas between California and Hawaii, the Great Pacific Garbage Patch (GPGP) is a significant ocean plastic accumulation zone. Recent research has revealed that the north-central Pacific Ocean's gyres have molded a floating island of plastic the size of the State of Taxes [14]. A key technological advancement was the use of synthetic fibers in fishing and aquaculture gear, yet gear losses became a large source of ocean plastic pollution. Ghost nets, also known as lost or discarded fishing nets, are a specific problem because they directly impact marine environments all around the world [15]. According to reports, by 2050, the world's oceans would have more pollutants, primarily plastic components, than the entire creatures in the oceans.

Non-biodegradable plastics, especially single-use ones, have lots of consequences on not only our health but also wildlife and the environment. They do not decompose or degenerate biologically. Agreeing to certain findings, laws simply based on the thickness of the plastic bags do not reduce their use. However, rules concentrated on banning the utilization of single-use plastic bags and imposing higher charges and levies on customers, resulting in a dramatic reduction in the use of plastic bags. Overall, findings demonstrate that public policy or laws can greatly influence attitude, perception, and behavior change toward biodegradable materials or ecofriendly activities [16].

Particularly single-use non-biodegradable plastics have negative effects on the environment, wildlife, and human health. They do not biologically decay or decompose. According to some research, legislation that only consider the thickness of plastic bags have little effect on the use of plastic bags. However, regulations focused on outlawing the use of single-use plastic bags and increasing user fees and levies, which caused a sharp decline in the use of plastic bags. Overall, the results show that laws or public policy can have a significant impact on changing people's attitudes, perceptions, and behaviors toward using biodegradable products or engaging in ecofriendly activities.

2.4   PLASTICS ARE NOT RECYCLED

Plastic has virtually surpassed every other man-made material in recent years, with production roughly expanding in recent years. For instance, with a lifespan of decades, construction accounts for half of all steel production. According to a study that was published in the peer-reviewed journal *Science Advances*, half of all plastic produced ends up in landfills in less than a year. Due to its durability, plastic has accumulated over time. The vast majority is being disposed of as waste, either in landfills or in the environment. That is to say, a large portion of it finally finds up in the oceans, which serve as the last sink. Not only "You can't manage what you don't measure," but also, "It's not simply that we make a lot; it's that we make more year after year" [17].

2.5   HUMAN BEHAVIOR

Raising public awareness of the immediate impact of plastic pollution and the health dangers of plastic itself is critical to educating the public about plastic management and consumption. Since plastic is regarded as an environmentally dangerous material, several projects aiming at transferring knowledge about the consequences of plastic pollution and consumption have been taking place. In addition, the relationship between culture and community behavior in terms of single-use plastic is the most important to be addressed.

Research shows that knowledge isn't the only thing that will change behavior. Usually, it'll require a lot more. The benefits shall be tangible, and the behavior to be doable and within their capabilities. One essential holdback is the near impossibility of avoiding single-use plastic. But it isn't just a case of people forgetting. It makes no difference if one carries his water bottle if there isn't anywhere to refill it. For others, it's a matter of personal preference that using their own products is safer. In many cases, the consumer has little choice except to accept plastic. And since it's so frequent, it becomes a matter of habit. We go to the supermarket, purchase some produce, place it in a plastic bag, and leave [18]. Plastic products have undoubtedly improved the comfort and ease of our lives. Largely utilized outcomes, on the other hand, offer a landfill nightmare, particularly when recycling alternatives are limited, and most of the material is non-biodegradable. Another benefit of plastic is the significant improvements in disease control have come from using single-use, sterile plastic products in medical fields because they reduce the possibility of cross-infection. Of course, these things need to be disposed of or recycled safely and properly. The first step toward changing the products and practices we take for granted despite these and other developments is to change our attitudes.

2.6   CIRCULAR ECONOMY FOR PLASTIC

A circular economy is an innovative approach to tackling plastic pollution that incorporates technical advancements, local creativity, commerce, business models, and job possibilities. Making plastics from plants instead of fossil fuels, redesigning products to use fewer materials, promoting recycling and reuse, and recycling plastic waste are all ways to maximize the positive effects of plastics while minimizing their negative effects on ecosystems, human health, and the environment. Local innovations that play an important part in the circular economy's ability to reduce plastic pollution, require marketing, investment, and before all that to believe it is an idea of worthiness to put any kind of effort.

Historically, cleaning-up events have been used to combat plastic pollution, but the only effective solution is to determine the key points or root causes... This kind of solution may be letting go of the current "take, make, waste" linear economy, which runs on massive amounts of cheap, readily available energy and other resources and generates disposable products. The circular economy, on the other hand, aims to keep resources in use for as long as possible, keeping as much value as possible, and then recovering and repurposing products and materials once their functional life has ended.

Its objective is to design waste materials out of the equation, leaving restoration and regeneration to take their place.

About 4% of the world's non-renewable oil and gas production is needed for plastics, and an additional 3%–4% is needed to power their production. Plastic is utilized in the packaging of disposable and other short-lived products that are discarded right after. We can conclude that our current plastic use is unsustainable based on these two data alone. Furthermore, vast amounts of waste plastics are found as garbage in landfills and natural habitats all over the world due to the durability of the polymers utilized. Recycling, which is one of the most dynamic solutions now available to alleviate the damage caused by plastic. It may reduce the oil usage, greenhouse gas, and the amount of waste that needs to be disposed of. Poor women and children in some cities, such as Bangladesh, gather used plastic materials and sell them to small commercial enterprises, who clean the waste and give it to specific industries for recycling to be again a new product sold on the market. As a result of the collection, processing, and trading of plastic waste, poor and small traders have found work. However, more research into the circular economy and the management of plastic pollution is required.

The extensive use along with the environmental consequences of synthetic plastics in product manufacture has reignited interest in biopolymers made from plant, animal, or microbial sources as a strong equivalent for a variety of applications. The real thing is creating new materials from abundant resources like food waste rather than especially needed materials, which come at a cost to the environment in any case. The number of raw materials that can be used has encouraged research that has been tremendously adopted by many entrepreneurs who create outstanding products using waste in recent years [19, 20].

2.7 RECYCLING VS. UPCYCLING

Nowadays, pollution, climate change, and destroyed ecosystem are matters of concern for international organizations and activists. There has been slow transition from linear economy of "take, make, and waste" to a circular economy of "make, use, and recycle." Governmental and private policies are becoming stricter in many nations, and the manufacturing industries are continually shifting to operate under these obligations. It is vital to lessen the negative effects of waste materials and their by-products on the environment, including contamination of the air, water, and greenhouse gases. In the future, repurposing the materials of waste in a way that consumes less energy than manufacturing them from scratch may be the way to go.

The process of breaking down waste and reusing it as raw materials to generate new products is referred to as recycling. In contrast, Reiner Pilz coined upcycling as "the method of recycling waste materials in their existing state without having to break them down into their original state" [21]. Although, clearly, this procedure uses less energy than typical recycling. It does, however, present its own difficulties.

2.7.1 WHAT EXACTLY IS RECYCLING?

Recycling is the process of creating products by turning waste into new raw materials. Interestingly, reduce, reuse, and recycle (3R process) have become popular. Production and consumer trends are gradually adopting this method. Millions of tons of waste are generated worldwide, of which only a few are recycled [22]. This shows that there is room for improvement in the recycling business. The main purpose of recycling was not to eliminate waste but to reduce its impact on the environment. For example, energy recovery is a type of recycling aimed at extracting energy by converting non-recyclable waste into heat, electricity, or fuel energy. It accounts for most of the recycling process around the world. This is usually a waste incinerator, which can cause harmful emissions as it is closely monitored, restricted, and regularly assessed.

2.7.2   AND UPCYCLING?

Upcycling is a type of recycling that is sometimes referred to as "creative recycling." It is not the same as recycling. Instead, upcycling, on the other hand, entails taking something that would otherwise be wasted and improving it in some way to make it useful again. This is an economically and ecologically interesting trial as it is becoming increasingly difficult to directly procure raw materials from natural resources. Materials such as aluminum are obtained by secondary recycling, not by primary sources from the mining industry. It is the process of reusing waste without breaking it down into basic forms to create higher quality or more valuable products. The term "upcycling" is new, but the philosophy behind it is not, it aims to reuse the waste in ways that increase its value and quality, rather than simply reusing it.

Widely recognized as two of the upcycling pioneers, Braungart and McDonough highlight the importance of considering future upcycling when designing products [23]. This principle is also known as "waste creation." Waste can be perfected using several strategies such as reuse planning, material optimization, and waste efficient sourcing. Waste disposal can be done from an upcycling perspective by creating products that are easy to assemble or disassemble and can include future redesigns as needed, faster and easier [24]. Assessing the environmental impact of a material is important and a common way of doing so is the Life Cycle Assessment (LCA). A "cradle-to-grave" kind of analysis that identifies the environmental impact of a product at every stage of its lifecycle, from obtaining raw materials through the product manufacturing to disposing of it at the end.

To create new sustainable materials for production, some companies, like Sabic, have even started fusing the recycling and upcycling processes. By upcycling post-consumer recycled polyethylene terephthalate (rPET) into more valuable polybutylene terephthalate (PBT) compound resins, Sabic can enhance the properties of the material while also extending the useful life of PET products and minimizing plastic waste [25].

2.8   RECYCLABILITY

High recyclability refers to a material's ability to be easily recycled, as well as the fact that its material qualities do not significantly depreciate from those of the original material. Some products are difficult to recycle or upcycle owing to their exceptional design/application or the nature of the materials used to produce them. For example, foam polystyrene is difficult and expensive to recycle. It's also difficult to recycle products with incompatible components because they're hard to separate. Plastics like HDPE, PET, and PVC are typically easier to recycle because they are commonly utilized as liquid containers or piping components. Other plastics, such as LDPE, are commonly used in food packaging, as squeezable tubes, which means they are regularly contaminated to the point where recycling them would be more energy-intensive than manufacturing them from scratch. Dual-material products, such as glossy paperboard used to make juice boxes, are also not easy to separate their components into the original state. Certain stuff that can't be recycled, instead, can be upcycled, and the other way around. And of what can be upcycled rather than recycled is the foam polystyrene, also known as Styrofoam, copper tubing and wires can easily be recycled rather than upcycled [26].

2.9   FROM BIOWASTE BURDEN TO USEFUL BIO-BASED PLASTIC

Biowaste makes up a significant amount of municipal solid waste (MSW). The constant generation of waste is causing management issues. Traditional waste discarding processes, such as incineration and landfilling, emit gases that may contribute to global warming. Plastic consumption is also fast expanding because of increased industrialization and population growth. Access to clean and green alternatives is critical for sustainable development to address this ever-increasing need. Researchers, scientists, governments, and stakeholders must all work together to make these technologies more feasible. We expect that the markets for agricultural by-products and recycled plastics will continue to expand.

Then, we're committed to aiding product designers and manufacturers in incorporating bio-based materials into high-quality, long-lasting materials that are also more likely to be preferred by the environment and people.

2.9.1 BIOPLASTICS

The modern world has a significant interest in sustainable alternatives to plastics and espouses any budding technology that is driving that growth. Traditional plastics are made from fossil fuels, while innovative plastics can also be made from renewable biomass. Bioplastic is simply plastic made from plant or other biological material rather than petroleum. It is also known as bio-based plastic. One can argue that every material either can or will undergo the process of biodegradation, which is theoretically very correct, but also not very accurate because practically some materials' behaviors are much slower than others, that some materials can take decades while others can take only a few days. That's why we consider some materials are biodegradable and others are not [26].

Bio-based plastics were used by humans in the industry for centuries long before the discovery of refining oil or harnessing the monomers that were obtained from it. For example, natural rubber, cellulose, and casein these materials were widely used bac —in the 18th and 19th centuries as resources for fibers used in functional objects such as carrying and packaging purposes [27].

Later, "Bakelite," the first totally synthetic material, was invented by Leo Baekeland—in 1907. This discovery was revolutionary. For the first time, human manufacture was unrestricted by natural constraints. There was only so much wood, metal, stone, bone, and horn that mother nature could provide. Humans, moreover, can now develop new materials. This opened the gate to a new era of manufacturing in which synthetic plastics are utilized every aspect of our daily life [28]. After many decades of this discovery, synthetic polymers became the first option when it comes to the production of various stuff with a market exceeding 600 billion USD [29]. That's because of low the price and the varied range of properties that can be obtained. But this of course comes with a heavy price and mainly on the environment.

The first patent presented a technical bioplastic was published—in 1947, by a small French company named Oraganico [30] but is now owned by Arkema [31]. They introduce Rilsan or Nylon 11 to the market which is a bio-based material made from castor beans and the polymerization of 11-aminoundecanoic acid. Such material has been found to have excellent mechanical properties and chemical resistance. Here's a quick yet not comprehensive summary of bioplastics' milestones:

- 1950—Amylomaize (corn with a higher than 50% amylose content) was successfully bred, and commercial bioplastics applications began to be explored.
- 1970—The environmental movement spurred more development in bioplastics.
- 1973— The surprising increase in the oil prices in the USA draws attention to the new possibilities of a plastic industry that is away from oil-based materials.
- 1989—Dr. Patrick R. Gruber succeeded to produce PLA from corn.
- 1992—Poirier, Dennis, Klomparens, Nawrath, and Somerville published a paper showing how polyhydroxybutyrate (PHB) can be made from the plant *Arabidopsis thaliana*.
- 2001—Nick Tucker used grass as a base to make bio-based material that can be used in plastic car parts.
- 2013—The synthesis of a bioplastic produced from waste blood and a bio-based crosslinking agent such as sugars or proteins has been granted a patent.
- 2018—First packaging made from a fruit, Jun Aizaki grew a fruit into a mold with the form of a cup and it worked.

Although the huge amount of research, the bio-based plastics with the current production lines are only on the order of 1% of the annual production amounts of plastic production [31]. Nevertheless, in the last years, this is changing because of multiple motives. Besides the research and development, the continuously varying fossil raw materials prices, the rise of environmental awareness, consumer preference, and some economics along with government policies toward "green procurement." All these reasons have contributed to increasing the market share of bioplastics with a growth rate of 17% annually. Moreover, it's expected to reach 30.9 billion USD by 2028 [32]. For the next years, various enhancements will be made in order to push the bioplastics to the level of usability of fossil-based plastics [33].

2.9.2 Its Applications

According to European Bioplastics, for the packaging of premium and branded products with specific needs as well as organic food, there is a significant demand for bioplastics. About 2.42 million tons of bioplastics were produced globally in 2021, with nearly 1.15 million tons of that volume going to the packaging sector.

Food packaging regulations are more different and sophisticated than those for other products. The package's primary functions are to store the product. The rate of change in quality, which includes both physical (mechanical damage during transit or storage, loss of crispness or appearance) and organoleptic changes (loss of taste, color, and odor), can be thought of as product protection [26]. Bioplastic used in food packaging is designed to protect food while also preserving its quality. These characteristics, as the mechanical and barrier properties of biopolymers, dependent on their structure, are crucial to tell how the packing material's properties change over time while in touch with the food. Bioplastics' end-use segments—presents:

- Packaging, bags for shopping.
- Waste collection bags that are compostable.
- Vegetable, fruit, meat, and egg trays and punnets.
- Catering service products that are disposable.

There are fundamental principles to verify that food packaging is functional, a set of variables listed below is mainly used to determine the best material for minimizing resources while maximizing shelf life (Table 2.1).

The continuous manufacturing of synthetic plastics will keep arising leading to more accumulation hence more waste—recycling and reuse are the only strategies that can work. Even though the challenges that face biopolymers, unlike synthetic plastics, they are sustaining their market growth. This generation of polymers has shown very promising results, with an advanced ability to overtake traditional plastic [34].

TABLE 2.1
Main Properties Standards Relating to Functionability of Bioplastics

Properties	Examples
Mechanical	Ductility
	Toughness
	Tensile strength
	Impact resistance
	Young's modulus
Physical	Transparency
	Water resistance
	Thermal resistance
	Barrier property (pore size)
Chemical	Chemical resistivity Fourier-transform infrared spectroscopy (FTIR)

2.10   INVEST IN NEW POSSIBILITIES

Bioplastics are being studied to wean society off fossil fuels while simultaneously addressing some of the environmental concerns linked to plastic waste. The synthesis and application of biowaste-based polymer, for example, are at the forefront of this science. Being 100% organic, bioplastic has the ability to decompose in just a matter of few weeks whether in the soil or sea. Moreover, today's techniques are innovative enough to produce bioplastics from the food waste or the unused organic substances. Researchers are concentrating their efforts on the use of such waste as a renewable feedstock for bioplastics. This waste mostly contains high-value compounds making it worthwhile to put it to new uses. Bioplastics made from starch are discovered to be easier to make than those made from other sources. Yet, the properties of such biofilms have some limitations [26]. Despite the green prospective that bioplastics have either in terms of carbon footprint or biodegradation, this class of materials is still needing research and laboratory work.

2.11   CONCLUSION AND WAY FORWARD

The demand for plastics is on a continuous rise; as a result, reducing plastic pollution will be one of the key difficulties. Reducing pollution to maintain human health and ecosystems would require proper legal measures and administration by relevant agencies. This can be accomplished by enforcing a strict regulatory system, encouraging users to separate their garbage, and increasing municipalities' capacity to collect as much solid waste as feasible in comparison to solid waste generation. Individual, family, community, and institutional levels of users' understanding of the significance of going green should be raised. Governments should encourage the plastic industry to reduce, reuse, and recycle. This would need technological and local innovation, as well as favorable public engagement and funding for such projects. It would be very appreciated if governments issued an order banning single-use plastic within the next few years as a means of controlling plastic production, processing, and trade. The government may take proactive approach, enlisting the help of consumers and stakeholders, this must be treated seriously right from the start.

It is critical to consider proper plastic management and pollution reduction in a holistic manner. With all its obstacles, it is tough and difficult to have a completely sustainable plastic industry, but each country can certainly develop a system for greener and cleaner plastic production that will help it achieve the Sustainable Development Goals (SDGs). Plastic pollution management can give a new route for adolescents and disadvantaged populations in both urban and rural areas to have access to decent labor, cleaner technologies, and low-pollution plastic because of now being rapid digital transformation. Local and global solutions would be required, as well as investment, effective innovation, and technology transfer.

The issues might be resolved by using materials that are biodegradable, bio-based, or both. Materials like that should be designed to ensure effective degradation while retaining their mechanical properties during the "use" phase. Applications out of these bags offer benefits since they can be degraded, eliminating the accumulation of plastics. The use of biodegradable plastics to mitigate environmental pollution due to leakage in open environments is another discussed advantage; currently, this seems optimistic!

ACKNOWLEDGMENTS

Author LH would like to express her thanks to **Dr. Irene Samy** for providing me opportunities and guidance.

Va bene, I would also love to thank **Ms. Giuseppina Miuli** who has significantly contributed to my scientific way of thinking, she is someone I am eagerly looking forward to her guidance due to her invaluable experience.

NOTES/THANKS/OTHER DECLARATIONS

Author LH appreciates **Medhat Benzoher**, they say you're gone, but I decided to bury this down too. I'm going to extend you forever. Or at least I'm going to try.

I'm grateful to **Malek Ghanem** and **Gabriel Eze** for their continuous support through all the ups and downs, thank you for being there.

REFERENCES

1. Rhodes, C.J. Plastic pollution and potential solutions. *Science Progress*, 101(3): 207–60, 2018.
2. Pietrelli, L., Pignatti, S., Fossi, M.C. Foreword-plastic pollution: A short and impressive story. *Rendiconti Lincei Scienze Fisiche e Naturali*, 29(4): 803–4, 2018.
3. Chen, Y., Awasthi, A.K., Wei, F., Tan, Q., Li, J. Single-use plastics: Production, usage, disposal, and adverse impacts. *Science of the Total Environment*, 52: 141772, 2021.
4. Rudin, A., Choi, P. Biopolymers. In *The Elements of Polymer Science and Engineering* (3rd ed, pp. 521–535). Oxford, Kidlington, Elsevier/AP, 2013.
5. Cecchini, C. Bioplastics made from upcycled food waste. Prospects for their use in the field of design. *The Design Journal*, 20(supp 1): S1596–610, 2017.
6. Narancic, T., Verstichel, S., Reddy Chaganti, S., Morales-Gamez, L., Kenny, S.T., De Wilde, B., Babu Padamati, R., O'Connor, K. Biodegradable plastic blends create new possibilities for end-of-life management of plastics but they are not a panacea for plastic pollution. *Environmental Science & Technology*, 52(18): 10441–52, 2018.
7. Alhazmi, H., Almansour, F.H., Aldhafeeri, Z. Plastic waste management: A review of existing life cycle assessment studies. *Sustainability*, 13(10): 5340, 2021.
8. Morris, B.A. *The Science and Technology of Flexible Packaging: Multilayer Films from Resin and Process to End-Use*. Oxford: William Andrew; 2017.
9. Selke, S.E., Hernandez, R.J. Packaging: Polymers in flexible packaging. *Encyclopedia of Materials: Science and Technology*, 6652–6, 2001.
10. Hou, P., Xu, Y., Taiebat, M., Lastoskie, C., Miller, S.A., Xu, M. Life cycle assessment of end-of-life treatments for plastic film waste. *Journal of Cleaner Production*, 201: 1052–60, 2018.
11. Chen, Y., Cui, Z., Cui, X., Liu, W., Wang, X., Li, X.X., Shouxiu, L. Life cycle assessment of end-of-life treatments of waste plastics in China. *Resources, Conservation and Recycling*, 146: 348–57, 2019.
12. Hopewell, J., Dvorak, R., Kosior, E. Plastics recycling: Challenges and opportunities. *Philosophical Transactions of the Royal Society B: Biological Sciences*, 364(1526): 2115–26, 2009.
13. Leal Filho, W., Salvia, A.L., Minhas, A., Paço, A., Dias-Ferreira C. The COVID-19 pandemic and single-use plastic waste in households: A preliminary study. *Science of the Total Environment*, 793: 148571, 2021.
14. Lebreton, L., Slat, B., Ferrari, F., Sainte-Rose, B., Aitken, J., Marthouse, R., Hajbane, S., Cunsolo, S., Schwarz, A., Levivier, A., Noble, K., Debeljak, P., Maral, H., Schoeneich-Argent, R., Brambini, R., Reisser, J. Evidence that the Great Pacific Garbage Patch is rapidly accumulating plastic. *Scientific Reports*, 8(1): 4666, 2018.
15. O'Hara, K., Iudicello, S., Bierce, R. *A Citizens Guide to Plastics in the Ocean: More Than a Litter Problem*. Washington, DC: Center for Marine Conservation; 1989.
16. Adeyanju, G.C., Augustine, T.M., Volkmann, S., Oyebamiji, U.A., Ran, S., Osobajo, O.A., Otitoju, A. Effectiveness of intervention on behaviour change against the use of non-biodegradable plastic bags: A systematic review. *Discover Sustainability*, 2(1): 13, 2021.
17. Geyer, R., Jambeck, J.R., Law, K.L. Production, use, and the fate of all plastics ever made. *Science Advances*, 3(7): e1700782, 2017.
18. Khoironi, A., Anggoro, S., Sudarno, S. Community behavior and single-use plastic bottle consumption. In *IOP Conference Series: Earth and Environmental Science*. IOP Publishing, 293(1), 012002, 2019.
19. Forrest, A., Giacovazzi, L., Dunlop, S., Reisser, J., Tickler, D., Jamieson, A., Meeuwig, J.J. Eliminating plastic pollution: How a voluntary contribution from industry will drive the circular plastics economy. *Frontiers in Marine Science*, 6: 627, 2019.
20. Syberg, K., Nielsen, M.B., Clausen, L.P.W., van Calster, G., van Wezel, A., Rochman, C., Hansen, S.F. Regulation of plastic from a circular economy perspective. *Current Opinion in Green and Sustainable Chemistry*, 29: 100462, 2021.

21. Xu, J., Gu, P. Five principles of waste product redesign under the upcycling concept. Proceedings of the 2015 International Forum on Energy, Environment Science, and Materials. 2015.

22. Calvo, S., Morales, A., Núñez-Cacho Utrilla, P., Guaita Martínez, J.M. Addressing sustainable social change for all: Upcycled-based social creative businesses for the transformation of socio-technical regimes. *International Journal of Environmental Research and Public Health*, 17(7): 2527, 2020.

23. McDonough, W., Braungart, M. *Cradle to Cradle: Remaking the Way We Make Things*. London: Vintage; 2009.

24. Cheshire, D. Designing Out Waste. In *The Handbook to Building a Circular Economy* (2nd ed, pp. 48–61), RUBS PUBNS LTD., 2021.

25. Sabic Introduces LNP(tm) ELCRIN(tm) IQ Upcycled Compounds to Extend the Useful Life of PET Bottles and Help Reduce Plastic Waste. SABIC. https://www.sabic.com/en/news/20177-sabic-introduces-lnp-elcrin-iq-upcycled-compounds-to-extend-useful-life-of-pet-bottles-and-help-reduce-plastic-waste, 2021.

26. Hamid, L., Samy, I. *Fabricating Natural Biocomposites for Food Packaging*. IntechOpen, 2021. doi: 10.5772/intechopen.100907.

27. Lackner, M. Bioplastics. *Kirk-Othmer Encyclopedia of Chemical Technology*, (6th ed, pp. 1–41), 2015. https://doi.org/10.1002/0471238961.koe00006.

28. Herzog, B., Kohan, M.I., Mestemacher, S.A., et al. Polyamides. In: *Ullmann's Encyclopedia of Industrial Chemistry* (pp. 1–47). Wiley-VCH, 2020. https://doi.org/10.1002/14356007.a21_179.pub4.

29. Plastic Market Size, Share & Trends Analysis Report By Product (PE, PP, PU, PVC, PET, Polystyrene, ABS, PBT, PPO, Epoxy Polymers, LCP, PC, Polyamide), By Application, By End-use, By Region, and Segment Forecasts, 2021–2028, Research and Market, https://www.researchandmarkets.com/reports/4751797/plastic-market-size-share-and-trends-analysis, 2021.

30. Furukawa, Y. High-performance polymers: Their origin and development. Raymond B. Seymour, Gerald S. Kirshenbaum. *Isis*, 78(4): 605–6, 1987.

31. Arkema Celebrates the 70th Anniversary of Bio-based Arkema Celebrates the 70th Birthday of Its Fflagship Rilsan(r) Polyamide 11 Brandnylon, Renewable Carbon. https://renewable-carbon.eu/news/arkema-celebrates-70th-anniversary-of-bio-based-arkema-celebrates-the-70th-birthday-of-its-flagship-rilsan-polyamide-11-brandnylon/, 2017.

32. Bioplastics Market Size, Share & Trends Analysis Report By Product (Biodegradable, Non-biodegradable), By Application (Packaging, Automotive & Transportation, Textile), By Region, And Segment Forecasts, 2021–2028, Grand View Research. https://www.grandviewresearch.com/industry-analysis/bioplastics-industry, 2021.

33. Brodin, M., Vallejos, M., Opedal, M.T., Area, M.C., Chinga-Carrasco, G. Lignocellulosics as sustainable resources for production of bioplastics – A review. *Journal of Cleaner Production*, 162: 646–64, 2017.

34. Narancic, T., O'Connor, K.E. Plastic waste as a global challenge: Are biodegradable plastics the answer to the plastic waste problem? *Microbiology*, 165(2): 129–37, 2019.

3 Synthetic Polysaccharides
Adored, Deplored and Ubiquitous

Rois Uddin Mahmud, Md. Raijul Islam,
Md. Abdur Rouf, and Md. Rubel Alam
BGMEA University of Fashion & Technology (BUFT)

Asif Mahmud Rayhan
Western Michigan University

Md Enamul Hoque
Military Institute of Science and Technology (MIST)

3.1 INTRODUCTION

Polysaccharides, which serve as structural components and energy storage for living cells, are the most abundant renewable polymers found in nature (Smith et al., 2020). Polysaccharides are complex carbohydrates that are an integral part of our diet and play a vital role in many biological processes. They are found in a variety of foods, such as fruits, vegetables, grains, and legumes, and are known for their high energy content, fiber content, and beneficial health effects (Li & Lin, 2021). It is one of the ubiquitous biopolymers with unique physical, chemical, and biological properties. As a biopolymer, it shows excellent characteristics and can be used in different sectors which lead to environmental and health benefits (AL-Oqla et al., 2022). It paves the way for the development of new adaptive materials for use in need-based applications. Monosaccharide building blocks are linked together by glycosidic linkage in a linear or highly branched order forming the backbone of polysaccharide (Li & Lin, 2021). Typically, more than 20 repeating units are joined to make polysaccharide. Starch, glycogen, and cellulose are the three most common natural polysaccharides.

To make a suitable polysaccharide, dozens of monosaccharides (homopolymer and copolymer) units can be combined. Starch, glycogen, and cellulose are examples of natural homopolymer-based polysaccharides that yield only glucose as the monosaccharide after complete hydrolysis. However, heteropolymers can also include amino sugars, sugar acids, and non-carbohydrate compounds in combination to monosaccharides. Heteropolymers, which are abundant in nature, include substances such as gums and pectins. Additionally, the glucose unit has a lot of hydroxyl groups ($-OH$), carboxyl acids ($-COOH$), or amines ($-NH_2$) covalently anchored on its periphery, offering a flexible platform for functionalization and post modification (Li & Lin, 2021).

However, in recent decades, synthetic polysaccharides, made by chemical or biological means, have become increasingly popular in various industries, including food and pharmaceuticals, construction, and energy. The biosynthesis polysaccharides can be produced by a myriad of glycosyltransferases (GTs). Synthetic polysaccharides are favored for their versatility, as they can be engineered to exhibit specific properties, such as high strength, viscosity, and stability, that make them useful in a range of applications. But chemical modifications can greatly hamper the structure-property relationships by the heterogeneity and by the random distribution of functional groups. But structural arrangements (e.g., monosaccharide components, positions of functional groups, modes of linkages) greatly confer the properties for particular applications. Polysaccharide isolation from natural resources often necessitates extensive purification and harsh treatments, resulting in polydispersity that is difficult to

DOI: 10.1201/9781003265054-3

analyze, reproduce, and control. However, synthesis of polysaccharides with multiple –OH groups in the laboratory is extremely difficult. However, the glycal assembly method is considered modern one for polysaccharide synthesis. This technique involves adding or removing additional sulfation, phosphorylation, acetylation, methylation, and other modifications to the glycans.

They are used as thickening agents, emulsifiers, and stabilizers in food, and as excipients in pharmaceuticals to improve the effectiveness of drugs. In construction, they are used as adhesives, binders, and insulation materials, while in energy, they are used as biofuels, lubricants, and hydraulic fluids. Its nontoxic biodegradable properties define the commercial applications in pharmaceuticals, photographic films, food products, and tertiary oil recovery. It has some superior uses in the areas of biomedical fields. Polysaccharides are used in many biomedical applications like cell identification, expansion, adherence, immune system, swelling and bacterial identifications, and many more (Wu et al., 2017). Carbohydrate-protein interactions (CPIs) and carbohydrate-carbohydrate interactions (CCIs) play a crucial role in regulating various biochemical processes such as cell differentiation, adhesion, proliferation, and immune responses (Yu & Delbianco, 2020). Despite the benefits of synthetic polysaccharides, their widespread use has led to criticism, as some of these compounds have been shown to have negative impacts on the environment and human health. For example, synthetic polysaccharides are not biodegradable and can persist in the environment for long periods, contributing to pollution and degradation of ecosystems. Additionally, some synthetic polysaccharides have been linked to adverse health effects, such as digestive problems, skin irritation, and allergic reactions. Given the adoration, depreciation, and ubiquity of synthetic polysaccharides, it is essential to understand their benefits, drawbacks, and overall significance in today's world. However, complex polysaccharides are difficult to understand at the molecular level for better applicability, limiting the viable reach. This chapter will provide a comprehensive overview of synthetic polysaccharides, exploring their properties, applications, and environmental and health impacts. The aim of this chapter is to provide an objective and balanced assessment of synthetic polysaccharides, highlighting the need for further research and regulation to ensure their safe and sustainable use.

3.2 SYNTHESIS AND EXPLORATION OF POLYSACCHARIDES

Saccharides units obtained in different natural carbohydrates from which isolation of saccharides is almost impracticable in many cases. Chemical synthesis of saccharides is now one of the interesting research parts by which different methods were developed recently. The regio- and stereoregulation in the polymeric chain is one of the most challenging parts of the synthesis of the synthetic polysaccharides unit which is very time-consuming. Different types of synthesis paradigm were used to make synthetic polysaccharide in various times. Currently, enzymatic synthesis, automated glycan assembly (AGA), one-pot glycosylation strategy, polymerization processes, and other synthetic techniques are being explored to facilitate the rapid creation of synthetic polysaccharides. Tailored synthetic materials, featuring precise compositions and modifications, have the potential to advance glycoscience and find applications in material science, biology, and nanotechnology. Their adaptability and diversity result in a variety of commercially exploited features. Chemical alterations enable the expansion of their uses (de Moura et al., 2015).

3.2.1 DIFFICULTIES IN SYNTHESIZING SYNTHETIC POLYSACCHARIDES

Preparing chemically synthesized monosaccharides can be time-consuming and frequently requires many stages to get the necessary long polysaccharide chain. Through the reaction of glycosylation, glycoside bonds are joined to form long-chain polymeric carbohydrates. Glycosylation processes are often challenging due to the severe conditions required, the low solubility of products and intermediates, and the formation of aggregates that can slow down the reactions (Delbianco et al., 2018). Stereo and regio-controlled chemistry in the long chain is another challenge in the synthesis. Significant effort has been expended to mechanistically comprehend the glycosylation reaction to predict its stereochemical consequence (Nigudkar & Demchenko, 2015). Since 1890, the Emil Fischer glycosylation process has been extensively used to synthesize specific glycosides and has undergone

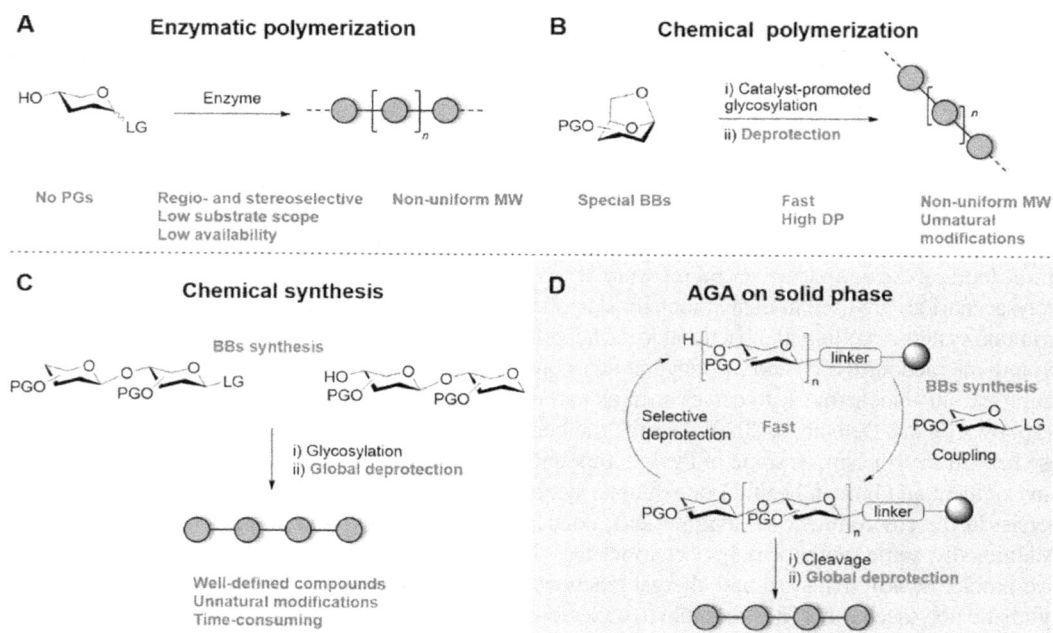

FIGURE 3.1 Explanation of the many techniques for the synthesis of polysaccharides. Polymerization techniques, enzymatic polymerization (a) and chemical polymerization (b); (c) chemical synthesis; and (d) AGA method (Fittolani et al., 2021).

significant optimization and modification (Haese et al., 2022). The synthesis of polysaccharides requires careful selection of monomeric or oligomeric building blocks, regio- and stereo-selective techniques for the formation of glycosidic linkages, precise control of glycan size, and the removal of protective groups. During lengthy polymerizations, loss of stereospecificity can occur, which significantly complicates the assembly process. But from the very beginning of 1981 to the date, many researchers applied many synthesis methods to create synthetic polysaccharides. Currently, several steps are being used to produce synthetic polysaccharides, mainly (i) AGA, (ii) enzymatic synthesis, (iii) chemical synthesis, and (iv) chemical polymerization, as shown in Figure 3.1.

3.2.2 Automated Glycan Assembly

Enzymes provide an advantage in glycosylation processes by allowing the use of exposed sugars as substrates and providing precise govern over regio- and stereoselectivity, making them an invaluable tool in the process. The efficiency of coupling is a critical factor in determining the maximum length of biopolymers that can be synthesized without requiring purification steps. To address this, automated synthesis techniques have been developed to ensure high coupling efficiencies, reduce cycle durations, and expand the limit of appropriate long-chain polysaccharides. These developments are partly inspired by the routine production of 200-mers in polynucleotide syntheses (Ma et al., 2012). 128-mer of polysaccharides are synthesized to date through the glycosylation process from the very beginning of 1981, a list of comparisons of AGA synthesis is shown in Table 3.1 (Zhu et al., 2020). The entitled strategy of monosaccharides through glycosylation which leads to polysaccharides has been taken by the AGA. Isolating a natural polysaccharide with an optimum length, content, and substitution is impossible due to microheterogeneity. Functional investigations of polysaccharides, which play necessary roles in various biological procedures such as bacterial infection, cell recognition, cell signaling, and viral entry, have often utilized heterogeneous and poorly defined polysaccharides, thereby neglecting their precise structure.

TABLE 3.1

Automated Polysaccharide Solid-Phase Synthesis Comparison (Zhu et al., 2020)

Biopolymer	Peptides	DNA	RNA	Polysaccharides
Introduced	1963	1981	1998	2001
Structure	Linear	Linear	Linear	Branched; regiocontrol required
Structure	Amide	Phosphodiester	Phosphodiester	Glycosidic bond
Stereogenic center	No	No	No	Yes; stereocontrol required
Capping	For selected sequences	Routinely	Routinely	Recently added
Coupling yield	99.5	>99.99	98.5	98.75
Length	50–100 mer	≈200	120	98.75

The linear 100-mer polymannoside was constructed using newly discovered linker and improved AGA procedures from variably protected monosaccharide building blocks. A branched 151-mer polymannoside was obtained through AGA-catalyzed convergent [31 + 30 + 30 + 30 + 30] block coupling of fully and partially protected polysaccharides (Joseph et al., 2020). In Figure 3.2, AGA assembly of polysaccharide concept is completely illustrated for the better understanding.

Figure 3.3a depicts a four-step synthetic cycle for the preparation of a glycoside, which involves a glycosylation, acidic wash, capping to mask any unreacted nucleophiles, and cleavage

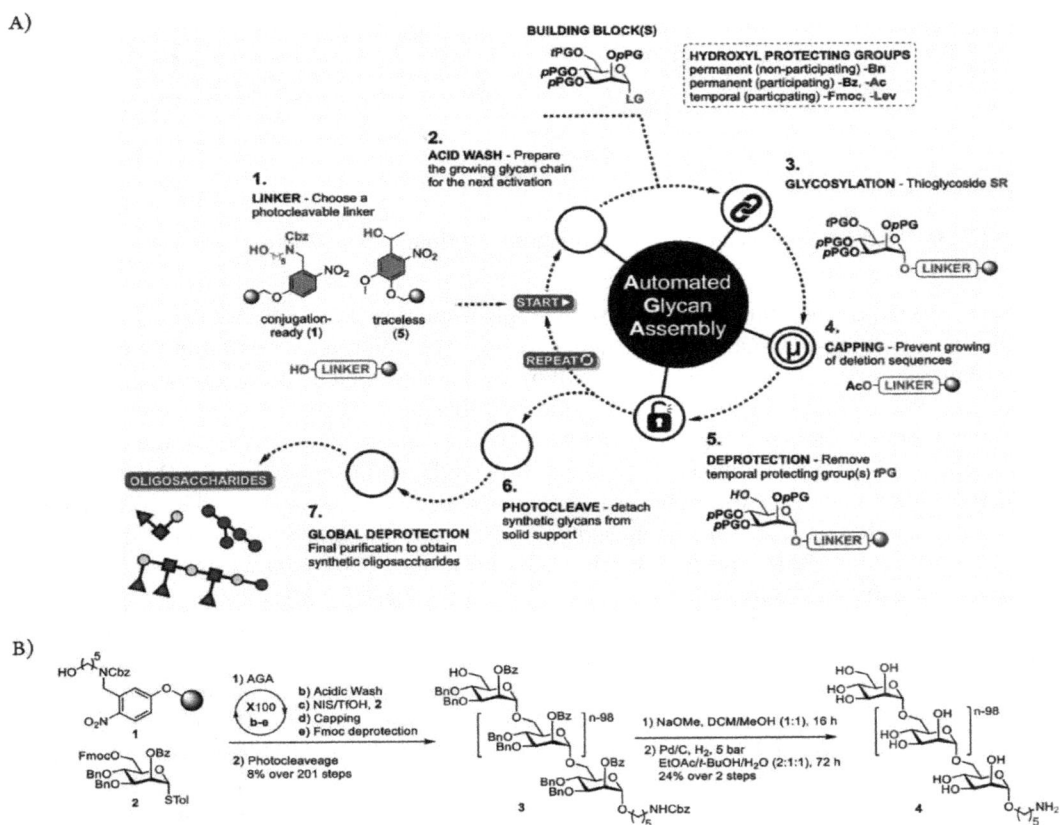

FIGURE 3.2 AGA assembly of polysaccharides (Joseph et al., 2020).

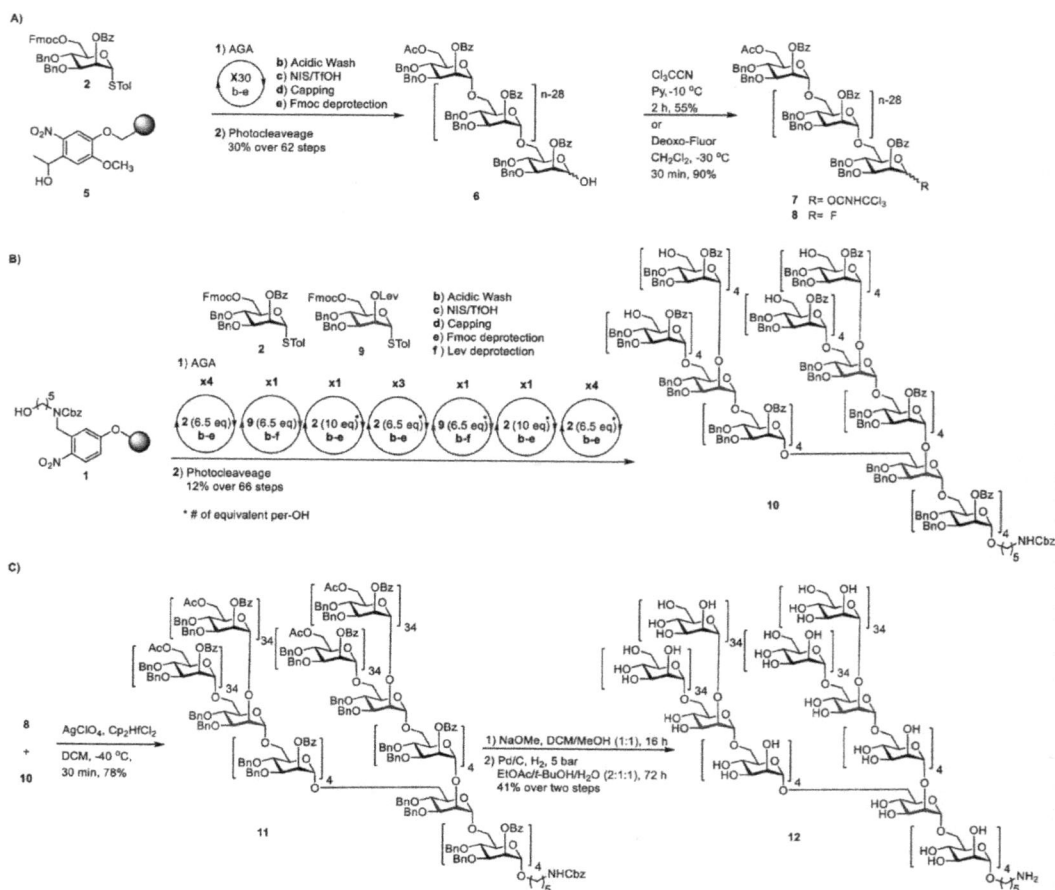

FIGURE 3.3 Synthesis of branched 151-mer polymannoside (Joseph et al., 2020).

of Fmoc carbonate in preparation for subsequent glycosylation. Figure 3.3b shows the synthesis of a 100-mer α-(1–6)-polymannoside using AGA and Merrifield resin containing a photolabile linker. The automated synthesizer runs coupling cycles that include an acidic wash, glycosylation using the second mannose thioglycoside building block, capping to obstruct any unreacted nucleophiles, and cleavage of the temporary Fmoc protecting group. By photocleavage, the 100-mer α-(1–6)-polymannoside 3 is released from the solid support, and 100-mer α-(1–6)-polymannoside 4 is created through a two-step global deprotection process. The AGA method was used to synthesize a 151-mer branched polymannoside through a block coupling strategy, as shown in Figure 3.3. In Figure 3.3a, the AGA of 30-mer glycosylating agents (7 and 8) is demonstrated using the AGA method. Merrifield resin 5 underwent 30 coupling cycles utilizing mannose thioglycoside building block 2, followed by photocleavage, yielding partially protected 30-mer 6. By treating polymannoside 6 with trichloroacetonitrile or deoxo-fluor, the matching 30-mer α-(1–6)-polymannoside donors 7 and 8 were generated. In Figure 3.3b, the block coupling strategy is shown. A branched 31-mer polymannoside acceptor 10, prepared using the AGA method, was coupled with the 30-mer α-(1–6)-polymannoside donors 7 and 8 using a 31 + 30 + 30 + 30 + 30 block coupling. The resulting product was a branched 151-mer polymannoside. In addition, Figure 3.3b shows the assembly of a 31-mer polymannoside acceptor 10. Four mannose thioglycoside building blocks 2 were incorporated, followed by the addition of branching building block 9. The first two parallel couplings with building block 2 required

10 equivalents to glycosylate the secondary C2-hydroxyl group and establish an α-1,2 linkage. After three more parallel glycosylations with building block 2, the procedure was repeated with the addition of two more branching points using building block 8, allowing for rapid expansion of the 31-mer polymannoside. After photocleavage and purification, 30 mg of the 31-mer polymannoside acceptor 10 was obtained. Figure 3.3c shows the successful combination of the 30-mer glycosylating agent 8 and the branched 31-mer 10 to produce a fully protected 151-mer polymannoside 11 with a yield of 78%. After the removal of all protecting groups through methanolysis and hydrogenolysis, a branched 151-mer polymannoside 12 was obtained with a yield of 1.8 mg. An important development in the field was the automated synthesis of straight and branched polysaccharides up to 100-mers using monosaccharides, which opened the door for the synthesis of bigger polysaccharides like the 151-mer. By acting as both glycosylating agents and glycosyl acceptors, the fully and partially protected polysaccharides made by AGA allow the synthesis of carbohydrate materials that contain both natural and synthetic monomers. This technology has enormous potential in both biological and material science applications, similar to the applications of the automated polynucleotide and polypeptide synthesis.

3.2.3 ENZYMATIC SYNTHESIS

Enzymes are advantageous in glycosylation reactions because they allow the use of exposed sugars as substrates and provide excellent govern over regioselectivity and stereoselectivity. They can polymerize monosaccharides or oligosaccharides that have a reactive leaving group to produce the desired polysaccharide. There are numerous types of enzymes, including as hydrolases, sucrases, phosphorylases, glycosynthases, and glycosyltransferases (Cobucci-Ponzano & Moracci, 2012; Danby & Withers, 2016; Hayes & Pietruszka, 2017; Mackenzie et al., 1998). Even though this method had several benefits, the range of substrates was restricted by the low availability of the enzymes and their high selectivity. The synthesis of synthetic polymers is generally hindered by the highly selective enzyme reactive site, which only accepts minor alterations. Additional difficulties with the enzymatic production of polysaccharides include low glycosylation yields and product hydrolysis (Wang & Huang, 2009). Due to the enzymes' inability to differentiate between acceptors of various lengths in the reaction mixture, homopolymers are frequently produced using this method as non-uniform samples. Since the beginning, numerous enzymes and substrates have been created with the goal of controlling the molecular organization of the finished product, increasing DP, or narrowing the dispersion of molecular weight (MW) (Hattori et al., 2012; Hiraishi et al., 2009). Kobayashi reported the first successful enzymatic cellulose synthesis in 1991 (Kobayashi et al., 1991). There are a lot of biocatalysts that have been shown to be able to selectively and quickly form glycosidic connections between saccharides without the time-consuming protection and deprotection procedures required in chemical synthesis (Smith et al., 2020). Enzymatic polymerizations may result in considerable amounts of distinct polysaccharides provided the necessary substrates and enzymes that are available in enough quantities. For polysaccharide synthesis in cell-free systems, three classes of enzymes can be used: (i) GTs, which transfer monosaccharides from activated sugar nucleotides to suitable acceptor substrates; (ii) glycoside hydrolases and glycosynthases, which are engineered glycoside hydrolases that use a glycoside hydrolysis's reverse reaction; and (iii) phosphorylases and sucrases, which use sugar-1 (Figure 3.4).

The hydrolytic breakdown of glycosidic bonds is catalyzed by the carbohydrate-active enzymes known as glycoside hydrolases (GHs) (Lombard et al., 2014; Davies & Henrissat, 1995). Due to their ability to degrade polysaccharides including cellulose, hemicellulose, and starch, these enzymes are important for the efficient valorization of plant biomass (Bornscheuer et al., 2014). They may be manufactured in vast quantities using bacterial expression methods and are frequently highly stable. Exo-glycosidases, which operate on terminal monosaccharides, are one type of GHs. Endoglycosidases, also known as endo-glycanases, hydrolyze internal glycosidic linkages (Ardèvol & Rovira, 2015). Additionally, they may be divided into two categories based on their

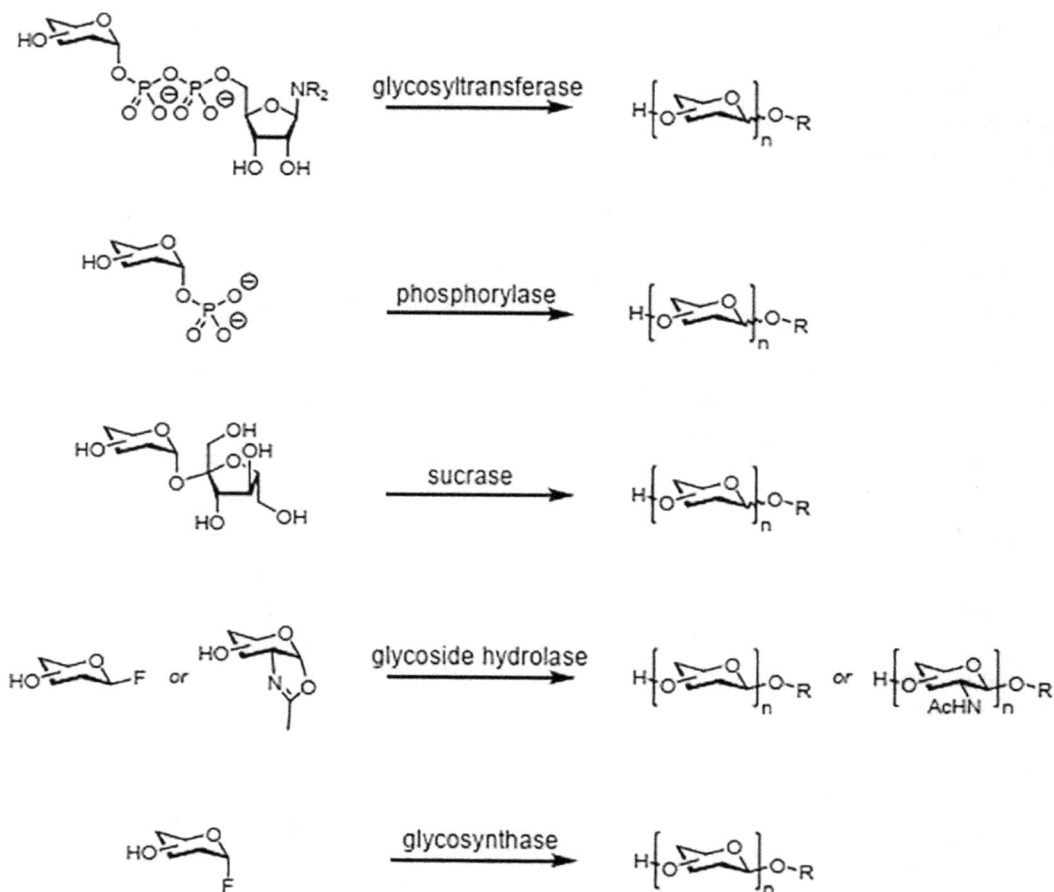

FIGURE 3.4 Typical families of enzymes investigated for the enzymatic polymerizations that produce polysaccharides and the processes they catalyze (Smith et al., 2020).

catalytic mechanism: retaining and inverting (Figure 3.5). By using a double-displacement method that entails forming an inverted covalent bond with the active site nucleophile, glycoside hydrolases can maintain the structure at the anomeric carbon. In contrast, inverting glycoside hydrolases employ a single displacement mechanism that involves a carbenium ion transition state and results in a net inversion of the anomeric center. Transglycosylation, a process in which GHs catalyzes the formation of a glycosidic link, is another capability. In retaining enzymes, the capacity to catalyze transglycosylation is more frequent.

When a reactive donor saccharide, such as a saccharide oxazoline or glycosyl fluoride, is present, significant amounts of glycosyl-enzyme precursors are produced that can preferentially react with the right acceptor substrates. Numerous other polysaccharides, such as glycosaminoglycans and derivatized celluloses, have been created in this fashion (Kadokawa, 2011). Nevertheless, the competing hydrolytic activity of the enzymes imposes an inherent restriction on the chain lengths and yields of polysaccharides produced by this method. To eliminate background hydrolysis activity and increase yields, the Withers group developed modified glycoside hydrolases, now known as "glycosylases," in which the catalytic nucleophile has been swapped out for a nonnucleophilic residue (Hayes & Pietruszka, 2017). Glycosidic bonds are generated extremely effectively when activated donors are combined with glycosylases that (i) have the opposite configuration from the original substrate and (ii) have a sufficient leaving group at the anomeric carbon.

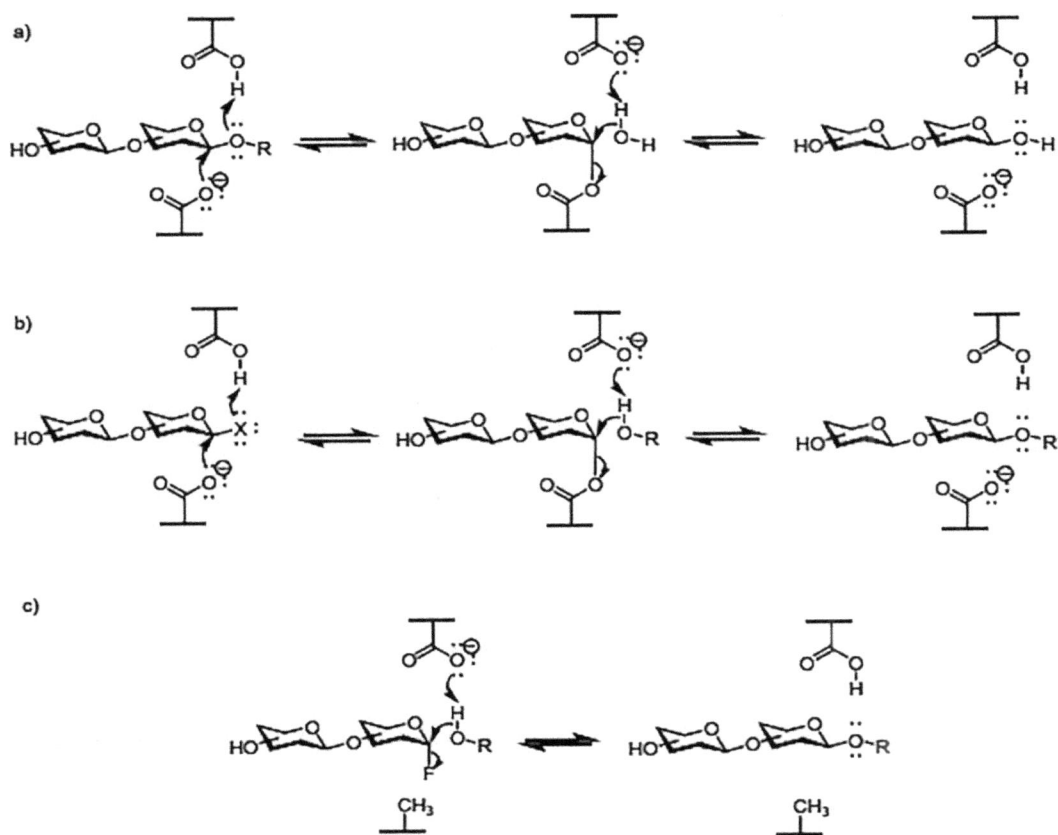

FIGURE 3.5 Pathways for (a) saccharide hydrolysis catalyzed by retaining glycosidases, (b) the formation of glycosidic bonds catalyzed by preserving glycosidase, and (c) the formation of glycosidic bonds catalyzed by glycosynthase (Smith et al., 2020).

3.2.4 CHEMICAL POLYMERIZATION

The passage highlights the advantages of chemical polymerization in achieving a broad range of cellulose structures, specifically through ring-opening polymerization (ROP) of orthoesters. ROP has been proven effective in producing (1–4) glucopyranan structures, with the first stereo-regular cellulose synthesized using 3,6-dibenzyl-protected 4 catalyzed by Ph3CBF4. This method also produced a protected polymer 14 with a productivity of 62% and a mean DP of 19.3 in just 2 hours. Selective transformation requires the presence of benzyl (Bn) groups. Several cellulose analogs have been produced using this technique, including 6-deoxy, 13C-labeled, L-GLC, and ethyl/methyl variants. The water solubility of the polymer can be adjusted by changing the concentration of methyl/ethyl. However, there is no control over the replacement pattern. For instance, 6-O-methyl and 6-O-ethyl celluloses were less water-soluble than heterogeneous polymers containing more ethyl groups. Although different BB ratios can alter the methyl/ethyl proportion, the replacement pattern cannot be controlled. The removal of benzyl protective groups can be a bottleneck, requiring several treatments at high temperatures and pressures. To overcome this issue, allyl groups can be used as an alternative, which can be removed within 4 hours using palladium chloride at 60°C (Figure 3.6).

The chemical production of glycans composed of [→4)-α-Rha-(1 → 3)-β- Man-(1 →] repeating units that are related to the O antigen of *Bacteroides vulgatus*, a frequent component of gut microbiota.

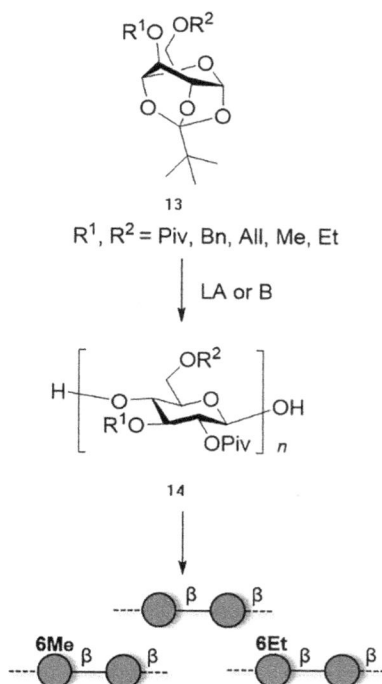

FIGURE 3.6 Lewis acid (LA) or base-enhanced cellulose synthesis by ROP from protected orthoester 4 (Fittolani et al., 2021).

We were able to synthesize a 128-mer glycan using the best mix of assembly techniques, protecting group arrangement, and glycosylation reaction (Zhu et al., 2020).

3.2.5 CHEMICAL SYNTHESIS

Chemical synthesis offers a significant advantage over polymerization methods because it allows complete rule over the polymer length, preventing non-uniform MW dispersions. This method also enables the creation of polymers with virtually any alteration pattern. By selectively removing strategically placed protective groups (PGs), it is possible to introduce branches or chemical changes into the polymer. The intense aggregation and insolubility of cello oligosaccharides, which have until now limited this process to just extremely short structures in low yields, have severely hindered this technology. To solve this problem, the products were transformed into acetate equivalents and studied similarly to how cellulose 20-mer made using a convergent process (Nishimura & Nakatsubo, 1996, 1997). Because of the creation of insoluble aggregates during the deprotection phase, which led to material waste during the purification process, the longest well-marked cellulose analog manufactured by chemical synthesis so far is a 12-mer (22, Figure 3.7), which was achieved with a 2% yield (Yu et al., 2019). AGA produced 22 as one of a number of modified compounds that were produced with substantially greater yields. By breaking up hydrogen-bond networks, especially the formation of insoluble clumps was decreased by strategically placed substituents, such as methyl, fluorine, and carboxymethyl groups. When comparing compounds with the same degree of substitution but different substitution patterns, significant differences were observed in the conformation of the molecule, such as the gyration radius and conformation of glycosidic bond, as well as in the aggregation behavior, including crystallinity and solubility. These differences emphasize the necessity of producing pure and well-furnished

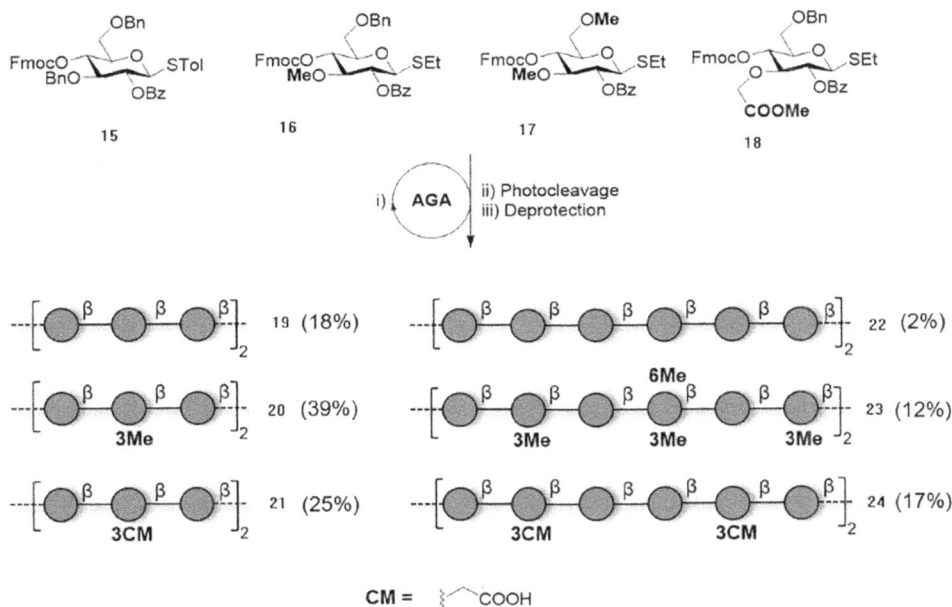

FIGURE 3.7 AGA of many cellulose analogs produced using BBs 6–9. Higher isolated yields were achieved as a result of changes that were strategically placed to prevent the development of insoluble aggregates. The coupling (NIS/TfOH), capping (Ac$_2$O), and Fmoc deprotection (piperidine or Et$_3$N) steps of the AGA cycle are all included (Fittolani et al., 2021).

polysaccharides to accurately characterize their structure and properties. When methyl groups were placed in blocks, for example, compounds with alternating methylation patterns created quasi-linear forms, whereas compounds with more curved geometries. As chemical synthesis provides for a high degree of exploitation flexibility, non-carbohydrate moieties can be utilized to modify the structure of the resulting molecules. Synthetic cellulose I was created by paralleling β-(1–4)-linked glucose chains with an anthraquinone molecule.

3.3 CHEMICAL STRUCTURE AND DIVERSIFICATION (CLASSIFICATION) OF SYNTHETIC POLYSACCHARIDES

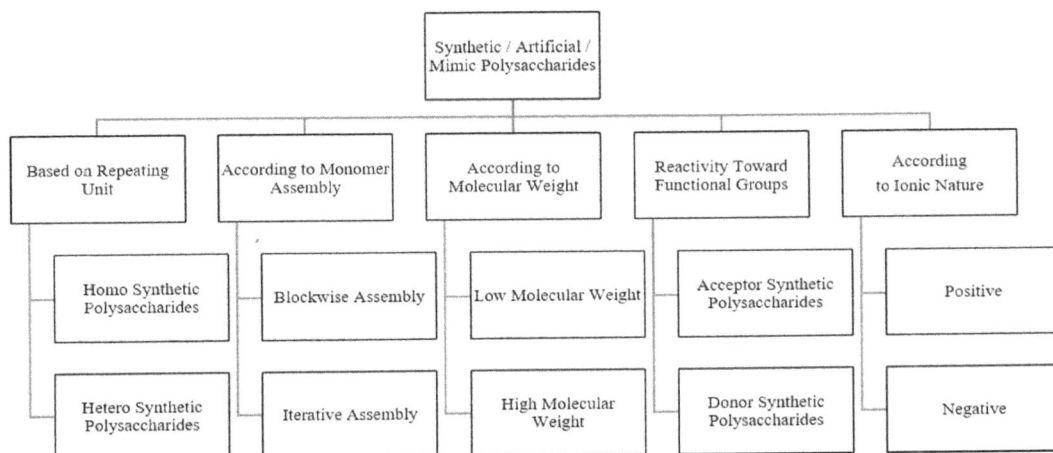

3.3.1 Homo Synthetic Polysaccharides

Homo synthetic polysaccharides are those polysaccharides that are made of a repetition of a mono-saccharide. For example – dextran, it is prepared by certain lactic acid bacteria from sucrose, the best-known *Leuconostoc mesenteroides*. It is a nontoxic polysaccharide and can form hydrogel cross-linking at that time need to use reagents like diisocyanates and epicholorohydrin. Dextran has some special qualities for which they are commercially very interested solubility, viscosity, thermal, and rheological properties. In Figure 3.8, the general pathways for carbohydrates and the synthesis of dextran in four different types of lactic acid bacteria: *Leuconostoc, Weissella, Lactobacillus*, and *Streptococcus*.

3.3.2 Hetero Synthetic Polysaccharides

Hetero synthetic polysaccharides are those that contain two or more different monosaccharide units, some heteropolysaccharides participate together with amino acid chains. Chitosan is one kind of heteropolysaccharides which have glucosamine (deacetylated monomer) and N-acetylglucosamine (acetylated monomer) and also they are linked through β-1,4-glycosidic bonds (Figure 3.9). By enzymatic or chemical deacetylation, chitin is derivative into chitosan. Due to having good bio-degradability and biocompatibility, it is easy to modify, and for this reason, it has extensive use in the biomedical (drug and vaccine) sector. Due to having antimicrobial quality, it is used in the food industry and tissue engineering.

3.3.2.1 Blockwise Assembly

Blockwise assembly refers to the process of synthesizing a polysaccharide by adding small, pre-synthesized building blocks, or blocks, together to form the final product. For examples of

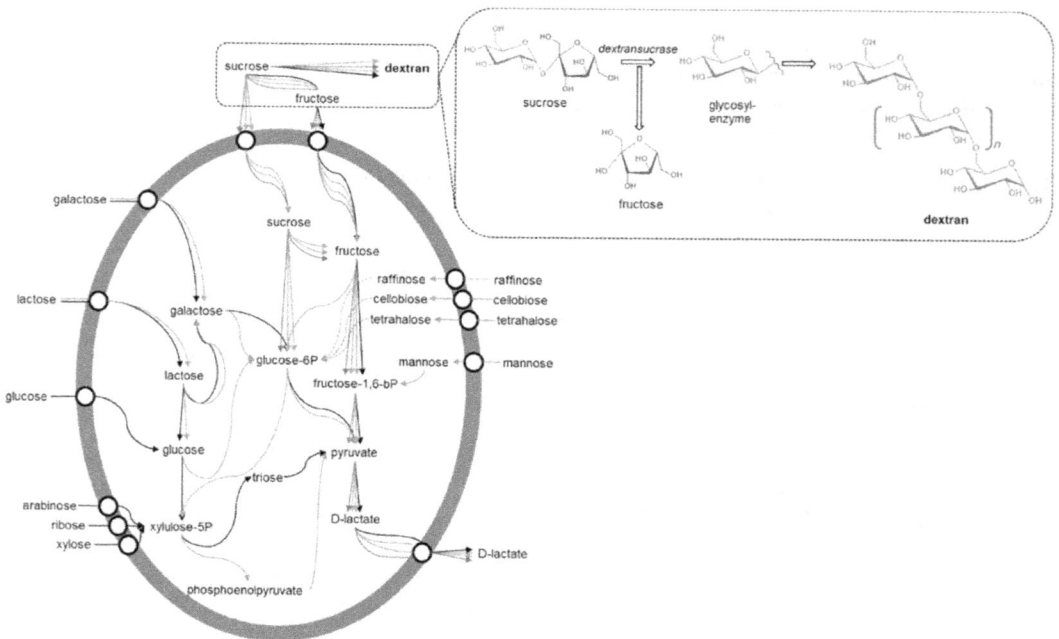

FIGURE 3.8 The general pathways of carbohydrates and the process of dextran synthesis in different lactic acid bacteria, including *Leuconostoc, Weissella, Lactobacillus*, and *Streptococcus* (Díaz-Montes, 2021).

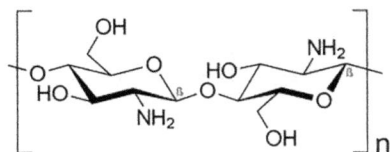

FIGURE 3.9 Chitosan, a linear polysaccharide.

the synthesis of oligo-β-(1–6)-glucosamines, blockwise assembly of oligosaccharide chains (Figure 3.10) is regarded to be a more dependable and scalable method.

3.3.2.2 Iterative Assembly

The process in which a growing polysaccharide chain is repeatedly modified until the desired structure is obtained. In iterative assembly (Figure 3.11), the polysaccharide chain is synthesized in a stepwise manner and the chemical structure is modified after each step, leading to the final desired product.

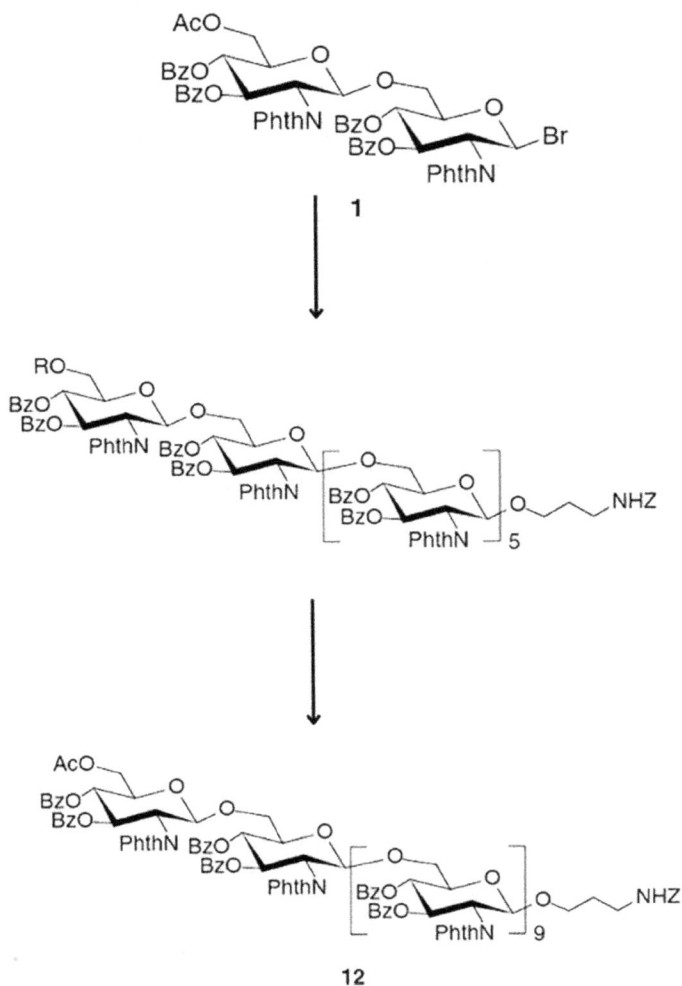

FIGURE 3.10 Blockwise assembling of oligosaccharide.

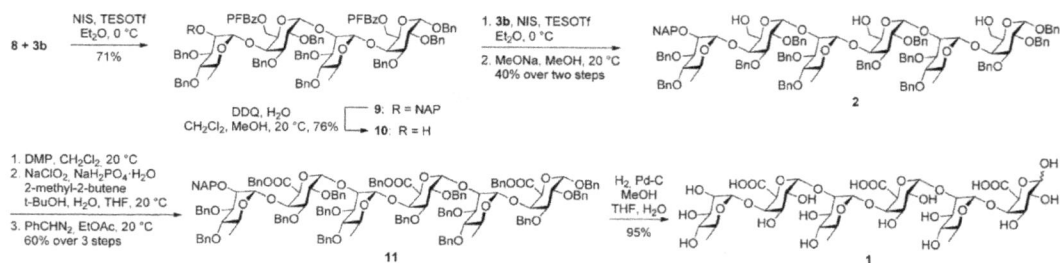

FIGURE 3.11 Iterative assembly of the hexa saccharide (Zakharova et al., 2013).

Iterative glycosylation allows two qualities in the chain -

- Stepwise elongation of polysaccharides
- Specific modifications of defined positions in the chain

3.3.3 ACCORDING TO MOLECULAR WEIGHT

3.3.3.1 Low Molecular Weight Synthetic Polysaccharide

Low molecular weight synthetic polysaccharide refers to a type of carbohydrate polymer composed of repeating units of simple sugars, with a relatively low molecular weight compared to other polysaccharides. These polysaccharides have a molecular weight ranging from a few hundred to a few thousand daltons.

For example, class 1 Dextrans, a low molecular weight polysaccharide, molecular weight is ≥ 1,000. Dextrans are synthesized by the action of leuconostoc mesenteric on sucrose, but it has a high molecular weight form too (Figure 3.12).

3.3.3.2 High Molecular Weight Synthetic Polysaccharide

High molecular weight synthetic polysaccharide refers to a type of carbohydrate polymer composed of repeating units of simple sugars, with a relatively high molecular weight compared to other polysaccharides. These polysaccharides have a molecular weight ranging from tens of thousands to millions of daltons. Examples of high molecular weight synthetic polysaccharides include chitosan, on average its molecular weight is 3,800–20,000 daltons.

- Chitosan (Figure 3.9)

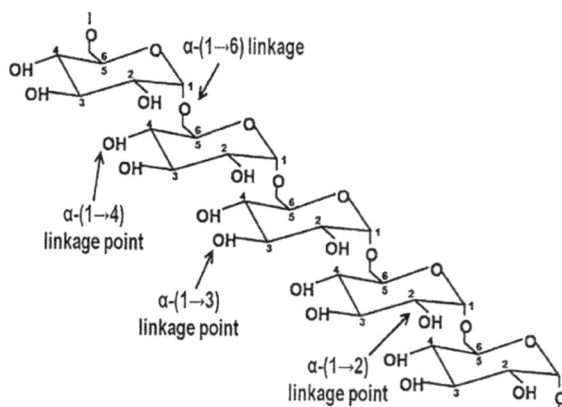

FIGURE 3.12 Class 1 dextrans (Al-Farga & Abed, 2016).

3.3.4 REACTIVITY TOWARD FUNCTIONAL GROUPS

3.3.4.1 Acceptor Synthetic Polysaccharides

Acceptor synthetic polysaccharide is a type of synthetic carbohydrate molecule that can accept chemical groups in a reaction, and they are referred to as "acceptors" because they have reactive functional groups, such as hydroxyl groups, that can accept chemical groups through chemical reactions such glycosylation (Figure 3.13).

3.3.4.2 Donor Synthetic Polysaccharides

Donor synthetic polysaccharides are a group of molecules that provides a sugar or other type of monomer unit that can be used in the formation of a glycosidic bond with another molecule (the "acceptor"). For example, glycosyl fluoride donor (Figure 3.14).

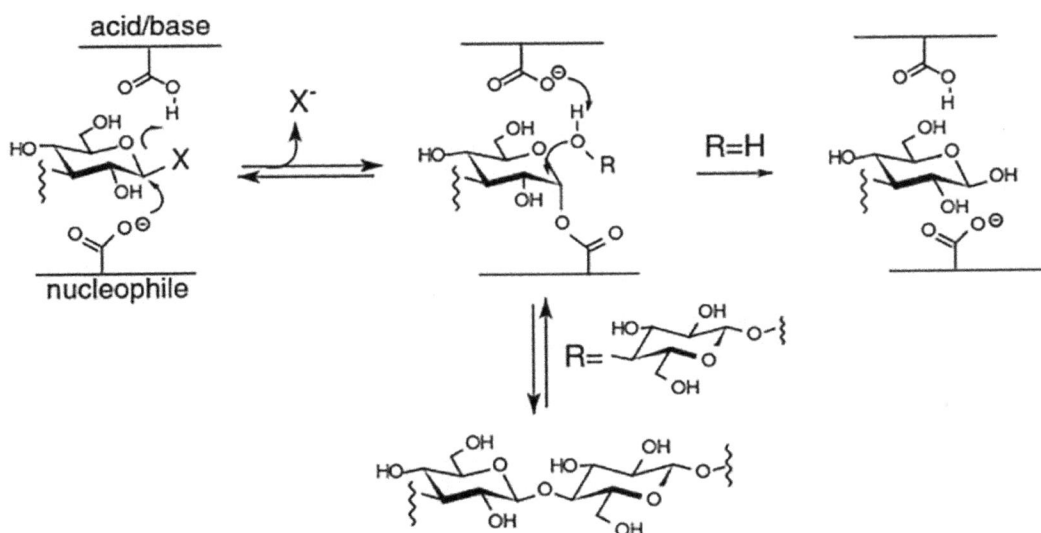

FIGURE 3.13 Glycosidases.

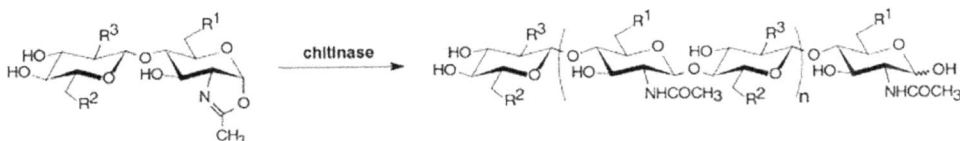

a	$R^1 = OH$	$R^2 = OH$	$R^3 = NHCOCH_3$
b	$R^1 = OH$	$R^2 = OH$	$R^3 = NH_2$
c	$R^1 = OH$	$R^2 = OH$	$R^3 = OH$
d	$R^1 = OH$	$R^2 = OH$	$R^3 = NHSO_3Na$
e	$R^1 = F$	$R^2 = OH$	$R^3 = NHCOCH_3$
f	$R^1 = OH$	$R^2 = F$	$R^3 = NHCOCH_3$
g	$R^1 = F$	$R^2 = F$	$R^3 = NHCOCH_3$

FIGURE 3.14 Activated glycosyl donors are oxazoline derivatives.

3.3.5 ACCORDING TO IONIC NATURE

3.3.5.1 Positively Charged Synthetic Polysaccharides

A positively charged polysaccharide refers to a polysaccharide molecule that has a positive charge because of the presence of more positive ions (e.g., cations) than negative ions (e.g., anions) in its chemical structure. This charge can influence the behavior and interactions of the polysaccharide with other molecules and can affect its solubility and stability in various environments. For example, there is only an alkaline polysaccharide containing a positive charge and that is chitosan.

3.3.5.2 Negatively Charged Synthetic Polysaccharides

Polysaccharides can carry a negative charge due to the presence of negatively charged functional groups such as carboxyl or sulfate groups. These negatively charged polysaccharides are called anionic polysaccharides, and they play important roles in various biological processes, cosmetics, and vaccines. For example, heparin has the highest negative charge density of any known biomolecule, and heparinoid is a chemically modified derivative of heparin (Figure 3.15).

3.4 FABRICATION TECHNIQUES OF SYNTHETIC POLYSACCHARIDES

Synthetic polysaccharide can fabricated in many ways for diversified uses. Some are mentioned below.

3.4.1 SOLUTION CASTING

The solution-casting method is based on the principle of Stokes' law (Das et al., 2018a). Membranes are made using the casting method of solution casting. The solution system's primary constituents are polymer and solvent although additional additives may also be used. When making the casting solution, the polymer we choose is quite important. In reality, the final membrane application is the only factor that determines whether the polymer is soluble in the chosen solvent at the proper concentration. Membranes with a porous structure typically result from low polymer concentrations, whereas membranes with a dense structure are produced by high polymer concentrations (Galiano, 2020). Solution casting is a common method for fabricating synthetic polysaccharides. It involves dissolving the polysaccharide in a suitable solvent to form a homogeneous solution. The solution is then cast onto a substrate, such as a glass or plastic slide, and allowed to dry. The solvent evaporates, leaving behind a thin film of polysaccharide material. One advantage of solution casting is that it can be easily scaled up to produce large amounts of polysaccharide films, making it a suitable method for large-scale production. Additionally, the solvent used for casting can influence the final morphology of the polysaccharide film, leading to improved properties such as increased

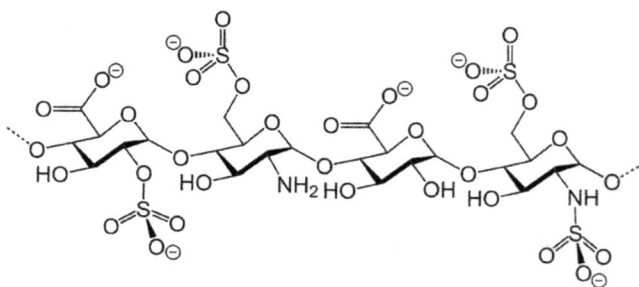

FIGURE 3.15 Heparin.

mechanical strength or enhanced stability. This material is versatile and can be utilized for many different applications, including but not limited to wound healing, tissue engineering, and drug delivery. However, solution casting also has some limitations, such as the need to carefully select solvents that are compatible with the polysaccharide material and the substrate, and the potential for the formation of defects in the film due to uneven drying or solvent evaporation. Additionally, some polysaccharides may not be soluble in commonly used solvents, making solution casting not suitable for these materials.

3.4.2 ELECTROSPINNING

A method for creating ultra-fine fibers from a polymer solution, electrospinning has the advantages of ease of use, cheap cost, a variety of raw material sources, and controlled process parameters (Rahimi et al., 2020). Lots of application need this method to prepare nanocomposites, nanofibers, etc. (Alam et al., 2023; Majumder et al., 2020). In electrospinning, a polymer solution is subjected to a high electric field, causing the solution to be drawn out into a fine fiber. The fiber is then collected onto a grounded collector. These fibers are referred to as nanofibers when they have diameters of less than 500 nm and are characterized by their large surface-to-mass ratio, high porosity, and mechanical performance; these properties have caught the interest of the scientific community (Mejía Agüero et al., 2022). The electric field and the solvent evaporation rate determine the diameter of the fiber produced. Synthetic polysaccharides, such as alginate, chitosan, and gelatin, can be used in electrospinning to create nanofibers with tunable properties (Rahimi et al., 2020).

A laboratory-scale electrospinning setup typically comprises four main parts, as depicted in Figure 3.1: syringe pump, a high-voltage supply, syringe, ground collector, and syringe. The process starts with an electrified pendant droplet held together by surface tension, which morphs into a cone shape (known as the Taylor cone) due to repulsion from surface charges. The cone then produces an electrified jet that moves in a straight line until it experiences a whipping instability. During this instability, the jet is stretched and thinned into fibers that have nanoscale diameters. The fibers then dry once they reach the ground collector. Overall, this process is well-suited for various applications and produces fibers with small diameters (Mokhena et al., 2000) (Figure 3.16).

Synthetic polysaccharides are attractive for electrospinning because they are biocompatible and biodegradable, making them compatible for biomedical applications such as wound healing and tissue engineering. Additionally, their properties, such as mechanical strength and surface charge, can be tailored by adjusting the spinning conditions and post-processing treatments.

Despite its potential applications, electrospinning of synthetic polysaccharides still faces some challenges, including low processing yields, limited fiber uniformity, and poor fiber stability. Further research is needed to overcome these challenges and to optimize the electrospinning process for these materials.

3.4.3 BLENDING

The most promising methods for producing polymer materials are the binary mixtures of natural polysaccharides (cellulose-chitin, cellulose-chitosan), as well as their mixtures with synthetic polymers (Rogovina & Vikhoreva, 2006). The blending of synthetic polysaccharides is a process where two or more polysaccharides are combined to create a new material with improved properties compared to the individual polysaccharides. The blend can be done through various methods, including physical blending, chemical modification, or a combination of both. Blending polysaccharides can offer several benefits, including improved mechanical strength, enhanced biocompatibility, and altered drug release behavior. For example, chitosan and alginate have been blended to

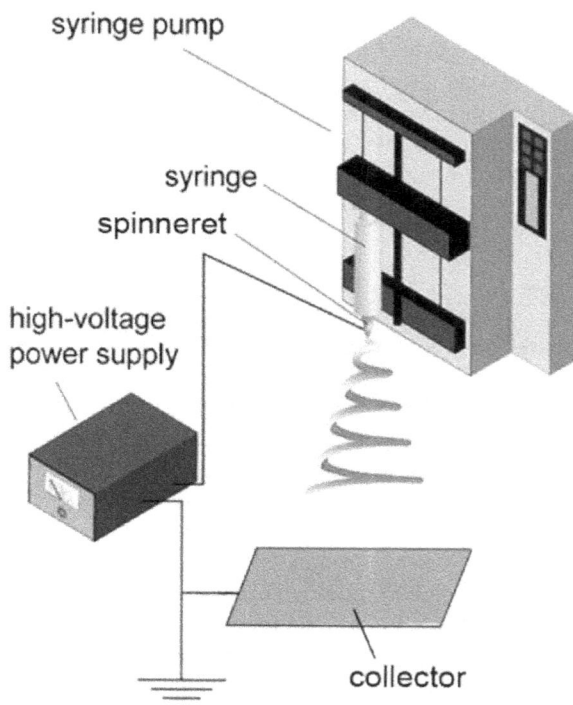

FIGURE 3.16 Typical electrospinning setup (Xue et al., 2017).

produce materials with improved mechanical properties and improved wound healing performance. The blending of polysaccharides can also lead to synergistic effects, where the properties of the blend are greater than the sum of its parts. Yet, the blending of polysaccharides can also present challenges, such as compatibility issues between the polysaccharides and the formation of a phase-separated structure. It is important to carefully consider the processing conditions, such as the blending ratio, temperature, and the presence of a solvent, to ensure the formation of a homogeneous blend.

3.4.4 LAYER-BY-LAYER ASSEMBLY

Using nanoscale polymer to change planar and spherical substrates, the layer-by-layer (LbL) assembly approach is a potential way to create highly conformal coatings. It has been used in the development of drug delivery systems, biosensors, enzyme immobilization, and cell adhesion (Tong et al., 2012). The technique offers many benefits, such as simplicity, adaptability, cheap operating costs, accurate structure control, and a high capacity for guest loading (Manuscript, 2013). This process is used to build up thin films by alternating depositions of oppositely charged materials, typically polymers or nanoparticles. In the case of polysaccharides, the layers can be built up by the sequential adsorption of polysaccharides with opposite charges onto a substrate. This allows for the creation of thin films with controlled thickness, composition, and surface charge. For example, if a positively charged polysaccharide is first adsorbed onto a substrate, it can then be followed by the adsorption of a negatively charged polysaccharide. This process can be repeated multiple times to build up the desired thickness. By varying the type and amount of polysaccharides used in each layer, the properties of the resulting film can be tailored to meet specific requirements. LbL assembly of polysaccharides has numerous potential applications, such as in the creation of protective coatings, and drug delivery systems, and in the design of biosensors and bioelectronics.

3.4.5 SOL-GEL PROCESSING

The sol-gel processing method, first described by Ebelman in the middle of the 18th century, was previously used to fabricate decorative and constructional materials, but it has since undergone extensive development for uses in the production of ceramics, glasses, catalysts, fibers, coatings, and composites (O'Brien et al., n.d.; Wright and Sommerdijk, 2011). This process is also used to fabricate nanocomposites (Das et al., 2018a). The sol-gel process generally entails the change of a solution from a liquid (usually colloidal) phase to a solid phase. The sol-gel process begins with the preparation of the sol solution. This is usually done by mixing a precursor, such as a metal alkoxide or a metal salt, with a solvent. The precursor is chosen based on the desired final product, as different precursors will yield different materials. The sol solution is then subjected to a chemical reaction, typically an acid-catalyzed hydrolysis that causes the precursor to form a gel-like structure. During the gelation process, the solvent evaporates and the gel structure forms. The gel structure can be further modified by controlling the conditions of the gelation process, such as temperature and evaporation rate, to produce materials with different structures and properties. Once the gel has formed, it can be subjected to further processing, such as drying and calcination, to obtain the final product. The drying process causes the solvent to continue to evaporate, causing the gel to shrink and become denser. Calcination, the heating of the material in a controlled environment, can be used to modify the structure of the material, improve its thermal stability, and remove any residual solvent.

This technique has many advantages over traditional fabrication methods, such as the ability to produce materials with high homogeneity, uniform pore sizes, and controlled porosity. It also allows for the incorporation of sensitive or reactive materials into the final product without damaging them, as the process is carried out at relatively low temperatures. The sol-gel fabrication of polysaccharides offers a multitude of possibilities due to the adaptable nature and versatility of the materials produced. These polysaccharides can be utilized in various fields, including biomedicine as drug carriers and tissue engineering scaffolds, sensing technologies, energy production as catalysts, environmental remediation, cosmetic products, biodegradable food packaging, and in membrane production for high selectivity and permeability. The combination of uniform structure, high surface area, and biocompatibility makes these materials a desirable option for numerous applications.

3.4.6 MICROFLUIDICS

The technology of manipulating fluid in channels of tens of micrometers-wide diameters is known as microfluidics. Microfluidics is a field of science and technology that deals with the manipulation and processing of small volumes of fluids, ranging from 10^{-9} to 10^{-18}L, through channels that have dimensions in the tens to hundreds of micrometers range (Whitesides, 2006). Microfluidic technology is a rapidly advancing field that provides various tools to manipulate small volumes of fluids, ranging from 10^{-9} to 10^{-18}L, and control chemical, biological, and physical processes essential for sensing. Lithographic techniques have facilitated the development of such tools, which can be integrated with electronic and optical components to construct functional sensors (Stroock, 2008). The microfluidic process typically starts with the preparation of two immiscible liquids, one of which contains the precursors and reagents, and the other containing the solvent. The two liquids are then mixed in a microfluidic device, and the resulting droplets are subjected to the chemical reactions necessary to synthesize the polysaccharide. After the reaction is complete, the droplets can be further processed to obtain the final product. This may involve drying, calcination, or other post-synthesis treatments to modify the properties of the polysaccharide. The final product can then be collected and analyzed to determine its structure and properties. The main advantage of the microfluidic approach is the high level of control it provides over the reaction conditions, including temperature, reaction time, and reactant concentration. This control enables the precise synthesis of polysaccharides with well-defined structures and properties. Additionally, the small droplet size in microfluidic systems allows for high-throughput synthesis, as many droplets can be processed simultaneously.

3.4.7 Photolithography

Typically, photolithography (PL) is used to create micro and nanostructures in a thin photoresist film by exposing it to light in a specific pattern (Nicaise et al., 2015). It entails applying a photoresist layer to a substrate and then subjecting it to UV light (Mustafa et al., 2020). The process has been explored using polysaccharides such as starch as the material to be patterned. Despite the environmentally friendly and biodegradable nature of polysaccharides, their use in photolithography is still in its early stages due to challenges such as low solubility and low thermal stability. However, ongoing research is aimed at developing new processing methods and materials to allow for the wider use of polysaccharides in photolithography as a sustainable alternative to traditional materials.

3.4.8 Microencapsulation

The microencapsulation process involves wrapping active substances in tiny capsules. This cutting-edge technology has been utilized in several industries, including cosmetics, pharmaceuticals, agrochemicals, and food, for encapsulating a variety of substances such as flavors, acids, oils, vitamins, and microorganisms. The success of microencapsulation relies on the appropriate selection of the coating material, the form of the core material, and the encapsulation technique (Silva et al., 2014). Synthetic polysaccharides, such as alginate or chitosan, can be used as coatings to encapsulate various substances, including drugs, flavors, and fragrances. The microencapsulation process involves droplet formation, coating, and drying, with properties such as size, stability, and release rate controlled by adjusting the coating solution and process conditions. Microencapsulation offers a flexible and versatile approach to delivering substances with potential applications in various industries.

3.5 VERSATILE APPLICATION OF SYNTHETIC POLYSACCHARIDES IN DIFFERENT FIELDS

3.5.1 Uses in Vaccine

Responses to infectious illnesses, inflammatory agents, and carcinogens are only some of the many challenges faced by the body, and the immune system plays a crucial part in each. Pharmaceutical companies are putting more of an emphasis on finding novel immune-stimulating substitutes that are both safe and effective, and polysaccharides are one of the finest options in this regard. This is due to the rise in the prevalence of cancer and infectious diseases as well as the shortcomings of several current medications, such as toxicity, resistance, and a lack of immune responses. Polysaccharides produced from traditional Chinese medicine have been shown to activate or control T cells and macrophages, increase interleukin activity, improve antibody levels, and control immunological function in the body (Yu et al., 2018). Polysaccharides were also shown to be a key modulator because they boost immunity via several pathways, including by activating macrophages, splenocytes, and thymocytes (Chen et al., 2019). Numerous efforts are being made to determine whether or not vaccines based on carbohydrates are viable options. Given the significance of polysaccharides in immunological cell–cell communication and tumor-associated carbohydrate antigen identification by the host immune system (Mohammed et al., 2020), usher in a new era in glycobiology and the availability of vaccinations (Zhang & Wang, 2015). Many vaccine formulations have used polysaccharide-based antigens, such as tumor-associated carbohydrate antigens and bacterial capsular polysaccharides (Morelli et al., 2011). Polysaccharides vaccines, including those for pneumonia and meningitis, have been on the market since the 1980s (Jennings, 1990; Nair, 2012). Due to the lacking of immunological memory and flip of class from IgM to IgG, traditional polysaccharide antigens, which are mainly pure polysaccharides capsular, have limitations such as a brief duration and a limited immunogenic response in newborn children and young children. Researchers have begun combining polysaccharide vaccines like diphtheria, tetanus toxins, with protein carriers that highly immunogenic, to overcome these drawbacks. These carriers interact with the immune system to increase the immunogenicity of the vaccine by eliciting a T cell-dependent response (Finn, 2004; Nikolaev & Sizova, 2011).

3.5.2 Applications in Biomedical Fields

Biomedical researchers have been using polysaccharides since the turn of the previous century (Muhamad et al., 2019). Numerous biomedical applications have investigated polysaccharides as a possible choice due to their many desirable characteristics, such as biocompatibility, non-immunogenicity, biodegradability, and increased stability and solubility (Boddohi & Kipper, 2010; Zhang & Wang, 2015). Some of such applications are shown in Table 3.2. Furthermore, polysaccharides are substances of choice and are employed in many biomedical and biotechnological (Usman et al., 2017) applications due to their availability of sources and affordable cost. Algal polysaccharides, for instance, have seen considerable usage in a variety of biological contexts, including wound treatment, regenerative medicine, and drug delivery regulation (De Jesus Raposo et al., 2015). This new family of biomaterials also has the advantage of being able to be made into hydrogels, like the hydrogels that have been successfully used to transport bone morphogenetic proteins and are loaded with heparin. It has been shown that polysaccharides can improve mechanical properties and make up for synthetic polymers' poor biological performance (D'Ayala et al., 2008; Boddohi & Kipper, 2010; Gemini et al., 2016; Cascone et al., 2001).

3.5.3 Drugs, Vaccine Delivery, and Tissues Engineering

Polysaccharides are gaining popularity in the domains of tissue engineering (Li et al., 2004; Ngwuluka, 2018), cosmetics (Bragd et al., 2004), and wound healing (D'Ayala et al., 2008; Muhamad et al., 2019) due to their use as drug carriers, bioactive materials, building blocks for drug delivery, and excipients to improve drug delivery. Natural polysaccharides are adaptable, meaning they could be used for anything from medicine and vaccine delivery to cosmetics

---

TABLE 3.2
Commercially Available Polysaccharides Drugs with Their Sources (Mohammed et al., 2021)

Source	Polysaccharides	Drugs	Biological Activity and Applications
Animal	Heparin	Heparin injection, heparin cream, low molecular weight heparin sodium gel, heparin sodium lozenge, and heparin sodium cream Chondroitin	Stabilize, distribute, and improve growth factors like FGF-2 (Boddohi & Kipper, 2010), have antiviral, anticoagulant, biosensor for thrombin (Finkenstadt, 2005), anti-inflammatory, and anti-angiogenic activities (Fedorov et al., 2013)
	Chondroitin sulfate	Tablets containing chondroitin sulfate, capsules containing chondroitin sulfate A sodium, and injections containing chondroitin sulfate	Activation of regulation of angiogenesis, interactions with matrix proteins, melanoma cell invasion, growth factors, and proliferation (Petit et al., 2006), coatings (Boddohi & Kipper, 2010), bacterial/viral infections (Yamada & Sugahara, 2008), morphogenesis, cell migration, and coatings (Boddohi & Kipper, 2010), and osteoarthritis (Chen et al., 2017)
	Hyaluronic acid	Injections of sodium hyaluronate and sodium hyaluronate ocular drops	Anti-arthritic (Gupta et al., 2019), drug carriers (Huang & Huang, 2018), osteoarthritis (Chen et al., 2017)
Plant	Astragalus PS	2-(Chloromethyl)-4-(4-nitrophenyl)-1,3-thiazole injection of astragalus	Antiviral (Xue et al., 2015), immunoregulatory (Yang et al., 2014), antitumor (Pu et al., 2016; Xue et al., 2015), and antioxidative (Pu et al., 2016)

(Continued)

TABLE 3.2 (*Continued*)

Commercially Available Polysaccharides Drugs with Their Sources (Mohammed et al., 2021)

Source	Polysaccharides	Drugs	Biological Activity and Applications
	Ginseng PS	Injections of ginseng polysaccharides	Hypoglycemic (Xie et al., 2004), immunostimulant (Lee et al., 2019), anti-inflammatory (Ullah et al., 2019)
	Fucoidan PS	Active component of a medication	Immune modulation, cancer inhibition, and pathogen inhibition (Fitton et al., 2015), cell proliferation and differentiation (Park et al., 2012), antioxidant (Wang et al., 2019), antiviral (Damonte et al., 2012), antitumor (van Weelden et al., 2019)
Microbial	Lentinan PS	Tablets containing lentinus edodes mycelia polysaccharides, lentinan injection, and lentinan capsules	Antitumor, hepatoprotective, immunologic activities, and antiviral
	Poria PS	Poria polysaccharide oral solution capsular	Immunomodulation, antioxidation, immunomodulation, anti-aging, anti-hepatitic, anti-diabetics, anti-hepatitic, antitumor, and anti-hemorrhagic fever
	Capsular PS	Group A and C meningococcal polysaccharide vaccine, polyvalent pneumococcal vaccine, and Vi polysaccharides typhoid vaccine	Vaccines, passive antibody therapies
	Dextran	Low molecule dextran, dextran 70 eye drops, and dextran 40 glucose injection	Applications on biotechnological field

and industrial purposes. Polysaccharides including gellan gum, xanthan gum, and scleroglucan, all of which are produced by microbes, have been the subject of substantial research for their potential use in drug administration (Alvarez-Lorenzo et al., 2013). By encapsulating the medication inside a nanoparticle carrier made of bioadhesive polysaccharides, drug uptake can be improved (Liu et al., 2008). Pectin, amylose, guar gum, inulin, chitosan, dextran, and chondroit-sulfateate are some of the other naturally occurring polysaccharides that have been studied for their potential as pharmaceutical excipients for colon-specific drug release (Chourasia & Jain, 2004). Chitin and chitosan are tissue-compatible polysaccharides that have been demonstrated to be effective in bone regeneration, tissue engineering, wound healing, and the transportation of medications and vaccinations (Chourasia & Jain, 2004; Vandamme et al., 2002). One of the most extensively utilized biopolymers for medication and vaccine administration is chitosan, which is depicted in Figure 3.9 (Luo & Wang, 2014). This is because a range of antigens can be encapsulated without harsh temperatures and organic solvents, preventing denaturalization and degradation (Arca et al., 2009) (Figure 3.17).

The potential of polysaccharide-based nanoparticles as nanometric carriers in the medication delivery system has recently seen an increase in research (Liu et al., 2008). Cationic polysaccharides, used as gene transfection vectors, are a novel class of non-viral gene delivery systems (Azzam et al., 2002). These polycations are created through a reductive amination reaction between primary amines and periodate-oxidized polysaccharides. While cross-linked polysaccharide hydrogels are employed in significant applications like tissue engineering, medication delivery systems. It is anticipated that the hyper-branched polymer would be a suitable carrier for gene delivery nanoparticles (Chen et al., 2019; Laurienzo, 2010).

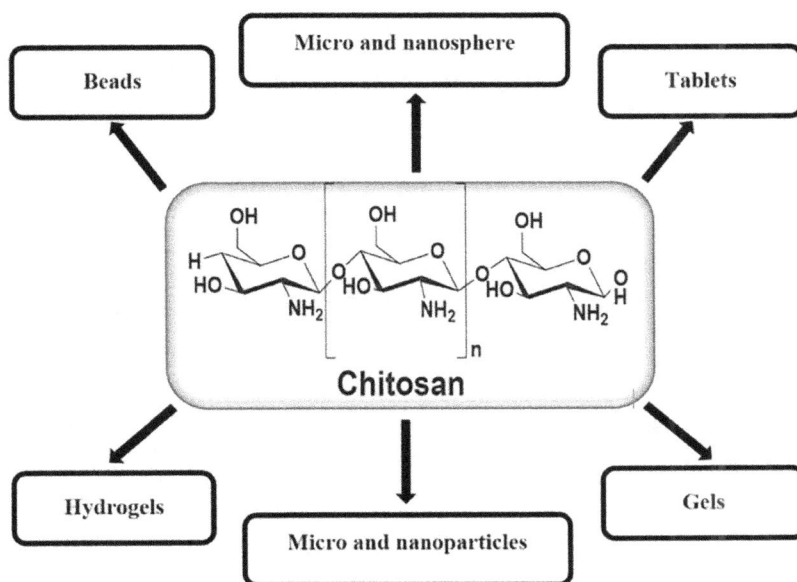

FIGURE 3.17 Chitosan-based drug delivery (Mohammed et al., 2021).

3.5.4 APPLICATION IN ANTITUMOR AND IMMUNOMODULATORY ACTIVITIES

Tumors are a primary cause of death globally (Zong et al., 2012), and numerous plants (Guo et al., 2019) and marine polysaccharides (De Jesus Raposo et al., 2015) have been shown to have anticancer properties. Several other types of natural polysaccharides, such as lentinan and schizophyllan (Chen et al., 2019), are potent antitumor agents. Antitumor activity and the ability to boost the efficacy of standard chemotherapeutic medicines are both hallmarks of polysaccharide-protein conjugates (Zong et al., 2012). This polysaccharide's immunosuppressive action against tumor growth has been demonstrated by numerous studies in recent years (Wu et al., 2019). Despite their unique chemical structures and conjugated components, mushrooms have long been valued for their ability to provide a variety of health benefits whether consumed or applied topically. Natural polysaccharides isolated from mushrooms have been the subject of substantial research due to their powerful anticancer and pharmacological activities (Ren et al., 2012; Zhang et al., 2007). Mushroom polysaccharides include *Ganoderma lucidum*, which showed strong in vitro immune activation and anticancer action on breast cancer cells (Lee et al., 2002; Zhao et al., 2010), and *Lentinus edodes*, which had a striking antitumor impact against subcutaneously implanted sarcoma (Paulsen, 2005). Mushroom polysaccharides have been shown to have anticancer effects (Pandya et al., 2019), and several mechanisms have been hypothesized for how they do so, including the inhibition of tumor growth, the stimulation of the immune system, and the induction of apoptosis in tumor cells. Mushroom polysaccharides are utilized in clinical trials to boost the efficacy of chemotherapeutic drugs while reducing their negative effects (Schepetkin & Quinn, 2006). The polysaccharides found in algae are also very significant due to their wide range of pharmacological properties, which include anticancer activity (Costa et al., 2010). Polysaccharides induce cytokine expression in effector cells like T lymphocytes, macrophages, B lymphocytes, natural killer cells, and cytotoxic T lymphocytes; these cytokines always exhibit antiproliferative activity, trigger apoptosis and differentiation in tumor cells shown in Figure 3.2, and secrete products like reactive oxygen intermediates, nitrogen, and interleukins (Meng et al., 2016; Seidi et al., 2018). Saccharide-coated nanoparticles have a longer half-life in the body's circulatory system. Furthermore, they are seen in high concentrations within tumors (Seidi et al., 2018) (Figure 3.18).

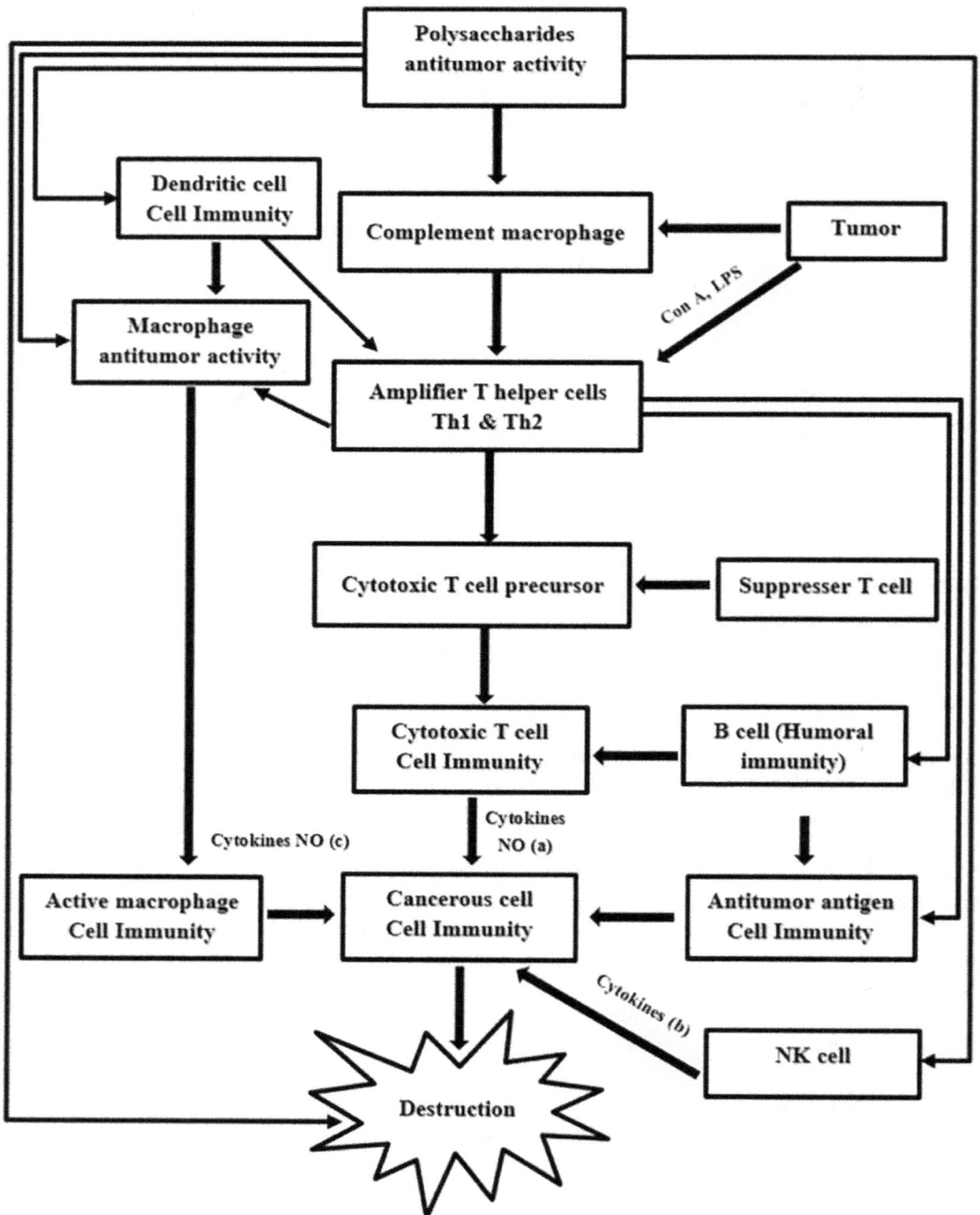

FIGURE 3.18 Probable immune system mechanism. The production of cytokines such IL-2, 3, 4, 6, 8, and 12 as well as TNF, IFN, and IL-1 by T cells was boosted by fungal polysaccharides. The NK cells' production of IL-1, 2, 3, 12, TNF-, and IFN- was boosted by cytokines B found in mushroom polysaccharides. The release of IL-6 and 8, IL-1, TNF, and IFN- from macrophages may be enhanced by cytokines C from fungal polysaccharides. LPS was employed as a general B cell, macrophage, and DC activator, while ConA was used as a specialized T cell activator (Meng et al., 2016; Mohammed et al., 2021).

3.5.5 Antioxidative Applications

Many diseases, including diabetes mellitus, cancer, neurodegenerative, serious tissue injuries, and inflammatory diseases, may be caused by reactive oxygen species (ROS) because they assault various macromolecules such as membrane lipids, DNA, and proteins. Since antioxidants can inhibit or lessen the harmful effects of these substances (Kardošová & Machová, 2006; Wijesekara et al., 2011) and boost health (Rocha De Souza et al., 2007), they are extremely helpful in this context. Because of the many negative effects that synthetic antioxidants, such as carcinogenesis and liver damage, have, a natural alternative would be quite useful. Plant-based polysaccharides have been shown to have potent antioxidant properties (Tseng et al., 2008), which may help shield the human body from the damaging effects of free radicals and reduce the severity of many diseases (Yuan et al., 2008). Since plant polysaccharides play such an important role in biology, scientists have spent a lot of time delving into their chemical properties to learn more about their potential uses in medicine, such as anticancer, immune-stimulating, and antioxidant actions (Jin, 2012). Polysaccharides, for example, have been shown to protect against and even reverse the damage to the heart and brain caused by free radicals (Chen et al., 2019); therefore, their presence as natural antioxidants that may scavenge these ROS is very helpful. *Hyriopsis cumingii* is just one example of a naturally occurring polysaccharide that has been studied for its antioxidative effects (Qiao et al., 2009). Polysaccharides isolated from the Chinese herb astragalus also exhibited strong antioxidant and anticancer action (Li et al., 2010). Anticoagulant/antithrombotic, antilipidemic, antiviral, and antioxidant actions are just some of the properties of the sulfated polysaccharides found in seaweed and red alga (Arad & Levy-Ontman, 2010; Jiao et al., 2011). For their antioxidant properties, sulfated polysaccharides found in seaweed, galactomannan, arabinogalactan, and pectic polysaccharides found in plants, and glucans and glycoproteins found in mushrooms have all been used (Yu et al., 2018).

Polysaccharides may have an anti-inflammatory effect through one of these mechanisms. For instance, TCM polysaccharides have anti-inflammatory activity primarily because they inhibit the expression of chemotactic factors and adherence factors as well as the activities of key enzymes involved in the inflammation process (Chen et al., 2016). Other polysaccharides inhibit inflammatory-related mediators like cytokines (IL-1β, IL-6, TNF-α) and NO (nitric oxide), and decreased the infiltration of inflammatory cells (Chen et al., 2016). Sulfated polysaccharides derived from algae show their anti-inflammatory effect by inhibiting the migration of leukocytes to sites of inflammation (Jiao et al., 2011).

3.5.6 Other Applications

3.5.6.1 Anti-inflammatory Activity

The anti-inflammatory effects of natural polysaccharides were proven experimentally (Muhamad et al., 2019) and led to their widespread usage in nanotechnology for the management of inflammatory disorders. One of these mechanisms may be responsible for polysaccharides' anti-inflammatory effect; for example, polysaccharides used in traditional Chinese medicine have anti-inflammatory properties primarily due to their capacity to inhibit the production of chemotactic and adhesion factors as well as the actions of important enzymes involved in the inflammation process (Chen et al., 2016). Algal sulfated polysaccharides inhibit leukocyte migration to inflammatory areas, which is how they exert their anti-inflammatory effects (Jiao et al., 2011; Ullah et al., 2019). Sulfated polysaccharides from algae demonstrate their anti-inflammatory impact differently than other polysaccharides, which have an inhibitory effect on inflammatory-related mediators including cytokines (IL-1β, IL-6, and TNF-α).

3.5.6.2 Hypoglycemic and Hypocholesterolemic Activities

Since the 1980s of the last century (Chen et al., 2019; Nie et al., 2018), several clinical investigations have been undertaken on polysaccharides for their hypoglycemic and hypocholesterolemic impact. Protecting against kidney impairment in people with type 2 diabetes, *Ganoderma atrium* polysaccharide also has the potential to cure hyperlipidemia, hyperglycemia, insulin resistance, and hyperinsulinemia.

Orally given insulinloaded dextran-chitosan nanoparticulate polyelectrolyte complex, for instance, has a greater bioavailability and a longer-lasting hypoglycemic impact (Ngwuluka, 2018). Proteins placed onto natural polysaccharides operate more steadily and have a longer therapeutic effect than proteins alone. Chitosan and kefran are two polysaccharides that have hypoglycemic and hypocholesterolemic properties (Nayak et al., 2019; Moradi & Kalanpour, 2019), as well as the sulfated polysaccharides that are extracted from *Bullacta exarate* (Yu et al., 2018). By modification of PPAR-mediated lipid metabolism, a dose-dependent hypoglycemic impact and improved insulin sensitivity can be achieved (Wu et al., 2019) were seen in mice administered a traditional Chinese medication comprising polysaccharides derived from *Tremella fuciformis* mushrooms.

3.5.6.3   Anticoagulant Activity

Anticoagulant activity is one of several polysaccharide qualities that has been examined in depth; sulfated polysaccharides like unfractionated and low molecular weight heparins are utilized as anticoagulant medications but have undesirable side effects such bleeding and thrombocytopenia (Costa et al., 2010). Antitumor, antioxidant, and anticoagulant actions are only some of the many biological effects of polysaccharides that have been demonstrated (Yu et al., 2018). These sulfated polysaccharides have anticoagulant action due in large part to their high sulfate concentration (De Jesus Raposo et al., 2013). Possible anticoagulant agents include plant-derived polysaccharides like pectin (Olennikov et al., 2015) and those acquired from marine sources like shellfish (shrimp, crab, squid, lobster, crayfish, etc.), marine macro-algae (seaweeds), marine fungus, microalgae, and corals (Mohan et al., 2019).

3.5.6.4   Antiviral Activity

It has been demonstrated that sulfated polysaccharides found from seaweeds have a suppressive effect on the replication of enveloped viruses such as the human immunodeficiency virus (HIV), herpes simplex virus (HSV), human cytomegalovirus, respiratory syncytial virus, and dengue virus (Jiao et al., 2011). This has been known since the 1950s when the polysaccharides's antiviral activity was first demonstrated. Sulfated exopolysaccharides (EPS) are produced by a variety of microalgal species and have a significant biological role as antiviral agents (De Jesus Raposo et al., 2013). Traditional Chinese medicine polysaccharides have been utilized for decades as antiviral drugs due to their proven ability to boost immunity by stimulating macrophages to increase their phagocytic capacity and secrete IL-2, IFN-, and antibodies (Chen et al., 2016).

3.6   SYNTHETIC POLYSACCHARIDES: FUTURE PROSPECTIVE, LIMITATION, AND CHALLENGES

3.6.1   Adored Synthetic Polysaccharides

Synthetic polysaccharides, also known as synthetic sugars, are a highly versatile class of materials that have garnered significant attention and adoration due to their various applications in the biomedical field. One of the most notable applications of synthetic polysaccharides is as hydrogels, which are highly absorbent materials that are widely used in wound dressings and drug delivery systems. Another important application is as mucoadhesive polymers, which have the ability to adhere to mucosal surfaces and have proven to be effective in oral drug delivery. Additionally, synthetic polysaccharides have been developed as biodegradable polymers, providing a more environmentally friendly alternative to traditional petroleum-based polymers. Furthermore, synthetic polysaccharides have also shown promise as materials for wound healing, with studies indicating that they can promote tissue regeneration and improve healing outcomes. The adoration for synthetic polysaccharides is driven by their versatility and potential for application in a range of biomedical fields, including drug delivery, wound healing, and regenerative medicine. These materials hold great potential for improving human health and well-being, making them a highly valued and adored component of the biomedical landscape.

3.6.2 DEPLORED SYNTHETIC POLYSACCHARIDES

Synthetic polysaccharides are being widely used in food additives, cosmetics, packaging and bio-medical applications. However, despite their widespread use, there are growing concerns about the impact of synthetic polysaccharides on both the environment and human health. One major concern is the contribution of synthetic polysaccharides to the microplastics crisis, which is causing significant harm to wildlife and marine life, as well as the potential for these particles to enter the food chain. Additionally, synthetic polysaccharides have been linked to toxicity and biocompatibility issues, with studies indicating that they may cause skin irritation and immune system suppression. The non-bio-degradable nature of synthetic polysaccharides also means that they persist in the environment for extended periods, leading to increased pollution and environmental harm. Despite these concerns, there is currently limited regulation and standardization in the production and use of synthetic poly-saccharides, making it difficult for consumers to make informed decisions about the products they use.

3.6.3 UBIQUITOUS SYNTHETIC POLYSACCHARIDES

Synthetic polysaccharides are employed as thickeners, emulsifiers, and stabilizers in the food indus-try to raise the caliber of food items. They are valued for their ability to enhance the texture, appearance, and stability of food, making them an attractive ingredient for food manufacturers. In pharmaceuticals, synthetic polysaccharides are used as excipients, substances added to drugs to improve their effectiveness. They play a critical role in controlling the release of drugs, improv-ing their solubility, and reducing their toxicity. In cosmetics, synthetic polysaccharides are used as moisturizers, emollients, and film-formers, providing a smooth, moisturized feel to the skin. They are also used as thickening agents and stabilizers to improve the consistency and shelf life of cosmetic products. In the textile industry, synthetic polysaccharides are used as fibers and yarns, providing strength, durability, and comfort to clothing and other textiles. They are also used as siz-ing agents to improve the stiffness, handling, and finish of textiles.

3.7 CONCLUSIONS

In conclusion, synthetic polysaccharides, or synthetic sugars, are a versatile class of materi-als with a wide variety of applications in various industries, including biomedical and cos-metic industries. These materials are adored for their versatility and potential to improve human health and well-being, as well as their ability to function as hydrogels, mucoadhesive polymers, and biodegradable polymers. However, there are also concerns about the negative impact of synthetic polysaccharides on the environment and human health, particularly with regard to microplastics and toxicity. This lack of regulation and standardization has led to calls for greater government oversight and action to ensure that synthetic polysaccharides are used in a safe and responsible manner. The use of synthetic polysaccharides in these industries highlights their versatility and ability to exhibit specific properties that make them useful in a range of applica-tions. However, it is important to note that the widespread use of these compounds has led to criticism, as some have been shown to have negative impacts on the environment and human health. This highlights the need for further research and regulation to ensure their safe and sus-tainable use in these industries.

Looking toward the future, there is great potential for synthetic polysaccharides to continue to play a significant role in the biomedical field, with ongoing research aimed at developing new applications and improving their safety and biocompatibility. At the same time, it is crucial to consider ethical considerations and ensure that these materials are produced and used in a safe and responsible manner. This may involve greater government oversight and regulation, as well as increased efforts to promote sustainability and reduce the environmental impact of synthetic polysaccharides.

In conclusion, synthetic polysaccharides are a versatile and valuable class of materials with both positive and negative aspects. It is essential to balance the potential benefits of these materials with the need to protect the environment and human health. Ultimately, the use of synthetic polysaccharides must be guided by ethical considerations and a commitment to responsible and sustainable practices. As a society, we must work together to ensure that these materials are used in a way that benefits both people and the planet.

REFERENCES

Alam, M. R., Alimuzzaman, S., Shahid, M. A., Fahmida-E-Karim, & Hoque, M. E. (2023). Collagen/Nigella sativa/chitosan inscribed electrospun hybrid bio-nanocomposites for skin tissue engineering. *Journal of Biomaterials Science, Polymer Edition*, *0*(0), 1–22. https://doi.org/10.1080/09205063.2023.2170139

Al-Farga, A., & Abed, S. (2016). Production of dextrans and their applications in human health and nutrition-Review. *European Academic Research*, 29, 3964–3988.

Al-Oqla, F. M., Alaaeddin, M. H., Hoque, M. E., & Thakur, V. K. (2022). Biopolymers and biomimetic materials in medical and electronic-related applications for environment-health-development nexus: Systematic review. *Journal of Bionic Engineering*, *19*(6), 1562–1577. https://doi.org/10.1007/s42235-022-00240-x

Alvarez-Lorenzo, C., Blanco-Fernandez, B., Puga, A. M., & Concheiro, A. (2013). Crosslinked ionic polysaccharides for stimuli-sensitive drug delivery. *Advanced Drug Delivery Reviews*, *65*(9), 1148–1171. https://doi.org/10.1016/j.addr.2013.04.016

Arad, S., & Levy-Ontman, O. (2010). Red microalgal cell-wall polysaccharides: Biotechnological aspects. *Current Opinion in Biotechnology*, *21*(3), 358–364. https://doi.org/10.1016/j.copbio.2010.02.008

Arca, H. Ç., Günbeyaz, M., & Şenel, S. (2009). Chitosan-based systems for the delivery of vaccine antigens. *Expert Review of Vaccines*, *8*(7), 937–953. https://doi.org/10.1586/erv.09.47

Ardèvol, A., & Rovira, C. (2015). Reaction mechanisms in carbohydrate-active enzymes: Glycoside hydrolases and glycosyltransferases. Insights from ab initio quantum mechanics/molecular mechanics dynamic simulations. *Journal of the American Chemical Society*, *137*(24), 7528–7547. https://doi.org/10.1021/JACS.5B01156/ASSET/IMAGES/LARGE/JA-2015-01156W_0016.JPEG

Azzam, T., Eliyahu, H., Raskin, A., Makovitzki, A., Barenholz, Y., Lineal, M., & Domb, A. J. (2002). Cationic polysaccharides as vectors for gene delivery. *American Chemical Society, Polymer Preprints, Division of Polymer Chemistry*, *43*(2), 671–672.

Boddohi, S., & Kipper, M. J. (2010). Engineering nanoassemblies of polysaccharides. *Advanced Materials*, *22*(28), 2998–3016. https://doi.org/10.1002/adma.200903790

Bornscheuer, U., Buchholz, K., & Seibel, J. (2014). Enzymatic degradation of (ligno)cellulose. *Angewandte Chemie International Edition*, *53*(41), 10876–10893. https://doi.org/10.1002/ANIE.201309953

Bragd, P. L., Van Bekkum, H., & Besemer, A. C. (2004). TEMPO-mediated oxidation of polysaccharides: Survey of methods and applications. *Topics in Catalysis*, *27*(1–4), 49–66. https://doi.org/10.1023/B:TOCA.0000013540.69309.46

Cascone, M. G., Barbani, N., Cristallini, C., Giusti, P., Ciardelli, G., & Lazzeri, L. (2001). Bioartificial polymeric materials based on polysaccharides. *Journal of Biomaterials Science, Polymer Edition*, *12*(3), 267–281. https://doi.org/10.1163/156856201750180807

Chen, L., Ge, M. D., Zhu, Y. J., Song, Y., Cheung, P. C. K., Zhang, B. B., & Liu, L. M. (2019). Structure, bioactivity and applications of natural hyperbranched polysaccharides. *Carbohydrate Polymers*, *223*, 115076. https://doi.org/10.1016/j.carbpol.2019.115076

Chen, Q., Shao, X., Ling, P., Liu, F., Han, G., & Wang, F. (2017). Recent advances in polysaccharides for osteoarthritis therapy. *European Journal of Medicinal Chemistry*, *139*, 926–935. https://doi.org/10.1016/j.ejmech.2017.08.048

Chen, Y., Yao, F., Ming, K., Wang, D., Hu, Y., & Liu, J. (2016). Polysaccharides from traditional Chinese medicines: Extraction, purification, modification, and biological activity. *Molecules*, *21*(12). https://doi.org/10.3390/molecules21121705

Chourasia, M. K., & Jain, S. K. (2004). Polysaccharides for colon targeted drug delivery. *Drug Delivery: Journal of Delivery and Targeting of Therapeutic Agents*, *11*(2), 129–148. https://doi.org/10.1080/10717540490280778

Cobucci-Ponzano, B., & Moracci, M. (2012). Glycosynthases as tools for the production of glycan analogs of natural products. *Natural Product Reports*, *29*(6), 697–709. https://doi.org/10.1039/C2NP20032E

Costa, L. S., Fidelis, G. P., Cordeiro, S. L., Oliveira, R. M., Sabry, D. A., Câmara, R. B. G., Nobre, L. T. D. B., Costa, M. S. S. P., Almeida-Lima, J., Farias, E. H. C., Leite, E. L., & Rocha, H. A. O. (2010). Biological activities of sulfated polysaccharides from tropical seaweeds. *Biomedicine and Pharmacotherapy*, *64*(1), 21–28. https://doi.org/10.1016/j.biopha.2009.03.005

Damonte, E., Matulewicz, M., & Cerezo, A. (2012). Sulfated seaweed polysaccharides as antiviral agents. *Current Medicinal Chemistry*, *11*(18), 2399–2419. https://doi.org/10.2174/0929867043364504

Danby, P. M., & Withers, S. G. (2016). Advances in enzymatic glycoside synthesis. *ACS Chemical Biology*, *11*(7), 1784–1794. https://doi.org/10.1021/ACSCHEMBIO.6B00340/ASSET/IMAGES/MEDIUM/CB-2016-00340M_0008.GIF

Das, R., Pattanayak, A. J., & Swain, S. K. (2018a). Polymer nanocomposites for sensor devices. In *Polymer-based Nanocomposites for Energy and Environmental Applications: A Volume in Woodhead Publishing Series in Composites Science and Engineering*. Elsevier Ltd. https://doi.org/10.1016/B978-0-08-102262-7.00007-6

Das, I., Sagadevan, S., Chowdhury, Z. Z., & Hoque, M. E. (2018b). Development, optimization and characterization of a two step sol-gel synthesis route for ZnO/SnO$_2$ nanocomposite. *Journal of Materials Science: Materials in Electronics*, *29*(5), 4128–4135. https://doi.org/10.1007/s10854-017-8357-5

Davies, G., & Henrissat, B. (1995). Structures and mechanisms of glycosyl hydrolases. *Structure (London, England : 1993)*, *3*(9), 853–859. https://doi.org/10.1016/S0969-2126(01)00220-9

D'Ayala, G. G., Malinconico, M., & Laurienzo, P. (2008). Marine derived polysaccharides for biomedical applications: Chemical modification approaches. *Molecules*, *13*(9), 2069–2106. https://doi.org/10.3390/molecules13092069

De Jesus Raposo, M. F., De Morais, A. M. B., & De Morais, R. M. S. C. (2015). Marine polysaccharides from algae with potential biomedical applications. *Marine Drugs*, *13*(5), 2967–3028. https://doi.org/10.3390/md13052967

De Jesus Raposo, M. F., De Morais, R. M. S. C., & De Morais, A. M. M. B. (2013). Bioactivity and applications of sulphated polysaccharides from marine microalgae. *Marine Drugs*, *11*(1), 233–252. https://doi.org/10.3390/md11010233

de Moura, F. A., Macagnan, F. T., & da Silva, L. P. (2015). Oligosaccharide production by hydrolysis of polysaccharides: A review. *International Journal of Food Science & Technology*, *50*(2), 275–281. https://doi.org/10.1111/IJFS.12681

Delbianco, M., Kononov, A., Poveda, A., Yu, Y., Diercks, T., Jiménez-Barbero, J., & Seeberger, P. H. (2018). Well-defined oligo- and polysaccharides as ideal probes for structural studies. *Journal of the American Chemical Society*, *140*(16), 5421–5426. https://doi.org/10.1021/JACS.8B00254/SUPPL_FILE/JA8B00254_SI_011.PDF

Díaz-Montes, E. (2021). Dextran: Sources, structures, and properties. *Polysaccharides*, *2*, 554–565. https://doi.org/10.3390/polysaccharides2030033

Fedorov, S. N., Ermakova, S. P., Zvyagintseva, T. N., & Stonik, V. A. (2013). Anticancer and cancer preventive properties of marine polysaccharides: Some results and prospects. *Marine Drugs*, *11*(12), 4876–4901. https://doi.org/10.3390/md11124876

Finkenstadt, V. L. (2005). Natural polysaccharides as electroactive polymers. *Applied Microbiology and Biotechnology*, *67*(6), 735–745. https://doi.org/10.1007/s00253-005-1931-4

Finn, A. (2004). Bacterial polysaccharide-protein conjugate vaccines. *British Medical Bulletin*, *70*, 1–14. https://doi.org/10.1093/bmb/ldh021

Fittolani, G., Tyrikos-Ergas, T., Vargová, D., Chaube, M. A., & Delbianco, M. (2021). Progress and challenges in the synthesis of sequence controlled polysaccharides. *Beilstein Journal of Organic Chemistry*, *17*(1), 1981–2025. https://doi.org/10.3762/bjoc.17.129

Fitton, J. H., Stringer, D. N., & Karpiniec, S. S. (2015). Therapies from fucoidan: An update. *Marine Drugs*, *13*(9), 5920–5946. https://doi.org/10.3390/md13095920

Galiano, F. (2020). Encyclopedia of membranes. *Encyclopedia of Membranes*, *2009*, 40872. https://doi.org/10.1007/978-3-642-40872-4

Gandini, A., Lacerda, T. M., Carvalho, A. J. F., & Trovatti, E. (2016). Progress of polymers from renewable resources: Furans, vegetable oils, and polysaccharides. *Chemical Reviews*, *116*(3), 1637–1669. https://doi.org/10.1021/acs.chemrev.5b00264

Guo, H., Zhang, W., Jiang, Y., Wang, H., Chen, G., & Guo, M. (2019). *Physicochemical, Structural, and Biological Properties of Polysaccharides from Dandelion.*

Gupta, R. C., Lall, R., Srivastava, A., & Sinha, A. (2019). Hyaluronic acid: Molecular mechanisms and therapeutic trajectory. *Frontiers in Veterinary Science*, *6*(JUN). https://doi.org/10.3389/fvets.2019.00192

Haese, M., Winterhalter, K., Jung, J., & Schmidt, M. S. (2022). Like visiting an old friend: Fischer glycosylation in the twenty-first century: Modern methods and techniques. *Topics in Current Chemistry*, *380*(4), 1–37. https://doi.org/10.1007/S41061-022-00383-9/FIGURES/14

Hattori, T., Ogata, M., Kameshima, Y., Totani, K., Nikaido, M., Nakamura, T., Koshino, H., & Usui, T. (2012). Enzymatic synthesis of cellulose II-like substance via cellulolytic enzyme-mediated transglycosylation in an aqueous medium. *Carbohydrate Research*, *353*, 22–26. https://doi.org/10.1016/J. CARRES.2012.03.018

Hayes, M. R., & Pietruszka, J. (2017). Synthesis of glycosides by glycosynthases. *Molecules 2017, 22*(9), 1434. https://doi.org/10.3390/MOLECULES22091434

Hiraishi, M., Igarashi, K., Kimura, S., Wada, M., Kitaoka, M., & Samejima, M. (2009). Synthesis of highly ordered cellulose II in vitro using cellodextrin phosphorylase. *Carbohydrate Research*, *344*(18), 2468–2473. https://doi.org/10.1016/J.CARRES.2009.10.002

Huang, G., & Huang, H. (2018). Application of hyaluronic acid as carriers in drug delivery. *Drug Delivery*, *25*(1), 766–772. https://doi.org/10.1080/10717544.2018.1450910

Müssig, J. (2010). Industrial applications of natural fibres: structure, properties and technical applications. 14–20 https://doi.org/10.1002/9780470660324.

Jennings, H. J. (1990). Capsular polysaccharides as vaccine candidates. *Current Topics in Microbiology and Immunology*, *150*(Finland 1979), 97–127. https://doi.org/10.1007/978-3-642-74694-9_6

Jiao, G., Yu, G., Zhang, J., & Ewart, H. S. (2011). Chemical structures and bioactivities of sulfated polysaccharides from marine algae. *Marine Drugs*, *9*(2), 196–233. https://doi.org/10.3390/md9020196

Jin, X. (2012). Bioactivities of water-soluble polysaccharides from fruit shell of Camellia oleifera Abel: Antitumor and antioxidant activities. *Carbohydrate Polymers*, *87*(3), 2198–2201. https://doi.org/10.1016/j.carbpol.2011.10.047

Joseph, A. A., Pardo-Vargas, A., & Seeberger, P. H. (2020). Total synthesis of polysaccharides by automated glycan assembly. *Journal of the American Chemical Society*, *142*(19), 8561–8564. https://doi.org/10.1021/jacs.0c00751

Kadokawa, J. I. (2011). Precision polysaccharide synthesis catalyzed by enzymes. *Chemical Reviews*, *111*(7), 4308–4345. https://doi.org/10.1021/CR100285V/ASSET/CR100285V.FP.PNG_V03

Kardošová, A., & Machová, E. (2006). Antioxidant activity of medicinal plant polysaccharides. *Fitoterapia*, *77*(5), 367–373. https://doi.org/10.1016/j.fitote.2006.05.001

Kobayashi, S., Kashiwa, K., Kawasaki, T., & Shoda, S. I. (1991). Novel method for polysaccharide synthesis using an enzyme: The first in vitro synthesis of cellulose via a nonbiosynthetic path utilizing cellulase as catalyst. *Journal of the American Chemical Society*, *113*(8), 3079–3084. https://doi.org/10.1021/JA00008A042/ASSET/JA00008A042.FP.PNG_V03

Laurienzo, P. (2010). Marine polysaccharides in pharmaceutical applications: An overview. *Marine Drugs*, *8*(9), 2435–2465. https://doi.org/10.3390/md8092435

Lee, D. Y., Park, C. W., Lee, S. J., Park, H. R., Seo, D. B., Park, J. Y., Park, J., & Shin, K. S. (2019). Immunostimulating and antimetastatic effects of polysaccharides purified from ginseng berry. *American Journal of Chinese Medicine*, *47*(4), 823–839. https://doi.org/10.1142/S0192415X19500435

Lee, I. H., Huang, R. L., Chen, C. T., Chen, H. C., Hsu, W. C., & Lu, M. K. (2002). Antrodia camphorata polysaccharides exhibit anti-hepatitis B virus effects. *FEMS Microbiology Letters*, *209*(1), 63–67. https://doi.org/10.1111/j.1574-6968.2002.tb11110.x

Li, Q., Williams, C. G., Sun, D. D. N., Wang, J., Leong, K., & Elisseeff, J. H. (2004). Photocrosslinkable polysaccharides based on chondroitin sulfate. *Journal of Biomedical Materials Research – Part A*, *68*(1), 28–33. https://doi.org/10.1002/jbm.a.20007

Li, R., Chen, W. C., Wang, W. P., Tian, W. Y., & Zhang, X. G. (2010). Antioxidant activity of astragalus polysaccharides and antitumour activity of the polysaccharides and siRNA. *Carbohydrate Polymers*, *82*(2), 240–244. https://doi.org/10.1016/j.carbpol.2010.02.048

Li, Z., & Lin, Z. (2021). Recent advances in polysaccharide-based hydrogels for synthesis and applications. *Aggregate*, *2*(2), 1–26. https://doi.org/10.1002/agt2.21

Liu, Z., Jiao, Y., Wang, Y., Zhou, C., & Zhang, Z. (2008). Polysaccharides-based nanoparticles as drug delivery systems. *Advanced Drug Delivery Reviews*, *60*(15), 1650–1662. https://doi.org/10.1016/j.addr.2008.09.001

Lombard, V., Golaconda Ramulu, H., Drula, E., Coutinho, P. M., & Henrissat, B. (2014). The carbohydrate-active enzymes database (CAZy) in 2013. *Nucleic Acids Research*, *42*(Database issue). https://doi.org/10.1093/NAR/GKT1178

Luo, Y., & Wang, Q. (2014). Recent development of chitosan-based polyelectrolyte complexes with natural polysaccharides for drug delivery. *International Journal of Biological Macromolecules, 64*, 353–367. https://doi.org/10.1016/j.ijbiomac.2013.12.017

Ma, S., Tang, N., & Tian, J. (2012). DNA synthesis, assembly and applications in synthetic biology. *Current Opinion in Chemical Biology, 16*(3–4), 260–267. https://doi.org/10.1016/J.CBPA.2012.05.001

Mackenzie, L. F., Wang, Q., Warren, R. A. J., & Withers, S. G. (1998). Glycosynthases: Mutant glycosidases for oligosaccharide synthesis. *Journal of the American Chemical Society, 120*(22), 5583–5584. https://doi.org/10.1021/JA980833D

Majumder, S., Sharif, A., & Hoque, M. E. (2020). Electrospun cellulose acetate nanofiber: Characterization and applications. In F. M. Al-Oqla & S. M. Sapuan (Eds.), *Advanced Processing, Properties, and Applications of Starch and Other Bio-Based Polymers* (pp. 139–155). Elsevier. https://doi.org/10.1016/B978-0-12-819661-8.00009-3

Mora, N. L., Hansen, J. S., Gao, Y., Ronald, A. A., Kieltyka, R., Malmstadt, N., & Kros, A. (2014). Preparation of size tunable giant vesicles from cross-linked dextran(ethylene glycol) hydrogels. *Chemical Communications, 50*(16), 1953–1955. https://doi.org/10.1039/C3CC49144G

Mejía Agüero, L. E., Saul, C. K., De Freitas, R. A., Rabello Duarte, M. E., & Noseda, M. D. (2022). Electrospinning of marine polysaccharides: Processing and chemical aspects, challenges, and future prospects. *Nanotechnology Reviews, 11*(1), 3250–3280. https://doi.org/10.1515/ntrev-2022-0491

Meng, X., Liang, H., & Luo, L. (2016). Antitumor polysaccharides from mushrooms: A review on the structural characteristics, antitumor mechanisms and immunomodulating activities. *Carbohydrate Research, 424*, 30–41. https://doi.org/10.1016/j.carres.2016.02.008

Mohammed, A. S. A., Naveed, M., & Jost, N. (2021). Polysaccharides: Classification, chemical properties, and future perspective applications in fields of pharmacology and biological medicine (a review of current applications and upcoming potentialities). *Journal of Polymers and the Environment, 29*(8), 2359–2371. https://doi.org/10.1007/s10924-021-02052-2

Mohammed, A. S. A., Tian, W., Zhang, Y., Peng, P., Wang, F., & Li, T. (2020). Leishmania lipophosphoglycan components: A potent target for synthetic neoglycoproteins as a vaccine candidate for leishmaniasis. *Carbohydrate Polymers, 237*(February), 116120. https://doi.org/10.1016/j.carbpol.2020.116120

Mohan, K., Ravichandran, S., Muralisankar, T., Uthayakumar, V., Chandirasekar, R., Seedevi, P., Abirami, R. G., & Rajan, D. K. (2019). Application of marine-derived polysaccharides as immunostimulants in aquaculture: A review of current knowledge and further perspectives. *Fish and Shellfish Immunology, 86*(November 2018), 1177–1193. https://doi.org/10.1016/j.fsi.2018.12.072

Mora, N. L., Hansen, J. S., Gao, Y., Ronald, A. A., Kieltyka, R., Malmstadt, N., & Kros, A. (2014). Preparation of size tunable giant vesicles from cross-linked dextran(ethylene glycol) hydrogels. *Chemical Communications, 50*(16), 1953–1955. https://doi.org/10.1039/C3CC49144G

Moradi, Z., & Kalanpour, N. (2019). Kefiran, a branched polysaccharide: Preparation, properties and applications: A review. *Carbohydrate Polymers, 223*, 115100. https://doi.org/10.1016/J.CARBPOL.2019.115100

Morelli, L., Poletti, L., & Lay, L. (2011). Carbohydrates and immunology: Synthetic oligosaccharide antigens for vaccine formulation. *European Journal of Organic Chemistry, 29*, 5723–5777. https://doi.org/10.1002/ejoc.201100296

Muhamad, I. I., Lazim, N. A. M., & Selvakumaran, S. (2019). Natural polysaccharide-based composites for drug delivery and biomedical applications. In M S Hasnain, A. K. Nayak (Eds.), *Natural Polysaccharides in Drug Delivery and Biomedical Applications*. Elsevier Inc. https://doi.org/10.1016/B978-0-12-817055-7.00018-2

Mustafa, F., Finny, A. S., Kirk, K. A., & Andreescu, S. (2020). Printed paper-based (bio)sensors: Design, fabrication and applications. In *Comprehensive Analytical Chemistry* (1st ed., Vol. 89). Elsevier B.V. https://doi.org/10.1016/bs.coac.2020.02.002

Nair, M. (2012). Protein conjugate polysaccharide vaccines: Challenges in development and global implementation. *Indian Journal of Community Medicine, 37*(2), 79–82. https://doi.org/10.4103/0970-0218.96085 https://doi.org/10.1021/JA953286U

Nayak, A. K., Ahmed, S. A., Tabish, M., & Hasnain, M. S. (2019). Natural polysaccharides in tissue engineering applications. In M. S. Hasnain, & A. K. Nayak (Eds.), *Natural Polysaccharides in Drug Delivery and Biomedical Applications*. Elsevier Inc. https://doi.org/10.1016/B978-0-12-817055-7.00023-6

Ngwuluka, N. C. (2018). Responsive polysaccharides and polysaccharides-based nanoparticles for drug delivery. In A. S. H. Makhlouf and N. Y. Abu-Thabit (Eds.), *Stimuli Responsive Polymeric Nanocarriers for Drug Delivery Applications: Volume 1: Types and Triggers*. Elsevier Ltd. https://doi.org/10.1016/B978-0-08-101997-9.00023-0

Nicaise, S. M., Amir Tavakkoli, K. G., & Berggren, K. K. (2015). Self-assembly of block copolymers by graphoepitaxy. In R. Gronheid, & P. Nealey (Eds.), *Directed Self-assembly of Block Co-polymers for Nano-manufacturing*. Elsevier Ltd. https://doi.org/10.1016/b978-0-08-100250-6.00008-0

Nie, S., Cui, S. W., & Xie, M. (2018). Practical applications of bioactive polysaccharides. *Bioactive Polysaccharides*, *2007*, 527–542. https://doi.org/10.1016/b978-0-12-809418-1.00011-3

Nigudkar, S. S., & Demchenko, A. V. (2015). Stereocontrolled 1,2-cis glycosylation as the driving force of progress in synthetic carbohydrate chemistry. *Chemical Science*, *6*(5), 2687–2704. https://doi.org/10.1039/C5SC00280J

Nikolaev, A. V., & Sizova, O. V. (2011). Synthetic neoglycoconjugates of cell-surface phosphoglycans of Leishmania as potential anti-parasite carbohydrate vaccines. *Biochemistry (Moscow)*, *76*(7), 761–773. https://doi.org/10.1134/S0006297911070066

Nishimura, T., & Nakatsubo, F. (1996). First stepwise synthesis of cellulose analogs. *Tetrahedron Letters*, *37*(51), 9215–9218. https://doi.org/10.1016/S0040-4039(96)02186-7

Nishimura, T., & Nakatsubo, F. (1997). Chemical synthesis of cellulose derivatives by a convergent synthetic method and several of their properties. *Cellulose*, *4*(2), 109–130. https://doi.org/10.1023/A:1018423503762/METRICS

Wright, J. D. & Sommerdijk, N. A. J. M. (2018). Sol-gel materials : Chemistry and applications. *Sol-Gel Material* doi:10.1201/9781315273808.

Olennikov, D. N., Kashchenko, N. I., Chirikova, N. K., Koryakina, L. P., & Vladimirov, L. N. (2015). Bitter gentian teas: Nutritional and phytochemical profiles, polysaccharide characterisation and bioactivity. *Molecules*, *20*(11), 20014–20030. https://doi.org/10.3390/molecules201119674

Pandya, U., Dhuldhaj, U., & Sahay, N. S. (2019). Bioactive mushroom polysaccharides as antitumor: An overview. *Natural Product Research*, *33*(18), 2668–2680. https://doi.org/10.1080/14786419.2018.1466129

Park, S. J., Lee, K. W., Lim, D. S., & Lee, S. (2012). The sulfated polysaccharide fucoidan stimulates osteogenic differentiation of human adipose-derived stem cells. *Stem Cells and Development*, *21*(12), 2204–2211. https://doi.org/10.1089/scd.2011.0521Paulsen, B. (2005). Plant polysaccharides with immunostimulatory activities. *Current Organic Chemistry*, *5*(9), 939–950. https://doi.org/10.2174/1385272013374987

Petit, E., Delattre, C., Papy-Garcia, D., & Michaud, P. (2006). Chondroitin sulfate lyases: Applications in analysis and glycobiology. *Advances in Pharmacology*, *53*(05), 167–186. https://doi.org/10.1016/S1054-3589(05)53008-4

Pu, X., Ma, X., Liu, L., Ren, J., Li, H., Li, X., Yu, S., Zhang, W., & Fan, W. (2016). Structural characterization and antioxidant activity in vitro of polysaccharides from angelica and astragalus. *Carbohydrate Polymers*, *137*, 154–164. https://doi.org/10.1016/j.carbpol.2015.10.053

Qiao, D., Ke, C., Hu, B., Luo, J., Ye, H., Sun, Y., Yan, X., & Zeng, X. (2009). Antioxidant activities of polysaccharides from Hyriopsis cumingii. *Carbohydrate Polymers*, *78*(2), 199–204. https://doi.org/10.1016/j.carbpol.2009.03.018

Rahimi, M., Noruzi, E. B., Sheykhsaran, E., Ebadi, B., Kariminezhad, Z., Molaparast, M., Mehrabani, M. G., Mehramouz, B., Yousefi, M., Ahmadi, R., Yousefi, B., Ganbarov, K., Kamounah, F. S., Shafiei-Irannejad, V., & Kafil, H. S. (2020). Carbohydrate polymer-based silver nanocomposites: Recent progress in the antimicrobial wound dressings. *Carbohydrate Polymers*, *231*(July), 3913–3931. https://doi.org/10.1016/j.carbpol.2019.115696

Ren, L., Perera, C., & Hemar, Y. (2012). Antitumor activity of mushroom polysaccharides: A review. *Food and Function*, *3*(11), 1118–1130. https://doi.org/10.1039/c2fo10279j

Rocha De Souza, M. C., Marques, C. T., Guerra Dore, C. M., Ferreira Da Silva, F. R., Oliveira Rocha, H. A., & Leite, E. L. (2007). Antioxidant activities of sulfated polysaccharides from brown and red seaweeds. *Journal of Applied Phycology*, *19*(2), 153–160. https://doi.org/10.1007/s10811-006-9121-z

Rogovina, S. Z., & Vikhoreva, G. A. (2006). Polysaccharide-based polymer blends: Methods of their production. *Glycoconjugate Journal*, *23*(7–8), 611–618. https://doi.org/10.1007/s10719-006-8768-7

Schepetkin, I. A., & Quinn, M. T. (2006). Botanical polysaccharides: Macrophage immunomodulation and therapeutic potential. *International Immunopharmacology*, *6*(3), 317–333. https://doi.org/10.1016/j.intimp.2005.10.005

Seidi, F., Jenjob, R., Phakkeeree, T., & Crespy, D. (2018). Saccharides, oligosaccharides, and polysaccharides nanoparticles for biomedical applications. *Journal of Controlled Release*, *284*(May), 188–212. https://doi.org/10.1016/j.jconrel.2018.06.026

Silva, P. T. da, Fries, L. L. M., Menezes, C. R. de, Holkem, A. T., Schwan, C. L., Wigmann, É. F., Bastos, J. de O., & Silva, C. de B. da. (2014). Microencapsulation: Concepts, mechanisms, methods and some applications in food technology. *Ciência Rural*, *44*(7), 1304–1311. https://doi.org/10.1590/0103-8478cr20130971

Smith, P. J., Ortiz-Soto, M. E., Roth, C., Barnes, W. J., Seibel, J., Urbanowicz, B. R., & Pfrengle, F. (2020). Enzymatic synthesis of artificial polysaccharides. *ACS Sustainable Chemistry and Engineering*, *8*(32), 11853–11871. https://doi.org/10.1021/acssuschemeng.0c03622

Stroock, A. D (2008). *Microfluidics. Opt. Biosens. Today Tomorrow*, 659–681, Elsevier Science. https://doi.org/10.1016/B978-044453125-4.50019-X

Tong, W., Song, X., & Gao, C. (2012). Layer-by-layer assembly of microcapsules and their biomedical applications. *Chemical Society Reviews*, *41*(18), 6103–6124. https://doi.org/10.1039/c2cs35088b

Tseng, Y. H., Yang, J. H., & Mau, J. L. (2008). Antioxidant properties of polysaccharides from Ganoderma tsugae. *Food Chemistry*, *107*(2), 732–738. https://doi.org/10.1016/j.foodchem.2007.08.073

Ullah, S., Khalil, A. A., Shaukat, F. & Song, Y. (2019). Sources, extraction and biomedical properties of polysaccharides. *Foods* (Basel, Switzerland) 8, 1–23.

Usman, A., Khalid, S., Usman, A., Hussain, Z., & Wang, Y. (2017). Algal polysaccharides, novel application, and outlook. In K. M. Zia, M. Zuber and M. Ali (Eds.), *Algae Based Polymers, Blends, and Composites: Chemistry, Biotechnology and Materials Science*. Elsevier Inc. https://doi.org/10.1016/B978-0-12-812360-7.00005-7

van Weelden, G., Bobi, M., Okła, K., van Weelden, W. J., Romano, A., & Pijnenborg, J. M. A. (2019). Fucoidan structure and activity in relation to anti-cancer mechanisms. *Marine Drugs*, *17*(1). https://doi.org/10.3390/md17010032

Vandamme, T. F., Lenourry, A., Charrueau, C., & Chaumeil, J. C. (2002). The use of polysaccharides to target drugs to the colon. *Carbohydrate Polymers*, *48*(3), 219–231. https://doi.org/10.1016/S0144-8617(01)00263-6

Wang, L. X., & Huang, W. (2009). Enzymatic transglycosylation for glycoconjugate synthesis. *Current Opinion in Chemical Biology*, *13*(5–6), 592–600. https://doi.org/10.1016/j.cbpa.2009.08.014

Wang, Y., Xing, M., Cao, Q., Ji, A., Liang, H., & Song, S. (2019). Biological activities of fucoidan and the factors mediating its therapeutic effects: A review of recent studies. *Marine Drugs*, *17*(3), 15–17. https://doi.org/10.3390/md17030183

Whitesides, G. M. (2006). The origins and the future of microfluidics. *Nature*, *442*(7101), 368–373. https://doi.org/10.1038/nature05058

Wijesekara, I., Pangestuti, R., & Kim, S. K. (2011). Biological activities and potential health benefits of sulfated polysaccharides derived from marine algae. *Carbohydrate Polymers*, *84*(1), 14–21. https://doi.org/10.1016/j.carbpol.2010.10.062

Wright, J. D., & Sommerdijk, N. A. J. M. (2011). Biosensors based on sol-gel-derived materials. In P. Ducheyne (Eds.), *Comprehensive Biomaterials* (Vol. 3). https://doi.org/10.1016/b978-0-08-055294-1.00118-5

Wu, Y., Xiong, D. C., Chen, S. C., Wang, Y. S., & Ye, X. S. (2017). Total synthesis of mycobacterial arabinogalactan containing 92 monosaccharide units. *Nature Communications*, *8*(1), 1–7. https://doi.org/10.1038/ncomms14851

Wu, Y. J., Wei, Z. X., Zhang, F. M., Linhardt, R. J., Sun, P. L., & Zhang, A. Q. (2019). Structure, bioactivities and applications of the polysaccharides from Tremella fuciformis mushroom: A review. *International Journal of Biological Macromolecules*, *121*, 1005–1010. https://doi.org/10.1016/j.ijbiomac.2018.10.117

Xie, J. T., Wu, J. A., Mehendale, S., Aung, H. H., & Yuan, C. S. (2004). Anti-hyperglycemic effect of the polysaccharides fraction from American ginseng berry extract in ob/ob mice. *Phytomedicine*, *11*(2–3), 182–187. https://doi.org/10.1078/0944-7113-00325

Xue, H., Gan, F., Zhang, Z., Hu, J., Chen, X., & Huang, K. (2015). Astragalus polysaccharides inhibits PCV2 replication by inhibiting oxidative stress and blocking NF-κB pathway. *International Journal of Biological Macromolecules*, *81*, 22–30. https://doi.org/10.1016/j.ijbiomac.2015.07.050

Xue, J., Xie, J., Liu, W., & Xia, Y. (2017). Electrospun nanofibers: New concepts, materials, and applications. *Accounts of Chemical Research*, *50*(8), 1976–1987. https://doi.org/10.1021/acs.accounts.7b00218

Yamada, S., & Sugahara, K. (2008). Potential therapeutic application of chondroitin sulfate/dermatan sulfate. *Current Drug Discovery Technologies*, *5*(4), 289–301. https://doi.org/10.2174/157016308786733564

Yang, M., Lin, H. B., Gong, S., Chen, P. Y., Geng, L. L., Zeng, Y. M., & Li, D. Y. (2014). Effect of astragalus polysaccharides on expression of TNF-α, IL-1β and NFATc4 in a rat model of experimental colitis. *Cytokine*, *70*(2), 81–86. https://doi.org/10.1016/j.cyto.2014.07.250

Yu, Y., & Delbianco, M. (2020). Synthetic polysaccharides. In A. P. Rauter, B. E. Christensen, R. Adamo (Eds.), *Recent Trends in Carbohydrate Chemistry: Synthesis, Structure and Function of Carbohydrates*. Elsevier Inc. https://doi.org/10.1016/B978-0-12-817467-8.00009-8

Yu, Y., Shen, M., Song, Q., & Xie, J. (2018). Biological activities and pharmaceutical applications of polysaccharide from natural resources: A review. *Carbohydrate Polymers*, *183*, 91–101. https://doi.org/10.1016/j.carbpol.2017.12.009

Yu, Y., Tyrikos-Ergas, T., Zhu, Y., Fittolani, G., Bordoni, V., Singhal, A., Fair, R. J., Grafmüller, A., Seeberger, P. H., & Delbianco, M. (2019). Systematic hydrogen-bond manipulations to establish polysaccharide structure-property correlations. *Angewandte Chemie International Edition*, *58*(37), 13127–13132. https://doi. org/10.1002/ANIE.201906577

Yuan, J. F., Zhang, Z. Q., Fan, Z. C., & Yang, J. X. (2008). Antioxidant effects and cytotoxicity of three purified polysaccharides from Ligusticum chuanxiong Hort. *Carbohydrate Polymers*, *74*(4), 822–827. https://doi. org/10.1016/j.carbpol.2008.04.040

Zakharova, A. N., Madsen, R., & Clausen, M. H. (2013). Synthesis of a backbone hexa saccharide fragment of the pectic polysaccharide rhamnogalacturonan I. *Organic Letters*, *15*(8), 1826–1829. https://doi. org/10.1021/ol400430p

Zhang, M., Cui, S. W., Cheung, P. C. K., & Wang, Q. (2007). Antitumor polysaccharides from mushrooms: A review on their isolation process, structural characteristics and antitumor activity. *Trends in Food Science and Technology*, *18*(1), 4–19. https://doi.org/10.1016/j.tifs.2006.07.013

Zhang, Y., & Wang, F. (2015). Carbohydrate drugs: Current status and development prospect. *Drug Discoveries & Therapeutics*, *9*(2), 79–87. https://doi.org/10.5582/ddt.2015.01028

Zhao, L., Dong, Y., Chen, G., & Hu, Q. (2010). Extraction, purification, characterization and antitumor activity of polysaccharides from Ganoderma lucidum. *Carbohydrate Polymers*, *80*(3), 783–789. https://doi. org/10.1016/j.carbpol.2009.12.029

Zhu, Q., Shen, Z., Chiodo, F., Nicolardi, S., Molinaro, A., Silipo, A., & Yu, B. (2020). Chemical synthesis of glycans up to a 128-mer relevant to the O-antigen of Bacteroides vulgatus. *Nature Communications*, *11*(1), 1–7. https://doi.org/10.1038/s41467-020-17992-x

Zong, A., Cao, H., & Wang, F. (2012). Anticancer polysaccharides from natural resources: A review of recent research. *Carbohydrate Polymers*, *90*(4), 1395–1410. https://doi.org/10.1016/j.carbpol.2012.07.026

4 Design and Structure of Polysaccharide-Based Nanoparticles
State of the Art

Tabassum Khan, Deepika Tiwari, and Nikita Sanap
SVKM's Dr. Bhanuben Nanavati College of Pharmacy

4.1 INTRODUCTION

Nanotechnology has recently accomplished wide interest due to its numerous applications in a variety of fields, like biomedical science and biotechnology [1]. The use of nanomaterials in pharmaceutical and biomedical research is currently gaining wider attention. NPs having diameter <100 nm have the ability to improve the patient compliance, site-specific drug delivery, and several advantages in biomedical and pharmaceutical industry including solubility of poorly water-soluble drugs [2]. Recently, nanotechnology is becoming more specialized in synthesizing nanoparticles having a wide range of physicochemical properties including surface charge, particle size, etc. [3].

As effective drugs delivered by nanoparticles have remarkable release/degradation mechanisms, the use of biopolymers like starch, cellulose, silk, gelatin, albumin, and chitosan (Ch) – in synthesizing nanoparticles offers several advantages as it is biocompatible, biodegradable, and least toxic as compared to synthetic NPs. Therefore, polysaccharide is coming to the notice to many researchers not only for their excellent physical and biological qualities but due to their different reactive group which is subjected to chemical modification [4, 5]. In advance, the fabrication of polysaccharide-based nanoparticles is particularly appropriate for preparing nanoparticles composed of polymeric backbone. Polysaccharide is composed of monosaccharide which is linked by *O*-glycosidic bonds to form polysaccharides. Polysaccharide obtained from different sources like algae (alginic acid and carrageenan), plants (cellulose, pectin, and guar gum), and animals like chitosan, hyaluronic, and chondroitin. Much research has been done up to this point on the fabrication and engineering of polysaccharide-based nanoparticles for medication, protein/peptide, and nucleic acid delivery system in biomedical applications [5]. They exhibit a wide variety of physical and chemical properties and exhibit a wide range of applications depending on the chemical structure [5, 6]. From the chemistry of polymer to material research to advance biomedical fields of application, polysaccharide nanoparticles (PSNPs) are extensively explored in numerous branches of fundamental and applied science. The principal behind PS is supramolecular self-assembly of polysaccharide that makes it possible to utilize it for the fabrication of inventive bio-nanomaterial. Polysaccharides are nontoxic, biocompatible, and often have beneficial bioactive properties.

4.2 NANOPARTICLES

Nanoparticles are entities, size ranging from 1 to 1,000 nm. Nanoparticles are considered as a link between macroscopic and microscopic structures. They exhibit the properties of both bulk materials and microscopic structures due to their unique size. The small particle size gives them the unique intrinsic property of a high surface-to-volume ratio. When in a free state, nanoparticles are highly mobile which causes a drastic decline in sedimentation rate. Nanoparticles are in a highly movable state when they are free, which leads to an extremely slow rate of sedimentation. Depending upon their use, they exhibit a wide range of configurations ranging from soft to hard materials depending and may possess quantum effects. The presence of quantum effects causes ultimate control over the surface energy of these particles that will result in control of initial protein adsorption to allow cellular interaction. Nanoparticles are differentiated from microparticles, fine particles, and coarse particles due to their small size that exhibits different physical and chemical properties encompassing other properties like colloidal, ultra-optical effect, and electrostatic properties [7]. Owing to Brownian motion, they do not sediment unlike colloidal particles. The properties of nanoparticles differ remarkably from those particles of significantly larger particle size of the same substance as the diameter of an atom commonly lies between 0.15 and 0.6 nm, a large portion of the nanoparticle's material lies within a few atomic diameters from its surface. Therefore, the properties of the surface layer seem to be prominent over those of the bulk material. Therefore, when a nanoparticle is dispersed in a medium possessing different composition, this effect becomes stronger, since there is a significant interaction between the surface of two materials. Polarized optical microscopy (POM), scanning electron microscopy (SEM), transmission electron microscopy (TEM), infrared atomic force microscopy (AFM), Raman spectroscopy, and zeta size analyzer are used for characterization and analysis of the morphology of nanoparticles.

4.2.1 Types of Nanoparticles

NPs are categorized on the basis of their morphology, size, physical, and chemical properties. Some of the common classes of NPs categorized on the basis of their physical and chemical characteristics are given below [8].

4.2.1.1 Carbon-based NPs

Fullerenes and carbon nanotubes (CNTs) form the two main classes of carbon-based NPs. Fullerenes contain nanomaterials that are composed of spherical hollow cage-like allotropic forms of carbon. Owing to their electrical conductivity, high strength, structure, electron affinity, and versatility, they have gained remarkable commercial interest. These materials possess arranged pentagonal and hexagonal carbon units in an arranged manner, each carbon being sp^2 hybridized. CNTs, on the other hand, are elongated, tubular structures, ranging 1–3 nm in diameter. These nanomaterials are metallic or semiconducting depending upon their diameter. Structurally, these nanomaterials resemble a sheet of graphite getting rolled upon itself. The rolled sheets are usually mono, double, or multi-walled and therefore they are named single-walled (SWNTs), double-walled (DWNTs), or multi-walled carbon nanotubes (MWCTs), respectively. They are prepared via two methods. One of the widely used methods involves the deposition of carbon precursors, i.e., atomic carbons that had been vaporized from graphite via laser or electric arc onto the metal particles. The other method used for synthesizing CNTs is chemical vapor deposition (CVD). CNTs are unique since they are thermally conductive throughout the length and show zero conductivity across the tube. Owing to their unique physicochemical and mechanical properties, these materials are employed in the form of pristine and also as nanocomposites for several commercial applications encompassing fillers, efficient gas adsorbents as remedies regarding environment-related issues, and as supporting medium for several inorganic and organic catalysts.

4.2.1.2 Metal Nanoparticles

Metal NPs are purely synthesized by precursors made up of metal. These nanoparticles are made by employing chemical, electrochemical, and photochemical methods. In chemical methods, chemical reducing agents are used to reduce the metal ion precursor present in the solution to get metal nanoparticles. Owing to their innovative optical properties, metal NPs find applications in many research areas, including imaging and detection of biomolecules, and environmental and bioanalytical applications. A sampling of SEM is done widely by using gold NPs to enhance the electronic stream and obtain SEM images of high quality.

4.2.1.3 Ceramic Nanoparticles

Ceramics NPs are inorganic nonmetallic solids, prepared by heat and continuous cooling, composed mainly of oxides, carbides, carbonates, and phosphates. They are mainly found in amorphous, dense, porous, polycrystalline, and hollow forms. These nanoparticles are chemically inert and highly resistant to heat. Therefore, these NPs are gaining interest from researchers owing to their wide usage in applications like catalysis, photocatalysis, photodegradation of dyes, imaging, etc.

4.2.1.4 Semiconductor Nanoparticles

Semiconductors are the materials that exhibit properties between metals and nonmetals due to which they find several applications. Due to the presence of wide band gaps, they show different properties on tuning. They find crucial applications in photocatalysis, photo optics, and electronic devices. For instance, the presence of significant band gaps and edge positions have demonstrated semiconductor NPs to be effective in water-splitting applications too [9].

4.2.1.5 Polymeric Nanoparticles

These are organic-based nanoparticles, collectively called polymer nanoparticles (PNP). They can be nanospherical or nanocapsules, based on the method of preparation. In the former, the matrix and the polymer are evenly dispersed, and they have a matrix-like structure [10]. In the latter case, the solid mass is surrounded by a polymeric shell and they possess core-shell morphology.

4.2.1.6 Lipid Based Nanoparticles

These NPs contain lipid moieties and are spherical in structure having diameter (10–100 nm). Lipid NPs are composed of a solid core and an external core which is stabilized by surfactants and emulsifiers. In the biomedical field, these nanoparticles have found applications in drug delivery and release of RNA for treating cancer [11].

4.3 POLYSACCHARIDE-BASED NANOPARTICLES

All living organisms have been subjected to polysaccharide-based nanoparticles. Also, they do not exhibit any toxic effects since they are biocompatible, biodegradable, and safe for various therapies. These nanoparticles are commonly synthesized by simple methods requiring mild temperature and pressure conditions as in ionotropic gelation and self-assembly of polyelectrolytes. They possess several advantages as they are safe, stable, and non-explosive and have natural availability compared to metal nanoparticles. They exhibit high stability, safety, non-explosive, and naturally accessible to the comparison of metal nanoparticles. Also, they are more stable as compared to peptide-based nanoparticles, as peptides are subjected to deformation and possess less risk of immunogenicity [12]. Polysaccharides-based nanoparticles are synthesized by self-assembly of nanoparticles as well as physical and chemical cross-linking. They find numerous applications in the therapeutic diagnosis of various diseases by employing imaging and therapeutic agents. Nanoparticles can be made from natural polymers by employing different polysaccharides such as alginic acid, dextran, chitosan, cyclodextrin, hyaluronic acid, and pullulan. These polymers are derived from different sources of plant origin (e.g., starch, guar gum), algal origin, microbial origin (e.g., dextran, gellant), and animal origin (chitosan, hyaluronic acid) [12]. These polymers exhibit some advantages over synthetic polymers as these are economical and find multiple applications in industries. Polysaccharide possesses many chemically reacting moieties in their structure that can be simply functionalized chemically and biochemically. Polysaccharide exhibits the presence of different hydrophilic groups such as hydroxyl, carboxyl, and amino groups in their structure that helps in enhancing bioadhesion with epithelia and mucous membranes by forming non-covalent bond and hence enhancing the bioavailability of many drugs [12].

Different types of polysaccharides that had been used to successfully synthesize nanoparticles according to the report are as follows:

Chitosan nanoparticles (CS-NPs)

Chitosan is the most commonly used polysaccharide for preparing nanoparticles since it is biocompatible and when given at a higher dose possesses no toxicity. It is gaining much wider attention from researchers since it has the ability to elongate the duration of drug residence at the absorption site and hence improve its bioavailability [1]. Rizvi et al. prepared chitosan nanoparticles loaded with carboplatin for treating breast cancer, showing improved anti-proliferative activity as compared to the native carboplatin. Li et al. synthesized doxorubicin-loaded CS-LDL (chitosan-low density lipoprotein) by self-assembly and evaluated it showing higher cell toxicity on gastric cancer SGC7901 cells.

Alginate nanoparticles (ALG-NPs)

Alginate is a linear polysaccharide that is polyanionic in nature, acquired from marine brown algae. Chemically, it is 1,4-linkage acid and β-D-mannuronic acid. Due to its mucoadhesive, biocompatibility, and biodegradability, it has several potential applications in the pharmaceutical and biomedical filed as drug delivery systems. Costha et al. developed a pH-sensitive sodium ALG iron oxide nanoparticle composite coated with bilayer hydroxyapatite by employing co-precipitation method. It was developed for loading of curcumin and 6-gingerol for targeted and controlled release of the same that showed improved drug release profile in a sustained and pH-controlled manner. Laraba-Djebari's group developed ALG nanoparticles loaded with scorpion venom toxin by ion gelation method for oral delivery of the vaccine. It was developed to provide protection to the venom against degradation caused by the biological environment and to improve its delivering efficiency to the APC (73).

Hyaluronic acid nanoparticles (HA-NPs)

Hyaluronic acid (HA), referred to as hyaluronan, forms the major part of the extracellular matrix, and it is found in abundant quantities in several organs, including connective, epithelial, and neural tissues. It is a linear polysaccharide that possesses chemical stability and the ability to imbibe water into it along with biocompatibility and biodegradability. Carboxylic acid present in the structure of HA enables efficient conjugation with other drugs and targeting ligands. Owing to the amazing properties possessed by HA, it finds potential application in pharmaceutical and medical science fields. Jeon et al. prepared a nano-sized drug delivery system for treating cancer. Amphililic hyaluronic acid conjugates were prepared by chemically conjugating hydrophobic 5β-cholanic acid to the backbone of HA. This amphiphilic HA conjugate enabled the formation of a stable nanoparticle by self-assembly in an aqueous media [1].

Dextran nanoparticles

Dextran is an unbranched polysaccharide, soluble in water, and widely utilized in medical products because of its significant and remarkable biocompatibility, biodegradability, non-antigenicity, and immunogenicity properties. Dextranspermine is an unbranched water-soluble polysaccharide and is extensively utilized in medical products because of its excellent biocompatibility, biodegradability, non-antigenicity, and immunogenicity properties. Dextranspermine conjugate-based nanoparticles were also prepared by researchers for delivering genes to leukemic cells and magnetic dextran spermine nanoparticles for targeting drug delivery to the brain [1].

4.4 DESIGN AND FUNCTIONALIZATION OF POLYSACCHARIDE-BASED NANOPARTICLES

4.4.1 Synthesis of Polysaccharide Nanoparticles

Several methods are reported in the literature for the synthesis of polysaccharide-based nanoparticles (Table 4.1).

TABLE 4.1
Synthetic Methods of Polysaccharide-Based Nanoparticles

S. No	Method	Merits	Demerits	Example	Ref.
1	Crosslinking and aggregation strategies	Highly applicable for polyelectrolyte complexation of charged polysaccharides	Lower chemical stability	Chitosan (CS), hyaluronic acid (HA), alginate	[5]
2	Nanoprecipitation	Fast, economical, and reproducible	Difficult to encapsulate water-soluble compounds	Dextran (DEX), Pullulan (PL)	[13]
3	Complex coacervation	Simple, does not need organic solvent	The physical properties of polyelectrolyte-complexes are affected by pH, temperature, and ionic strength	Oleoyl-carboxymethyl-chitosan (OCMCS), HA	[13]
5	Emulsion based	Easy to scale up	Large volume of water is eliminated due to water-soluble drug leakage into saturated aqueous external phase	Doxorubicin-loaded HA NPs	[14]
6	Ionotropic gelation	Free from organic solvent	Due to poor mechanical strength, particle can be disintegrate	CSNPs	[15]

4.5 STRUCTURE AND DESIGN OF POLYSACCHARIDE-BASED NANOPARTICLES

4.5.1 CHITOSAN

Chitosan polysaccharide is cationic and found in nature. It is made up of *N*-acetyl-2-deoxy-D-glucopyranose and 2-amino-2-deoxy-D-glucopyranose linked by (1,4)-β-glycosidic bonds [4, 16]. Chitosan is formed by the removal of alkyl group from chitin, and different molecular weights of chitosan can be made with varying degrees of deacetylation while it shows various physicochemical and biological characteristics on the basis of chitosan synthesis [17]. Various chitosan modifications have been done to enhance the target specificity and bioavailability. As drug delivery vectors, a range of chitosan derivatives have been developed, including sugar-bearing, carboxyalkyl, and quaternized chitosan has been invented as a drug delivery vector [18] (Figure 4.1).

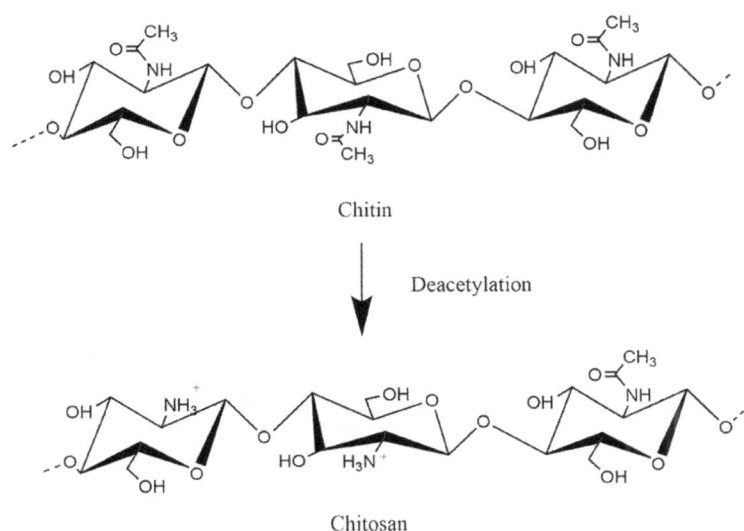

Chitin

Deacetylation

Chitosan

FIGURE 4.1 Deacetylation of chitosan from chitin.

4.5.2 Functionalization of Chitosan

Chitosan has functional groups that are strategically placed to provide polysaccharide with unique qualities and attributes due to the existence of an amino group at the C-2 position of the glucosamine. Then amino group represents its cationic nature having wound healing, antibacterial action, and its mucoadhesive, making it an excellent carrier for drug delivery. Chitosan has pKa 6.5, and it is soluble in acidic solution and insoluble water solution. The chitosan is protonated and polycationic in nature and forms various anions such as lipids, proteins, DNA, alginate, and pectin. The physicochemical properties, such as toxicity and solubility, depend on the degree of dealkylation. The functionalization of the hydroxyl group and the amino group as *N*-modified, *O*-modified, and *N*,*O*-modified chitosan resulting enhance biological activity. The antimicrobial activity, solubility, and its mucoadhesive properties can be increased by quaternized, *N*-alkyl-/*N*-benzylchitosan and phosphorylated chitosan. By using electrophilic reactants like alkyl halide, acids and isocyanides, nonselective *N*, *O*-modified chitosan derivatives, and selective *O*-modified chitosan derivative are synthesized, wherein the acid protonated amino group which leaves, the −OH group to undergo the reaction, and by securing a hydroxyl functional group, *N*-modified derivatives are obtained [19] (Figure 4.2).

4.5.3 Synthesis of Chitosan Nanoparticles

Chitosan nanoparticles are reported to be synthesized by the following two methods:

1. Deprotonation of amino group
2. Ionic crosslink method

Sodium tripolyphosphate (TPP) is used as a crosslinking agent to enhance intermolecular bonding between negative phosphate in TPP structure and positive charges of chitosan amino group. Chitosan powder incorporated in 1% acetic acid solution was allowed for magnetic stirring for 24 hours [20]. For amino deprotonation, chitosan was used in concentrations 0.9 g/100 g and 1.5 g/100 g having pH 3.5–6.7 with 6 M NaOH. In the ionic crosslinking method, triphosphate (TPP)

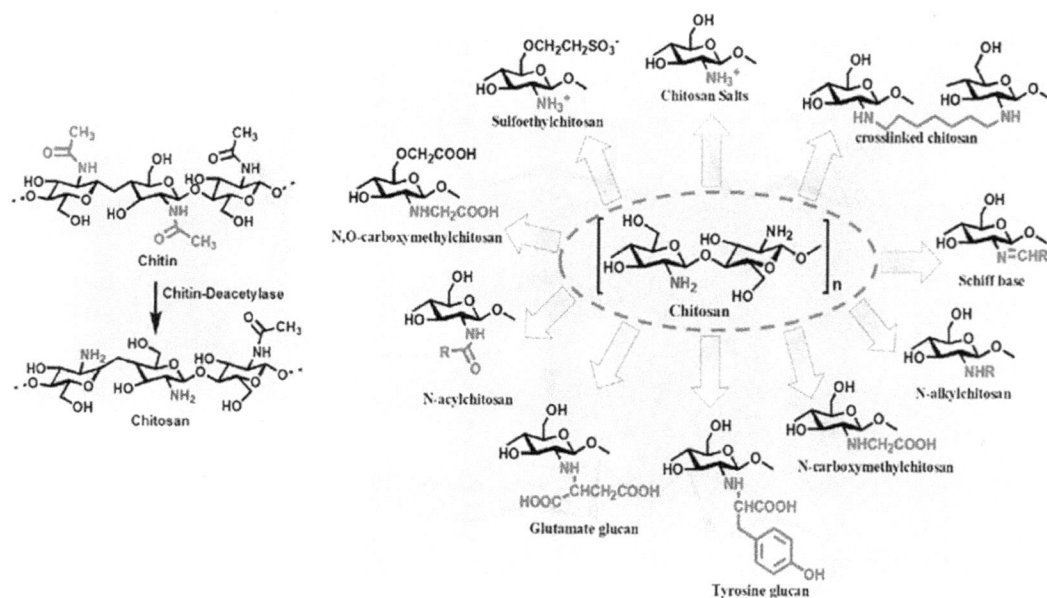

FIGURE 4.2 Structures of some functionalized chitosan derivatives [19].

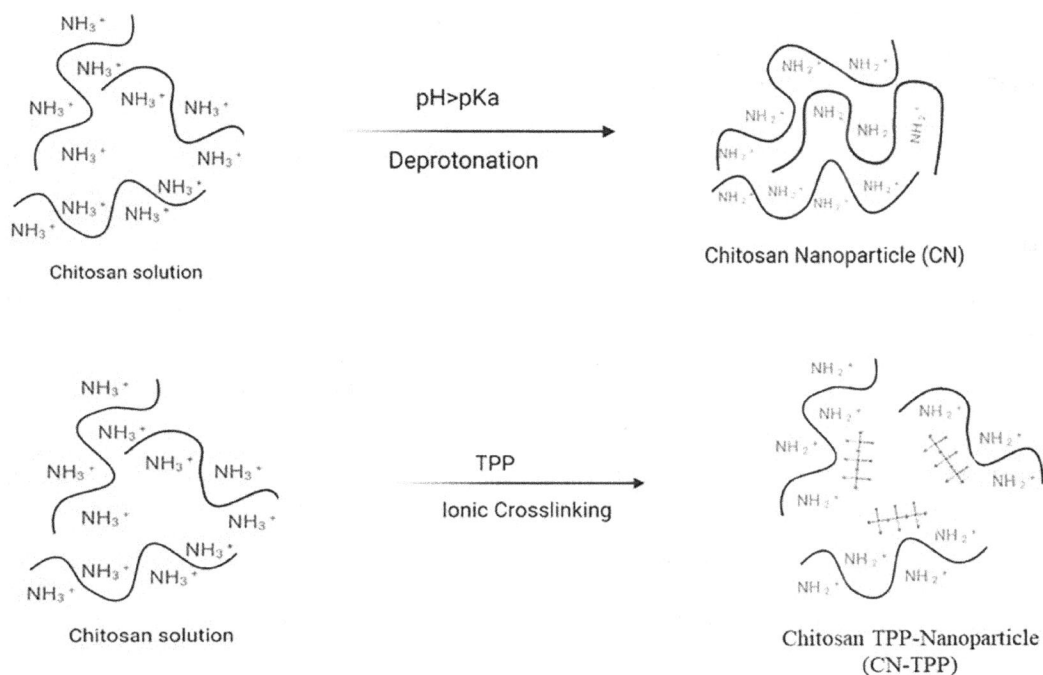

FIGURE 4.3   Deprotonation of amino group of chitosan and ionic crosslinking between chitosan and TPP.

aqueous solution having pH 8 is added in chitosan solution having pH 3.5 [21]. The final concentration of solution is 0.9 g/100 g and 1.5 g/100 g with pH values 4.34 and 5.16, respectively, resulting in the 3 : 1 mass ratio of CS:TPP. The final nanoparticle is designed as 0.9CN-TPP and 1.5CN-TPP, respectively [5] (Figure 4.3).

4.5.4   Synthesis of Chitosan-Dicarboxylic Acid NPs

One gram chitosan in 5.58 mmol glucosamine was added and dissolved in 50 mL HCL (0.37% v/v) aqueous solution. Certain amount of anhydride (phenyl succinic or phthalic) was incorporated and dissolved in pyridine, and a chitosan solution was added dropwise and stirred. To maintain the pH of the reaction mixture, some quantity of NaOH was added and the reaction was continued for 40 minutes. Then precipitate was washed with absolute alcohol and acetone and allowed for drying for 48 hours in a hot air oven for 35°C [22].

4.5.5   Galactosylated Chitosan Nanoparticles

4.5.5.1   Synthesis (GACHNPs)

Chitosan was coupled with the lactobionic acid by carbodiimide reaction. When a small quantity of NHS (1.80 g) is dissolved in 20 mL of HCl buffer solution at a pH of 4.7, then the carboxylic group of lactobionic acid (2.80 g) is activated. Then a small quantity of chitosan (1.25 g was dissolved in 200 mL of HCl 90.1 M) buffer solution. Then the lactobionic acid solution and the chitosan solution were mixed by using a magnetic stirrer for 72 hours. Then the final product was distilled in a dialysis tube to purify galactosylated chitosan. Then the final product was dried by using lyophilization [23] (Table 4.2).

TABLE 4.2
Polysaccharide-Based Nanoparticles

Polysaccharide-Based Polymer	Drug	Uses	Advantages	Ref
Chitosan	Carboplatin	Breast cancer	Enhanced solubility and bioavailability with controlled release	[12]
Chitosan	Curcumin	Breast cancer	Improved solubility and systemic bioavailability	[24]
β-CD adamantine-grafted hyaluronic acid	Paclitaxel	Cancer	Dual stimuli responsive nanoparticles with enhanced solubility and targeting ability	[25]
β-CD and folic acid	Platinum-based agent LA-12	Cancer	The permeability of anticancer agent across the membrane was enhanced	[25]
Sodium alginate and hydroxyapatite	Curcumin and 6-gingerol	Cancer	Better sustained drug release profile after a period of time in a pH-controlled manner	[26]
Alginate	Scorpion venom toxin	Vaccine	Good immunotherapeutic efficiency by toxins	[26]
Glutamic acid modified alginate	Zidovudine	Antiviral	Haemocompatible and cyto-compatible with sustained release of drug	[27]
Alginate and chitosan	Naringenin	Antihyperglycemic	Less or no toxicity and secured biocompatible and biodegradable polymeric vehicle	[28]
Hyaluronic acid and 4-carboxybenzaldehyde	Doxorubicin	Cancer	Improve the targeting ability and antitumor activity toward HeLa cells	[28]
Enhanced targeting ability and antitumor activity toward HeLa cells	Paclitaxel	Cancer	Dual-targeting capacity and strong antitumor efficacy	[28]
Carboxymethyl pullulan and N-trimethyl CS chloride, CS glutamate, CS chloride	–	Vaccine	Potential carrier for nasal vaccination	[29]
Carboxymethyl dextran and lithocholic acid	Doxorubicin	Cancer	Higher cytotoxicity against cancer cells	[30]

4.6 PROPERTIES OF POLYSACCHARIDE-BASED NANOPARTICLES

Polysaccharide-based NPs prepared by ionic crosslinking have the ability to efficiently encapsulate labile biomacromolecular drugs encompassing peptides and proteins without loss of activity. Alonso and coworkers successfully incorporated chitosan/tripolyphosphate (CS/TPP) nanoparticles for delivering biomacromolecular drugs and other low molecular weight drugs (Mw), such as doxorubicin (DOX) [31–32]. High physical stability was shown by CS/TPP nanoparticles, and successful encapsulation was achieved for both plasmid DNA and dsDNA oligomers (20 mers), irrespective of the molecular mass of CS. Increasing the amount of deacetylation of CS (75.5%–92%) leads to slight improvement in the encapsulation efficiency [33]. Calcium-alginate nanoparticles prepared by ionic cross-linking showed high efficiency in transfection [34]. Nanoparticles prepared by polyelectrolyte complexation (PEC) have been demonstrated to be highly biocompatible. The molecular weight of CS greatly affected the particle sizes of the CS-DNA PECs, and high efficiency in transfection was achieved when CS with an Mw of 40–84 kDa was employed [35]. Kim et al. synthesized nanocomplex by employing anionic HA and cationic PEGylated tumor necrosis factor (TNF)-related apoptosis inducing ligand (TRAIL), a potential therapeutic protein for cancer and rheumatoid arthritis (RA) [36]. Enhanced in vivo stability and a remarkable therapeutic effect against RA compared to native poly(ethylene glycol) was achieved in the prepared HATRAIL nanocomplex. Kwon's research group established hydrophobically modified glycol CS (HGC), which has the ability to form self-assembled nanoparticles, by the means of chemical conjugation of 5β-cholanic acid to glycol CS. These HGC nanoparticles achieved numerous advantageous characteristics such as serum stability, deformability, and rapid uptake by tumor cells. In vivo results showed these nanoparticles to exhibit prolonged blood circulation and remarkable tumor specificity for delivering various anticancer drugs [31].

4.7 APPLICATIONS OF POLYSACCHARIDE-BASED NANOPARTICLES

Owing to the biocompatibility and biodegradability of polysaccharide NPs, they have remarkable advantages and prospective applications in biomedicine and biomaterials engineering. In recent years, applications in drug delivery and tissue engineering have also been studied and investigated owing to their favorable properties such as mucoadhesiveness and cellular uptake. For example, Park and coworkers investigated photodynamic therapy by employing chitosan CS-based nanoparticles. Here, the reactive oxygen species sensitive thioketal linker was utilized for conjugating photosensitizer pheophorbide A with CS. In the aqueous environment, the self-assembly of the amphiphilic CS-conjugate into nanoparticles (NPs) occurred, leading to the self-quenching of the photosensitizer. Near IR was used for the activation of photosensitizer. The mechanism of circulation of polymeric drug or gene-loaded nanoparticles has also been studied recently. The drug or gene-loaded nanoparticles are injected into the body that crosses the epithelial barrier, circulates in the blood vessels, and reaches the target site. The nanoparticles being more accessible to small openings and pores in tissues get accumulated at the target site, such as tumors, due to the enhanced penetration and retention effect (EPR). The application of drug or gene-loaded nanoparticles overcomes and prevents the adverse effect caused to the tissue by delivering the drug systemically. Also, polysaccharide NPs have been employed to enhance the solubility of drugs that are poorly soluble in water, deliver active drug molecules to the target site, and facilitate controlled and sustained release of drugs and genes. Polysaccharide NPs based on theranostic systems have also been fabricated as described by Lee and coworkers, e.g., fluorescence imaging and drug delivery, magnetic resonance imaging (MRI), and near-IR fluorescence imaging and delivery of drugs. Yameens and coworkers developed a novel enzyme-responsive DEX-based oligoester NPs by crosslinking, for the controlled release of 5-fluorouracil, an anticancer drug with improved release kinetics and decreased cell viability.

Polymer nanoparticles containing polysaccharide shells are also widely used as nanocarriers. For example, Xu et al. incorporated CS oligosaccharides, forming a cationic hydrophilic shell surrounding the anionic hydrophobic PLGA NPs. It was investigated from the results that the core-shell NPS formed showed enhanced water solubility and improved stability as compared to PLGA NPs without a CS oligosaccharide shell. The developed nanoparticles thus have application in transporting and delivering hydrophobic drugs and other nutraceuticals. Seabra and workers developed an alginate-based system for delivering NO and Ag NPs. Alginate nanoparticles were made by gelation and Ca^{2+} precipitation. Ag NPs were developed by utilizing $AgNO_3$, tea extract, and mercaptosuccinic acid, and later, it was encapsulated into the alginate NPs. The particles, loaded with NO through nitrosation of mercaptosuccinic, were demonstrated for antibacterial properties. It was investigated that the developed alginate-based NPs showed antibacterial activity against three bacterial strains. It displayed the synergistic effect of NO release and Ag NPs, without being toxic to Vero cells.

Tissue engineering utilizes the principle of engineering in life sciences and develops biological substitutes that help restore, maintain, and improve functioning of tissue. From the genesis of scaffolds, hydrogels, and particulate systems ranging from biocompatible polymer, it also includes growth factors which are soluble-secreting proteins that play an important role in tissue regeneration. Controlled release of therapeutic factors as in the case of drug delivery enhances the efficiency of tissue engineering. Polysaccharide-based nanoparticles have been found to be a suitable material as a carrier since it has the ability to protect the protein-based growth factor, showing a versatile release profile with less possible side effects. Mandal and coworkers synthesized dual growth factor loaded CSNPs (chitosan nanoparticles) by incorporating epidermal growth factor (EGF), fibroblast growth factor (FGF), and CS. The developed CSNPs demonstrated successful delivery of growth factors from CSNPs, enhanced in vitro fibroblast proliferation, controlled release, and nontoxicity. These results paved the way for further application of these dual growth factors loaded nanoparticles in the tissue engineering domain. Nanotechnology by utilizing polysaccharide-based NPs has also been applied in the field of cosmetics. Several research studies are ongoing regarding the application of polysaccharide-based NPs for improving and enhancing the properties of cosmetic products, using cosmetic ingredients for stabilizing pickering emulsion, and delivering useful substances. An emulsion is known to be a dispersion of two liquids that are immiscible with one another. It consists of a stabilizer for the stability of the interfacial layer. The formed emulsion, in the absence of an interfacial stabilizer, becomes unstable leading to phase separation that causes creaming of the emulsion. In order to overcome the issue, surfactants and solid NPs are employed in the emulsion to lower the interfacial tension between the two liquid phases, thereby improving the stability of the emulsion. The resulting emulsion stabilized by incorporating solid NPs is quoted as pickering emulsions. Solid NPs have been proven to be a better alternative for stabilizing emulsion over surfactants. Since surfactants used as a stabilizer improve both kinetic and thermodynamic stability of the emulsion, they are associated with several limitations such as toxicity, high cost, and issues in recovery. Whereas polysaccharide NPs used as solid NPs are environment friendly and biologically compatible. Pickering emulsion also provides improved stability preventing coalescence and produces a long-term durable stabilized emulsion that has the potential application in cosmetics.

Polysaccharide-based NPs have also been employed as nanocarriers in cosmeceuticals since they are biodegradable. For example, cosmeceutical vitamin A alcohol, i.e., retinol, has been successfully encapsulated into CSNPs that demonstrated improved water solubility. Apart from skin care, polysaccharide NP-based nanocarriers have also been applied in hair care. Gelfuso and coworkers synthesized MXS minoxidil sulfate (MXS)-loaded CSNPs for targeted delivery to hair follicles. MXS has significant applications in treating androgenentic conditions i.e. alopecia in males as well as females, but these are also associated with several limitations such as low solubility in water, high irritancy, and side effects substantiated with its potent antihypertensive activity. Therefore, the targeted delivery of minoxidil to hair follicles has always been a topic for interest to the researchers. MXS-loaded CSNPs developed by Gelfuso et al. were investigated to accumulate into the hair follicles, demonstrating sustained release of MXS. Hence, MXS-loaded CSNPs were found to be an easy and promising delivery system to enhance the topical treatment of alopecia [29]. Also, there are several potent applications of polysaccharide-based NPs in the food and agricultural field. Packaging form a vital part in the food industry since it imparts protection against physical and chemical damages during storage and transportation, thereby improving shelf life and providing ease in handling. Many researchers investigated the improvement in physical and chemical properties, including mechanical strength and thermal stability in the food packaging material by incorporating CSNPS. For example, for the development of sustainable packaging material, polysaccharide NPs were incorporated that demonstrated lower water vapor permeability. Sun and coworkers investigated the corn starch-based packaging film by incorporating taro-derived selenium nanoparticles (SNPs). The resulting packaging film demonstrated significantly higher tensile strength which was the result of strong interaction between taro SNPs and films [2].

4.8   ADVANTAGES AND LIMITATIONS

The advantages associated with polysaccharide-based NPs are as follows:

1. Significant less toxic reagents involved in the preparation process
2. The preparation and synthesis involve simple steps

3. Optimization enables improvement in the yield and retention efficiency
4. Formation of nanoscale preparation composed of mucoadhesive polysaccharides such as chitosan with certain drugs increases the residence time in the body
5. Biodegradable enabling enhanced absorption via the epithelial barrier
6. Enables targeted delivery of hydrophobic drugs by getting incorporated in self-assembled amphiphilic polysaccharides
7. Improves the permeability, bioavailability, and retention of hydrophobic drugs [2]

Disadvantages associated with polysaccharide-based NPs are as follows:

1. Molecular imaging and targeted therapeutics have shown interesting results, but the conclusive performance of these nanocarriers is yet to be established and studied in clinical trials.
2. The clinical translation of these novel nanocarriers due to their inherent complexity is the greatest challenge.
3. Significant challenges are faced during the optimization of nanoparticle properties for improving targeting and reducing non-specific tissue residence.

4.9 CONCLUSION AND FUTURE PERSPECTIVE

As reviewed above, several polysaccharide-based nanoparticles had been synthesized and prepared for therapeutic drug, genes, and cosmeceutical delivery. This gives insight for the development of more polysaccharide-based nanoparticles for drug delivery systems [37]. Physicochemical properties, drug loading ability, in vitro toxicity, and in vivo tests have been explored for these polysaccharide NPs. There is a need for more investigational research for cases of specific interaction of these nanoparticles with human organs, tissues, and cells. The effect brought by them on metabolism along with a broad range of applications in drug delivery, etc. is yet to be explored in the future. These nanoparticles have the potential to overcome unmet clinical needs in the form of personalized medicines. Until now, the research on polysaccharide-based nanoparticles had been restricted to academic institutions. Therefore, there is a need for medical translation of nanoparticles, their nanoscale manufacturing, scale-up, and regulatory requirements.

REFERENCES

1. W. Khan, E. Abtew, S. Modani, and A. J. Domb, "Polysaccharide based nanoparticles," *Isr. J. Chem.*, vol. 58, no. 12, pp. 1315–1329, 2018, doi: 10.1002/ijch.201800051.
2. J. Zhang, P. Zhan, and H. Tian, "Recent updates in the polysaccharides-based nano-biocarriers for drugs delivery and its application in diseases treatment: A review," *Int. J. Biol. Macromol.*, vol. 182, pp. 115–128, 2021, doi: 10.1016/j.ijbiomac.2021.04.009.
3. P. Rabl, S. J. Kolkowitz, F. H. L. Koppens, J. G. E. Harris, P. Zoller, and M. D. Lukin, "A quantum spin transducer based on nanoelectromechanical resonator arrays," *Nat. Phys.*, vol. 6, no. 8, pp. 602–608, 2010, doi: 10.1038/nphys1679.
4. M. S. Huh *et al.*, "Polysaccharide-based nanoparticles for gene delivery," *Top. Curr. Chem.*, vol. 375, no. 2, pp. 1–19, 2017, doi: 10.1007/s41061-017-0114-y.
5. Z. Liu, Y. Jiao, Y. Wang, C. Zhou, and Z. Zhang, "Polysaccharides-based nanoparticles as drug delivery systems," *Adv. Drug Deliv. Rev.*, vol. 60, no. 15, pp. 1650–1662, 2008, doi: 10.1016/j.addr.2008.09.001.
6. S. Mizrahy, and D. Peer, "Polysaccharides as building blocks for nanotherapeutics," *Chem. Soc. Rev.*, vol. 41, no. 7, pp. 2623–2640, 2012, doi: 10.1039/c1cs15239d.
7. C. Torres-Torres, A. López-Suárez, B. Can-Uc, R. Rangel-Rojo, L. Tamayo-Rivera, and A. Oliver, "Collective optical Kerr effect exhibited by an integrated configuration of silicon quantum dots and gold nanoparticles embedded in ion-implanted silica," *Nanotechnology*, vol. 26, no. 29, p. 295701, 2015, doi: 10.1088/0957-4484/26/29/295701.

8. J. Zhang, P. Zhan, and H. Tian, "Recent updates in the polysaccharides-based nano-biocarriers for drugs delivery and its application in diseases treatment: A review," *Int. J. Biol. Macromol.*, vol. 182, pp. 115–128, 2021, doi: 10.1016/j.ijbiomac.2021.04.009.

9. G. M. Souza, K. C. de O. Vieira, L. V. Naldi, V. C. Pereira, and L. K. Winkelstroter, *Nanotechnology for Advances in Medical Microbiology.* 2021. Available at: http://www.springer.com/series/16324.

10. M. Rai, M. Patel, and R. Patel, "Nanotechnology in Medicine Toxicity and Safety Edited by," 2022, Accessed: Apr. 01, 2022. Online.. Available: www.wiley.com.

11. A. Husen, *Introduction and Techniques in Nanomaterials Formulation.* INC, 2020.

12. W. Khan, E. Abtew, S. Modani, and A. J. Domb, "Polysaccharide based nanoparticles," *Isr. J. Chem.*, vol. 58, no. 12, pp. 1315–1329, 2018, doi: 10.1002/ijch.201800051.

13. C. Gavory, A. Durand, J. L. Six, C. Nouvel, E. Marie, and M. Leonard, "Polysaccharide-covered nanoparticles prepared by nanoprecipitation," *Carbohydr. Polym.*, vol. 84, no. 1, pp. 133–140, 2011, doi: 10.1016/j.carbpol.2010.11.012.

14. B. Tao and Z. Yin, "Redox-responsive coordination polymers of dopamine-modified hyaluronic acid with copper and 6-mercaptopurine for targeted drug delivery and improvement of anticancer activity against cancer cells," *Polymers (Basel)*, vol. 12, no. 5, 2020, doi: 10.3390/POLYM12051132.

15. J. Yang, S. Han, H. Zheng, H. Dong, and J. Liu, "Preparation and application of micro/nanoparticles based on natural polysaccharides," *Carbohydr. Polym.*, vol. 123, pp. 53–66, 2015, doi: 10.1016/j.carbpol.2015.01.029.

16. M. Rinaudo, "Chitin and chitosan: Properties and applications," *Prog. Polym. Sci.*, vol. 31, no. 7, pp. 603–632, 2006, doi: 10.1016/j.progpolymsci.2006.06.001.

17. J. Berger, M. Reist, J. M. Mayer, O. Felt, and R. Gurny, "Structure and interactions in chitosan hydrogels formed by complexation or aggregation for biomedical applications," *Eur. J. Pharm. Biopharm.*, vol. 57, no. 1, pp. 35–52, 2004, doi: 10.1016/S0939-6411(03)00160-7.

18. Y. S. Wang *et al.*, "Self-assembled nanoparticles of cholesterol-modified O-carboxymethyl chitosan as a novel carrier for paclitaxel," *Nanotechnology*, vol. 19, no. 14, 2008, doi: 10.1088/0957-4484/19/14/145101.

19. J. Jhaveri, Z. Raichura, T. Khan, M. Momin, and A. Omri, "Chitosan nanoparticles-insight into properties, functionalization and applications in drug delivery and theranostics," *Molecules*, vol. 26, no. 2, 2021, doi: 10.3390/molecules26020272.

20. S. H. Yuk, K. Choi, K. Kim, and I. C. K. C. Kwon, "In Vivo Targeted Delivery of Nanoparticles for Theranosis," *Acc. Chem. Res.*, vol. 44, no. 10, doi: 1018-10281018, 2011.

21. H. Q. Mao *et al.*, "Chitosan-DNA nanoparticles as gene carriers: Synthesis, characterization and transfection efficiency," *J. Control. Release*, vol. 70, no. 3, pp. 399–421, 2001, doi: 10.1016/S0168-3659(00)00361-8.

22. R. M. Saeed, I. Dmour, and M. O. Taha, "Stable chitosan-based nanoparticles using polyphosphoric acid or hexametaphosphate for tandem ionotropic/covalent crosslinking and subsequent investigation as novel vehicles for drug delivery," *Front. Bioeng. Biotechnol.*, vol. 8, no. January, pp. 1–21, 2020, doi: 10.3389/fbioe.2020.00004.

23. C. Wang, Z. Zhang, B. Chen et al, "Design and evaluation of galactosylated chitosan/graphene oxide nanoparticles as a drug delivery system," *J. Colloid Interface Sci.*, vol. 516, pp. 332–341, 2018, doi: 10.1016/j.jcis.2018.01.073.

24. F. Croisier and C. Jérôme, "Chitosan-based biomaterials for tissue engineering," *Eur. Polym. J.*, vol. 49, no. 4, pp. 780–792, 2013, doi: 10.1016/j.eurpolymj.2012.12.009.

25. M. E. Brewster, and T. Loftsson, "Pharmaceutical applications of cyclodextrins. 1. Drug solubilization and stabilization," *J. Pharm. Sci.*, vol. 85, no. 10, pp. 1017–1025, 1996.

26. C. M. Magin *et al.*, "Cyclodextrin-based host–guest supramolecular nanoparticles for biomedical applications," *Chem. Rev.*, vol. 98, no. 5, pp. 187–230, 2014, doi: 10.1016/S1369-7021(10)70058-4.

27. M. S. Hasnain and A. K. Nayak, "Drug delivery using interpenetrating polymeric networks of natural polymers: A recent update," *Polymeric and Natural Composites*, 2022, Available: https://doi.org/10.1016/j.jddst.2021.102915.

28. K. Y. Lee and D. J. Mooney, "Alginate: Properties and biomedical applications," *Prog. Polym. Sci.*, vol. 37, no. 1, pp. 106–126, 2012, doi: 10.1016/j.progpolymsci.2011.06.003.

29. A. Plucinski, Z. Lyu, and B. V. K. J. Schmidt, "Polysaccharide nanoparticles: From fabrication to applications," *J. Mater. Chem. B*, vol. 9, no. 35, pp. 7030–7062, 2021, doi: 10.1039/d1tb00628b.

30. M. Swierczewska, H. S. Han, K. Kim, J. H. Park, and S. Lee, "Polysaccharide-based nanoparticles for theranostic nanomedicine," *Adv. Drug Deliv. Rev.*, vol. 99, pp. 70–84, 2016, doi: 10.1016/j.addr.2015.11.015.

31. G. Saravanakumar, D.-G. Jo, and J. H. Park, "Polysaccharide-based nanoparticles: A versatile platform for drug delivery and biomedical imaging," *Curr. Med. Chem.*, vol. 19, no. 19, pp. 3212–3229, 2012, doi: 10.2174/092986712800784658.

32. N. Csaba, M. Köping-Höggård, and M. J. Alonso, "Ionically crosslinked chitosan/tripolyphosphate nanoparticles for oligonucleotide and plasmid DNA delivery," *Int. J. Pharm.*, vol. 382, no. 1–2, pp. 205–214, 2009, doi: 10.1016/j.ijpharm.2009.07.028.

33. Y. Xu, and Y. Du, "Effect of molecular structure of chitosan on protein delivery properties of chitosan nanoparticles," *Int. J. Pharm.*, vol. 250, no. 1, pp. 215–226, 2003, doi: 10.1016/S0378-5173(02)00548-3.

34. J. O. You, and C. A. Peng, "Calcium-alginate nanoparticles formed by reverse microemulsion as gene carriers," *Macromol. Symp.*, vol. 219, pp. 147–153, 2004, doi: 10.1002/masy.200550113.

35. T. Ishii, Y. Okahata, and T. Sato, "Mechanism of cell transfection with plasmid/chitosan complexes," *Biochim. Biophys. Acta Biomembr.*, vol. 1514, no. 1, pp. 51–64, 2001, doi: 10.1016/S0005-2736(01)00362-5.

36. Y. J. Kim *et al.*, "Ionic complex systems based on hyaluronic acid and PEGylated TNF-related apoptosis-inducing ligand for treatment of rheumatoid arthritis," *Biomaterials*, vol. 31, no. 34, pp. 9057–9064, 2010, doi: 10.1016/j.biomaterials.2010.08.015.

37. N. Amreddy, A. Babu, R. Muralidharan, A. Munshi, and R. Ramesh, "Polymeric nanoparticle-mediated gene delivery for lung cancer treatment," *Top. Curr. Chem.*, vol. 375, no. 2, p. 35, 2017.

5 Polysaccharides as Adhesive
Sweet Solutions to Sticky Situations

Jeffy Joji, Swetha K.S., Anila Antony, and Neetha John
Central Institute of Petrochemicals Engineering & Technology

5.1 INTRODUCTION

Polymers in adhesive technology feature a major contribution within global marketing. The phenol-formaldehyde resins, polyesters, elastomers, polyacrylates, and polyurethanes are the backbone for polymers in adhesive applications.[1] While these polymers are highly functional and have a big selection of applications, the sources of the polymers are non-renewable and depleting petrochemical sources. Also, a variety of toxic chemicals are used for the synthesis of those adhesives. As an example, in the synthesis of PU adhesive, toxic monomers like methylene diphenyl diisocyanate (MDI) are used. Likewise, for phenol-formaldehyde resins, the usage of formaldehyde is important. These adhesives thereby are toxic to animals and other organisms.

Bioadhesives or the employment of biopolymers as an adhesive are very prominent nowadays as they are from natural sources, renewable, non-toxic, and price effective. Polysaccharides, nucleic acids, polyesters, polythioesters, polyoxoesters, proteins, polyisoprenoids, and polyphenols are the key examples of bioadhesives. Polysaccharides have been used for a long term as an adhesive in many applications. Polysaccharides are polymeric carbohydrates bounded together by glycosidic linkage with an enormous number of hydroxyl group. Polysaccharides, which are the foremost abundant natural biopolymer, have many physical, chemical, and biological characteristics. Monosaccharides act as the constituent element, and glycosidic linkages form the building block of those polymers, which successively regulates the heterogeneity and complication of the polysaccharides. Polysaccharides are the biggest constituent of biomass. It is said that over 90% of the carbohydrate mass in nature is within the sort of polysaccharides. Polysaccharides originate from a range of sources—from the farm, the forest, generally all plants, the ocean, shells of aquatic organisms, microbes, and eventually by chemical alteration of natural polysaccharides. The event of polysaccharide adhesives is of utter significance because of their structural mutability and various intra- and intermolecular interactions.

This review mainly focuses on nine differing types of polysaccharide-based adhesive. It briefly describes the sources, synthesis routes, properties, and also advantages. As we move forward, the review gives more information about the mechanism of adhesive formation and testing techniques. Toward the last part, the review ends with the possible application areas of polysaccharide adhesives.

5.2 TYPES OF POLYSACCHARIDES-BASED ADHESIVE

5.2.1 Chitosan-Based Adhesive

Chitosan is one among the interesting bio-based polymers such as protein, tannin, and lignin which are used as adhesives.[2] The presence of OH– groups within the chitosan causes interaction with other chemical functions, which provides the adhesive with more cohesive strength. The properties such as biodegradability, biocompatibility, and good mechanical properties make them a decent material

DOI: 10.1201/9781003265054-5

for adhesive application. It is the second most abundant natural polysaccharide on earth. In 1859, it was found that chitin is often made water soluble by chemical modification. The chitin was altered chemically and was called chitosan by Hoppe-Seyler.[3] Chitosan possesses two sorts of monomers 2-amino-2-deoxy-β-D-glucopyranose and 2-acetamido-2-deoxy-β-D-glucopyranose, which contains an acetamido group and an amino group.[4] The main dissimilarity within the structure of chitosan and chitin is that in second position of chitosan, acetamide group is substituted (Figure 5.1).

The source of chitosan is a natural source like shrimp, crab, and lobster.[5] It is also found on the exoskeleton of insects and fungal cell walls.[6] Every shellfish waste doesn't seem to be a source of chitin.[7] The content of chitin varies from source to source, like oyster contains 4%–6% of chitin while blue crab contains 14% of chitin.[8] The percentage of chitin depends on the proportions of minerals, proteins, carotenoids, and species.[8]

The countries such as Japan, the US, India, Poland, and Australia produce chitin and chitosan commercially.[9] The annual production of chitin is approximately 1×10^{10} to 1×10^{12} tons in nature.[10] The high amount of production of chitin in nature makes it a less costly and mostly obtainable biopolymer.[11] Chitosan is produced from chitin by the enzymatic deacetylation of chitin.[12] The property within which the chitosan strongly depends on molecular chain orientation and regular packing. Crystalline of chitosan is much less than chitin, which makes it more dissolvable and available to reagents.

The material which is employed to affix two surfaces by means of strong adhesion is thought as adhesive.[13] The adhesion is guaranteed by the diffusion of adhesive within the gap of the surface before it gets solidified. The adsorption theory gives the mechanism of adhesion. The theory contains interatomic and intermolecular forces formed between adhesive molecules and molecules at the surface of adherend.

An adhesive in the liquid state is characterized by viscosity, surface tension, and diffusion of the adhesive on the surface of a material. To achieve a decent molecular interaction, the surface tension of the adhesive should be equal to the material surface energy. Comparing the surface tension of metals and polymers (metal Aluminum-1134 m Nm^{-1} and polymers Polyethylene 35.76 m Nm^{-1}),[14,15] chitosan at low concentration has a surface tension of 64 m Nm^{-1}, which decreases after an extended time.[16] The chitosan solution with more concentration has lower surface tensions. These key factors, the high dispersive part, and lower surface tension make the chitosan spread easily in every kind of material. It shows Newtonian behavior and shear-thinning behavior at not up to 0.25% (w/v) and above this value.[17] With increasing concentration, the viscosity of chitosan has been increased, and with temperature, it decreased.[18] The viscosities of classical and artificial adhesives vary significantly. For instance, in the case of phenol formaldehyde, the viscosity range between

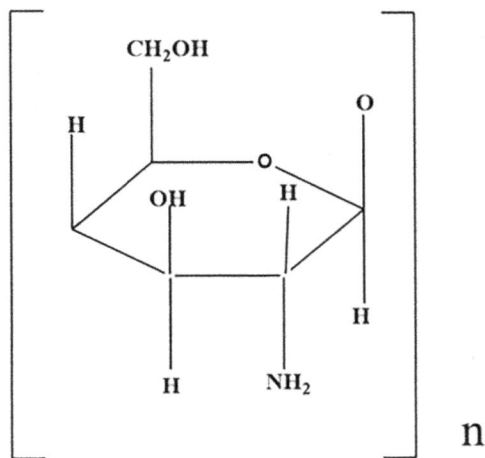

FIGURE 5.1 Structure of chitosan.

60 and 2825 MPa s, and polysaccharide adhesive range from 40 to 400 Pa s. The chitosan adhesive includes a larger value which makes it suitable as an adhesive. It also has an excellent bonding property.[19] The glass transition temperature and the temperature of decomposition are two values that characterize the adhesive thermal resistance. The thermal decomposition temperature of chitosan is 250°C, which makes it useful as an adhesive in temperatures above the room temperature.[19,20]

The chitosan adhesive found application mainly in biomedical adhesives, which mainly focus on purposes like surgery and substitute for traditional drug delivery systems.[21] It is also employed in skin wound closure because of its hemostatic properties.[22] It is a wonderful mucoadhesive (swollen stage) that might adhere to hard and soft tissues.[23] The films made of chitosan are used in wound healing and tissue repair.[24–26] It is also used as a wood adhesive.

5.2.2 CELLULOSE BASED ADHESIVE

Cellulose is the most abundant biopolymer in the world, and it's a polysaccharide consisting of β-1,4 linked D-glucose units. They're synthesized from a sizable amount of living organisms starting from the bacterium *Dictyostelium discoideum* to larger trees.[27] The primary reason cellulose is used as paper and other composite materials are due to its easy adhesion. Cellulose could be a linear molecule, and they are bonded by intramolecular and intermolecular hydrogen bonds and also Van der Waals force. Cellulose has a strong affinity to water-containing molecules. The hydrogen bonding that occurs when wetted fibers dry and comes in contact with one another is primarily responsible for the binding of cellulose fibers in paper.[28] The advantage of cellulose is that a variety of hydroxyl groups present on cellulose molecule leads to the formation of different derivatives, and these derivatives successively give comparable adhesive properties to cellulose. On the opposite hand, the most problem that lies within cellulose is that the OH groups of cellulose have likely less chance to break making the polymer insoluble in solvents. As a result, cellulose undergoes etherification and esterification to make derivatives that have excellent adhesive properties.

5.2.2.1 Cellulose Modification

Esterification and etherification are the two main chemical modifications of cellulose. Esterification is the reaction of mixing an organic acid with an alcohol to produce an ester and water. Common esterified cellulose includes nitrocellulose, cellulose ester, and cellulose ester butyrate. On the opposite hand, etherification is a process of dehydration of alcohols to provide ethers. Methyl cellulose, ethyl cellulose, and carboxy methyl cellulose are the samples of etherified cellulose.

5.2.2.2 Esterification

A German-Swiss scientist named Schonbein found the procedure for esterification for the primary time. He synthesized cellulose nitrate by a reaction of cellulose, acid, and aqua fortis. The formed material was too inflammable that its usage was restricted, and its first use was as smokeless gunpowder. The particular synthesis of cellulose nitrate involves the reaction between cellulose and nitric acid, where sulfuric acid acts as a catalyst giving cellulose nitrate and water. The foremost common application of cellulose nitrate is general-purpose household cement. When applied to a substrate, the solution immediately loses all its solvent to make strong, moisture-resistant, transparent films (Figure 5.2).[29]

Cellulose acetate, on the other side, despite being the most important ester derivative of cellulose, is not widely used as an adhesive. It is synthesized from the chemical reaction of cellulose and acetic anhydride, catalyzed by sulfuric acid.[30] Then the triacetate is partially hydrolyzed to achieve the desired degree of substitution (DS). Lately, a variety of other synthesis methods for cellulose ester has been performed.[31] Cellulose acetate butyrate is another esterified cellulose and has more advantages than cellulose acetate. These include good spraying ability, chemical stability, good chlorine tolerance, and cold-crack and moisture resistance, and it acts as a stable carrier dispersing medium for pigments.[32] Though there is a number of synthesis methods used, the foremost common method of production may be a reaction between acetic anhydride and butyric acid or propionic acid catalyzed by sulfuric acid.

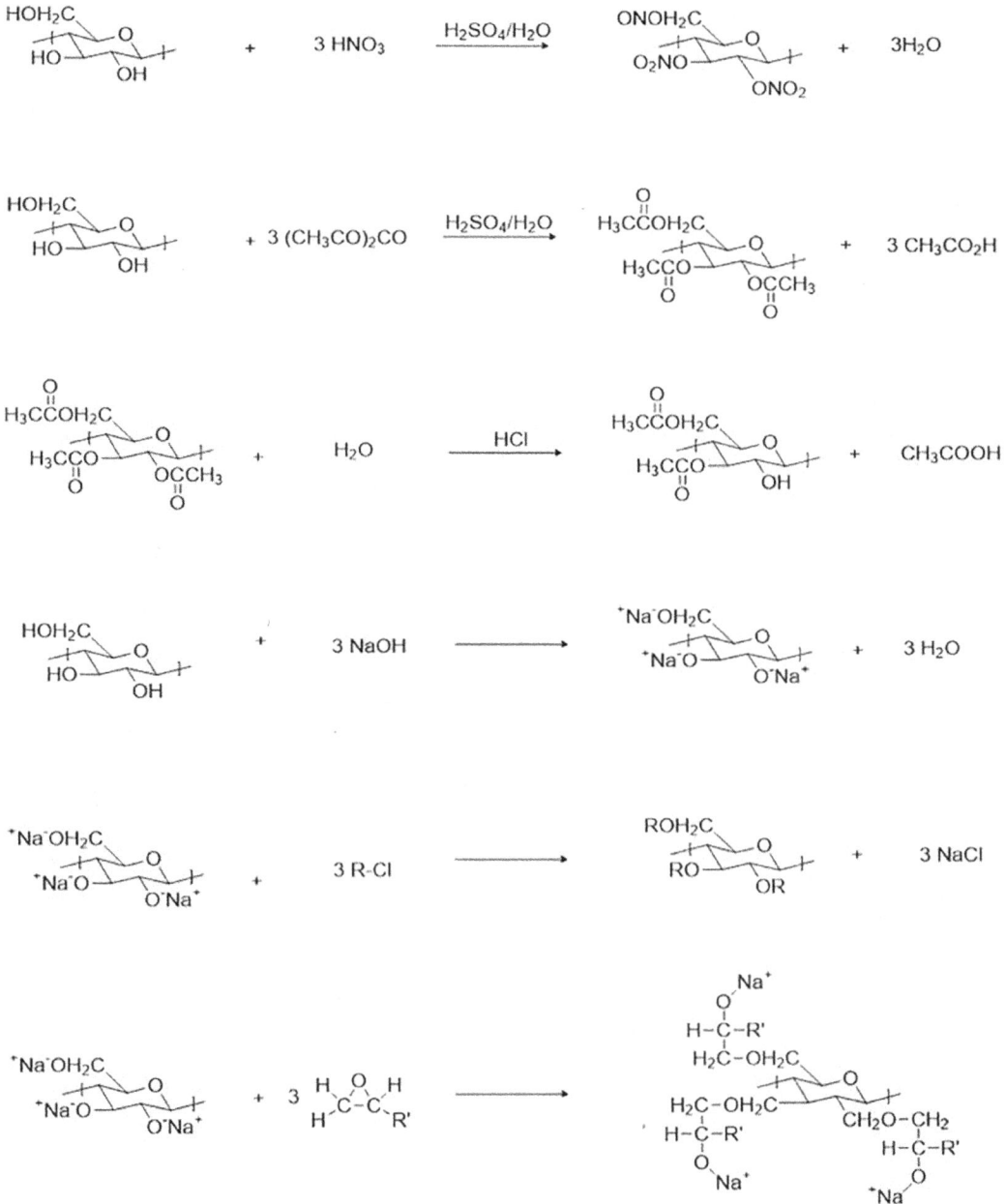

FIGURE 5.2 Significant reactions of cellulose.

5.2.2.3 Etherification

One of the earliest used etherified cellulose is methyl cellulose. They are mainly used for sizing and paper coating to give oil resistance, as a binder in ceramics, as a non-tinting paste for wallpaper, and in adhesives for leather drying.[33] Ethyl cellulose is another type. Its synthesis method is similar to methyl cellulose. It is obtained by a reaction between ethyl chloride and cellulose. As a result of their soluble nature, the films usually have a tendency to swell little or no within the presence of water. In adhesive application, ethyl cellulose can either be applied as a solvent or in a hot melt form. Carboxy methyl cellulose is the most typically and widely used etherified cellu-lose. It's also called sodium salt and is very soluble in water because of its ionic nature. They have significant use in the food industry as a thickener as a result of their non-toxicity.[34] In addition, a variety of research on carboxy methyl cellulose as an oral delivery system and other pharmaceuti-cal applications is taking place.[35]

5.2.3 STARCH-BASED ADHESIVE

Starch is a biodegradable, abundant, renewable, and inexpensive polymer. Moreover, it is the second most abundant natural polymer. The abundance range comes after cellulose and is on the market mainly from the roots, corn, stalks rice, wheat, potato, and tapioca. Starch could be a polysaccha-ride, consisting of amylose and amylopectin. Linear amylose and branched amylopectin are the two main components of starch, representing its simplest forms. Lipids, proteins, ash, and moisture are constituents of starch in low amounts. The amylose and amylopectin content in a starch depends on its source, and most starches contain 20%–30% amylose, although different varieties can contain quite 80% amylose (Figure 5.3).[36]

Starch in water cannot act as adhesive as starch is so tightly bound in granules. So, for starch to be used as an adhesive, modification is important. The branches of amylose and amylopectin are opened using suitable methods like acidic/alkali treatment, heat treatment, and oxidation.[37]

In heat treatment, starch is simply heated or cooked until gelation occurs. Usually, the gelation temperature for starch lies between 55°C and 75°C. But just in the case of starch, a phenomenon called retrogradation occurs. It is a process where suspensions of amylose and high-amylose starches have a bent to harden and become solid upon cooling.[38] In alkali treatment, gelation temperature may be lowered when alkalis like sodium hydroxide are added.[39] Acid treatment is completed by heating the starch to 45°C–55°C using suitable mineral acids and then neutralizing it with a base. Oxidized starch is additionally commercially available. There is a variety of oxidizing agents used, but aqueous alkaline hypochlorite treatment is the most typically preferred. Oxidized starch will be more viscous, stable, and has an excellent range of adhesion. Hypochlorite is taken into account as one of the foremost efficient oxidants.[40]

5.2.4 PULLULAN-BASED ADHESIVE

Pullulan, a linear glucosic polysaccharide, is a water-soluble polymer synthesized by the polymor-phic fungus *Aureobasidium pullulans*.[41] A low degree of hydrogen bonding in the crystal form makes it water soluble. It's an exopolysaccharide and encompasses a large choice of applications in numer-ous sectors ranging from food and pharmaceutical industries to biomedical applications.[42] Pullulan, a linear α-D-glucan, contains malt trioseas subunits and is connected by (1–6)-α-D-glycosidic

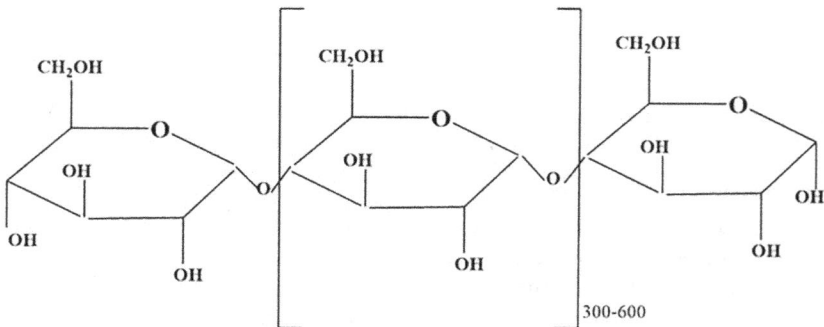

FIGURE 5.3 Structure of starch.

linkages. Because it contains a unique linkage pattern, pullulan shows exemplary physical properties, like fiber and film formation, adhesion capacity, etc. Due to these advantages, pullulan has long been employed in various applications like food adhesive, blood plasma alternatives, and cosmetic additives. One of the most features of pullulan is that it can be chemically altered to decrease the water solubility or to extend the pH sensitivity by functionalizing the polymer by various means.[43,44]

The characteristics that make it an excellent choice of adhesive is that pullulan can easily form fibers, compression moldings, and oxygen-proof films. The introduction of various groups in pullulan helps in modifying its properties thereby helping in numerous adhesive applications.

5.2.5 LEVAN-BASED ADHESIVE

Levan is a polysaccharide that is found in some plants and microorganisms. They comprise mostly fructosyl residues linked by β-2,6-carbon[45] and are mainly synthesized from sucrose or molasses. Some of the unique characteristics which make this polymer different from others are strong adhesivity, their property to self-assemble into spherical colloids in water, very low intrinsic viscosity, and biocompatibility. Because of its attractive properties, it is having various applications in medicine, beauty products, food, and other industries.[46] Compared to other natural polymers, it has higher shear and tensile strength. The polymer has all the properties of the material needed for an adhesive like thermal stability, tensile strength, and reversibility.

Levan is a natural and biodegradable adhesive that acts as a binder in the manufactured wood, foundry sand cores,[47] fiberglass, and bare aluminum.[48] It is also used as a bioadhesive in engineering smart scaffolds, drug delivery systems, tissue engineering, and wound-healing applications.[49] Adhesion can be explained as bonding between the surfaces of two different substances. In the case of Levan, the hydroxyl group along with Van der Waal forces keeps the adhesion. Metals create strong adhesive bonds with Levan than the polymers. The main adhesive application of Levan adhesive is membranes and films fabricated from Levan for medical applications such as drug delivery systems, implants, sensors, and optoelectronics. Adhesive Levan can bond aluminum films, similar to a wood adhesive.

5.2.6 DEXTRAN-BASED ADHESIVE

Dextran is a natural polymer that has attained great attention due to its natural resources, nontoxicity, and good water-soluble ability. It is a bacterial-derived polysaccharide of glucose in which they are linked mainly by – (1–6) type.[50] It is synthesized directly by dextransucrases and by polymerizing the glucose of the sucrose in dextran. In industry, it is produced through *Leuconostoc mesenteroides* NRRLB512.[51] The dextran is classified into two categories based on the length of the chains like simple dextran having a molecular weight greater than 40 kDa and oligodextran having a molecular weightless than 40 kDa.[52,53]

It has been employed in biomedical fields by chemically modifying dextran into hydrogels via crosslinking.[54] Synthetic dextran hydrogels are utilized in tissue adhesives and vascular tissue engineering.[55,56] Aldehyde dextran together with a series of amine-PEG act as a controlled biocompatible tissue adhesive.[57] The dextran-based bioadhesive can crosslink at a fast rate and may adhere with a very good strength. They can't adhere to the surface of gels because of the high swelling ratio.

5.2.7 XANTHAN BASED ADHESIVE

Xanthan gum is an extracellular heteropolysaccharide consisting of D-glucose, D-mannose, D-glucuronic acid, pyruvic acid which has acetal linkages, and *O*-acetyl repeating unit.[58] It is a pentasaccharide derived from bacteria and fungi.[59] Xanthan is synthesized commercially by fermentation. It is used in a large number of applications like in eatables, beauty aids, drugs, and medicines.[60,61] When combined with other polysaccharides like starch, we get a variety of films and hydrogels having different properties.[62–64] One of the main disadvantages lies within its hydrophilicity making its application limited to adhesives.

Therefore, improving the property, xanthan gum is oxidized which decreases hydrophilicity.[65] Xanthan gum reacts with chitosan efficiently.[66] The combined adhesive of chitosan with oxidized dextran has good bonding strength, which finds application in surgery for bone or tissue gluing.[67]

It found application in toothpastes and gels as an ideal stabilizer. It is also used as an emulsion stabilizer and water binder in creams and lotions. Hypoallergenic and benign nature of xanthan makes it useful in pharmaceutical applications. Sialorrhea was treated using thiolated xanthan gum. The thiolation was done using L-cysteine which releases tannin in buccal mucosa. This functionalization of xanthan gum resulted in improved adherence on the buccal mucosa.[68]

5.2.8   Gum Arabic

Gum arabic aka gum Acacia is a tree gum efflux. The gum is oozed out from Acacia Senegal or Acacia Seyal trees. This has been commercially used since ancient times as paints for hieroglyphic inscriptions by Egyptians. The main component of gum arabic is polysaccharides along with glycoproteins and oligosaccharides. It's neutral or slightly acidic and composed of 1,3-linked β-D-galactopyranosyl units. L-Arabinose, L-rhamnose, and D-glucuronic acid have also been found as components of this gum. They are environmentally friendly, harmless, and have low carbon footprints. It's easily soluble, excellent emulsifier, highly binding, and a really good stabilizing agent. As a result, it's heavily in usage in food, pharmaceutical, textile, and cosmetic industries.[69] It also has antimicrobial properties and helps in demineralizing teeth; it will increases its usage in dentistry.[70] Gum arabic is generally a good adhesive for soft substrates.[71] As gum arabic is a polyelectrolyte, the viscosity of the solution drops when electrolytes are present as a result of charge screening and at low pH when the carboxyl groups get undissociated.[72]

5.2.9   Gellan Gum

Gellan gum is a hydrophilicpolymer derived from bacteria. It's an anionic polysaccharide that consists of tetra saccharide as a repeating unit, and it can withstand temperature around 120°C. Gellan gum can be produced artificially by fermenting sugar with bacteria. It works as a plant-based alternative to gelatin. It is also similar to xanthan gum. It is used as an additive for food which is used to bind and is used as a stabilizer for processed foods. It is also a gelling agent and gives jelly like consistency to food. It is also used in medical and pharmaceutical application for tissue regeneration, bone repair, and drug manufacturing.

5.3   MECHANISM OF ADHESION

Adhesives have been used for many centuries and are still a popular alternative to mechanical joints in engineering applications. The binding strength of an adhesive to an adherend can be described as a sum of various forces such as mechanical, physical, and chemical and also involves several theories. The basic principle behind the adsorption theory of adhesion is that adhesion occurs once two materials come into contact at the molecular level. This can be physisorption or chemisorption. The primary or secondary forces are more than enough supply to provide a high bond strength along with good wettability. A clean surface, ideally, zero contact angles between the adherend and adhesive is required to have adequate wettability, and it greatly depends on interfacial tension. Thus, the contact angle can be calculated as follows:

$$\gamma_{lg}Cos\theta = \gamma_{sg} - \gamma_{sl}$$

where γ is the interfacial tension, θ denotes contact angle, and s, l, and g are solid, liquid, and gas, respectively. The mechanical theory involves interlocking, where the adhesive penetrates the cavities or pores on the surface of the material. As the roughness of the exterior area increases, the mechanical adhesion also enhances. One of the indicators used to identify roughness is the root mean square, R_{rms}, as follows:[73]

$$R_{rms} = \sqrt{\frac{\sum_i (h_i - h)^2}{n}}$$

where h_i is the ith sampling point feature height value, h is the mean height value, and n is the number of data points. In the 1920s McBain and Hopkins' stated: "that a good joint must result whenever a strong continuous film of partly embedded adhesive is formed in situ." The mechanical theory was also used to explain the adhesion of all other metallic substrates. Further, in the 1940s, Deryagin introduced the electrostatic theory which reveals the interactive electron transfer between adherend and adhesive. In this theory, the adhesive-substrate system is considered comparable to a parallel plate condenser. In addition, diffusion theory by Voyutskii put forward a representation for polymer-to-polymer adherence. This theory explains the adhesion of polymers that are mutually capable of mixing by considering the dependence of time and molecular weight on adhesion.[74]

5.3.1 Mechanical Tests for Adhesion Strength

A wide range of testing methods have been performed to find out how strong are the adhesives, and this includes the test for tension, compression, cleavage, shear, and peel stresses. Their primary use is to induce shear stress to find the stability of an adhesive by evaluating the properties like tensile strength, durability, deformability, and brittleness.[75] Besides that, lap tests, such as thin-lap shear, double-lap shear, strap joint, and thick adhered shear, are most familiar among mechanical tests for adhesive joints. From the standpoint of specimen preparation and testing, a single-lap or thin-lap joint test is pretty simple. However, these tests are aware of the limitations of accurately determining the design parameters of the joint, as the force is given at the end. To overcome these disadvantages, a new mechanical test was put forward: the double-lap test. Even though the test is expensive, it gives the shear strength of adhesives directly. The shape of thin adherend tails and end fillets is helpful to control the point of stress maximum. Moreover, the tension pull tests determine the force needed to stretch the adhesive from the surface, and the peel test measures the resistance of the adhesive-adherend system to force peel loading (Figure 5.4).[76]

5.4 APPLICATIONS OF POLYSACCHARIDE-BASED ADHESIVE

5.4.1 Biomedical and Pharmaceutical Applications

The biocompatibility, biodegradability, antimicrobial activity, non-toxicity, and low-cost properties make polysaccharide-based adhesives a suitable candidate for several biomedical applications. In drug delivery and biological hemostasis application, chitosan is used as a promising bio adhesive. A mucoadhesive neuronanoemulsion based on N,N,N-trimethyl chitosan and flaxseed oil was reported for application of drug delivery directly from nose to brain in Parkinson's disease treatment.[77] In 2020, Zhang et al. prepared hydroxybutyl chitosan and diatom-biosilica based porous composite sponge with effective hemorrhage control. The composite sponge can shortened the clotting time up to 70% than that of control by the quick absorption of plasma, increasing hemocyte and platelet concentrations and thereby triggering the blood coagulation.[78] Furthermore, the macroporous scaffold from methacrylate dextran and aminoethyl methacrylate was used as cell adhesive in tissue engineering. The cell adhesion and neurite outgrowth was improved by the covalent modification with extracellular matrix-derived peptides.[79] Recently, Singh et al. reported composites of bacterial cellulose and carbene as bioadhesive in oral cavity applications.[80]

The polysaccharides such as levan, pullulan, chitosan, and starch have found notable applications in the pharmaceutical field. Levan is used to producing temporary coating or bandage and

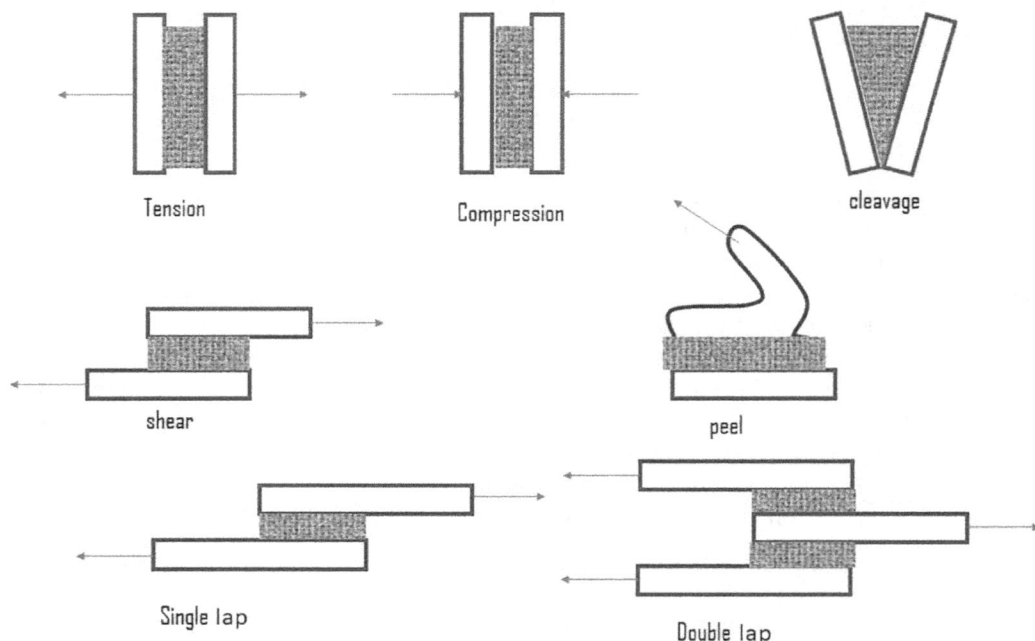

FIGURE 5.4 Various modes of mechanical testing.

tablet binder since it is an adhesive of very strength and a water-soluble film-forming material.[81] The in-vitro studies of bioadhesive films of pullulan-polyacrylamide blends show the potential of this material in immediate-release formulations for dermal applications.[82] The substituted starches such as carboxymethyl starch, starch acetate, and crosslinked starch has wide application in oral tablets to control drug release (Figure 5.5).[83]

5.4.2   INDUSTRIAL APPLICATIONS

The polysaccharide-based adhesives are used as binders and thickening agents in wood, paper, and textile industries. The extracellular polysaccharide-based wood adhesive exhibits improved performance in shear strength and moisture resistance.[84] In the wood industry, modified polysaccharides significantly improve the adhesion properties. In this regard, starch-based wood adhesive with silica nanoparticles enhances both the bonding strength and water resistance compared to starch-based adhesive without silica nanoparticles.[85] Besides that, the chitosan-phenolics system is also used as a strong wood adhesive.[86] In the paper industry, adhesives are used for repairing holes and gaps, or consolidating paper. Due to high chemical stability for artificial aging, low color alteration, depolymerization, and pH changes, starch base-adhesives are commonly used in paper consolidation.[87]

5.4.3   FOOD AND PACKAGING APPLICATION

In the preservation of spiced and flavored coatings in oil roasting of nutmeat products, the dextrin-based food-grade adhesives are used along with slight amount of xanthan, carboxymethyl cellulose, or its mixtures. Besides that, the dried sap of gum arabic has significant use in many foodstuffs applications as a stabilizer, binder, and shelf-life enhancer.[75] Adhesives are inevitable in packaging. The polysaccharide-based adhesives, mainly starch and its derivatives, come in the classification of water-based adhesives for packaging application. Due to low-cost, biodegradability, and availability

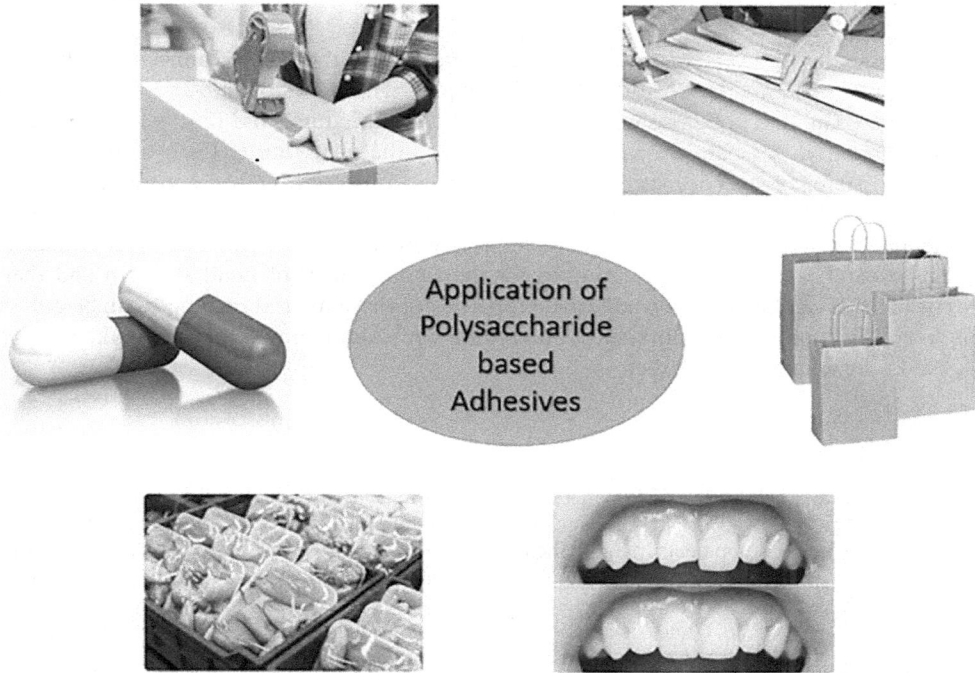

FIGURE 5.5 Application of polysaccharide-based adhesive.

in good quality grades, starch adhesives are remarkably used for corrugated board making, paper bag manufacturing, and also for tube winding. In fast-drying applications, dextrin adhesives, derivatives of starch, are widely used. It is also available in different viscosities, and modifications are also possible, which makes it suitable for applications such as bonding paper-based materials, tube winding, and high-speed labeling (Table 5.1).[88]

TABLE 5.1
List of Polysaccharides and Their Applications

Polymer	Application	Reference
Starch	• Paper adhesive • Corrugating adhesive • Wood adhesive • Pharmaceutical industries	84,89
Chitosan	• Biomedical • Wood adhesive	77,86
Dextrin	• Packaging • Carton sealing • Bag manufacturing	88,89
Cellulose	• Biomedical and pharmaceutical • Agricultural	75
Pullulan	• Pharmaceutical • Paper adhesive • Cosmetics	82,89
Levan	• Bandage and tablet coating	81
Gum arabic	• Food and packaging	75

5.5 CONCLUSION

Polysaccharides have developed an increasing application in the field of adhesives. The ease of availability, low cost, and biodegradability have made them more popular. Polysaccharides such as starch, chitosan, and cellulose which are obtained from natural and renewable resources are the right alternative for petroleum-based adhesives in the near future. Each of these bioadhesives poses different properties such as non-toxicity, biocompatibility, and antimicrobial activity, along with good bond strength. The wide range of applications of polysaccharide-based adhesives includes the biomedical, pharmaceutical, packaging, cosmetics, wood, and paper industries. Detailed research and studies are required for the improvement of bond strength and thermal properties of these adhesives. The adhesive properties of these natural polymers can be enhanced through modification by cross-linking or combining with additives and this can make them more attractive in the commercial market.

REFERENCES

1. Singh, M.; Kadian, S.; Manik, G. Polymers in Adhesive Applications, *Encyclopedia of Materials: Plastics and Polymers* **2021**, *4*, 370–381. Elsevier, USA. https://doi.org/10.1016/B978-0-12-820352-1.00124-3.
2. Pizzi, A. Recent Developments in Eco-Efficient Bio-Based Adhesives for Wood Bonding: Opportunities and Issues. *Journal of Adhesion Science and Technology* **2006**, *20* (8), 829–846. https://doi.org/10.1163/156856106777638635.
3. Bai Qu ; Yangchao Luo. Chitosan-based hydrogel beads: Preparations, modifications and applications in food and agriculture sectors -A review. *International Journal of Biological Macromolecules* **June 2020**, *152*, 427–448, Elsevier, USA.
4. Rinaudo, M. Chitin and Chitosan: Properties and Applications. *Progress in Polymer Science* **2006**, *31* (7), 603–632. https://doi.org/10.1016/j.progpolymsci.2006.06.001.
5. Ramos Berger, L. R.; Montenegro Stamford, T. C.; de Oliveira, K. Á. R.; de Miranda Pereira Pessoa, A.; de Lima, M. A. B.; Estevez Pintado, M. M.; Saraiva Câmara, M. P.; de Oliveira Franco, L.; Magnani, M.; de Souza, E. L. Chitosan Produced from Mucorales Fungi Using Agroindustrial By-Products and Its Efficacy to Inhibit Colletotrichum Species. *International Journal of Biological Macromolecules* **2018**, *108*, 635–641. https://doi.org/10.1016/j.ijbiomac.2017.11.178.
6. Hamed, I.; Özogul, F.; Regenstein, J. M. Industrial Applications of Crustacean By-Products (Chitin, Chitosan, and Chitooligosaccharides): A Review. *Trends in Food Science & Technology* **2016**, *48*, 40–50. https://doi.org/10.1016/j.tifs.2015.11.007.
7. Campana, Fº. S. P.; Signini, R. Efeito de Aditivos Na Desacetilação de Quitina. *Polímeros* **2001**, *11* (4), 169–173. https://doi.org/10.1590/S0104-14282001000400006.
8. Kaur, S.; Dhillon, G. S. Recent Trends in Biological Extraction of Chitin from Marine Shell Wastes: A Review. *Critical Reviews in Biotechnology* **2015**, *35* (1), 44–61. https://doi.org/10.3109/07388551.2013.798256.
9. Varun, T. K.; Senani, S.; Jayapal, N.; Chikkerur, J.; Roy, S.; Tekulapally, V. B.; Gautam, M.; Kumar, N. Extraction of Chitosan and Its Oligomers from Shrimp Shell Waste, Their Characterization and Antimicrobial Effect. *Vet World* **2017**, *10* (2), 170–175. https://doi.org/10.14202/vetworld.2017.170-175.
10. Ferren, L. G.; Ward, R. L.; Campbell, B. J. Monoanion Inhibition and 35Cl Nuclear Magnetic Resonance Studies of Renal Dipeptidase. *Biochemistry* **1975**, *14* (24), 5280–5285. https://doi.org/10.1021/bi00695a008.
11. Gerhardt, R.; Farias, B. S.; Moura, J. M.; de Almeida, L. S.; da Silva, A. R.; Dias, D.; Cadaval, T. R. S.; Pinto, L. A. A. Development of Chitosan/Spirulina Sp. Blend Films as Biosorbents for Cr^{6+} and Pb^{2+} Removal. *International Journal of Biological Macromolecules* **2020**, *155*, 142–152. https://doi.org/10.1016/j.ijbiomac.2020.03.201.
12. Yang, Z.; Chai, Y.; Zeng, L.; Gao, Z.; Zhang, J.; Ji, H. Efficient Removal of Copper Ion from Wastewater Using a Stable Chitosan Gel Material. *Molecules* **2019**, *24* (23), 4205. https://doi.org/10.3390/molecules24234205.
13. McBain, J. W.; Hopkins, D. G. On Adhesives and Adhesive Action. *Journal of Physical Chemistry* **1925**, *29* (2), 188–204. https://doi.org/10.1021/j150248a008.
14. Aqra, F.; Ayyad, A. Surface Tension (ΓLV), Surface Energy (ΓSV) and Crystal-Melt Interfacial Energy (ΓSL) of Metals. *Current Applied Physics* **2012**, *12* (1), 31–35. https://doi.org/10.1016/j.cap.2011.04.020.

15. Kurek, M.; Guinault, A.; Voilley, A.; Galić, K.; Debeaufort, F. Effect of Relative Humidity on Carvacrol Release and Permeation Properties of Chitosan Based Films and Coatings. *Food Chemistry* **2014**, *144*, 9–17. https://doi.org/10.1016/j.foodchem.2012.11.132.

16. Dmitrenko, M. E.; Penkova, A. V.; Kuzminova, A. I.; Ermakov, S. S.; Roizard, D. Investigation of Polymer Membranes Modified by Fullerenol for Dehydration of Organic Mixtures. *Journal of Physics: Conference Series* **2017**, *879*, 012010. https://doi.org/10.1088/1742-6596/879/1/012010.

17. Boukhelata, N.; Taguett, F.; Kaci, Y. Characterization of an Extracellular Polysaccharide Produced by a Saharan Bacterium Paenibacillus Tarimensis REG 0201M. *Annals of Microbiology* **2019**, *69* (2), 93–106. https://doi.org/10.1007/s13213-018-1406-3.

18. Owczarz, P.; Ziółkowski, P.; Dziubiński, M. The Application of Small-Angle Light Scattering for Rheo-Optical Characterization of Chitosan Colloidal Solutions. *Polymers* **2018**, *10* (4), 431. https://doi.org/10.3390/polym10040431.

19. Umemura, K.; Inoue, A.; Kawai, S. Development of New Natural Polymer-Based Wood Adhesives I: Dry Bond Strength and Water Resistance of Konjac Glucomannan, Chitosan, and Their Composites. *Journal of Wood Science* **2003**, *49* (3), 221–226. https://doi.org/10.1007/s10086-002-0468-8.

20. Umemura, K.; Kawai, S. Modification of Chitosan by the Maillard Reaction Using Cellulose Model Compounds. *Carbohydrate Polymers* **2007**, *68* (2), 242–248. https://doi.org/10.1016/j.carbpol.2006.12.014.

21. Khanlari, S.; Dubé, M. A. Bioadhesives: A Review: Bioadhesives: A Review. *Macromolecular Reaction Engineering* **2013**, *7* (11), 573–587. https://doi.org/10.1002/mren.201300114.

22. Shah, N. V.; Meislin, R. Current State and Use of Biological Adhesives in Orthopedic Surgery. *Orthopedics* **2013**, *36* (12), 945–956. https://doi.org/10.3928/01477447-20131120-09.

23. Dash, M.; Chiellini, F.; Ottenbrite, R. M.; Chiellini, E. Chitosan-A Versatile Semi-Synthetic Polymer in Biomedical Applications. *Progress in Polymer Science* **2011**, *36* (8), 981–1014. https://doi.org/10.1016/j.progpolymsci.2011.02.001.

24. Chatelet, C. Influence of the Degree of Acetylation on Some Biological Properties of Chitosan Films. *Biomaterials* **2001**, *22* (3), 261–268. https://doi.org/10.1016/S0142-9612(00)00183-6.

25. Barton, M. J.; Morley, J. W.; Mahns, D. A.; Mawad, D.; Wuhrer, R.; Fania, D.; Frost, S. J.; Loebbe, C.; Lauto, A. Tissue Repair Strength Using Chitosan Adhesives with Different Physical-Chemical Characteristics: Chitosan Adhesives with Different Physical-Chemical Characteristics. *Journal of Biophotonics* **2014**, *7* (11–12), 948–955. https://doi.org/10.1002/jbio.201300148.

26. Jayakumar, R.; Menon, D.; Manzoor, K.; Nair, S. V.; Tamura, H. Biomedical Applications of Chitin and Chitosan Based Nanomaterials: A Short Review. *Carbohydrate Polymers* **2010**, *82* (2), 227–232. https://doi.org/10.1016/j.carbpol.2010.04.074.

27. Saxena, I. M.; Brown, R. M. Biosynthesis of Cellulose. In *Progress in Biotechnology*; Elsevier, 2001; Vol. 18, pp 69–76. https://doi.org/10.1016/S0921-0423(01)80057-5.

28. Gardner, D. J.; Oporto, G. S.; Mills, R.; Samir, M. A. S. A. Adhesion and Surface Issues in Cellulose and Nanocellulose. *Journal of Adhesion Science and Technology* **2008**, *22* (5–6), 545–567. https://doi.org/10.1163/156856108X295509.

29. Mattar, H. *Nitrocellulose: Structure, Synthesis, Characterization, and Applications*. https://doi.org/10.18576/wefej/010301.

30. Gralén, N. Cellulose and Cellulose Derivatives (High Polymers, Volume V, 2nd Ed. In 3 Parts). Emil Ott and Harold Spurlin, Coeditors; Asst. Editor, Mildred W. Grafflin. Interscience, New York-London, 1954. Part I, XVI+509 pp. Part II, VIII+546 pp. $12.00 Each. *Journal of Polymer Science* **1955**, *18* (89), 443–444. https://doi.org/10.1002/pol.1955.120188920.

31. Cheng, H. N.; Dowd, M. K.; Selling, G. W.; Biswas, A. Synthesis of Cellulose Acetate from Cotton Byproducts. *Carbohydrate Polymers* **2010**, *80* (2), 449–452. https://doi.org/10.1016/j.carbpol.2009.11.048.

32. Edgar, K. J.; Buchanan, C. M.; Debenham, J. S.; Rundquist, P. A.; Seiler, B. D.; Shelton, M. C.; Tindall, D. Advances in Cellulose Ester Performance and Application. *Progress in Polymer Science* **2001**, *26* (9), 1605–1688. https://doi.org/10.1016/S0079-6700(01)00027-2.

33. Melissa, G.; Baumann, D.; Conner, A. H. Carbohydrate Polymers as Adhesives. In *Handbook of Adhesive Technology*, Marcel Dekker, Inc., New York, 299; 1994.

34. Ergun, R.; Guo, J.; Huebner-Keese, B. Cellulose. In *Encyclopedia of Food and Health*; Elsevier, 2016; pp 694–702. https://doi.org/10.1016/B978-0-12-384947-2.00127-6.

35. Javanbakht, S.; Shaabani, A. Carboxymethyl Cellulose-Based Oral Delivery Systems. *International Journal of Biological Macromolecules* **2019**, *133*, 21–29. https://doi.org/10.1016/j.ijbiomac.2019.04.079.

36. Whistler, R. L.; BeMiller, J. N. *Starch: Chemistry and Technology*, 3rd ed.; Food science and Technology, International Series; Academic Press: London, 2009.

37. Cornejo-Ramírez, Y. I.; Martínez-Cruz, O.; Del Toro-Sánchez, C. L.; Wong-Corral, F. J.; Borboa-Flores, J.; Cinco-Moroyoqui, F. J. The Structural Characteristics of Starches and Their Functional Properties. *CyTA – Journal of Food* **2018**, *16* (1), 1003–1017. https://doi.org/10.1080/19476337.2018.1518343.
38. Yu, H.; Cao, Y.; Fang, Q.; Liu, Z. Effects of Treatment Temperature on Properties of Starch-Based Adhesives. *BioRes* **2015**, *10* (2), 3520–3530. https://doi.org/10.15376/biores.10.2.3520-3530.
39. Qin, Y.; Zhang, H.; Dai, Y.; Hou, H.; Dong, H. Effect of Alkali Treatment on Structure and Properties of High Amylose Corn Starch Film. *Materials* **2019**, *12* (10), 1705. https://doi.org/10.3390/ma12101705.
40. Gadhave, R. V.; Mahanwar, P. A.; Gadekar, P. T. Starch-Based Adhesives for Wood/Wood Composite Bonding: Review. *Open Journal of Polymer Chemistry* **2017**, *07* (02), 19–32. https://doi.org/10.4236/ojpchem.2017.72002.
41. Cheng, K.-C.; Demirci, A.; Catchmark, J. M. Pullulan: Biosynthesis, Production, and Applications. *Applied Microbiology Biotechnology* **2011**, *92* (1), 29–44. https://doi.org/10.1007/s00253-011-3477-y.
42. Filiz Yangılar, P. O. Y. *Pullulan: Production and Usage in Food Industry*, Semantic Scholar.
43. Abhilash, M.; Thomas, D. Biopolymers for Biocomposites and Chemical Sensor Applications. In *Biopolymer Composites in Electronics*; Elsevier, Netherlands, 2017; pp 405–435. https://doi.org/10.1016/B978-0-12-809261-3.00015-2.
44. Lochhead, R. Y. The Use of Polymers in Cosmetic Products. In *Cosmetic Science and Technology*; Elsevier, 2017; pp 171–221. https://doi.org/10.1016/B978-0-12-802005-0.00013-6.
45. Arvidson, S. A.; Rinehart, B. T.; Gadala-Maria, F. Concentration Regimes of Solutions of Levan Polysaccharide from Bacillus Sp. *Carbohydrate Polymers* **2006**, *65* (2), 144–149. https://doi.org/10.1016/j.carbpol.2005.12.039.
46. Feng, J.; Gu, Y.; Quan, Y.; Zhang, W.; Cao, M.; Gao, W.; Song, C.; Yang, C.; Wang, S. Recruiting a New Strategy to Improve Levan Production in Bacillus Amyloliquefaciens. *Scientific Reports* **2015**, *5* (1), 13814. https://doi.org/10.1038/srep13814.
47. Zikmanis, P.; Brants, K.; Kolesovs, S.; Semjonovs, P. Extracellular Polysaccharides Produced by Bacteria of the Leuconostoc Genus. *World Journal of Microbiology and Biotechnology* **2020**, *36* (11), 161. https://doi.org/10.1007/s11274-020-02937-9.
48. Combie, J.; Steel, A.; Sweitzer, R. Adhesive Designed by Nature (and Tested at Redstone Arsenal). *Clean Technologies and Environmental Policy* **2004**, *6* (4). https://doi.org/10.1007/s10098-004-0244-0.
49. Mihailescu, I. N.; Bigi, A.; Gyorgy, E.; Ristoscu, C.; Sima, F.; Oner, E. T. Biomaterial Thin Films by Soft Pulsed Laser Technologies for Biomedical Applications. In *Lasers in Materials Science*; Castillejo, M., Ossi, P. M., Zhigilei, L., Eds.; Springer Series in Materials Science; Springer International Publishing: Cham, 2014; Vol. 191, pp 271–294. https://doi.org/10.1007/978-3-319-02898-9_11.
50. Nowakowska, M.; Zapotoczny, S.; Sterzel, M.; Kot, E. Novel Water-Soluble Photosensitizers from Dextrans. *Biomacromolecule s***2004**, *5* (3), 1009–1014. https://doi.org/10.1021/bm034506w.
51. Díaz-Montes, E. Dextran: Sources, Structures, and Properties. *Polysaccharides* **2021**, *2* (3), 554–565. https://doi.org/10.3390/polysaccharides2030033.
52. Heinze, T.; Liebert, T.; Heublein, B.; Hornig, S. Functional Polymers Based on Dextran. In *Polysaccharides II*; Klemm, D., Ed.; Advances in Polymer Science; Springer: Berlin Heidelberg, 2006; Vol. 205, pp 199–291. https://doi.org/10.1007/12_100.
53. Foster, J. H.; Killen, D. A.; Jolly, P. C.; Kirtley, J. H. Low Molecular Weight Dextran in Vascular Surgery: Prevention of Early Thrombosis Following Arterial Reconstruction in 85 Cases. *Annals of Surgery* **1966**, *163* (5), 764–770. https://doi.org/10.1097/00000658-196605000-00013.
54. Wondraczek, H.; Elschner, T.; Heinze, T. Synthesis of Highly Functionalized Dextran Alkyl Carbonates Showing Nanosphere Formation. *Carbohydrate Polymers* **2011**, *83* (3), 1112–1118. https://doi.org/10.1016/j.carbpol.2010.09.013.
55. Liu, Y.; Chan-Park, M. B. Hydrogel Based on Interpenetrating Polymer Networks of Dextran and Gelatin for Vascular Tissue Engineering. *Biomaterials* **2009**, *30* (2), 196–207. https://doi.org/10.1016/j.biomaterials.2008.09.041.
56. Liu, Y.; Chan-Park, M. B. A Biomimetic Hydrogel Based on Methacrylated Dextran-Graft-Lysine and Gelatin for 3D Smooth Muscle Cell Culture. *Biomaterials* **2010**, *31* (6), 1158–1170. https://doi.org/10.1016/j.biomaterials.2009.10.040.
57. Artzi, N.; Shazly, T.; Baker, A. B.; Bon, A.; Edelman, E. R. Aldehyde-Amine Chemistry Enables Modulated Biosealants with Tissue-Specific Adhesion. *Advanced Matererials* **2009**, *21* (32–33), 3399–3403. https://doi.org/10.1002/adma.200900340.
58. Jansson, P.; Kenne, L.; Lindberg, B. Structure of the Extracellular Polysaccharide from Xanthomonas Campestris. *Carbohydrate Research* **1975**, *45* (1), 275–282. https://doi.org/10.1016/S0008-6215(00)85885-1.

59. Nur Hazirah, M. A. S. P.; Isa, M. I. N.; Sarbon, N. M. Effect of Xanthan Gum on the Physical and Mechanical Properties of Gelatin-Carboxymethyl Cellulose Film Blends. *Food Packaging and Shelf Life* **2016**, *9*, 55–63. https://doi.org/10.1016/j.fpsl.2016.05.008.

60. Soares, R. M. D.; Lima, A. M. F.; Oliveira, R. V. B.; Pires, A. T. N.; Soldi, V. Thermal Degradation of Biodegradable Edible Films Based on Xanthan and Starches from Different Sources. *Polymer Degradation and Stability* **2005**, *90* (3), 449–454. https://doi.org/10.1016/j.polymdegradstab.2005.04.007.

61. Katzbauer, B. Properties and Applications of Xanthan Gum. *Polymer Degradation and Stability* **1998**, *59* (1–3), 81–84. https://doi.org/10.1016/S0141-3910(97)00180-8.

62. Tako, M.; Teruya, T.; Tamaki, Y.; Ohkawa, K. Co-Gelation Mechanism of Xanthan and Galactomannan. *Colloid and Polymer Science* **2010**, *288* (10–11), 1161–1166. https://doi.org/10.1007/s00396-010-2242-6.

63. Lopes, L.; Andrade, C. T.; Milas, M.; Rinaudo, M. Role of Conformation and Acetylation of Xanthan on Xanthan-Guar Interaction. *Carbohydrate Polymers* **1992**, *17* (2), 121–126. https://doi.org/10.1016/0144-8617(92)90105-Y.

64. Bryant, C. Influence of Xanthan Gum on Physical Characteristics of Heat-Denatured Whey Protein Solutions and Gels. *Food Hydrocolloids* **2000**, *14* (4), 383–390. https://doi.org/10.1016/S0268-005X(00)00018-7.

65. Hemar, Y.; Tamehana, M.; Munro, P. A.; Singh, H. Viscosity, Microstructure and Phase Behavior of Aqueous Mixtures of Commercial Milk Protein Products and Xanthan Gum. *Food Hydrocolloids* **2001**, *15* (4-6), 565–574. https://doi.org/10.1016/S0268-005X(01)00077-7.

66. Argin-Soysal, S.; Kofinas, P.; Lo, Y. M. Effect of Complexation Conditions on Xanthan-Chitosan Polyelectrolyte Complex Gels. *Food Hydrocolloids* **2009**, *23* (1), 202–209. https://doi.org/10.1016/j.foodhyd.2007.12.011.

67. Hoffmann, B.; Volkmer, E.; Kokott, A.; Augat, P.; Ohnmacht, M.; Sedlmayr, N.; Schieker, M.; Claes, L.; Mutschler, W.; Ziegler, G. Characterisation of a New Bioadhesive System Based on Polysaccharides with the Potential to Be Used as Bone Glue. *Journal of Materials Science: Materials in Medicine* **2009**, *20* (10), 2001–2009. https://doi.org/10.1007/s10856-009-3782-5.

68. Laffleur, F.; Michalek, M. Modified Xanthan Gum for Buccal Delivery-A Promising Approach in Treating Sialorrhea. *International Journal of Biological Macromolecules* **2017**, *102*, 1250–1256. https://doi.org/10.1016/j.ijbiomac.2017.04.123.

69. Patel, S.; Goyal, A. Applications of Natural Polymer Gum Arabic: A Review. *International Journal of Food Properties* **2015**, *18* (5), 986–998. https://doi.org/10.1080/10942912.2013.809541.

70. Ali, B. H.; Ziada, A.; Blunden, G. Biological Effects of Gum Arabic: A Review of Some Recent Research. *Food and Chemical Toxicology* **2009**, *47* (1), 1–8. https://doi.org/10.1016/j.fct.2008.07.001.

71. Katiyar, V.; Tripathi, N. Functionalizing Gum Arabic for Adhesive and Food Packaging Applications. *SPE Plastics Research Online* **2017**. https://doi.org/10.2417/spepro.006897.

72. Phillips, G. O., Ed. *Handbook of Hydrocolloids*, 2nd ed. Woodhead Publishing in Food Science, Technology and Nutrition; CRC: Boca Raton, FL, 2009.

73. Patel, A. K.; Mathias, J.-D.; Michaud, P. Polysaccharides as Adhesives. *Reviews of Adhesion and Adhesives* **2013**, *1* (3), 312–345. https://doi.org/10.7569/RAA.2013.097310.

74. da Silva, L. F. M., Öchsner, A., Adams, R. D., Eds. *Handbook of Adhesion Technology*; Springer International Publishing: Cham, 2018. https://doi.org/10.1007/978-3-319-55411-2.

75. Ali, A.; Rehman, K.; Majeed, H.; Khalid, M. F.; Akash, M. S. H. Polysaccharide-Based Adhesives. In *Green Adhesives*; Inamuddin, Boddula, R., Ahamed, M. I., Asiri, A. M., Eds.; Wiley, Beverly, MA, 2020; pp 165–180. https://doi.org/10.1002/9781119655053.ch8.

76. Duncan, B.; Crocker, L. *Review of Tests for Adhesion Strength*; NPL Report MA TC(A)67.

77. Pardeshi, C. V.; Belgamwar, V. S. N,N,N-trimethyl Chitosan Modified Flaxseed Oil Based Mucoadhesive Neuronanoemulsions for Direct Nose to Brain Drug Delivery. *International Journal of Biological Macromolecules* **2018**, *120*, 2560–2571. https://doi.org/10.1016/j.ijbiomac.2018.09.032.

78. Zhang, K.; Li, J.; Wang, Y.; Mu, Y.; Sun, X.; Su, C.; Dong, Y.; Pang, J.; Huang, L.; Chen, X.; Feng, C. Hydroxybutyl Chitosan/Diatom-Biosilica Composite Sponge for Hemorrhage Control. *Carbohydrate Polymers* **2020**, *236*, 116051. https://doi.org/10.1016/j.carbpol.2020.116051.

79. Lévesque, S. G.; Shoichet, M. S. Synthesis of Cell-Adhesive Dextran Hydrogels and Macroporous Scaffolds. *Biomaterials* **2006**, *27* (30), 5277–5285. https://doi.org/10.1016/j.biomaterials.2006.06.004.

80. Singh, J.; Tan, N. C. S.; Mahadevaswamy, U. R.; Chanchareonsook, N.; Steele, T. W. J.; Lim, S. Bacterial Cellulose Adhesive Composites for Oral Cavity Applications. *Carbohydrate Polymers* **2021**, *274*, 118403. https://doi.org/10.1016/j.carbpol.2021.118403.

81. *Polysaccharides for Drug Delivery and Pharmaceutical Applications*; Marchessault, R. H., Ravenelle, F., Zhu, X. X., Eds.; ACS Symposium Series; American Chemical Society: Washington, DC, 2006; Vol. 934. https://doi.org/10.1021/bk-2006-0934.

82. Vishwanath, B.; Shivakumar, H. R.; Sheshappa, R. K.; Ganesh, S.; Prasad, P.; Guru, G. S.; Bhavya, B. B. In-Vitro Release Study of Metoprolol Succinate from the Bioadhesive Films of Pullulan-Polyacrylamide Blends. *International Journal of Polymeric Materials* **2012**, *61* (4), 300–307. https://doi.org/10.1080/00 914037.2011.584227.

83. Hong, Y.; Liu, G.; Gu, Z. Recent Advances of Starch-Based Excipients Used in Extended-Release Tablets: A Review. *Drug Delivery* **2016**, *23* (1), 12–20. https://doi.org/10.3109/10717544.2014.913324.

84. Haag, A. P.; Geesey, G. G.; Mittleman, M. W. Bacterially Derived Wood Adhesive. *International Journal of Adhesion and Adhesives* **2006**, *26* (3), 177–183. https://doi.org/10.1016/j.ijadhadh.2005.03.011.

85. Wang, Z.; Gu, Z.; Hong, Y.; Cheng, L.; Li, Z. Bonding Strength and Water Resistance of Starch-Based Wood Adhesive Improved by Silica Nanoparticles. *Carbohydrate Polymers* **2011**, *86* (1), 72–76. https://doi.org/10.1016/j.carbpol.2011.04.003.

86. Peshkova, S.; Li, K. Investigation of Chitosan-Phenolics Systems as Wood Adhesives. *Journal of Biotechnology* **2003**, *102* (2), 199–207. https://doi.org/10.1016/S0168-1656(03)00026-9.

87. Borges, I. da S.; Casimiro, M. H.; Macedo, M. F.; Sequeira, S. O. Adhesives Used in Paper Conservation: Chemical Stability and Fungal Bioreceptivity. *Journal of Cultural Heritage* **2018**, *34*, 53–60. https://doi.org/10.1016/j.culher.2018.03.027.

88. Emblem, A.; Hardwidge, M. Adhesives for Packaging. In *Packaging Technology*; Elsevier, Netherlands, 2012; pp 381–394. https://doi.org/10.1533/9780857095701.2.381.

89. Baumann, M. G. D.; Conner, A. H. Carbohydrate Polymers as Adhesives. In *Handbook of Adhesive Technology*; World CAT, Marcel Dekker, New York, 1994; pp 299–313.

6 Carbohydrate-Based Therapeutics

Evolution from Wellness Pursuit to Medical Treatment

Shradha S. Tiwari
Annasaheb Dange College of B. Pharmacy

Surendra G. Gattani
Swami Ramanand Teerth Marathwada University

Bhasha Sharma
University of Delhi

Md Enamul Hoque
Military Institute of Science and Technology (MIST)

6.1 INTRODUCTION

Carbohydrates are the important ingredient of the dietary component. Carbohydrates are the complex biomolecules that exist abundantly around us. It can be found profusely on the cell surface of eukaryotes and prokaryotes. Carbohydrates are an integral part of living system and are involved in various biological processes, cellular interactions, and signaling to various cellular molecules and cell surface receptors. This biomolecule plays an important role in various diseases. Carbohydrates are involved in different imperative biological process, including cell identification events, signal transduction, intercellular adhesion, cellular differentiation, molecular recognition, and embryonic development. These molecules serves as the building blocks of genetic materials like DNA and RNA. Carbohydrates are immediate source of energy; moreover, these are involved in a variety of biological events, including reproduction, immune responses, inflammation, building macromolecules, signal transmission, signal transduction, and infection. Cell recognition is accelerated through cell surface saccharides, which is involved in drug development. Lectin is a carbohydrate-binding protein present on the cell surface, and saccharides bind to that site in various diseased states. Oligosaccharides are located on the cell exteriors, playing a crucial role in controlling the interactions of cells with other cells in the extracellular environment and with various molecules. As carbohydrates are involved in cellular recognition and signaling, these are used as therapeutics and diagnostics in various areas like infectious diseases, cancer, inflammation, immunology, neurodegenerative disease, vascular diseases, diabetes, and arthritis. Novel carbohydrates-based therapeutics have wide scope and applications. Carbohydrates exhibit significant biological importance as biologically active molecule. In addition, they act as starting materials for various therapeutic agents like D-glucose.[1,2]

Carbohydrates are involved in formation of genetic materials, supporting the structure of organisms. A large variety of monosaccharide and oligosaccharide residues are attached through glyosidic linkages to form fundamental glycoconjugates, glycoproteins, glycolipids, and glycosylated natural products. Natural and semisynthetic glycoconjugates have been used as anticancer and antimicrobial drugs. Carbohydrates possess outstanding characteristics like biocompatible, biodegradable, hydrophilic, nonimmunogenic, and nontoxic. Carbohydrate-based microparticles and nanoparticles have grabbed huge attention due to biomedical applications as it is being used for bioimaging, biosensing, and as targeted and controlled drug delivery. Polysaccharides are biocompatible substance having suitable rigidity and functionality, forming polymeric biomaterials that have extensive applications in drug delivery, biomedical, tissue engineering, etc. Carbohydrate-based polymers like β-cyclodextrins can be used as solubilizing agents in various dosage forms. Carbohydrates have been extensively explored in the development and discovery of new drug. Marine carbohydrates are significant biological macromolecules that broadly exist in marine algae and animals. Examples include alginate, carrageenan, porphyran, fucoidan, ulvan, agarose, and chitosan.[3,4]

6.2 CLASSIFICATION OF CARBOHYDRATES

Carbohydrates are huge source of energy, available in the form of starch, sugar, legumes, fruits, vegetables, and fibers. Carbohydrates, based on the number of sugar moieties present, are mainly classified as monosaccharides (single sugar unit), disaccharides (two carbohydrate units), oligosaccharides (three to nine carbohydrate units), and polysaccharides (more than nine carbohydrate units). Monosaccharides and oligosaccharides residues are bridged by glyosidic linkages to form crucial glycoconjugates, including glycoproteins, glycolipids, and glycosylated natural products. Polysaccharides are also called as glycans.[5]

Polysaccharides containing similar monosaccharides are homopolysaccharides, and polysaccharides containing different monosaccharides are heteropolysaccharides. Based on the number of carbon atoms present, monosaccharides are further classified as tetrose, pentose, hexose, nanose, etc. Naturally occurring sugar acids (monosaccharides) are another important class of carbohydrates. Aldonic acid, uronic acid, glucuronic acid, ascorbic acid, and sialic acid are some physiologically important sugar acids. Sialic acids are diverse monosaccharides that can mediate various pathological and physiological processes. Pathogens of diseases like human influenza A, *Helicobacter pylori*, and *Vibrio cholerae* attach to sialic acid present on human cell surface, thus regulating immune response and ligand selection.[6,7] Imino sugar is another important carbohydrate class, obtained from natural plants. Imino sugar inhibits selective glycosyl transferases involved in cancer and microbial disease, thus it can be used as therapeutic agent.[8]

6.3 CARBOHYDRATES AS THERAPEUTICS

6.3.1 Glycoconjugates as Therapeutic Agents

Carbohydrates exhibit diversity in structure, property, and function that facilitate to synthesize novel carbohydrate-based therapeutic agents or make significant improvements to advance the activity of existing drugs.[8] Carbohydrates-based drugs are of natural origin, obtained from marine sources, animal extracts, plant origin, bacterial sources, etc. There are some examples of natural origin carbohydrate-based therapeutics. For instance, Ancer 20 is used in the management of leukopenia, and it is derived from different tubercle bacillus. Heparin is the oldest carbohydrate-based drug, isolated from porcine intestinal mucosal tissue, and used as an anticoagulant drug. It is complex mixture of glycosaminoglycan (GAG) polysaccharides. Hyaluronic acid is natural oligosaccharide that has been used extensively, especially in ophthalmology. Acemannan is complex polysaccharide isolated from aloe vera used in wound healing. Acarbose is obtained from *Actinoplanes utahensis* bacterial species and is used in the management of Type II diabetes mellitus. Sodium hyaluronate is obtained from animal source and is used as an antiarthritic agent (Figure 6.1).[4,9]

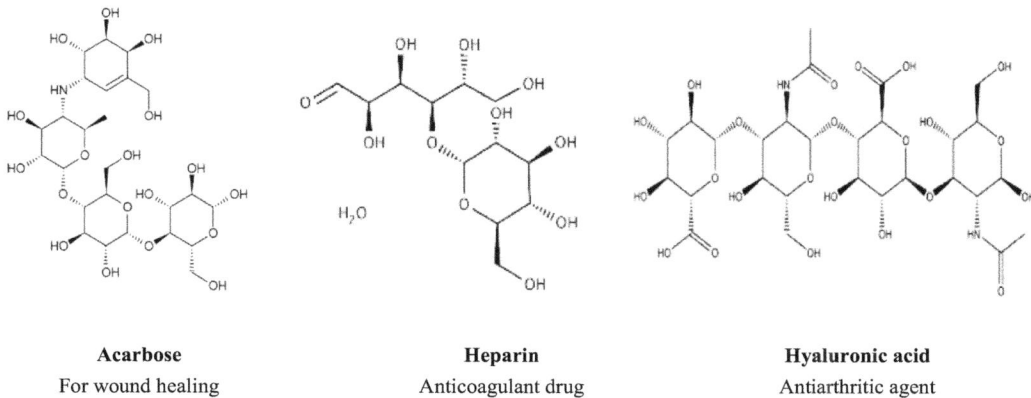

Acarbose
For wound healing

Heparin
Anticoagulant drug

Hyaluronic acid
Antiarthritic agent

FIGURE 6.1 Structures of drugs used in wound healing, anticoagulant drug, and anti-arthritic agent.

6.3.2 CARBOHYDRATE-BASED ANTIBIOTICS

Antibiotics are the chemical compounds produced by microorganisms that interfere with metabolic process and precisely inhibit the expansion of or kill microorganisms. Carbohydrate antibiotics are microbial metabolites, and these can be isolated from bacteria, fungi, plants, algae, etc. Carbohydrate-based antibiotics are mostly fungal and bacterial origin. Carbohydrate-based antibiotics are microbial metabolites having antifungal and antibacterial properties with carbohydrate component in chemical structure. There are three types of carbohydrate-based antibiotics, like aminoglycosides, macrolides, nucleoside analogs, and glycosylated aromatic structures.

First antibiotics group includes carbohydrates linked glycosidically with cyclitols or aminocyclitols, i.e., aminoglycoside antibiotics which consist of gentamicin, amikacin, tobramycin, streptomycin, and kanamycin. Second group includes oligosaccharides in which monosaccharides are linked glycosidically as well as with one or more orthoester linkages, i.e., orthosomycins antibiotics which consist of flambamycin, hygromycin, destomycins, avilamycins, etc. Third group include carbohydrates linked glycosidically with non-carbohydrate group of antibiotic like macrolide antibiotics, e.g., erythromycin, clarithromycin, azithromycin, nystatin, etc. Macrolide antibiotics have broad-spectrum antibacterial activities, and they are used in the management of respiratory tract infections associated with patients allergic to penicillin.[1,2] Carbohydrates are linked to nucleotide moiety, giving rise to antibiotics such as liposidomycin, tunicamycin, mureidomycins, etc.

Aminoglycoside antibiotics are highly potent drugs that inversely bind to small subunit of bacterial ribosome 30 S and thereby inhibit protein synthesis. These are used in the treatment of aerobic, gram positive bacterial infection. Streptomycin has been used in the treatment of tuberculosis. Streptomycin is an effective against gram positive as well as gram negative bacteria. Amikacin and kanamycin are mainly effective in the treatment of multidrug-resistant tuberculosis. Kanamycin is used in the management of infections caused by *Escherichia coli*, Proteus, and Acinetobacter species.[4]

Nucleoside antibiotics inhibit biosynthesis of peptidoglycan attached on the outer plasma membrane of bacterial cell wall. Mildiomycin is a peptidyl nucleoside antibiotic produced by an actinomycete that obstructs the fungal peptidyl-transferase center of the larger ribosomal subunit and acts as antifungal agent. Among lipopeptide antibiotics, ramoplanin is glycolipodepsipeptide produced by *Actinoplanes* spp. It is new class of antimicrobial agent antibiotic that interferes with peptidoglycan production and thereby blocks bacterial cell wall biosynthesis. It is used in the treatment for multiple antibiotic-resistant *Clostridium difficile* infection of the gastrointestinal tract (Figure 6.2).[10,11]

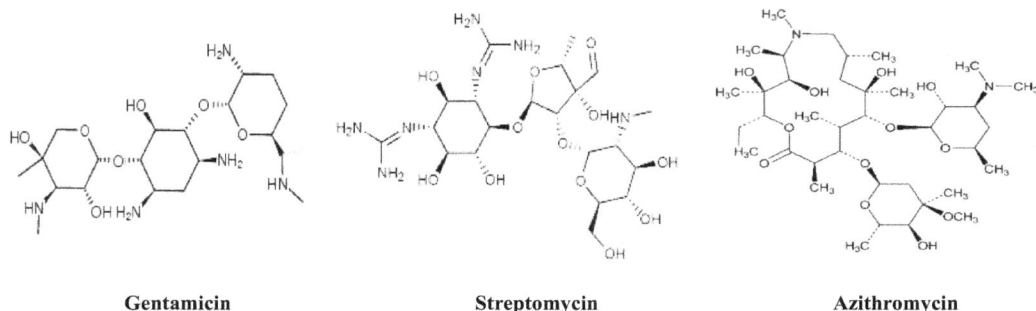

Gentamicin   **Streptomycin**   **Azithromycin**

FIGURE 6.2   Structure of antibiotics.

6.3.3 CARBOHYDRATE-BASED VACCINE

Vaccination is important to fight against infectious diseases. Pneumovax was the first polysaccharide-based vaccine, commercially launched by Merck in 1983. Avery and Heidelberger in the 1920s have discovered carbohydrate-based vaccines, containing capsular polysaccharides (CPS) for *Streptococcus pneumoniae*. Carbohydrate-based vaccines have been discovered, developed, and effectively used against a variety of infectious diseases, including cancer, hepatitis, meningitis, pneumonia, influenza, and viral diseases. Carbohydrate antigens are present on the bacterial cell surface in the form of glycans, and these glycans have unique complex structures that distinguish them from mammalian glycans. Thus glycan become promising target for vaccines and biomarkers. Polysaccharides and glycoconjugates present on cell surface of bacteria, fungi, viruses, and parasites act as important antigenic targets for vaccine design and development.[12]

Quimi-Hib® is carbohydrate-based synthetic vaccine used for *Haemophilus influenza* Hib type b, and it consists of synthetic capsular polysaccharide antigen from Hib coupled with carrier protein tetanus toxoid (TTox). Glycosyl phosphatidyl inositol based antimalarial vaccines have been developed. *Neisseria meningitides* vaccines are available in various brands including Menveo (GSK), Menactra® (Sanofi Pasteur), and Nimenirix® (Pfizer). Provenge® sipuleucel-T, an autologous cellular immunotherapy, used in the treatment of late-stage prostate cancer vaccine. Vaccination is used to prevent pneumococcal disease. Pneumococcal polysaccharide vaccine PPSV10, PPSV13, PPSV23, marketed under various brands Synflorix®, Prevnar13®, and Pneumovax®23, respectively, are used in the prevention of infection caused by *Streptococcus pneumoniae* bacteria.

Typhoid fever is the severe systemic infectious disease caused by *S. typhi*. Typhoid Vi polysaccharide vaccine, known as Typhim Vi, was developed by Sanofi pharma to prevent typhoid fever.[12,13]

Tumor-associated carbohydrate antigen (TACA) is one of the promising targets for the development of anticancer vaccines. Carbohydrates alone cannot provoke T-cell-dependent immune responses, which are necessary for cancer therapy. TACA linked with carrier protein was used to improve immunogenicity. TACA coupled with T-cell peptide multiepitopes form multicomponent glycoconjugate vaccines. GM2 ganglioside (melanoma-specific antigen) can be attached with immunostimulant keyhole limpet hemocyanin (KLH) to form an anticancer vaccine.[14] Heptavalent vaccine consists of seven TACAs: globohexaosylceramide (globo-H), GM2, Lewis-y, MUC1–32 (mucin 1–32 cancer antigen aa), sTn(c), TF(c), and Tn(c) conjugated with KLH, an immunomodulator. This vaccine possibly encourages the production of IgG and IgM antibodies and an antibody-dependent cell-mediated cytotoxicity (ADCC) against tumors expressing these antigens.[15,16]

6.3.4 CARBOHYDRATE-BASED ANTICANCER THERAPEUTICS

Glycosylation plays an essential role in cancer malignancy expansion and development. Glycans are distinctively expressed on the surface of tumor cells, and these glycans have been targeted for the discovery and development of anticancer therapeutic, diagnostic, and preventive treatments, including vaccines magnetic nanoparticles and monoclonal antibodies. Synthetic oligosaccharides have been used in the development of therapeutic vaccines for cancer. Peptide and proteins or nucleic acid-based therapeutics are the most promising anticancer biopharmaceuticals.[17,18]

Various clinical studies and attempts of vaccination with carbohydrate-based therapeutics have been proven unsuccessful, mostly due to the glycan-associated immune tolerance. However, novel monoclonal antibodies are specially designed to identify cancer-bound carbohydrates and provoke tumor cell killing. Several carbohydrate-based magnetic nanocomposites and nanoparticles have been developed for safe and effective treatment and diagnosis of cancer. These carbohydrate-based magnetic nanoparticles show various advantages such as enhanced stability, higher biocompatibility, and lower toxicity. In various research studies, TACAs, primarily expressed on cancer tissues, have been found by National Institutes of Health as significant biomarkers of cancer prognosis and for therapeutics design. Several preclinical and clinical studies have demonstrated that naturally acquired, actively induced, or passively administered antibodies against TACAs are competent to remove circulating tumor cells and micro-metastases in cancer patients.[18]

TACA alone is poorly immunogenic to induce a T-cell dependent immune response, which is very essential for cancer therapy. TACAs can be ideal candidates to bring out an antitumor immune response; TACAs were used as conjugate with T-cell stimulating protein carriers, like KLH, tetanus toxoid (TT), bovine serum albumin (BSA), and diphtheria toxin (CRM197). TACA were coupled with various polysaccharides like Toll-like receptor 2 (TLR2) ligand, zwitterionic polysaccharide (PS A1), and T-cell peptide epitopes for the development of fully synthetic, self-adjuvating, multi-component vaccines for cancer.[19]

Cancer cells contain a glucose transporters (GLUTs) and lectins on its membrane surface, and these can transport and bind carbohydrate moieties, respectively. Increased energy in proliferation of cancer cells is satisfied by GLUTs, and these allow for an increased uptake of glucose at a superior rate than normal cell. This occurrence is commonly referred to as the Warburg effect. Thus sugar-based targeted drug delivery have been designed and developed. Several cytotoxic drugs, including glufosfamide, chlorambucil, busulfan, docetaxel, and paclitaxel, have been glycoconjugated and found to be less toxic to normal cells than the parent aglycons.

Some plasma tumor markers, including prostate-specific antigen (PSA), cancer antigen 125 (CA125), and alpha-fetoprotein (AFP), have been clinically used in United States for diagnosis of cancer in early stage. Recently, carbohydrate-based noninvasive diagnosis cancer tools, like lectin binding, metabolic oligosaccharide engineering (MOE) imaging technology, and glycan microarrays, have been used for tumor screening.[20]

Mifamurtide is synthetic immunomodulator used in the treatment for osteosarcoma, a type of bone cancer. It is derivative of muramyl dipeptide (MDP). It is naturally occurring immune stimulatory component of cell walls found in *Mycobacterium* species (Figure 6.3).[21]

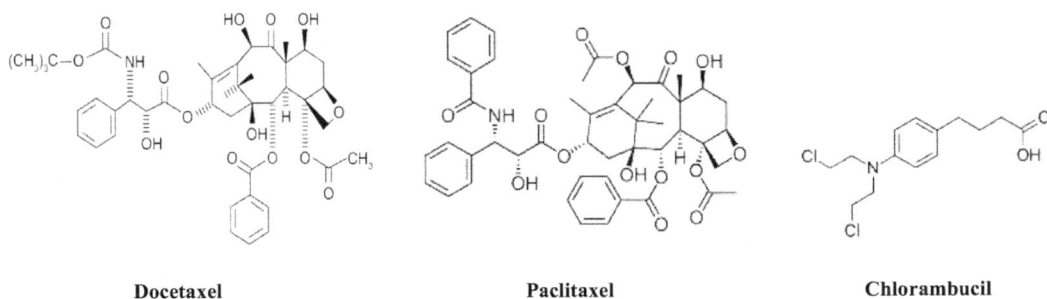

Docetaxel   Paclitaxel   Chlorambucil

FIGURE 6.3 Structure of anticancerous drugs.

6.3.5 CARBOHYDRATE-BASED DIAGNOSIS

In many diseases, there is change in the expression of oligosaccharides and glycoproteins. Glycan-based diagnostic have been developed for various human diseases. Diabetes type I is recognized at late stage, after getting autoimmune damage of pancreatic cells. Magnetic resonance imaging can be used to detect diabetes type I in early stage. Glycans are involved in various cellular events with significant implications for therapeutic purposes. The glycan-based plasmonic sensor were developed for prostate cancer diagnosis. Lamarre et al. have developed antibodies against the tumor-associated carbohydrate antigen (Tn), or TACA, that develop early in carcinogenesis, making them a remarkable alternative as a target for prostate cancer diagnostics.[22] Carbohydrates are crucial biomolecules that can act as markers in the recognition of immune systems. Carbohydrates exhibit antimicrobial, antimicrobial, and antifungal properties. Various serum glycoprotein biomarkers, including carcino-embryonic antigen (CEA), carbohydrates antigen 19-9 (CA19-9) and 125 (CA125), alpha-fetoprotein (AFP), and PSA, have been discovered to be useful in the preliminary detection and diagnosis of colon, ovarian, and prostate cancers, respectively.[23]

Specific lectins have been screened as potential carbohydrate tumor biomarkers in the diagnosis of cancer. Various lectin proteins, including *Artocarpus integrifolia* agglutinin (AIA), *Amaranthus caudatus* agglutinin (ACA), *Arachis hypogea* agglutinin (AHA), *Vicia villosa* lectin (VVL), *Artocarpus integrifolia* agglutinin (AIA), *Grionia simplicifolia* agglutinin I (GSA I), and *Ulex europaeus* agglutinin I (UEA I), are able to recognize Tn, TF, and STn alteration of CA125 antigens. Glycan microarray technology can detect antibodies against specific antigens like GOBO-H in patients' blood serum, and this provides glycan protein interaction.

Positron emission tomography (PET) scan can also be used for early detection of cancer, as it depends on enhanced concentration of 2-fluoro deoxy glucose concentration in tumor cells. Nowadays, metabolic oligosaccharide engineering technology (MOE) have been used for the diagnosis of cancer.[24]

Aridol (mannitol inhalation powder) is a sugar alcohol indicated for diagnosing asthma to assess bronchial hyper responsiveness.[25] Regadenoson is an adenosine analogue and a coronary vasodilator used in radionuclide myocardial perfusion imaging (MPI) as a diagnostic agent.[26]

6.3.6 CARBOHYDRATE-BASED ANTIVIRAL DRUGS

Viral pathogens and diseases are one of the most life threatening diseases among human history. Around 60% of pandemic diseases are viral diseases, including pathogenic avian influenza (HPAI) A (H5N1) in 2009–2010, Ebola virus in West Africa in 2014–2016, recently happening coronavirus disease 2019 (COVID-19), etc. Acquired immune deficiency syndrome (AIDS) is one of the serious viral diseases caused by human immunodeficiency virus. Hepatitis liver infection is caused by the hepatitis B virus (HBV). Carbohydrate-based drugs give unique advantages in the antiviral drug development. Nucleotides and nucleosides affect and disrupt the replication of viral DNA and RNA. Idoxuridine, vidarabine, ribavirin, entecavir, telbivudine, clevudine, and sofosbuvir are examples of nucleotide or nucleoside antiviral drugs. Entecavir is carbocyclic nucleoside-based antiviral drug developed by Bristol-Myers Squibb (BMS) pharmaceuticals. It is primarily used in the treatment of HBV and HIV. Entecavir is deoxyguanosine triphosphate analogue nucleoside, and it inhibits reverse transcription, HBV-DNA polymerase replication, and transcription in the viral replication process. Vidarabine, a nucleoside analog, interferes in the synthesis of viral DNA. Telbivudine is l-nucleoside analogue of thymidine developed by Idenix and Novartis. It is used in treatment of chronic HBV. It interacts with viral polymerase and inhibits viral replication and results in termination of DNA synthesis.[27,28]

Ribavirin interferes on the synthesis of viral mRNA, and it acts on DNA and RNA viruses. It is the only approved nucleoside analogue for the treatment of HCV. Remedesivir, Azvudine, Molnupiravir, Carragelose, Peramivir, and Maribavir are the examples of carbohydrate-based antiviral drugs.

Remdesivir is an antiviral drug that is a nucleoside analogue developed for the treatment of ebola virus. Molnupiravir was found helpful in reducing nasopharyngeal COVID-19 virus and viral RNA, with high safety and tolerability.

Zanamivir is the first neuraminidase inhibitor developed by the Australian biotech firm Biota Holdings. It is used in the treatment of influenza A and B (Figure 6.4).[26–28]

6.3.7 CARBOHYDRATE-BASED ANTIDIABETIC AGENTS

Diabetes is metabolic disorder affecting many people across the world. DM Type I is commonly treated with insulin replacement therapy, while type II DM is treated with oral hypoglycemic drugs. Diabetes mellitus type II is non-insulin-dependent diabetes. If the diabetes is not cured properly, it may lead to severe complications.[1,2] Diabetes mellitus is a most common chronic metabolic disorder characterized by hyperglycemia that results from insulin resistance, or no insulin secretion, or both. SGLT II inhibitors reduce reabsorption of glucose, and thereby increase urinary glucose excretion by approximately 60–80 g per day and improve hyperglycemia. SGLT II inhibitors are the newest class of drugs for type 2 diabetes. These reduce the reabsorption of renal-filtered glucose back into the bloodstream, thereby leading to increased glucose in the urine. SGLT II inhibitors and carbohydrate-based glycosidase inhibitors are major class of antidiabetic. Carbohydrate-based antidiabetic drugs can be developed by natural or synthetic process.[29,30]

6.3.7.1 Carbohydrate-Based Glycosidase Inhibitors

Glycosidase plays an important role in oligosaccharides biosynthesis in cell organelles. Glycosidase enzymes are responsible for hydrolysis of glycoside linkages in the digestion of carbohydrates and polysaccharides. α-Glycosidase inhibitors lower blood glucose levels by preventing hydrolysis of polysaccharide, delaying carbohydrate metabolism. Antidiabetic drugs for the treatment and management type II diabetes can be developed using this glycosidase inhibitors mechanism. Carbohydrates from diet get converted into simple sugars or monosaccharides by α-glycosidase enzyme at intestinal cell linings, and they get absorbed through intestine. α-Amylase and α-glucosidase enzymes are accountable for carbohydrates metabolism.[29–32]

Thus alpha glycosidase inhibitors prevent carbohydrate digestion and thus reduce blood glucose level. α-Glycosidase inhibitors are classified as irreversible glucosidase inhibitors, reversible α-glucosidase inhibitors, and sucrose inhibitors. Acarbose, emiglitate, voglibose, and miglitol are reversible α-glucosidase inhibitors of antidiabetic drugs.[31,32]

6.3.7.2 Carbohydrate-Based Sodium-Glucose Co-Transporter-II (SGLT II) Inhibitors

In the kidney, approximately 180 g of glucose is excreted per day in primitive urine through the glomerular filtration. Most of the glucose present in the primitive urine is reabsorbed by SGLTII and SGLT I present in the proximal tubule, and usually glucose is not excreted in the urine. SGLTII is responsible

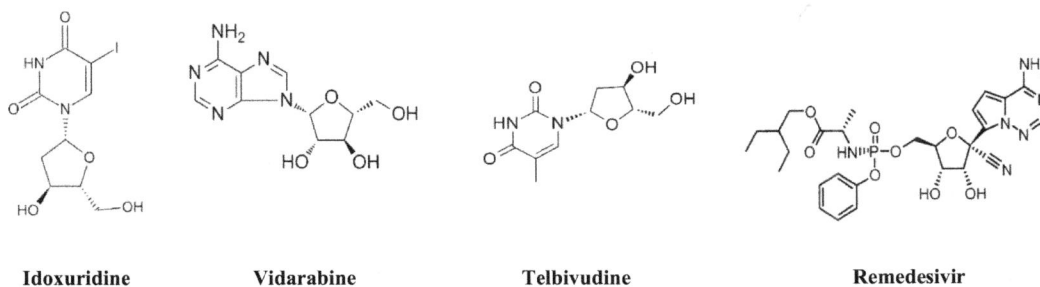

| Idoxuridine | Vidarabine | Telbivudine | Remedesivir |

FIGURE 6.4 Structure of antiviral drugs.

for the 90% of glucose reabsorbtion. SGLT II inhibitors reduce renal tubular reabsorbtion of glucose and thereby increases urinary glucose excretion by approximately 60–80 g per day and controls hyperglycemia. SGLT II inhibitors are the newest class of drugs for type II diabetes. These reduce the reabsorbtion of renal-filtered glucose back into the bloodstream, thereby increases glucose in the urine.[32]

Recently, dapagliflozin, canagliflozin, empagliflozin, sotagliflozin, ipragliflozin, tofogliflozin, and luseogliflozin are the new class of SGLT II inhibitors.[33]

Phlorizin, remogliflozin, sergliflozin, and dapagliflozin are initially synthesized O-aryl glucopyranosides that act as SGLT II inhibitors. Dapagliflozin was developed by companies BMS and AstraZeneca. Empagliflozin was developed by Boehringer Ingelheim and Eli Lilly Company. Canagliflozin hemihydrate (CNZ) is the first gliflozin, a selective sodium glucose co-transporter (SGLT II) inhibitor approved for the management of type II diabetes under the brand name "Invokana" by Janssen pharmaceuticals (Figure 6.5).[32,33]

6.3.7.3 Other Carbohydrate-Based Antidiabetics

Naturally found, plant-derived flavonoids like rutin and quercetin have potential as hyperglycemic agents for the treatment of diabetes.

6.3.8 Carbohydrate-Based Central Nervous System Drugs

Carbohydrates have been proved promising drugs for the treatment of central nervous system (CNS) disorders. Carbohydrate-based nanoparticles have been used as noninvasive technique for drug delivery to the central nervous system. Sodium oligomannate (GV-971) is an oral oligosaccharide derived from marine algae used in the treatment of Alzheimer's disease (AD). It is developed by Shanghai Green Valley Pharmaceuticals. Sodium oligomannate is derived from marine brown algae β-D-(1,4)-polymannuranate containing 2–10 sugars. In clinical trial studies, it has been proved effective and safe in the treatment of AD. Sugammadex is neuromuscular reversal drug used as reversal of neuromuscular blocking agents (NBA). It is novel synthetic ϒ-cyclodextrin molecule developed by Merck Sharp & Dohme. It is used in reversal of neuromuscular blockade induced by rocuronium bromide and vecuronium bromide in general anesthesia during surgery.[34,35] Topiramate is monosaccharide D-fructose derivative and carbonic anhydrase inhibitor found naturally. It is used in the management of seizures in epilepsy and to prevent migraines.[36]

6.4 CARBOHYDRATES IN CARDIOVASCULAR DISEASES

Digoxin is one of the oldest cardiovascular drugs used for the management of atrial fibrillation and the symptoms of heart failure. Digoxin slows heart rate and improves prognosis in patient with heart failure. Acetyldigoxin is prodrug of digoxin that shows enhanced absorption, efficacy, and tolerance. Methylated derivative of digoxin, medigoxin, shows superior bioavailability.

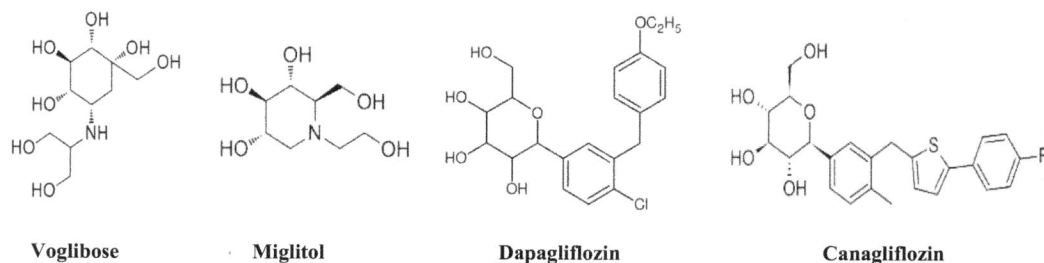

| Voglibose | Miglitol | Dapagliflozin | Canagliflozin |

FIGURE 6.5   Structure of inhibitors.

Ouabain is cardiac glycoside obtained from seeds of the *Acokanthera schimperi* and *Strophanthus gratus* plants. It is used in the treatment of congestive heart failure. Tribenoside is a glucofuranoside derivative, is an anti-inflammatory drug, and reduces vascular permeability and causes vasoconstriction to treat hemorrhoids.

Cangrelor is a nucleoside analogue and an intravenous, reversibly-binding platelet P2Y12 receptor antagonist. It shows antithrombotic activity and used in the treatment of myocardial infarction.[37]

Ticagrelor is antiplatelet agents and used in the prevention and treatment of thromboembolism in patients with acute coronary syndrome.[38]

Fondaparinux sodium is synthetic anticoagulant agent consisting of five monomeric sugar units and a *O*-methyl group. It is structurally similar to polymeric glycosaminoglycan heparin, and heparan sulfate is used as antithrombotic agent. Tinzaparin sodium is a low-molecular-weight heparin used in the treatment of deep-vein thromboembolism.[39]

6.5 CARBOHYDRATES FOR BIOMEDICAL APPLICATIONS

Carbohydrates are abundantly found in nature and have the affinity to form supramolecular network, thus polysaccharides or carbohydrate-based natural polymers have found promising applications in biomedical field. Carbohydrate polymers like starch, chitosan, cellulose, cyclodextrin, alginate, dextran, galactose, and derived polymers can be used for nanocomposites formation. Nanocomposites have been designed with nanoparticles (metal, metal oxide), nanocarbons (carbon nanotube, graphene, graphene oxide), metal, and inorganic nanofillers. It has several biomedical applications, including antibacterial, wound healing, inorganic nanofillers, tissue engineering, implants, biosensing, cosmetics, and drug delivery system.[40]

Chitosan and its derivatives like *N,N,N*-trimethyl-chitosan, *O*-carboxymethyl-*N,N,N*-trimethyl-chitosan, and *N,O*-carboxymethyl-chitosan have been used widely for wound healing in veterinary medicines. Chitosan-based nanocomposites exhibit antibacterial properties.[41] Polysaccharides drug conjugate have been used in anticancer drug delivery system. Carbon nanocomposites prepared with carbon nanotubes and biopolymers like starch, cellulose, chitosan have been used in drug delivery. Alginates in the form of hydrogel, nanogel and nanoparticles is used in anticancer drug delivery. Various natural polysaccharides, including alginates, cyclodextrin, chitosan, pullulan, hyaluronic acid, guar gum, pectin, dextran, and cellulose, have been used in anticancer drug delivery systems.[42]

Glycosaminoglycans like hyaluronic acid or hyaluronan are extensively found on the mammalian cell surface as well as in the extracellular matrix. Therefore, they are widely used in biomedical studies as drugs or carriers in drug delivery systems. Fluorescent hyaluronic acid-iodixanol nanogels were used in targeted X-ray computed tomography (CT) imaging and chemotherapy.[43] Hyaluronic acid-coated chitosan nanoparticles loaded with chemotherapeutic drug 5-fluorouracil were designed as tumor-targeting drug delivery system. These nanoparticles were found to enhance drug accumulation in tumor cells and significantly improve antitumor efficiency (Figure 6.6).[44]

6.6 CARBOHYDRATES IN GENE THERAPY

In gene therapy, genetic material (DNA, short interfering RNA-siRNA or antisense molecule) is introduced into body to replace defective or missing genetic material for the treatment of disease or an abnormal medical condition. Gene sequences introduced into cells for gene transfection in the form of plasmid DNA or oligonucleotides, that interfere with common gene expression either at the transcriptional level (triplex DNA), or at the translational level (antisense DNA or short-interfering RNA). Viral and nonviral vectors have been typically used for gene transfection. Biodegradable nanoparticles and other nanostructures can be used for nonviral gene delivery.

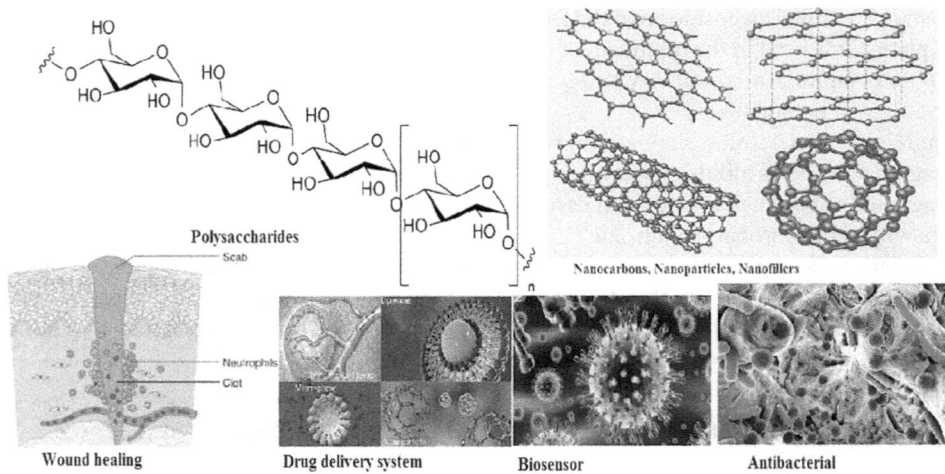

FIGURE 6.6 Carbohydrates for biomedical applications.

Various polysaccharides including chitosan, alginates, pullulan, hyaluronic acid, dextran, and their derivatives have been extensively used as polymeric backbones for the development of nanoparticles, which can be used as gene delivery carriers. Polysaccharides have been discovered in biomedicines as gene delivery carriers.[45,46] Nanoparticles, liposomes, hydrogels, matrix, and nanoparticles can be fabricated using cyclodextrins, pectins, guar gum for encapsulating oligonucleotides, DNA, and siRNA. Nanoparticle-based polysaccharides are the most important candidates of nonviral vectors for gene delivery systems.

Perfect gene delivery system requires an appropriate vector competent to protect and facilitate the genes to reach the target cells. Polysaccharide nanoparticles offer safe, efficient, and controllable methods for gene delivery.[46,47]

Chen et al. have formulated nanoparticles decorated with carbohydrate-based targeting polymers like mannose, galactose, dextran, galactose, etc. for efficient macrophage-targeted gene therapy. Study revealed that carbohydrate-based polymers showed strong entrapment of mRNA and pDNA with a higher encapsulation efficiency.[48]

6.7 OTHER CARBOHYDRATE-BASED THERAPEUTICS AND ADJUVANTS

Xyloglucan (XG) is a polysaccharide obtained from tamarind seeds. It is a natural hemicellulose that consists of main chain of glucan backbone with xylose and galactose side chains. Xyloglucan exhibits distinct mucoadhesive and gelling properties, and it can be used in drug delivery system.[49] Chitin and its derivative chitosan are antibacterial, biocompatible, degradable, and nontoxic biomaterials. Chitin and chitosan nanofibers are used for promoting bone regeneration and repairing defects. Chitosan has several biomedical applications, including bacterial inhibition, drug delivery, and wound healing.[50] Nano-carbohydrates have been used for biosensing, cell imaging, and screening in genetics, biotechnology, and medicines. Glycosylated gold nanoparticles, glycosylated quantum dots, and self-assembled glyconanoparticles can be used for imaging, therapeutics, and biodiagnostic devices.[51] Carragelose is sulfated polymer obtained from red seaweed (Chondrus crispus). It is an algae-based Iota-carrageenan type polysaccharide drug. Carragelose is used as antiviral drug for the treatment of common cold, influenza, and covid-19 disease[52] (Table 6.1, Figures 6.7 and 6.8).

TABLE 6.1

Other Carbohydrate-Based Therapeutics[26,49–58]

Drug	Descriptions	Use
Cangrelor	Nucleoside analog	Used in the treatment of cardiovascular diseases
Lactulose and lactitol	Synthetic disaccharide composed of galactose and fructose	Laxative agent for the treatment of chronic constipation
Sucralfate	Aluminum salt of sulfated sucrose	Used in the treatment of the GIT diseases, including duodenal ulcers, gastritis, peptic ulcer, stress ulcer, gastro-esophageal reflux disease, and dyspepsia
Teniposide	Derivatives of podophyllotoxin containing a carbohydrate unit	Used in the treatment of leukemia, lymphoma, small cell lung cancer
Miglustat	Iminosugar mimicking glucose	For the treatment of Gaucher's disease.
Sodium hyaluronate	It is sodium salt of hyaluronic acid	Antiarthritic agent
Topiramate	Sulfamate substitutes monosaccharide	For the treatment of epilepsy, Lennox–Gastaut syndrome, migraines
Auranofin	Carbohydrate-containing gold complex	Antirheumatic agent
Fucoidan	Sulfated polysaccharide obtained from brown algae (*Fucus vesiculosus*)	Anti-coagulation and Anti-thrombosis agent
Heparin sulfate	Sulfated linear polysaccharide, consisting of 1,4-glycosidic bonds between D-glucosamine (GlcN) and D-glucuronic acid (GlcA) or l-iduronic acid (IdoA) units	Used as an anticoagulant drug in treating cardiopulmonary bypass, extracorporeal membrane oxygenation, hemodialysis
Ticagrelor	Ticagrelor	Antiplatelet effect
Diquafosol tetrasodium	Second-generation uridine nucleotide	Used in dry eye disease
Paromomycin	Aminoglycoside antibiotic	Oral drug effective for the treatment of infections caused by intestinal protozoa and leishmaniasis
Fidaxomicin	Narrow spectrum macrocyclic Antibiotic	For the treatment of diarrhea associated with *Clostridium difficile* bacterial infection
Sialic acid	High carbon sugar with a complex carbon skeleton and neuraminidase inhibitors	For the treatment of influenza

6.8 CONCLUSION

Carbohydrates are essential biomolecules universally present and execute diverse biological roles in living system. Carbohydrates are integral part of living system and are involved in various biological process. Carbohydrates are structurally diverse molecules that have been extensively explored in the development and discovery of new drug.

Carbohydrates have been used as excipients, solubilizing agents in drug delivery systems. Carbohydrates have been used as biomarkers for diagnosis of various diseases like cancer, and they have huge biomedical applications. Carbohydrate-based vaccines have been found to be a promising approach for the treatment of cancer or infections caused by viruses, bacteria, fungi, protozoan parasites, and helminths. Natural and synthetic carbohydrate-based therapeutics have been significantly used for the treatment and management of diabetes, cancer, viral diseases, arthritis, cardiovascular diseases, microbial diseases, malaria, CNS-related diseases, etc. Carbohydrate-based therapeutics have been designed as nanomaterials for targeded and controlled drug delivery systems. Carbohydrate-based drugs and therapeutics are supreme candidates for current and future treatment to achieve wellness.

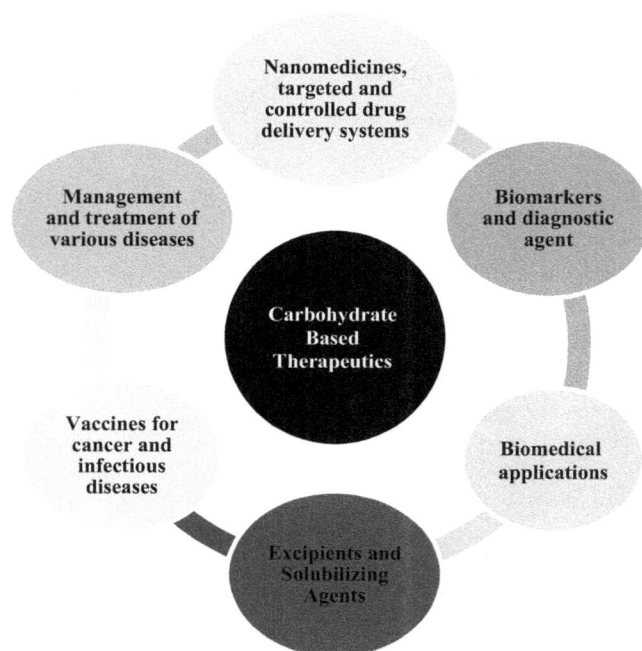

FIGURE 6.7   Applications of carbohydrate-based therapeutics.

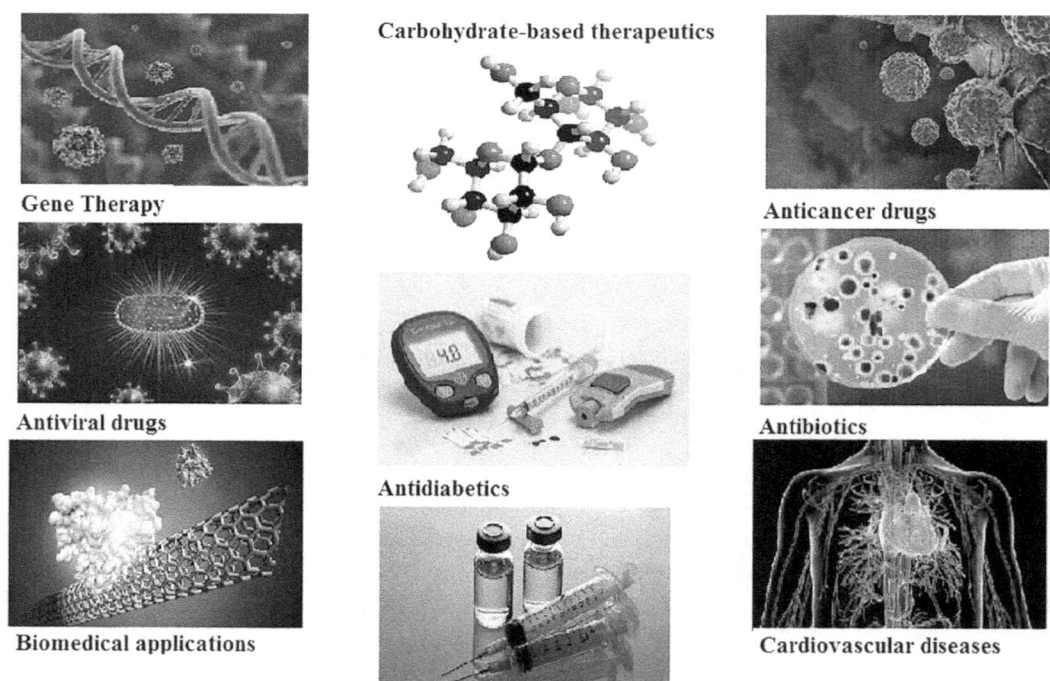

FIGURE 6.8   Carbohydrate-based therapeutics.

LIST OF ABBREVIATION

BSA	Bovine serum albumin
DNA	Deoxyribonucleic acid
GLUTs	Glucose transporters
KLH	Keyhole limpet hemocyanin
MOE	Metabolic oligosaccharide engineering technology
PSA	Prostate-specific antigen
TACAs	Tumor-associated carbohydrate antigens
TT	Tetanus toxoid

REFERENCES

1. Mishra S, Upadhaya K, Mishra KB, Shukla AK, Tripathi RP, Tiwari VK. Carbohydrate-Based Therapeutics: A Frontier in Drug Discovery and Development, Editor(s): Atta-ur-Rahman, *Studies in Natural Products Chemistry*, Elsevier, Volume 49, 2016, pp. 307–61, ISSN 1572-5995, ISBN 9780444636010.
2. Cipolla L, Araújo AC, Bini D, Gabrielli L, Russo L, Shaikh N. Discovery and design of carbohydrate-based therapeutics. *Expert Opin Drug Discov.* 2010 Aug;5(8):721–37.
3. Sunasee R, Adokoh CK, Darkwa J, Narain R. Therapeutic potential of carbohydrate-based polymeric and nanoparticle systems. *Expert Opin Drug Deliv.* 2014 Jun;11(6):867–84.
4. Fernández-Tejada A, Cañada FJ, Jiménez-Barbero J. Recent developments in synthetic carbohydrate-based diagnostics, vaccines, and therapeutics. *Chemistry.* 2015 Jul 20;21(30):10616–28. doi: 10.1002/chem.201500831.
5. Cummings JH, Stephen AM. Carbohydrate terminology and classification. *Eur J Clin Nutr.* 2007 Dec;61(Suppl 1):S5–18.
6. Stuetz AE, *Iminosugars as Glycosidase Inhibitors: Nojirimycin and Beyond*, vol. 14, Wiley-VCH Verlag GmbH, 1999, pp. 22–24.
7. Varki A. Sialic acids in human health and disease. *Trends Mol Med.* 2008 Aug;14(8):351–60. doi: 10.1016/j.molmed.2008.06.002.
8. Horne G, Wilson FX. Therapeutic applications of iminosugars: current perspectives and future opportunities. *Prog Med Chem.* 2011;50:135–76. doi: 10.1016/B978-0-12-381290-2.00004-5.
9. Tiwari V. K. (ed.) *Carbohydrate in Drug Discovery and Development Synthesis and Application*, Elsvier, 2020.
10. Brandish PE, Kimura KI, Inukai M, Southgate R, Lonsdale JT, Bugg TD, Antimicrob. *Agents Chemother.* 1996;40:1640–4.
11. Thomson JM, Lamont IL. Nucleoside analogues as antibacterial agents. *Front Microbiol.* 2019 May 22;10:952.
12. Miljkovic M. Carbohydrate-Based Antibiotics, In *Carbohydrates* (pp. 469–486), Springer, New York, NY, 2010. doi: 10.1007/978-0-387-92265-2-14
13. Mettu R, Chen CY, Wu CY. Synthetic carbohydrate-based vaccines: challenges and opportunities. *J Biomed Sci.* 2020 Jan 3;27(1):9.
14. Zhu J, Warren JD, Danishefsky SJ. Synthetic carbohydrate-based anticancer vaccines: the memorial Sloan-Kettering experience. *Expert Rev Vaccines.* 2009 Oct;8(10):1399–413.
15. Feng D, Shaikh AS, Wang F. Recent advance in tumor-associated carbohydrate antigens (TACAs)-based antitumor vaccines. *ACS Chem Biol.* 2016 Apr 15;11(4):850–63.
16. Hütter J, Lepenies B. Carbohydrate-Based Vaccines: An Overview, Editor(s) Lipenies B, *Methods in Molecular Biology*, Humana Press, New York, 2015, pp. 1–10, 1331.
17. Ranjbari J, Mokhtarzadeh A, Alibakhshi A, Tabarzad M, Hejazi M, Ramezani M. Anti-cancer drug delivery using carbohydrate-based polymers. *Curr Pharm Des.* 2018 Feb 12;23(39):6019–32.
18. Hossain F, Andreana PR. Developments in carbohydrate-based cancer therapeutics. *Pharmaceuticals (Basel).* 2019 Jun 4;12(2):84
19. Jin KT, Lan HR, Chen XY, Wang SB, Ying XJ, Lin Y, Mou XZ. Recent advances in carbohydrate-based cancer vaccines. *Biotechnol Lett.* 2019 Jul;41(6–7):641–50.
20. Shende P, Shah P. Carbohydrate-based magnetic nanocomposites for effective cancer treatment. *Int J Biol Macromol.* 2021 Apr 1;175:281–93.

21. Frampton JE. Mifamurtide: a review of its use in the treatment of osteosarcoma. *Paediatr Drugs*. 2010 Jun;12(3):141–53.
22. Lamarre M, Tremblay T, Bansept MA, Robitaille K, Fradet V, Giguère D, Boudreau D. A glycan-based plasmonic sensor for prostate cancer diagnosis. *Analyst*. 2021 Nov 8;146(22):6852–60. doi: 10.1039/d1an00789k.
23. Namikawa T, Kawanishi Y, Fujisawa K, Munekage E, Iwabu J, Munekage M, Maeda H, Kitagawa H, Kobayashi M, Hanazaki K. Serum carbohydrate antigen 125 is a significant prognostic marker in patients with unresectable advanced or recurrent gastric cancer. *Surg Today*. 2018;48:388–94.
24. Dube DH, Bertozzi CR. Metabolic oligosaccharide engineering as a tool for glycobiology. *Curr Opin Chem Biol*. 2003;7:616–25
25. Fernández-Tejada A., Javier Cañada F, Jiménez-Barbero J. Recent developments in synthetic carbohydrate-based diagnostics, vaccines, and therapeutics. *Chemistry*. 2015;21(30):10616–28.
26. Xin C. et al., Carbohydrate-based drugs launched during 2000–2021. *Acta Pharm Sin B*. doi: 10.1016/j.apsb.2022.05.020.
27. Osborn HM, Evans PG, Gemmell N, Osborne SD. Carbohydrate-based therapeutics. *J Pharm Pharmacol*. 2004 Jun;56(6):691–702. doi: 10.1211/0022357023619.
28. Bertozzi CR, Kiessling LL. Chemical glycobiology. *Science*. 2001 Mar 23;291(5512):2357–64.
29. Shyangdan DS, Uthman OA, Waugh N. SGLT-2 receptor inhibitors for treating patients with type 2 diabetes mellitus: a systematic review and network meta-analysis. *BMJ Open*. 2016;6(2):e009417.
30. Seufert J, Laubner K. Outcome-Studien zu SGLT-2-Inhibitoren. *Internist (Berl)*. 2019;60(9):903–911.
31. Saisho Y. SGLT2 inhibitors: the star in the treatment of type 2 diabetes? *Diseases*. 2020;8(2):14.
32. Rang H, Dale M, Ritter J, Moore P. *Pharmacology* (9th edition), Churchill Livingstone, 2003, pp. 408–19.
33. Tiwari S, Wadher S, Fartade S, Vikhar C. Gliflozin a new class for type-II diabetes mellitus: an overview. *IJPSR*. 2019;10(9), 4070–7.
34. Syed YY. Sodium oligomannate: first approval. *Drugs*. 2020 Mar;80(4):441–4. doi: 10.1007/s40265-020-01268-1. Erratum in: *Drugs*. 2020 Feb 24;
35. Nag K, Singh DR, Shetti AN, Kumar H, Sivashanmugam T, Parthasarathy S. Sugammadex: a revolutionary drug in neuromuscular pharmacology. *Anesth Essays Res*. 2013 Sep–Dec;7(3):302–6.
36. Shank RP, Gardocki JF, Streeter AJ, Maryanoff BE. An overview of the preclinical aspects of topiramate: pharmacology, pharmacokinetics, and mechanism of action. *Epilepsia*. 2000; 41:3–9.
37. Jiang H, Qin X, Wang Q, Xu Q, Wang J, Wu Y, Chen W, Wang C, Zhang T, Xing D, Zhang R. Application of carbohydrates in approved small molecule drugs: a review. *Eur J Med Chem*. 2021 Nov 5;223:113633.
38. Kabil MF, Abo Dena AS, El-Sherbiny IM. Ticagrelor. *Profiles Drug Subst Excip Relat Methodol*. 2022;47:91–111.
39. Neely JL, Carlson SS, Lenhart SE. Tinzaparin sodium: a low-molecular-weight heparin. *Am J Health Syst Pharm*. 2002 Aug 1;59(15):1426–36.
40. Gim S, Zhu Y, Seeberger PH, Delbianco M. Carbohydrate-based nanomaterials for biomedical applications. *Wiley Interdiscip Rev Nanomed Nanobiotechnol*. 2019 Sep;11(5):e1558. doi: 10.1002/wnan.1558.
41. Seidi F, Khodadadi Yazdi M, Jouyandeh M, Dominic M, Naeim H, Nezhad MN, Bagheri B, Habibzadeh S, Zarrintaj P, Saeb MR, Mozafari M. Chitosan-based blends for biomedical applications. *Int J Biol Macromol*. 2021 Jul 31;183:1818–50.
42. Yadav N, Francis AP, Priya VV, Patil S, Mustaq S, Khan SS, Alzahrani KJ, Banjer HJ, Mohan SK, Mony U, Rajagopalan R. Polysaccharide-drug conjugates: a tool for enhanced cancer therapy. *Polymers (Basel)*. 2022 Feb 27;14(5):950.
43. Zhu Y, Wang X, Chen J, Zhang J, Meng F, Deng C, Cheng R, Feijen J, Zhong Z. Bioresponsive and fluorescent hyaluronic acid-iodixanol nanogels for targeted X-ray computed tomography imaging and chemotherapy of breast tumors. *J Control Release*. 2016 Dec 28;244(Pt B):229–39.
44. Wang T, Hou J, Su C, Zhao L, Shi Y. Hyaluronic acid-coated chitosan nanoparticles induce ROS-mediated tumor cell apoptosis and enhance antitumor efficiency by targeted drug delivery via CD44. *J Nanobiotechnol*. 2017 Jan 10;15(1):7.
45. Wasupalli GK, Verma D. 3-Polysaccharides as Biomaterials, Editor(s): Thomas S, Balakrishnan P, Sreekala MS, In *Woodhead Publishing Series in Biomaterials, Fundamental Biomaterials: Polymers*, Woodhead Publishing, 2018, pp. 37–70.
46. Huh MS, Lee EJ, Koo H, Yhee JY, Oh KS, Son S, Lee S, Kim SH, Kwon IC, Kim K. Polysaccharide-based nanoparticles for gene delivery. *Top Curr Chem (Cham)*. 2017 Apr;375(2):31.

47. Serrano-Sevilla I, Artiga Á, Mitchell SG, De Matteis L, de la Fuente JM. Natural polysaccharides for siRNA delivery: nanocarriers based on chitosan, hyaluronic acid, and their derivatives. *Molecules*. 2019 Jul 15;24(14):2570.

48. Chen Q, Gao M, Li Z, Xiao Y, Bai X, Boakye-Yiadom KO, Xu X, Zhang XQ. Biodegradable nanoparticles decorated with different carbohydrates for efficient macrophage-targeted gene therapy. *J Control Release*. 2020 Jul 10;323:179–90.

49. Pardeshi CV, Kulkarni AD, Belgamwar VS, Surana SJ.7-Xyloglucan for Drug Delivery Applications, Editor(s): Thomas S, Balakrishnan P, Sreekala MS, In *Woodhead Publishing Series in Biomaterials, Fundamental Biomaterials: Polymers*, Woodhead Publishing, 2018, pp. 143–69.

50. Tao F, Cheng Y, Shi X, Zheng H, Du Y, Xiang W, Deng H. Applications of chitin and chitosan nanofibers in bone regenerative engineering. *Carbohydr Polym*. 2020 Feb 15;230:115658.

51. Jebali A, Nayeri EK, Roohana S, Aghaei S, Ghaffari M, Daliri K, Fuente G. Nano-carbohydrates: synthesis and application in genetics, biotechnology, and medicine. *Adv Colloid Interface Sci*. 2017 Feb;240:1–14.

52. Figueroa JM, Lombardo ME, Dogliotti A, Flynn LP, Giugliano R, Simonelli G, Valentini R, Ramos A, Romano P, Marcote M, Michelini A, Salvado A, Sykora E, Kniz C, Kobelinsky M, Salzberg DM, Jerusalinsky D, Uchitel O. Efficacy of a nasal spray containing iota-carrageenan in the postexposure prophylaxis of COVID-19 in hospital personnel dedicated to patients care with COVID-19 disease. *Int J Gen Med*. 2021 Oct 1;14:6277–86.

53. Cao X, Du X, Jiao H, An Q, Chen R, Fang P, Wang J, Yu B. Carbohydrate-based drugs launched during 2000−2021, *Acta Pharm Sin B*. doi: 10.1016/j.apsb.2022.05.020.

54. Alexopoulos D, Pappas C, Sfantou D, Lekakis J. Cangrelor in percutaneous coronary intervention: current status and perspectives. *J Cardiovasc Pharmacol Ther*. 2018 Jan;23(1):13–22.

55. Candelli M, Carloni E, Armuzzi A, Cammarota G, Ojetti V, Pignataro G, Santoliquido A, Pola R, Pola E, Gasbarrini G, Gasbarrini A. Role of sucralfate in gastrointestinal diseases, *Panminerva Med*. 2000;42:55–9.

56. Keating GM. Cangrelor: a review in percutaneous coronary intervention. *Drugs*. 2015 Aug;75(12):1425–34.

57. Jiang H, Qin X, Wang Q, Xu Q, Wang J, Wu Y, Chen W, Wang C, Zhang T, Xing D, Zhang R. Application of carbohydrates in approved small molecule drugs: a review. *Eur J Med Chem*. 2021 Nov 5;223:113633.

58. Tiwari VK, Mishra RC, Sharma A, Tripathi RP. Carbohydrate based potential chemotherapeutic agents: recent developments and their scope in future drug discovery. *Mini Rev Med Chem*. 2012;12:1497–519.

7 Starch-Based Advanced Materials and Their Applications

Aiswarya P. R. and Sabu Thomas
Mahatma Gandhi University

7.1 INTRODUCTION

Plastic is becoming one of the most dangerous environmental issues, as the production of a new variety of plastic products is rapidly increasing day by day. The low degradable behavior of plastic is the main reason that makes plastic a harmful material. Therefore, waste plastic disposal became a threat to nature and living beings. There is a strong urge to replace conventional non-degradable plastic materials with other biopolymer materials, which are easy to degrade. The biopolymer industry is advancing since biodegradable plastic is a good solution to environmental problems. As biopolymers are also able to offer better properties similar to conventional plastic materials, the demand is increasing drastically. Natural polymer materials like starch, cellulose, collagen, lignin, etc. are tremendously used for commercial uses. Among them, starch is a promising material for a lot of applications due to its renewability, availability, biodegradability, and low cost. In addition, starch and derivatives of starch provide varieties of properties favorable for advanced material applications [1]. Desired properties of materials are obtained by starch modification or coupling starch with other polymeric materials. The derivatives of starch include starch processed by different modifications like physical, chemical, or mechanical modifications, starch-based blends and composites, and so on [2]. But the degradability of the materials should not be lost because of such modification processes and also the additives must not be toxic because there are a lot of edible applications such as food packaging and coatings. Because of this reason, natural edible fillers and reinforcing agents like natural fibers, starch, cellulose crystals, etc. are to be used in starch-based materials. The self-reinforcement technique, that is, reinforcing starch with modified starch particles, is also widely used by several industries [3]. The main fields where the use of starch and starch-based materials was established were electronics, drug delivery, pharmaceuticals, structural materials, packaging, etc. [2]

7.2 WHY STARCH IS A PROMISING MATERIAL FOR ADVANCED APPLICATIONS

Starch is produced naturally by the polymerization of glucose during photosynthesis in plants. It is a source of energy and is mainly used as food. It has a lot of industrial applications due to its different desirable properties. It is considered a favorable material for various purposes because of the basic advantages of a natural biodegradable material [2]. Also, it has various unique properties such as water retention, water dissolution, gelatinization, ease of modification, etc. that make it a promising material in the industrial field [1]. On the other hand, some undesirable properties, which reduce the extensive use of starch for industrial applications, are the hydrophilic nature, brittleness, retrogradation, and thermal degradation. Therefore, the functional −OH group of starch is modified to mitigate that limitations [2]. The basic structure of starch is shown in Figure 7.1.

DOI: 10.1201/9781003265054-7

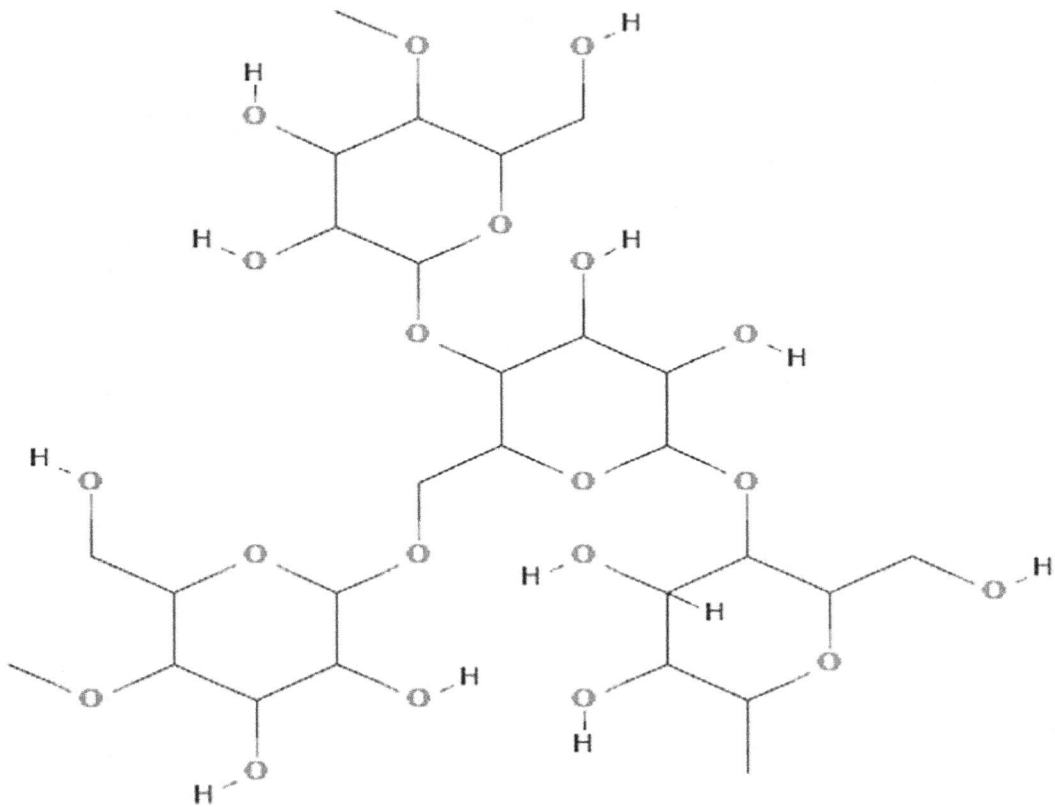

FIGURE 7.1 Basic structure of the starch molecule. By Palanisamy et al. [4] is licensed under CC, reprinted from Polymers, MDPI.

The structure and composition of starch are responsible for its physical and chemical properties. Starch is made up of amylose, a linear glucose chain that has an α-1,4-glucosidase bond, and amylopectin, a branched glucose chain that has branching at the α-1,6-position. The ratio of amylose and amylopectin varies in different botanical sources of starch. It is differed mainly due to climate conditions, location of cultivation, soil properties, etc. Another component present in starch is phosphorous, a non-carbohydrate component that affects its gel strength, lucidity, and solubility depending on the foreign material with which it bonds [2].

7.3 STARCH-BASED ADVANCED MATERIALS

The hydrophilic nature of starch, brittleness, retrogradation, and thermal degradation has limited its extensive use for industrial polymer applications that require mechanical integrity. Extensive use of starch is limited to industrial applications which require mechanical integrity. Hydrophilicity, brittleness, retrogradation, thermal degradation, etc. are the properties that hinder its mechanical strength. It has limited mechanical integrity since it will degrade under relatively low temperatures and cannot be melt-processed. These poor mechanical strengths, thermal stability, and poor solubility of starch limited its applications in polymer-based industries. To overcome this crisis, starch is subjected to several modifications. Several kinds of research are still being conducted for starch modifications. Many studies have relieved that conducting modifications on starch not only improves the mechanical properties but also helps to achieve various other functional properties. Figure 7.2 shows the evolution of starch-based materials.

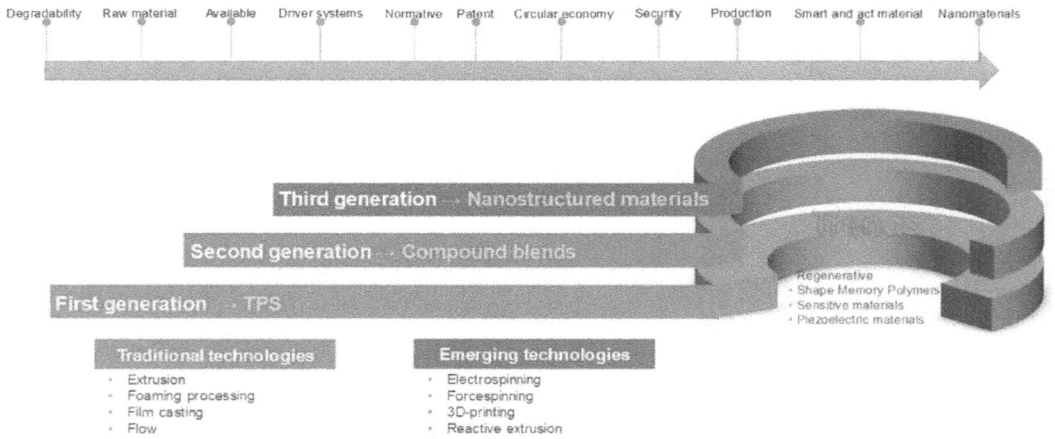

Degradability   Raw material   Available   Driver systems   Normative   Patent   Circular economy   Security   Production   Smart and act material   Nanomaterials

Third generation — Nanostructured materials

Second generation — Compound blends

First generation — TPS

Regenerative
· Shape Memory Polymers
· Sensitive materials
· Piezoelectric materials

Traditional technologies
· Extrusion
· Foaming processing
· Film casting
· Flow

Emerging technologies
· Electrospinning
· Forcespinning
· 3D-printing
· Reactive extrusion

FIGURE 7.2   Evolution of starch-based materials. By García-Guzmán et al. [5] is licensed by CC, reprinted from Polysaccharides, MDPI.

7.3.1   MODIFICATION OF NATIVE STARCH

7.3.1.1   Thermo-Plasticization

Starch is not a thermoplastic material; therefore, it should be converted into a thermoplastic so that it becomes an ideal material for commercial applications. Therefore, starch is converted into thermoplastic with or without shear stress in the presence of plasticizers like water or polyols in the heat range 90°C–180°C. Gelatinization of starch is the reason for this conversion. This is an irreversible process by which the starch granules lose their crystalline nature. In the process, the amylopectin double helix unwinds and amylose gets separated from amylopectin. The crystalline order is lost mainly because of the penetration of water into the granules. The poor mechanical and higher hydrophilic nature of thermoplastic starch make it unsuitable for applications. To overcome these drawbacks, several modifications are conducted. Incorporating reinforcements, chemical modifications, and blending with other polymers are some of those methods to improve the properties of starch-based materials [5].

7.3.1.2   Chemical Modification of Native Starch

Chemical modification of starch is mainly carried out by incorporating functional groups through the process of reduction, substitution, or crosslinking [5]. In the substitution method, the hydroxyl groups present in the starch chain are changed into acetyl group or similar other groups as per the requirement and thus altering the molecular structure of starch. Several researchers reported that esterification of starch with organic acids and acid anhydrides improves several properties of starch and is used for packaging application [6]. Starch esterification is a nucleophilic substitution process. Commonly used organic acids for starch esterification are formic acid, acetic acid, maleic acid, propanoic acid, and butanoic acid [5]. Acylated starch prevents retrogradation. Gelatinized starch on cooling the dissociated amylose chains show a tendency to realign. This reassociating tendency is called retrogradation. Thus, during low-temperature storage, trapped water is appeared at the surface of the material and may cause other damage. Succinate starches provide properties like low gelatinization temperature, the ability to swell in cold water to increase viscosity, and good filming properties [7].

Modification of starch by reduction consists of producing more functional groups hence reactivity increases. Starch modified by reduction or oxidation showed reduced swelling capacity and solubility, and showed increased pasting viscosity [8]. By oxidation, the pyranosidic ring structure is opened, and reactive functional groups are formed in the framework. Commonly oxidizing agents used for this technique are hydrogen peroxide, oxygen, ozone, bromine, chromic acid, permanganate, nitrogen dioxide, etc. [7]. Crosslinking is a method of covalently interconnecting linear or branched chains, resulting in improved hydrophobicity and structural stability of starch granules. Citric acid is an inexpensive and nontoxic crosslinking agent, and it reacts with the hydroxyl group in starch and produces intermolecular covalent bonds [9]. Other common crosslinking agents are sodium trimetaphosphate (STMP), sodium tripolyphosphate (STPP), epichlorohydrin (ECH), and phosphorus chloride ($POCl_3$) [8].

7.3.1.3 Physical Modification of Native Starch

Starch is physically modified by different treatments under the action of heat, pressure, or mechanics, in the dry or wet state, including fractionation. Physical modifications by nonthermal treatment are a highly demanding method since the amount of energy required for the process is very less, and it is environment friendly. And the enhancement of material properties by these modification techniques is very satisfying. It mainly helps to change physical properties, and in some cases, it is reported that chemical properties can also be changed. Obtaining chemical changes in starch without using a chemical modification technique is an interesting result. In a work on waxy wheat starch, physical properties like gelatinization and crystallinity are influenced by hydrostatic pressure treatments [10]. Research suggested that plasma treatments of starch affected the crosslinking of starch molecules and altered the melting temperature. On treating starch films with hexamethyldisiloxane (HMDSO) cold plasma, it is reported that chemical changes including substitution and crosslinking have occurred. C–OH bonds present in the starch were reduced, and there was an increase C–Si bonds, which is a hydrophobic blocking group and resulted in increased barrier property of the film. These changes resulted in thermal transitions like melting temperature (Tm) and enthalpy (ΔH) of the material. The overall result suggested that high crosslinking and hydrophobic blocking group (C–Si) improved the water repellency of the film and enhanced the performance [11]. In pharmaceutical industries, pre-gelatinized starch is commonly used to improve its flowability and enhance the disintegration and hardness of the starch. Therefore, the amount of pre-gelatinized starch used for the production of tablets will be much less than unmodified starch [7]. Starches are mainly used as excipients in drug production.

7.3.2 STARCH-BASED MATERIALS WITH FILLER/REINFORCEMENT

Lignocellulosic fibers as reinforcement are mainly studied to improve TPS properties. Different types of organic lignocellulosic fibers are being subjected to this study. Some examples include green coconut, hemp, cellulose fibers, sisal, nanofibers, and microfibers from recycled paper, polysaccharide-based nanocrystals, and microcrystalline cellulose [7]. Adding lentil flour fiber in thermoplastic starch films showed an improved resistant character which is favorable property to use as the best packaging material [12]. Similarly, starch-based form composite material is prepared using different types of cellulosic fibers such as cellulose, wood pulp fiber, and municipal solid waste fiber, which help in enhancing the tensile strength without affecting the foam density of the material [13]. Starch-based films filled with cellulose fibers also contributed to an enhancement in resistance. This is because of the decrease of starch chain mobility caused by the starch-cellulose interaction [14]. Most of the studies concluded that the addition of 2.5%–3.0% of fibers improved the mechanical resistance of films [3] because of the formation of the nonhomogeneous structure of films, resulting in brittle composites. The differences in processing method and length of fibers added to films have an important effect on the mechanical properties. The nature and characteristics of fibers vary with the difference in processing techniques. Soy-based biocomposite kenaf fiber is prepared with extrusion injection molding, compression molding, and injection molding. In this study, it is found that the compression molding process is beneficial to both the thermal and mechanical properties of the composites [15].

7.3.3 Starch Blends with Biodegradable Polymers

Nature is in high demand of limiting petroleum-based materials. Therefore, it is relevant to find a substitute at least for the short term using materials for applications like packaging and disposal. Even though starch is a degradable natural material, it is limited to use commercially because of its water solubility. Several researchers reported that native starch can be converted into thermoplastic starch by plasticization and restructuring. But plasticized starch also shows an unfavorable hydrophilic nature. As a solution for this, plasticized starch is blended with another biodegradable polymer material and desired properties can be acquired [3]. The blending of polymer is very easy and cheap than synthesizing a new macropolymer. But the drawback is the non-miscibility of two different polymers due to their unique chemical structures [16]. Starch blending with polymers like polycaprolactone (PCL), polylactic acid (PLA), polyvinyl alcohol (PVA), poly(butylene adipate-co-terephthalate) (PBAT), chitosan, etc. is reported in several articles.

7.3.3.1 Polycaprolactone (PCL)

Polycaprolactone is a biodegradable aliphatic polyester. Comparatively, PCL has a low melting point (60°C) and thermal transition temperature around −60°C [7]. Researchers worked on PCL as a good biodegradable material. But blending thermoplastic starch (TPS) with PCL did not contribute to any significant changes in tensile strength and elongation of break. Reports suggested that the mechanical properties of PCL/TPS blends depend on the type of starch used [17]. Using gelatinized starch for blending showed some favorable properties in blends like increased fluidity, increased absorption of water, reduced crystallinity of PCL, and maintained good dispersion of blend. In some cases, the mechanical and thermal properties of PCL became worse when blended with starch, mainly due to poor dispersion of the two phases [7].

7.3.3.2 Poly(Butylene Adipate-Co-Terephthalate) (PBAT)

PBAT is an aromatic-aliphatic copolyester and biodegradable. The major properties of PBAT are similar to polyethylene. But the usage of PBAT is reduced due to its high cost. PBAT/TPS blends are studied by using Nisin and EDTA as compatibilizers for active packaging applications and reported that the nisin improved the compatibility and smoothness of films, while EDTA increased the hydrophilicity and oxygen barrier of films [5]. Another work reported that PBAT/TPS film containing 6% sodium benzoate and potassium sorbate extended the shelf-life of packaged food by delaying microbial growth [17]. Another work using maleate PBAT and citric acid on PBAT/TPS films improved strain at break.

7.3.3.3 Poly(Lactic Acid) (PLA)

Polylactic acid is an aliphatic polyester suitable to use as packaging material, and it is categorized as a Generally Recognized as Safe (GRAS) by the Food and Drug Administration (FDA) [7]. PLA is comparatively a high-cost degradable material, and starch is considered a suitable material to blend with PLA to reduce the cost and enhance biodegradation properties. PLA is hydrophobic and TPS is hydrophilic, hence the compatibility of the material may be less due to the limited barrier and mechanical properties [18]. Recently, several studies have been conducted to improve the compatibility of PLA and TPS [19]. PLA has a wide range of physical properties. Poor mechanical and thermal properties of PLA limit its usage in industrial material applications. Stereo-complexation introduced a new way to maintain the range of properties of PLA. The results revealed that they have favorable properties to be used as biodegradable packaging material [20]. Reduced hydrophilicity of TPS is shown when TPS is produced from modified graft starches which results in the enhancement of compatibility with PLA. Here PLA-gMTPS act as a compatibilizer by reducing the surface tension of the TPS phase during blending [7]. But an excess amount of coupling agent will act as a plasticizer and will lead to a decrease in the maximum tensile strength. PLA/TPS blends were studied with different starch sources. The weight proportion of PLA: TPS is fixed at 50:50, and two different types of starch, that is, native cassava starch and acetylated starch with different polarities of acetylation. Acylated starch blends showed a better melt flow ability, higher storage modulus, and thermal stability. By increasing the degree of acetylation, the water barrier properties of blends also increased. These improved properties made this blend a promising material for industrial application [21].

7.3.3.4 Poly (Vinyl Alcohol) (PVA)

PVA is a synthetic polymer commonly used in the textile industry due to its high solubility in water. Even though it is a comparatively expensive polymer, it has good chemical resistance and high mechanical properties. Disadvantages such as a limited barrier and thermal properties limit its wide application ranges. But since it is highly compatible with other materials, it is often mixed with other polymers to obtain the desired properties [7]. TPS/PVA films provide high mechanical properties and low degradation when compared to TPS films. Biodegradable sago starch film is prepared and studied using different amounts of silica. An optimal amount of silica provided high tensile strength but low elongation at break (2 wt%). This study also concluded that increasing amounts of silica created a network structure in the plastic film, and it resulted in a higher water resistance [22]. Another study reported that adding 25% of polyvinyl alcohol (PVA) into bioplastic made from cassava starch produced bioplastic with higher tensile strength than the bioplastic without the addition of PVA. Fifty percent addition of PVA showed a higher decomposition temperature than the bioplastic without PVA addition, which is evidence of improved thermal property. When 100% PVA is added, elongation is also slightly increased [23].

In some cases, nanofibers were added to improve the hydrophobic nature of the TPS/PVA films. Several reports suggest that adding nanofibers or nanoparticles can enhance the water absorption of the film. Due to the addition of nanoparticles, aggregations, voids, and cavities are produced in the film, thus the morphology of the material becomes imperfect [24]. PVA is a polymer containing –OH functional group in its backbone. Therefore, the blend synthesized from starch with the biodegradable –OH bond in PVOH is highly degradable. Studies conducted on PVOH/starch blends suggest that the biodegradability of the material increased with the increase in the amount of starch. In addition, these blends are highly polar. They can be easily molded into various shapes since it is thermoplastic and water soluble. It is a good material to make films, and due to its thermal stability, it can be used as a container. Biodegradable agents like *Pseudomonas* sp. degrade the starch-PVOH material into carbon dioxide and water [1].

7.3.3.5 Chitosan

Chitosan/TS blends are commonly used due to their mechanical properties. Intermolecular hydrogen bonding between NH_3^+ of the chitosan backbone and OH– of the starch is the reason for the mechanical properties of such blends. Using glycerol as a plasticizer reduces the tensile strength, but the elongation of break increases due to plasticization. Glycerol increases the interaction between starch and chitosan by forming hydrogen bonding, and the mobility of glycerol itself is reduced. In addition, the water vapor transmission rate is reduced due to the presence of chitosan, and the crystalline structure is depressed due to the presence of starch [25].

7.3.3.6 Others

Starch blends with other polysaccharides show more favorable mechanical and thermal properties than the polymers alone. Pectin is a nontoxic complex heteropolysaccharide with a linear backbone of α-(1–4)-linked polygalacturonic acid residues and neutral side chain sugars [7]. The galacturonic acid can be partially methoxylated or amidated. Unmodified starch/pectin blends can form films with mechanical properties like synthetic polymer films. A novel hydrocolloid pectin-starch (30/70) films with bioactive extracts are synthesized by crosslinking pectin-starch gels with poly(ethylene glycol) diglycidyl ether (PEGDGE). The films obtained had shown an adequate water-uptake capability, ranging from 100% to 160%. PEGDGE also inhibits the disintegration of the pectin-starch films [25]. Another work reported a novel method for synthesizing green hydrogels by the crosslinking of polyvinyl alcohol (PVA) and aliphatic dicarboxylic acid [26].

7.3.4 TPS Blends with Synthetic Polymers

Bio-based products are those which are partially or fully made from renewable resources. The materials based on TPS with other synthetic polymers contribute to an increase in the performance properties of starch-based materials. In this way, plastic waste can be reused and better quality products are obtained [7].

7.3.4.1 Polypropylene (PP)

Polypropylene (PP) is an excellent material for producing TPS blends. It is a linear hydrocarbon polymer and one of the most widely used polyolefins produced. PP and TPS are incompatible due to their different polarity; TPS is strongly polar, while PP is nonpolar. Therefore, a suitable compatibilizer is used, such as maleic anhydride grafted polypropylene (MA-g-PP), to improve the interfacial adhesion of the composite phases. When maleic anhydride is grafted into PP (PP-g-MA), covalent bonds are formed with the hydrophilic starch. That is why maleic anhydride is called the most acceptable and economical compatibilizer for PP/TPS blends [7]. PP/TPS blends with maleic anhydride as compatibilizers with different concentrations were studied to understand the changes in properties of the material in different maleic anhydride concentrations. They revealed that increasing the amount of compatibilizer up to 20 wt% improved the mechanical properties, rheological behavior, and morphology [27]. Another study on PP/TPS blend with C^{14}, C^{16}, and C^{18} carboxylic acids as compatibilizers concluded that the addition of carboxylic acids proved to be equivalent or better compatibilizers when compared to maleic anhydride polypropylene and no compatibilizer. Also, the adhesion between both polymer phases was improved and the tensile strength, elongation, and impact strength increased with the addition of carboxylic acids [28].

7.3.4.2 Natural Rubber (NR)

TPS films with natural, modified rubbers show interesting behavior. The molecular interactions and crosslinking of rubbers with starch or polymers resulted in changes in the mechanical properties of the films. In some cases, the addition of rubber improves the tensile strength; however, in other cases, it results in a soft material. The influence of glycerol added to the TPS is very important in this case. Some reports suggested that a high amount of rubber causes the separation of phases or agglomeration in the mix, resulting in a total change in the matrix of materials. Thus, the blend becomes brittle. As a result, the properties become worse. Therefore, less amount of rubber is favorable in most of the cases. Several reports mentioned that, by using different types of rubber, materials with different properties are obtained. For example, the interaction of epoxidized rubber is compared with natural rubber and found that epoxidized rubber has shown better interaction [29]. Another similar work studied the interaction of chitosan and starch with epoxidized natural rubber. They found that the chemical interaction of chitosan with the epoxy groups improved the mechanical properties of the blend [30]. The hydroxyl groups in the three-dimensional network of lignin in the rubberwood sawdust interact with starch to act as an interfacial compatibilizer and produce a higher tensile strength [31].

7.3.4.3 Polyethylene (PE)

Polyethylene is a simple polymer with a lineal chain backbone that is widely used for different industrial applications. Commercially available polyethylenes are of different densities: low (PLD) and high (PHD). The PHD has properties like high rigidity and resistance, and the advantages of the use of PLD include great flexibility, good resistance to impact, and high resistance at high temperatures. The researchers found that LDP with rice starch showed low compatibility, making it difficult to modulate the tensile strength due to the presence of both crystalline and amorphous regions [5].

7.3.4.4   Poly (Vinyl Chloride) (PVC)

PVC is a polymer that is commonly used by various industries for fittings, electrical insulation, synthetic plastic products, floor coverings, etc. The strong mechanical and thermal properties, resistance to chemicals, low water absorption, etc. are the properties that made PVC a demanding material for commercial applications. Studies revealed that the degradability of PVC can be improved by the addition of starch. Starch-based PVC material is easily degraded because of the presence of carbon chains, which attract biodegradable agents. Oxidation takes place when the microorganisms broke the carbon chain by releasing carbon dioxide and water. After the oxidation, a weak salt product with a chlorine atom is formed which is not harmful to the environment [1].

7.3.5   Starch-Based Composite Materials

Better mechanical and functional properties are obtained when inorganic and organic reinforcement is incorporated into the starch matrix. Studies revealed that a very low concentration of reinforcement is sufficient for getting desirable properties. Achieving high dispersion, finding suitable plasticizers, controlling the interfacial strength of the inorganic reinforcement in the biopolymer matrix, etc. are the main challenges faced in the preparation of nanocomposites. Therefore, organic cationic compatibilizer is widely used for getting suitable dispersion [3]. Water is a plasticizer for starch-based materials. Nowadays, polysaccharide materials at the nanoscale are used for improving the mechanical properties of starch-based materials since they are edible and compatible. For example, cellulose and starch nanocrystals exhibit improved tensile strength and Young's modulus of starch-based materials. SEM images of starch films are represented in Figure 7.3. It clearly showed good compatibility of a starch matrix with both crystals because of the same chemical units present in both components (glucose) [32]. SEM clearly showed very good compatibility between the starch matrix and the two crystals due to the same chemical unit (glucose).

Cellulose crystals provide higher thermal stability, mechanical properties, and better processibility. On the other hand, starch crystals exhibited higher protection against UV radiation. The starch composites prepared using nanocrystals of starch or cellulose are not only biodegradable but also edible, making them suitable for use as food packages or coatings [3].

Inorganic reinforcements into starch-based composites include clays (montmorillonite) (MMT) and bentonite [33], nano-clay [34], nano-silica (SiO_2) [35], and metal oxides [36]. The incorporation of bentonite and hectorite clay strengthens the polymeric structure and increases the compatibility of starch/PVA blends [31]. Adding nano-SiO_2 as a nanofiller to TPS films contributed to better tensile strength and Young's modulus [37]. Adding clay to potato starch/hectorite nanocomposite films exhibited enhanced mechanical and biodegradable properties. Clay as the nanofiller in a corn starch film showed the highest absorption and best antibacterial properties. In this case, the swelling increased with time and decreased with the addition of plasticizer and hectorite clay [38]. Corn starch—lithium perchlorate ($LiClO_4$)—nano-silica composites exhibited a high-water repellent property.

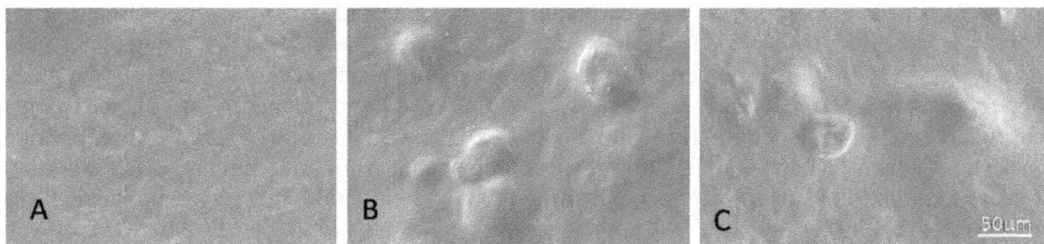

FIGURE 7.3   Surface SEM images of the starch film: (a) pure starch film; (b) containing starch crystals; (c) containing cellulose crystals. By Amjad Ali et al. [32], reprinted from Composites Part B, Elsevier.

This is due to the network structure formed by nano-silica in the corn starch matrix, which is preventing the water molecules from dissolving. The incorporation of natural mineral fibers like talc improves the properties like low water vapor and oxygen permeability, making the material suitable for packaging application [12].

7.4 APPLICATIONS OF STARCH-BASED ADVANCED MATERIALS

Overexploitation of natural resources and increased climate changes lead industries to search for more environmentally friendly materials. Plastic is used widely by all industries because of its low cost of production, its higher versatility, and good barrier properties to water and gas permeability, as well as its mechanical and optical properties. But the accumulation of this waste affected the natural ecosystem. Therefore, there is high demand for pollution-free approaches in every industry. More than 40% of plastic which is produced is used for packaging applications. The packaging industry is now looking forward to replacing such non-degradable materials. Biopolymer is the best solution to overcome this situation because it has a lot of similar characteristics to synthetic polymer materials. It is easily available and has high biocompatibility, and it is nature friendly. In most of the reported works, common sources of starch like wheat, rice, potato, corn, etc., are used. But due to the increased demand for biodegradable advanced materials, several works included non-conventional starch sources such as pehuen [39], pea, lentil, faba bean [40], chickpea [41], chestnut [42], jackfruit seed [43], and avocado seed [44]. As already mentioned, in different botanical sources, the properties of starch will be different.

7.4.1 APPLICATION IN THE FOOD INDUSTRY

Food packaging industries are in a crucial stage of replacing non-biodegradable polymer materials with sustainable and biodegradable materials. Developing new innovative material which is biodegradable and at the same time can preserve food for a long time is challenging. Nowadays, usage of bioplastic as packaging materials increased abundantly due to its biodegradability, biocompatibility, and low cost of starch-based materials [5].

The development of functional packaging materials mainly focuses on:

    i. Improving the materials by mixing different materials like micro/nanostructures so that they provide a better barrier to oxygen and water vapor.
    ii. Adding bioactive substances like antioxidant or antibacterial agents

Active packaging includes oxygen scavengers to decrease fat oxidation, ethylene scavengers to minimize fruit and vegetable ripening, humidity, and odor absorbers [5].

The fish and meat on normal preservation produce a color change and off-odor and flavor. The color change is due to the oxidation of fats and leads to the decoloration of pigments like myoglobin, carotenoids, etc. Odor and flavor changes occur due to rancidity, resulting from lipid oxidation. Nutrients like vitamin E, β-carotene, and ascorbic acid are lost due to oxidization. To overcome these limitations, vacuum packaging is used. Even this method is not completely satisfying since it will not fully remove oxygen. Therefore, it became necessary to develop a novel packaging material that is highly efficient in preventing oxidation and quality loss. Thus, oxygen scavengers were introduced to eliminate residual oxygen. Oxygen scavengers are incorporated in sachets, films, or labels, and they prevented food from contamination. It is reported that thermoplastic starch films are efficient in controlling lipid oxidation in food [7]. Thermoplastic films are prepared by the incorporation of low-density polyethylene and green tea. The work proved that the lipid oxidation was significantly reduced due to the hydrophobic nature of films [45]. Using antioxidant extracts in the edible cassava starch films showed an improved UV-blocking properties (Figure 7.4) [46].

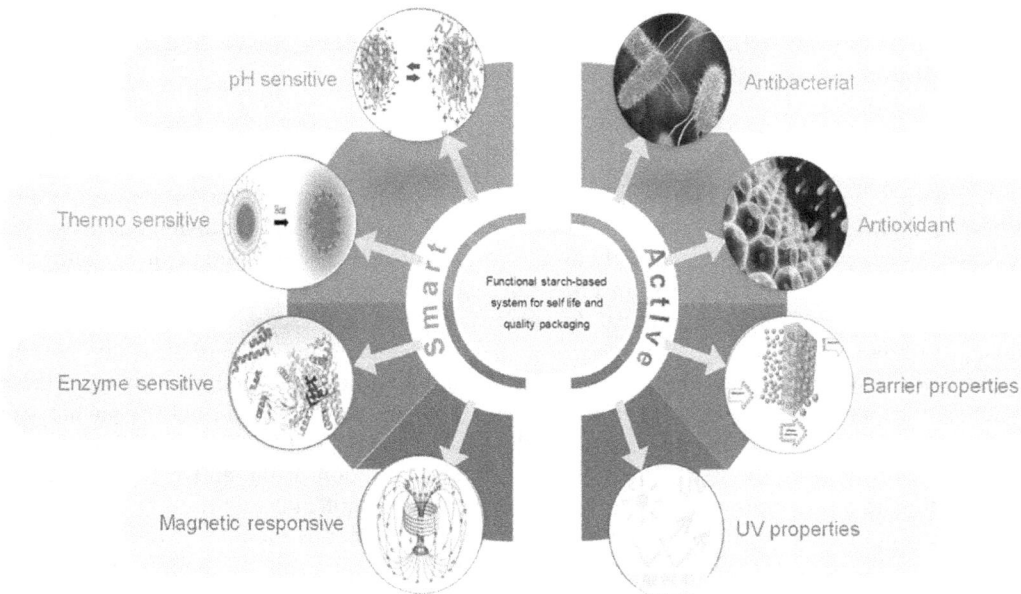

FIGURE 7.4   Active and smart starch-based food packaging. By García-Guzmán et al. [5] is licensed under CC, reprinted from Polysaccharides, MDPI.

Active packaging deals with improving food safety and quality. Hence, the most importance is given to the preparation of packaging materials by using active films having antibacterial, antioxidant, and barrier properties. Smart packaging helps inform the consumer about the kinetic changes, which are related to the quality of food or the environment it contains. This is to ensure food quality and to minimize losses [5]. Smart packaging films based on a change of color are developed by adding pH-sensitive and responsive indicators like anthocyanins, betacyanins, curcumin, etc. in the biopolymer-based matrix [7]. These materials are prepared by blending with polymers such as PVA, PLA, carrageenan, chitosan, etc., and they respond to the magnetic field or have enzyme-responsive characteristics. Recently, natural pigments are actively been used as indicators. In a work, blueberry residues as a visual pH indicator are added to the corn starch-cellulose matrix to monitor fish spoilage. The material exhibited good pH sensitivity and thermal stability [47]. Similarly, cassava starch films with anthocyanins showed high pH sensitivity over a wide pH range, and potato starch-based films with anthocyanins displayed the color difference at pH 1–12 and helped to the spoilage of pork [48]. Starch films with chitosan nanoparticles were studied as a packaging material for cherry tomatoes, and it is found that the growth of microorganisms was effectively inhibited compared with pure starch film. Adding cinnamon oil to the starch matrix also showed reduced microbial growth and protected the fruit from further contamination during transport and storage [49]. Tapioca starch film with anthocyanin-rich bay laurel berry extracts is synthesized and studied. The color of the film is changed from purple to red when it is exposed to hydrogen peroxide gas, and when exposed to ammonia, it quickly changed to blue and then became olive [50]. Nutmeg oil, ZnO NP, and ham extract are incorporated into starch/PVA films. These films showed pH-sensitive and antibacterial properties. Purple extract turned cherry red when pH become acidic, while an alkaline pH changed from brownish yellow to light green at neutral pH [51]. Similarly, PVA/starch/propolis/procyanidin rosemary extract films exhibited a change in color depending on pH; reddish to blue under acidic pH, blue under neutral pH, and yellow under alkaline pH [52].

Starch-based materials are undoubtedly a promising material for food packaging applications. Even though it is biodegradable and sustainable, drastic production and usage are still not established due to its limited optical properties, barrier resistance, and mechanical properties compared with the conventional non-biodegradable polymeric materials. Studies are active to overcome such limitations [5].

7.4.2 PHARMACEUTICAL AND BIOMEDICAL APPLICATION

Starch is widely used in the pharmaceutical industry since it has fundamental properties such as morphology, particle size, shape, surface area, porosity, density, gelling, viscosity, and its white, soft, smooth dryness, etc. Starch is mostly used as excipients [7]. The excipient can be defined as a substance other than an active drug or pro-drug that is included in the manufacturing process or is contained in finished pharmaceutical dosage forms. It can be categorized as binders, disintegrants, and glidants in tablets [2]. Starch became an attractive material in biomedical applications because of its availability, abundance, low cost, biodegradability, and biocompatibility. Starch can act as a dispersing agent to uniformly distribute the drug particles and bind the particles to form loose agglomerates to form compacted tablets for oral ingestion [7]. Starch is also used as disintegrants in the drug industry, which helps in the assimilation of the active components of the drug into the body. Hydrogen bonds present in the starch molecules with other constituents of the tablets bind them together. But in the presence of an aqueous medium (water), the water is absorbed and results in the rupture of hydrogen bonding. This results in the loosening of the tablet, and it breaks apart, which are then absorbed by the body [2]. Modified starch like pre-gelatinized starch is better than native starch for biomedical applications. Due to pre-gelatinization, the starch structure is partially opened exposing a greater number of hydroxyl groups and thereby forming hydrogen bonds between the other constituents in the tablets [2]. Starch is modified to improve flow, disintegration, and direct-compression properties, thereby enhancing the demand for starch in industrial applications. Pre-gelatinized starch has excellent wettability, easy dispersion in cold water, high viscosity, moisture sorption, and swelling [7].

Commonly in tablets excipients will be at a higher concentration than that of the active drug. The role of the excipient is to increase the mass of the pharmaceutical form and dilute the drug. Therefore, in such cases, diluents are the excipients, and the diluents must possess enough flow-through characteristics [7]. Tablets with low crushing strength and fast disintegration can be made by using acetylated starch as a binder. Acetylated starch has high granular size, solubility, swelling power, and water absorption capacity [7]. Paracetamol-loaded starch-citrate nanoparticles are studied with the degree of substitution (DS) ranging from 0.11 to 0.90 and found that they had low toxicity and could be promising pH-responsive nanocarriers for targeted drug delivery [53]. An important property that is needed for a material used for drug delivery is much-adhesivity. It is reported that crosslinked thiolated starch coated magnetite nanoparticles for sustained drug release observed enhanced cell viability, mucoadhesive, and drug release properties [54]. Hydrogels are three-dimensional polymer networks that have high absorption capacity and can be used for drug delivery. The drug is loaded into the porous network, and a controlled release of the drug is achieved. Similarly, cross-linked starch can be used as a drug delivery system since it can form hydrogels. Starch has a wide range of applications in the pharmaceutical and biomedical fields due to its biocompatibility characteristics, low immunogenicity, and versatility of chemical and physical modifications. The studies suggested that products synthesized from such biodegradable materials exhibited very less side effects.

7.4.3 APPLICATIONS OF STARCH IN WATER TREATMENT

Water treatment commonly involves two types of purification: removing undissolved solid contaminants and removing the dissolved contaminants. Different kinds of filtration techniques are employed successfully to remove solid contaminants. Similarly, there are a lot of well-known purification techniques for removing dissolved contaminants. But these techniques involve various chemicals that are toxic to both humans and the environment. And replacing such techniques with environment-friendly techniques is a major challenge. Starch is an ideal material for sustainable application since it is nontoxic, easily available, low cost, and can be easily subjected to different modifications. –OH functional groups present on the starch structure can target and trap the contaminant particles. Various nanoparticles are also used for decontamination of water, and in such cases, starch can act as a dispersing agent or stabilizer due to its ability to be soluble even in elevated temperatures or simply because of its gelation characteristics. Better dispersion is needed for the efficient decontamination of water [2]. Fe–Pd nanoparticles are prepared with and without starch as a dispersing agent for the dichlorination of trichloroethene (TCE) hydrocarbon. Morphology is analyzed using transmission electron microscopy and found that Fe–Pd nanoparticles without starch were agglomerated and Fe–Pd nanoparticles with starch are efficiently dispersed. The dispersion is achieved by the interaction of starch and iron. As a result, the dichlorination rate was greater for starch-Fe–Pd than for the particles without the stabilization of starch [55].

7.4.4   APPLICATION OF STARCH IN POROUS FOAM STRUCTURES

Metal and ceramic foam structures are widely used for applications even at elevated temperatures and pressures. These materials are manufactured mainly by the replication method. The polymer foam structure is synthesized, and the space in the foam is filled with ceramics and allowed to solidify. The porous ceramic structure is obtained by burning the polymer foam. The drawback is the porous structure of the ceramic form is dependent on the polymeric foam used for the preparation. Another method of foam production is agitating a suspension containing ceramic particles and a foaming agent. By removing the liquid and sintering, a porous structure is obtained. In this method also, there is no any way to control the structure formed by the process. The structure of the form is responsible for the properties. Therefore, obtaining desirable property is not possible. In addition, dispersing agents and chemicals used for the process are expensive and not nature friendly [2]. Starch can be used as a pore former in the production of ceramic and metallic foam structures. An optimal quantity of starch is added to the suspension containing ceramic powder and heated under constant stirring. Starch granules start to swell by the uptake of water in the suspension. And thus, ceramic powder is consolidated into a porous structure. The pore structure and size are also dependent on the type of starch used, the size and amount of starch granules, and the rate of swelling of starch granules. The higher the swelling, the larger will be the pore structure, and the higher the amount of starch also leads to the increase in pore size. At the desirable size, the starch is burned out and the structure is dried out [2].

7.4.5   APPLICATION OF STARCH IN SELF-HEALING POLYMERIC MATERIALS

Polymer materials that are hard and brittle will catastrophically fail when cracks develop. Elastomers can be incorporated into the polymer to stop the further propagation of the crack. But studies found that adding this kind of phase affects the mechanical properties like tensile and flexural properties of the material. Controlling cracks without diminishing the properties became a demand and lead to the development of self-healing polymers [2]. Self-healing polymers contain crosslinked formaldehyde-based microcapsules shells, which are well-dispersed within them. The microcapsules are allowed to survive the process of dispersion within the polymer phase without rupture. These microcapsules contain highly reactive liquids (healant), such as epoxies, glycidyl methacrylate (GMA), and isocyanates. These liquids readily react with multiple reactive functional groups on polymer chains. At the time of the development of cracks, multiple microcapsules containing healant are ruptured. The healant flows into the cracks, and crosslinking takes place by bridging the gap produced due to the crack [2]. Gelatinized waxy maize starch (WMS) is loaded with poly(d,l-lactide-coglycolide) microcapsules, and the effect is studied. Gelatinization of starch activates the −OH functional group of the starch and reacts during self-healing crosslinking. The result showed that the healing property increased as the number of microcapsules increased [7].

7.4.6 OTHER ADVANCED APPLICATIONS OF STARCH AND ITS DERIVATIVE MATERIALS

Starch and starch-based materials are used for various applications, such as electronics, photonics, energy, sensors, and super-hydrophobic surfaces. Superhydrophobic films, coatings, and papers have attracted tremendous attention in a wide range of sectors. Hydrophobic papers are developed by coating two layers of different solutions on the surface of the paper. The first layer is gelatinized starch, which contains enzymes, sizing agents, crosslinkers, and aluminum sulfate as pH adjusters, and the second layer was a suspension of hexamethyldisilazane treated silica nanoparticles (HMDSSiNPs) in ethanol. The second layer is sprayed immediately after applying the first layer, and ethanol is allowed to evaporate. The hydrophobic nature is obtained due to the bonds that are created between the two layers. The contact angle analysis revealed that the developed paper is super-hydrophobic with a contact angle of 162°. The effect of the starch layer is studied by the submersion test. The paper with and without the starch layer is submerged in water for 2 minutes. The paper with both layers was dry after 2 minutes, but the paper without starch was wet, and the water penetrated through the fibers of the paper. Also, the paper with a starch layer has increased mechanical durability. The paper is treated with a very thin layer of materials; therefore, the visual appearance of the paper was not affected [3].

Starch has a good optical property and is used for applications in photonics. Due to low oxygen permeability, it is shown to have greater photostability. Gelatinized starch is doped with rhodamine 6G water and cast on the glass to form a layer. When it is photoexcited, a random lasing effect was observed due to the random formation of starch granules. Compared with other biomaterials, starch has exhibited a better photonic property [56]. Starch-based aerogel is an advanced biodegradable material with low density and high-specific surface area and is used for different potential applications. The procedures for fabricating the aerogel are starch gelatinization, retrogradation, organic solvent exchange, and superfacial CO_2 drying. The properties of aerogels are described by the parameters like density, pore size, surface area, crystallinity, thermal stability and conductivity, mechanical properties, digestibility, etc. These properties of starch aerogels are achieved by varying preparation conditions, starch modifications, and the addition of other ingredients. Starch aerogels have diverse applications such as encapsulation and controlled release of bioactive compounds, packaging applications, tissue engineering, thermal insulation, CO_2 adsorption, etc. [57].

7.5 CONCLUSION

The real importance of a sustainable and ecofriendly lifestyle will be going to get noticed when this life is about to end, and that day, we will be ready to pay any amount to save our life. However, money can do nothing and cannot buy life. The advantage of sustainable ecofriendly materials is that they are safe for us and future generations too. The use of biodegradable and natural materials can mitigate the current environmental and climate issues. Starch gained interest as a natural polymer that is biodegradable and easy to modify and blend with other materials to produce desirable properties to be used in different industrial applications. By varying the type and degree of modification, the properties of the final product can be varied easily. Therefore, starch and its derivatives became a very promising material for a lot of applications. Using starch as a self-healing material became a better alternative for conventionally used toxic chemical healants. Starch has become a promising material for drug delivery because of its degradability and biocompatibility. It is also easily available and less expensive. The pharmaceutical industry utilizes it as disintegrants, binders, dispersants, and lubricants, making use of its properties like bendability and gel-ability. Similarly, various fields are utilizing their favorable properties for numerous potential applications. Even though starch is at the forefront of sustainable materials, its competitive use as food is a limitation for extensive use in commodity applications. Even though, more areas of application of starch and derivatives are still in research.

REFERENCES

1. Mohd Amin, A. M., Mohd Sauid, S. & Ku Hamid, K. H. Polymer-starch blend biodegradable plastics: An overview. *Adv. Mater. Res.* **1113**, 93–98 (2015).
2. Ogunsona, E., Ojogbo, E. & Mekonnen, T. Advanced material applications of starch and its derivatives. *Eur. Polym. J.* **108**, 570–581 (2018).
3. Jiang, T., Duan, Q., Zhu, J., Liu, H. & Yu, L. Starch-based biodegradable materials: Challenges and opportunities. *Adv. Ind. Eng. Polym. Res.* **3**, 8–18 (2020).
4. Palanisamy, C. P., Cui, B., Zhang, H., Jayaraman, S. & Muthukaliannan, G. K. A comprehensive review on corn starch-based. *Polymers (Basel)* **12**, 2161 (2020).
5. García-Guzmán, L. *et al.* Progress in starch-based materials for food packaging applications. *Polysaccharides* **3**, 136–177 (2022).
6. Kaur, B., Ariffin, F., Bhat, R. & Karim, A. A. Progress in starch modification in the last decade. *Food Hydrocoll.* **26**, 398–404 (2012).
7. Garcia, M. A. V. T., Garcia, C. F., & Faraco, A. A. G. Pharmaceutical and biomedical applications of native and modified starch: A review. *Starch-Stärke* **72**(7–8), 1900270 (2020).
8. Vanier, N. L., El Halal, S. L. M., Dias, A. R. G. & da Rosa Zavareze, E. Molecular structure, functionality and applications of oxidized starches: A review. *Food Chem.* **221**, 1546–1559 (2017).
9. Qin, Y. *et al.* Effects of citric acid on structures and properties of thermoplastic hydroxypropyl amylomaize starch films. *Materials (Basel).* **12**, 1–13 (2019).
10. Hu, X. P., Zhang, B., Jin, Z. Y., Xu, X. M. & Chen, H. Q. Effect of high hydrostatic pressure and retrogradation treatments on structural and physicochemical properties of waxy wheat starch. *Food Chem.* **232**, 560–565 (2017).
11. Sifuentes-Nieves, I. *et al.* Influence of gelatinization process and HMDSO plasma treatment on the chemical changes and water vapor permeability of corn starch films. *Int. J. Biol. Macromol.* **135**, 196–202 (2019).
12. López, O. V., Castillo, L. A., García, M. A., Villar, M. A. & Barbosa, S. E. Food packaging bags based on thermoplastic corn starch reinforced with talc nanoparticles. *Food Hydrocoll.* **43**, 18–24 (2015).
13. Glenn, G. M. *et al.* Cellulose fiber reinforced starch-based foam composites. *J. Biobased Mater. Bioenergy* **1**, 360–366 (2008).
14. Menzel, C. Improvement of starch films for food packaging through a three-principle approach: Antioxidants, cross-linking and reinforcement. *Carbohydr. Polym.* **250**, 116828 (2020).
15. Liu, W., Drzal, L. T., Mohanty, A. K. & Misra, M. Influence of processing methods and fiber length on physical properties of kenaf fiber reinforced soy based biocomposites. *Compos. Part B Eng.* **38**, 352–359 (2007).
16. Schwach, E. & Avérous, L. Starch-based biodegradable blends: Morphology and interface properties. *Polym. Int.* **53**, 2115–2124 (2004).
17. Hubackova, J. *et al.* Influence of various starch types on PCL/starch blends anaerobic biodegradation. *Polym. Test.* **32**, 1011–1019 (2013).
18. Palai, B., Biswal, M., Mohanty, S. & Nayak, S. K. In situ reactive compatibilization of polylactic acid (PLA) and thermoplastic starch (TPS) blends; synthesis and evaluation of extrusion blown films thereof. *Ind. Crops Prod.* **141**, 111748 (2019).
19. Zhou, L., Zhao, G., Feng, Y., Yin, J. & Jiang, W. Toughening polylactide with polyether-block-amide and thermoplastic starch acetate: Influence of starch esterification degree. *Carbohydr. Polym.* **127**, 79–85 (2015).
20. Li, Z., Tan, B. H., Lin, T. & He, C. Recent advances in stereocomplexation of enantiomeric PLA-based copolymers and applications. *Prog. Polym. Sci.* **62**, 22–72 (2016).
21. Noivoil, N. & Yoksan, R. Compatibility improvement of poly(lactic acid)/thermoplastic starch blown films using acetylated starch. *J. Appl. Polym. Sci.* **138**, 1–16 (2021).
22. Ismail, H. & Zaaba, N. F. The mechanical properties, water resistance and degradation behavior of silica-filled sago starch/PVA plastic films. *J. Elastomers Plast.* **46**, 96–109 (2014).
23. Syamani, F. A. *et al.* Characteristics of bioplastic made from modified cassava starch with addition of polyvinyl alcohol. *IOP Conf. Ser. Earth Environ. Sci.* **591**, 1–10 (2020).
24. Heidarian, P., Behzad, T. & Sadeghi, M. Investigation of cross-linked PVA/starch biocomposites reinforced by cellulose nanofibrils isolated from aspen wood sawdust. *Cellulose* **24**, 3323–3339 (2017).

25. Liu, H., Adhikari, R., Guo, Q. & Adhikari, B. Preparation and characterization of glycerol plasticized (high-amylose) starch-chitosan films. *J. Food Eng.* **116**, 588–597 (2013).
26. Carreño, G. *et al.* Sustained release of linezolid from prepared hydrogels with polyvinyl alcohol and aliphatic dicarboxylic acids of variable chain lengths. *Pharmaceutics* **12**, 1–17 (2020).
27. Raee, E., Avid, A. & Kaffashi, B. Effect of compatibilizer concentration on dynamic rheological behavior and morphology of thermoplastic starch/polypropylene blends. *J. Appl. Polym. Sci.* **137**, 1–8 (2020).
28. Martins, A. B. & Santana, R. M. C. Effect of carboxylic acids as compatibilizer agent on mechanical properties of thermoplastic starch and polypropylene blends. *Carbohydr. Polym.* **135**, 79–85 (2016).
29. Cai, Z. *et al.* The modification of properties of thermoplastic starch materials: Combining potato starch with natural rubber and epoxidized natural rubber. *Mater. Today Commun.* **26**, 101912 (2021).
30. Jantanasakulwong, K. *et al.* Reactive blending of thermoplastic starch, epoxidized natural rubber and chitosan. *Eur. Polym. J.* **84**, 292–299 (2016).
31. Sarkar, A. *et al.* Preparation of novel biodegradable starch/poly(vinyl alcohol)/bentonite grafted polymeric films for fertilizer encapsulation. *Carbohydr. Polym.* **259**, 117679 (2021).
32. Ali, A. *et al.* Preparation and characterization of starch-based composite films reinfoced by polysaccharide-based crystals. *Compos. Part B* (2017) doi:10.1016/j.compositesb.2017.09.017.
33. Campos-Requena, V. H. *et al.* Thermoplastic starch/clay nanocomposites loaded with essential oil constituents as packaging for strawberries – In vivo antimicrobial synergy over Botrytis cinerea. *Postharvest Biol. Technol.* **129**, 29–36 (2017).
34. Javanbakht, S. & Namazi, H. Solid state photoluminescence thermoplastic starch film containing graphene quantum dots. *Carbohydr. Polym.* **176**, 220–226 (2017).
35. Liu, Y., Fan, L., Mo, X., Yang, F. & Pang, J. Effects of nanosilica on retrogradation properties and structures of thermoplastic cassava starch. *J. Appl. Polym. Sci.* **135**, 1–9 (2018).
36. Liu, Q. *et al.* Enhanced dispersion stability and heavy metal ion adsorption capability of oxidized starch nanoparticles. *Food Chem.* **242**, 256–263 (2018).
37. Lendvai, L., Apostolov, A. & Karger-Kocsis, J. Characterization of layered silicate-reinforced blends of thermoplastic starch (TPS) and poly(butylene adipate-co-terephthalate). *Carbohydr. Polym.* **173**, 566–572 (2017).
38. Khodaeimehr, R., Peighambardoust, S. J. & Peighambardoust, S. H. Preparation and characterization of corn starch/clay nanocomposite films: Effect of clay content and surface modification. *Starch/Staerke* **70**, 1–12 (2018).
39. Castaño, J., Rodríguez-Llamazares, S., Carrasco, C. & Bouza, R. Physical, chemical and mechanical properties of pehuen cellulosic husk and its pehuen-starch based composites. *Carbohydr. Polym.* **90**, 1550–1556 (2012).
40. Li, L. *et al.* Characteristics of pea, lentil and faba bean starches isolated from air-classified flours in comparison with commercial starches. *Food Chem.* **276**, 599–607 (2019).
41. Yniestra Marure, L. M., Núñez-Santiago, M. C., Agama-Acevedo, E. & Bello-Perez, L. A. Starch characterization of improved chickpea varieties grown in Mexico. *Starch/Staerke* **71**, 1–29 (2019).
42. Shubeena *et al.* Effect of acetylation on the physico-chemical properties of Indian Horse Chestnut (Aesculus indica L.) starch. *Starch/Staerke* **67**, 311–318 (2015).
43. Zhang, Y. *et al.* Jackfruit starch: Composition, structure, functional properties, modifications and applications. *Trends Food Sci. Technol.* **107**, 268–283 (2021).
44. Macena, J. F. F., de Souza, J. C. A., Camilloto, G. P. & Cruz, R. S. Physico-chemical, morphological and technological properties of the avocado (Persea americana mill. cv. hass) seed starch. *Cienc. Agrotecnol.* **44**, (2020).
45. Panrong, T., Karbowiak, T. & Harnkarnsujarit, N. Thermoplastic starch and green tea blends with LLDPE films for active packaging of meat and oil-based products. *Food Packag. Shelf Life* **21**, 100331 (2019).
46. Piñeros-Hernandez, D., Medina-Jaramillo, C., López-Córdoba, A. & Goyanes, S. Edible cassava starch films carrying rosemary antioxidant extracts for potential use as active food packaging. *Food Hydrocoll.* **63**, 488–495 (2016) doi:10.1016/j.foodhyd.2016.09.034.
47. Teixeira, A., Pereira-júnior, V. A., Silva-pereira, M. C. & Stefani, R. Chitosan/corn starch blend films with extract from Brassica oleraceae (red cabbage) as a visual indicator of fish deterioration. *LWT – Food Sci. Technol.* **61**, 1–5 (2014) doi:10.1016/j.lwt.2014.11.041.
48. Andretta, R., Luchese, C. L., Tessaro, I. C. & Spada, J. C. Development and characterization of pH-indicator films based on cassava starch and blueberry residue by thermocompression. *Food Hydrocoll.* **93**, 317–324 (2019) doi:10.1016/j.foodhyd.2019.02.019.

49. Díaz-Galindo, E. P., Nesic, A., Bautista-Baños, S., Dublan García, O., & Cabrera-Barjas, G. Corn-starch-based materials incorporated with cinnamon oil emulsion: Physico-chemical characterization and biological activity. *Foods* **9**(4), 475 (2020).

50. Yun, D. *et al.* Development of active and intelligent films based on cassava starch and Chinese bayberry (Myrica rubra Sieb. et Zucc.) anthocyanins. *RSC Adv.* **9**, 30905–30916 (2019) doi:10.1039/c9ra06628d.

51. Jayakumar, A. *et al.* Starch-PVA composite films with zinc-oxide nanoparticles and phytochemicals as intelligent pH sensing wraps for food packaging application. *Int. J. Biol. Macromol.* **136**, 395–403 (2019).

52. Mustafa, P. *et al.* PVA/starch/propolis/anthocyanins rosemary extract composite films as active and intelligent food packaging materials. *J. Food Saf.* **40**, 1–11 (2019) doi:10.1111/jfs.12725.

53. Chin, S. F., Romainor, A. N., Pang, S. C., Lee, B. K., & Hwang, S. S. pH-responsive starch-citrate nanoparticles for controlled release of paracetamol. *Starch - Stärke* **71**(9-10), 1800336 (2019).

54. Saikia, C., Hussain, A., Ramteke, A., Sharma, H. K. & Maji, T. K. Crosslinked thiolated starch coated Fe_3O_4 magnetic nanoparticles : Effect of montmorillonite and crosslinking density on drug delivery properties. *Starch - Starke* **66**, 1–12 (2014) doi:10.1002/star.201300277.

55. He, F. & Zhao, D. Preparation and characterization of a new class of starch-stabilized bimetallic nanoparticles for degradation of chlorinated hydrocarbons in water. *Environ. Sci. Technol.* **39**, 3314–3320 (2005).

56. Cyprych, K., Sznitko, L. & Mysliwiec, J. Starch : Application of biopolymer in random lasing. *Org. Electron.* **15**, 2218–2222 (2014).

57. Zhu, F. Starch based aerogels: Production, properties and applications. *Trends Food Sci. Technol.* **89**, 1–10 (2019).

8 Chitin and Chitosan Derivatives to Proffer New Functional Materials

Abdellah Halloub, Raji Marya, Hamid Essabir,
Rachid Bouhfid, and Abou el kacem Qaiss
Moroccan Foundation for Advanced Science, Innovation and Research
Mohammed VI Polytechnic University

8.1 INTRODUCTION

Polysaccharides are widely distributed throughout the biosphere, serving a variety of essential functions in human lives, including energy storage and structural materials [1]. Cellulose, lignin, hemicellulose, pectin, chitin, and keratin are earth's most abundant organic polymers [2,3]. Chitin, on the other hand, has a far a smaller number of uses than cellulose. This is caused by a different factor, including the scarcity of natural chitin structures that may be employed with little processing and the polysaccharide's poor solubility qualities. As a result, most of the obtained chitin undergoes substantial alkaline deacetylation to yield chitosan. Indeed, this kind of polysaccharide is a combination of β(1→4)N-acetyl-d-glucosamine with d-glucosamine connected units [4]. Because of essential aspects like biodegradability and biocompatibility, as well as its mucoadhesive and harmless character, chitosan owns a huge interest from a variety of applications fields for example the food industry, pharmacy, biotechnology, and biomedicine [5].

Natural polymer modification is a recurring topic in materials research that led to the formation of new derivatives with distinct properties. Polysaccharide modification can be done in a variety of ways. Chitosan is susceptible to chemical modifications at free amino groups from deacetylated units at C2, as well as hydroxyl groups (–OH) at C3 and C6 [6]. Chitosan has restricted applicability because of its poor solubility in most organic solvents. In fact, the backbone of chitosan should have a molecular modification to expand its range of applications. The inclusion of functional groups in the molecular structure enables various modification approaches [7]. After chemical modification, physical connections (inorganic composites and polyelectrolyte complexes), and other different approaches such as the utilization of avidin-biotin interaction, it appears to have a lot of potential applications, especially in the biomedical industry. Biocompatibility, biodegradability, antimicrobial ability, anticancer capacity, wound-healing ability, planned drug delivery, gene therapeutic, biosensing capacity, anticoagulation capacity, and many other intrinsic properties of chitosan have drawn a lot of attention to it for use as a scaffold material in tissue engineering, membranes, nanoparticles, and many other applications [8]. Apart from the optimism and reputation it has in the medical field, it is also being researched in other industrial fields like pulp and paper, cosmetics, and water purification [9].

8.2 PROCESSING OF CHITOSAN

Crab and shrimp shells are the most common sources of chitosan. This polymer is acid-treated and then alkaline-treated in industrial processing to eliminate calcium carbonates and proteins, respectively. To remove pigments and other contaminants, decolorization and purifying processes are sometimes added to the process. Decalcification or demineralization is the procedure of calcium carbonate elimination, whereas deproteinization is the procedure of proteins removing [10]. Using

DOI: 10.1201/9781003265054-8

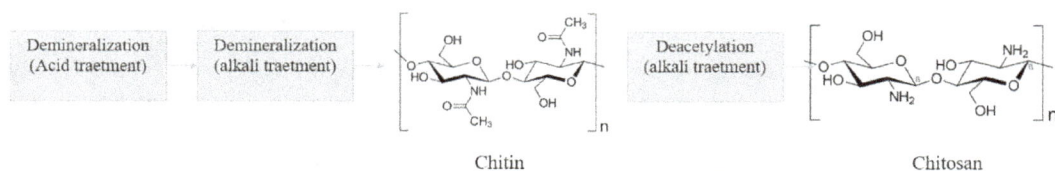

Chitin   Chitosan

FIGURE 8.1 Chitin and chitosan extraction process.

different several such as sodium hypochlorite, hydrogen peroxide, and acetone, the decolorization procedure primarily removes pigments such astaxanthin and β-carotene [7]. Deacetylation is commonly used to produce chitosan from chitin (Figure 8.1). Chitin can be converted into its deacetylated form, chitosan, or even partially fragmented by the demineralization processing in an acidic environment. Among the used acids, sulfuric acid, nitric acid, formic acid, acetic acid, and especially dilute hydrochloric acid are preferred. After that, the material is filtered and rinsed before being dried overnight. Afterward, the dried material is alkaline treated with a dilute NaOH solution in order to perform a deproteinization process. Then, the material is washed to eliminate surplus quantities of utilized NaOH and dried to produce chitin. To produce the chitosan, the resultant chitin is then deacetylated by removing the acetyl groups. In the existence of a solution of concentrated sodium hydroxide (NaOH), this reaction occurs at an elevated temperature. The purification stage is then carried out to bring the chitosan to neutrality and remove any impurities, resulting in the desired chitosan.

Usually, the chitin and the chitosan isolation method can be divided into two classifications: the chemical and the biological extraction techniques. Indeed, the chemical technique has several disadvantages: it is not an environmentally friendly process as it involves the use of harmful acids and powerful bases. The other standpoint is that the mineral and protein that have been solubilized cannot be utilized as nutrition. That is why, in recent years, biological techniques have gotten a lot of attention. For the demineralization and deproteinization phases, this biological technique uses lactic acid-producing bacteria and bacterial proteases, respectively. In an enzymatic approach, chitin deacetylase is utilized to deacetylate chitin [11].

Several published papers contain precise research on the chitin and chitosan isolation that exist in various sources using those isolation techniques. Narguess et al. report the extraction of chitin and chitosan that exists in a variety of Egyptian insects. Demineralization was performed via HCl solution (1.0 M), and proteins were removed using an alkaline treatment via sodium hydroxide solution (1.0 M) at 100°C for 8 hours. The chitin was treated with 50% NaOH for 8 hours at 100°C [12]. Further, Haripriya et al. report demineralization and deproteinization processes in the manufacture of chitin and chitosan that exists in *Penaeus monodon* shrimp waste with HCl solution (1.0 M) and NaOH aqueous solution (3.0 M) over 75 minutes at lab temperature. The deacetylation of chitin was performed at 90°C over 50 minutes using a 50% sodium hydroxide solution in a 1:50 ratio [13]. Another paper reported that the shell of Antarctic krill, *Euphausia superba*, was processed via HCl aqueous solution (1.7 M) at room temperature over 6 hours and subsequently via NaOH aqueous solution (2.5 M) at 70°C over 60 minutes. Using 1% potassium permanganate, the pigments were removed [14].

However, several studies reported the utilization of the biological method to extract the chitin and the chitosan. Fatemeh et al. report a biological approach for extracting chitosan from prawn waste. The lactic acid fermentation approach, which resulted in demineralization and deproteinization of prawn waste, was applied for 4 and 6 days using *Pseudomonas aeruginosa* bacteria. Then, the extracted chitin was treated with 50% sodium hydroxide for chemical deacetylation [15]. In other study, and using improved enzymatic deproteinization, Islem et al. produce chitin and chitosan from shrimp shells. Proteases were used to deproteinize the shrimp wastes at 60°C for 6 hours, utilizing couple enzymes, the Bromelain with the Alcalase. After that, the chitin was mixed with a NaOH aqueous solution (12.5 M) solution and heated to 140°C for 4 hours [16].

8.3   CHITOSAN MODIFICATIONS FOR NEW FUNCTIONAL MATERIALS

As previously noted, chitosan has several unique features that make it a viable option for a variety of applications; nevertheless, because it is insoluble in most usual solvents, it is anticipated to give much-reduced efficiency in such fields. The unaltered forms of chitosan, on the other hand, encompass most commercial uses. Many other types of modified chitosan are presently in use, and many more are being investigated. This modified chitosan was created utilizing self-assembly, sol-gel process, ultrasonication, and microcontact printing, among other ways or procedures.

Ionic gelation, micro emulsion, and spray drying procedures have all been used to produce chitosan nanocarriers; however, molecular self-assembly is the most advantageous since it provides the needed functional group's organized molecular structure without requiring any additional modifications. There are two types of self-assembly methods [17]:

- Mono-component systems: where only the chitosan molecules are engaged in self-assembly;
- Multicomponent systems: where other compounds, as well as chitosan, can be involved.

Hydrogen bonding, ionic bonds [18], hydrophobic interaction [18], and van der Waals contact of molecules have all had a positive impact on molecular self-assembly [17]. The composite impact of such forces becoming the self-assembly leading power. The self-assembly process was used to make a bio-nanocomposite film based on chitosan with montmorillonite hybrid blocks. Another prominent approach for modifying chitosan is sol-gel. New composite macroporous scaffolds were generated by combining a polymer solution with bioactive glazier in the presence of a sol-gel precursor solution. Sol-gel procedures were applied to examine bioartificial polymeric hybrids by combining PVA solution and chitosan with bioactive glazier reagents [19]. Despite this, the resulting hybrid displayed superior mechanical, morphological, and cell survival characteristics. Another study employed gelation techniques to embed chopped silk fibers and electrospun silk fibers in chitosan/glycerophosphate to generate hyaline cartilage regeneration scaffolds [20]. The sol-gel approach was also used for vanadium, molybdenum, and chromium oxonions adsorb from aqueous solution using chitosan and silica nanocomposites [21].

A membrane of chitosan was created using the phase inversion technique of immersion-precipitation as a possible wound dressing. Herein, an acetic acid solution (0.5 wt%) was applied as the solvent, where NaOH (2 wt%)-Na_2CO_3 (0.05 wt%) solution was applied as the non-solvent. A casting procedure was used to complete the preparation. Pervaporation at 50°C for 10–15 minutes was applied to adapt the thickness of the skin surface and the porosity of the sponge-like sublayer. This asymmetric membrane has good antibacterial activity, as well as oxygen permeation and fluid drainage capabilities. Controlling evaporative water loss is also a benefit of this membrane. The porous cellular structure owned by the membrane is due to this chitosan modification process [22].

Microcontact printing [23] is a method for making nanostructured macromolecules including dendrimers, conducting polymers, and peptides. This method of modification, on the other hand, applied in conjunction with others to produce nanofibers in a wide variety of sizes. Nanofibers production has recently drawn attention to the mechanical treatment technique of modification [24].

8.4   APPLICATION FIELD

Food packaging, antibacterial wound healing, tissue engineering, medication release, water treatment, and bleeding control are just a few of the uses for chitin and chitosan in the industrial and medical industries.

8.4.1   CHITIN AND CHITOSAN FUNCTIONAL MATERIALS IN FOOD PACKAGING ACTIVITIES

Since chitosan films and coatings are bioactive, biodegradable, and biocompatible, they have been intensively explored for food preservation in recent decades. However, in order to be mass manufactured at a cheap cost, it is importance to improve their thermal, mechanical, and water barrier properties [25]. Several studies have proven chitosan films' mechanical, thermal, and barrier characteristics, allowing them to be utilized for dry food packing. Furthermore, these materials' functional features, like their excellent antibacterial and antioxidant abilities, have led to their identification as viable synthetic polymer alternatives in the food industry [26]. Further, in comparison to chitosan standalone films, chitosan blends with other polysaccharides, like pectin, starch, cellulose, and alginate [27–29], microbial polysaccharides [30,31] and proteins, such as gelatin [31] and whey proteins [32], have shown improvements in mechanical characteristics, greater moisture permeability, and start to become more and more water insoluble. This is attributed to electrostatic interactions of chitosan's protonated amino groups with the side-chain groups in the other polymer that had a negative charge in the specific pH [27,33]. Several papers noted challenges in the full dissolution at least with one of the used polymers under specified settings and the creation of highly insoluble compounds within polymers during the manufacturing of blends. This restriction can be solved by bilayer systems, which have greater water vapor barrier characteristics than mixed films [31,34].

Indeed, incorporating nanoscale fillers (e.g., nanocellulose) into films made of chitosan, which could create an interaction with polymeric chains, is one method for addressing the hydrophilic nature of chitosan's inherent problems, such as inadequate water resistivity, low barrier, and mechanical characteristics [35,36]. Because of their great compatibility with chitosan, cellulose nanofibers (CNFs) and nanocrystalline cellulose (NCCs) are attractive reinforcements for chitosan in order to create ecofriendly biofilms with desired characteristics. The strong contact between chitosan molecules and nanocellulose with significant length-to-diameter ratios, caused by electrostatic association and hydrogen bonding, results in the construction of an interactive network structure, which increases the crystallinity of the films [37,38]. As a result, chitosan/nanocellulose composites offer a huge range of potential applications [39,40]. Further, nanocrystalline cellulose has been employed as a reinforcing component for chitosan-guar gum [41], gelatin-chitosan, and starch-chitosan biocomposites [42]. In the previous studies, a translucent, heat-resistant biopolymer-based nanocomposite with improved mechanical and barrier characteristics was created. That new group of risk-free, harmless, sustainable, and ecofriendly chitosan/nanocellulose films could be replacing petroleum-based polymers as a food packaging material.

Currently, the discovery of active biocompounds that give superior antibacterial and antioxidant capabilities to edible films is now the focus of scientific study in the chitosan active packaging field [43–45]. Furthermore, more research has been published to see how the addition of those chemicals impacts the film's mechanical characteristics [46–48]. Currently, films are exposed to various food matrices to examine their effect on the food's organoleptic properties depending on the time [36,49–51]. Lekjing et al. [51] investigated the impacts of chitosan/clove oil on the lifetime and properties of cooked pork sausages. They found that the association of those two ingredients limited microbe expansion, slowed the oxidation of fatty acid, and extended the lifetime of cooked pork sausages by more than 6 days. However, at the outset of the storage period, there were some detrimental effects on odor and taste qualities. In comparable studies, adding ginger and rosemary essential oils to poultry flesh decreased the oxidation progresses [36]. Souza et al. [49] improved that chitosan films with rosemary showed great antimicrobial behavior versus *Bacillus cereus* (decrease of 7.2 log) and *Salmonella enterica* in in vitro studies (decrease of 5.3 log). In summary, bioactive compounds combined in chitosan films have shown excellent promise in elongating lifetime and preserving food property, as well as reducing postharvest fungus and foodborne bacteria in the food chain.

8.4.2 Chitin and Chitosan Functional Materials in Wound Healing Activities

Polymers are frequently applied in the form of a film or an ointment. When exposed to UV light, a chitosan derivative forms a hydrogel, which has lately been presented as a soft tissue biological glue. Usually, hydrogel efficiently protects a wound by securely attaching two parts of skin together to speed-up healing and closure. Silver sulfadiazine was used in the developed bilayer chitosan wound dressing to help reduce wound infection. It should both heal and preserve the wound from bacterial infection to be an effective dressing. Various chitosan-based polymers have been employed in this purpose because they recover without leaving scars. Chitooligomers appear to be involved in improved collagen fibril integration into the extracellular matrix, and they improve vascularization. In comparison to alternative material dressings, chitosan hydrogels had the ability to deliver therapeutic payloads to the desired area.

In a model of corneal alkali burn wound for rabbit, a combination processing of corneal epithelial cells/carboxymethyl chitosan/gelatin/hyaluronic acid blended membranes improved corneal wound healing and restored normal structure, paving the way for the investigation of a novel approach to corneal epithelial reconstruction and treatment via tissue engineering techniques [52]. In an in vitro model, Kratz and others studied the impact of heparin/chitosan biocomposite on the energizing of wound re-epithelialization on human skin. The outcome was considerable; however, the efficacy solely depends on the heparin concentration in the heparin-chitosan gel [53]. In the UV-irradiation presence, Ishihara et al. employed photocross-linkable chitosan hydrogel and discovered that full-thickness wound healing was accelerated [54]. Obara et al. enhanced it further by incorporating fibroblast growth factor-2 to have a faster healing procedure [55].

8.4.3 Chitin and Chitosan Functional Materials in Water Treatment Process

Many organic contaminants have been discovered in a variety of water sources, and industrial effluents are outflows from several industries. Pesticides, biphenyls, hydrocarbons, phenols, fertilizers, plasticizers, oils, detergents, greases, medicines, and other chemicals are among them [56]. As a result, industrial wastewater is a severe environmental issue. When spilled into rivers and lakes, it poses a threat to water quality. In addition, organic and inorganic impurities are thoroughly eliminated to achieve ever-higher environmental quality standards. The use of nontoxic biodegradable biopolymers like chitin and chitosan in wastewater treatment is becoming more widely recognized [57]. They can aggregate and precipitate at neutral or alkaline pH due to their polycationic characteristics. Furthermore, the lengthy polymer chain may make it easier for the polymer to contact the polluted medium [58]. Chitosan can be employed as an excellent adsorbent material for wastewater contaminants' removal because of its backbone's amino ($-NH_2$) and hydroxyl ($-OH$) groups. As a result, chitosan has a substantial advantage over the other polysaccharides (such as starch or cellulose), where its chemical structure permits it special changes to build biocomposites for specific applications. The reactive groups can create composites with a variety of chemicals. Kaolinite, bentonite, oil palm ash, magnetite, montmorillonite, zeolites, polyurethane, and other adsorbents have lower absorbed wastewater contaminants and withstand an acidic condition than chitosan. Instead, the cationic charge present in chitosan can neutralize and flocculate anionic suspended colloidal particles in wastewaters, lowering chemical oxygen demand, chlorides, and turbidity [85]. Chitosan could regenerate, and its environmentally benign character permits it to be used in adsorption processes [59]. Chitosan is applied as a coagulant/flocculant for contaminated wastewaters, in heavy metal or metalloid adsorption (Cu(II), Cd(II), Pb(II), Fe(III), Zn(II), Cr(III)), and for the elimination of dyes from industrial wastewater, as well as organic pollutants like organic oxidized, organochloride pesticides, or fatty and oil pollutants.

Activated carbon was made from the empty fruit bunch of an oil palm and employed as a filler in the fabrication of a PEG diglycidyl ether crosslinked chitosan/activated carbon composite films. Because of the strong adherence between the matrix and filler, the film had a high adsorption potential for Cd^{+2} [60]. In fact, chitosan films enhanced with cellulose extracted from oil palm empty fruit bunch were properly manufactured with an improvement in the mechanical properties. This improvement is attributed to the hydrogen bonds formed between chitosan and cellulose particles in the composite film, leading to an excellent adhesiveness of the matrix-filler interface and a homogeneous structure of biocomposite film. Cadmium ions were removed from aqueous solutions using this technique. The chitosan/cellulose biocomposites film holds a lot of promise as an affordable, high-performance adsorbent for removing Cd^{2+} that exists in water [37]. Hydroxysodalite/chitosan biocomposites are produced from aluminum waste using a hydrothermal process and successfully used to remove Ni(II) and Pb(II) ions from contaminated water [61]. Table 8.1 represents the recent applications and properties for chitin and chitosan derivatives to proffer new functional materials.

TABLE 8.1
Recent Applications and Properties of Chitin and Chitosan Derivatives

Form	Schematic Representation	Applications	Properties	Ref
Hydrogels	OH⁻ diffusion / mold / gel / gelation front / solution / 1 cm	Bone tissue engineering	• Complex dynamic living tissue • Undergoes regrowth • Self-repair	[62,63]
		Drug delivery	• pH sensitivity • Biocompatibile • Enzymatic biodegradability • Polycationic nature	[64–66]
		Waste water treatment	• Renewable • Biodegradable • Environmentally friendly • Resource • Hydrophilic biopolymer with high reactivity • Cationic	[67–69]
Membranes	Hydrogen gas / Anode / Chitin sheet / Cathode / Air (Oxygen gas) / Current collector / Seal	Fuel cells	• Proton conductor in the humidified condition	[70–72]

(Continued)

TABLE 8.1 (*Continued*)
Recent Applications and Properties of Chitin and Chitosan Derivatives

Form	Schematic Representation	Applications	Properties	Ref
Membranes		Wound healing	• Structural similarities with the epidermal and dermal layers of the human skin • Biocompatibility • Antibacterial • Hemostatic • Healing properties	[73–75]
Fibers		Wound dressing	• Antibacterial • Hemostatic properties	[76–78]

(*Continued*)

TABLE 8.1 (*Continued*)
Recent Applications and Properties of Chitin and Chitosan Derivatives

Form	Schematic Representation	Applications	Properties	Ref
Fibers		Tissue engineering	• Biocompatibility • Antibacterial • Pore size • Mechanical performances • Surface properties • Porosity	[79–81]
Scaffolds		Oil/water separation	• Positive charge • Biocompatibility • Osteoconductivity • Biodegradability	[82–84]

(Continued)

TABLE 8.1 (Continued)

Recent Applications and Properties of Chitin and Chitosan Derivatives

Form	Schematic Representation	Applications	Properties	Ref
Scaffolds		Drug delivery	• Biocompatibility • Biodegradability • Antibacterial activity	
Scaffolds		Regenerative dentistry	• Biomaterial-based aerogels • Biodegradability • Antibacterial • Renewable	

TABLE 8.1 (*Continued*)
Recent Applications and Properties of Chitin and Chitosan Derivatives

Form	Schematic Representation	Applications	Properties	Ref
Film		Smart packaging	• Biomaterial-based aerogels • Biodegradability • Antibacterial • Renewable	[85]

8.5 CONCLUSION

Chitosan, because of its inherent properties, could be used in a several industries field, in particular biomedical industry, water treatment, pulp, paper, and so on. Nevertheless, chitosan has some constraints as well. In recent years, the modification of chitosan's structure has gained much more attention. Several methods of modifying chitosan have been discussed in this chapter. In addition to its use in different biomedical fields, chitosan has properties that make it a promising biomaterial. There is hope and conviction that it will become a highly valuable biomaterial in the future, and the numerous supplies and distinctive features point in that direction.

REFERENCES

1. A. Halloub *et al.*, "Stable smart packaging betalain-based from red prickly pear covalently linked into cellulose/alginate blend films," *Int J Biol Macromol*, vol. 234, p. 123764, Feb. 2023, doi: 10.1016/J.IJBIOMAC.2023.123764.
2. L. Hu *et al.*, "Design, synthesis and antimicrobial activity of 6-*N*-substituted chitosan derivatives," *Bioorg Med Chem Lett*, vol. 26, no. 18, pp. 4548–4551, 2016, doi: 10.1016/j.bmcl.2015.08.047.
3. A. Halloub *et al.*, "Intelligent food packaging film containing lignin and cellulose nanocrystals for shelf life extension of food," *Carbohydr Polym*, vol. 296, p. 119972, Nov. 2022, doi: 10.1016/j.carbpol.2022.119972.
4. P. Zou *et al.*, "Advances in characterisation and biological activities of chitosan and chitosan oligosaccharides," *Food Chem*, vol. 190, no. 12, pp. 1174–1181, 2016, doi: 10.1016/j.foodchem.2015.06.076.
5. S. Kumari and R. Kishor, *Chitin and Chitosan: Origin, Properties, and Applications*. INC, 2020. doi: 10.1016/B978-0-12-817970-3.00001-8.
6. R. F. Bombaldi de Souza, F. C. Bombaldi de Souza, A. Thorpe, D. Mantovani, K. C. Popat, and Â. M. Moraes, "Phosphorylation of chitosan to improve osteoinduction of chitosan/xanthan-based scaffolds for periosteal tissue engineering," *Int J Biol Macromol*, vol. 143, pp. 619–632, 2020, doi: 10.1016/j.ijbiomac.2019.12.004.
7. S. Kumari, P. Rath, A. Sri Hari Kumar, and T. N. Tiwari, "Extraction and characterization of chitin and chitosan from fishery waste by chemical method," *Environ Technol Innov*, vol. 3, pp. 77–85, 2015, doi: 10.1016/j.eti.2015.01.002.
8. J. I. Lozano-Navarro *et al.*, "Antimicrobial, optical and mechanical properties of chitosan-starch films with natural extracts," *Int J Mol Sci*, vol. 18, no. 5, p. 997, 2017, doi: 10.3390/ijms18050997.
9. M. R. Hossain, A. K. Mallik, and M. M. Rahman, "Fundamentals of Chitosan for Biomedical Applications," *Handbook of Chitin and Chitosan*, Sreerag Gopi, Sabu Thomas, Anitha Pius (eds.), pp. 199–230, Elsevier, 2020. doi: 10.1016/b978-0-12-817966-6.00007-8.
10. Y. S. Puvvada, S. Vankayalapati, and S. Sukhavasi, "Extraction of chitin from chitosan from exoskeleton of shrimp for application in the pharmaceutical industry," *Int Curr Pharm J*, vol. 1, no. 9, pp. 258–263, 2012, doi: 10.3329/icpj.v1i9.11616.
11. I. Hamed, F. Özogul, and J. M. Regenstein, "Industrial applications of crustacean by-products (chitin, chitosan, and chitooligosaccharides): A review," *Trends Food Sci Technol*, vol. 48, pp. 40–50, 2016, doi: 10.1016/j.tifs.2015.11.007.
12. N. H. Marei, E. A. El-Samie, T. Salah, G. R. Saad, and A. H. M. Elwahy, "Isolation and characterization of chitosan from different local insects in Egypt," *Int J Biol Macromol*, vol. 82, pp. 871–877, 2016, doi: 10.1016/j.ijbiomac.2015.10.024.
13. H. Srinivasan, V. Kanayairam, and R. Ravichandran, "Chitin and chitosan preparation from shrimp shells Penaeus monodon and its human ovarian cancer cell line, PA-1," *Int J Biol Macromol*, vol. 107, no. Part A, pp. 662–667, 2018, doi: 10.1016/j.ijbiomac.2017.09.035.
14. Y. Wang *et al.*, "Crystalline structure and thermal property characterization of chitin from Antarctic krill (Euphausia superba)," *Carbohydr Polym*, vol. 92, no. 1, pp. 90–97, 2013, doi: 10.1016/j.carbpol.2012.09.084.
15. F. Sedaghat, M. Yousefzadi, H. Toiserkani, and S. Najafipour, "Bioconversion of shrimp waste Penaeus merguiensis using lactic acid fermentation: An alternative procedure for chemical extraction of chitin and chitosan," *Int J Biol Macromol*, vol. 104, pp. 883–888, 2017, doi: 10.1016/j.ijbiomac.2017.06.099.
16. I. Younes, O. Ghorbel-Bellaaj, R. Nasri, M. Chaabouni, M. Rinaudo, and M. Nasri, "Chitin and chitosan preparation from shrimp shells using optimized enzymatic deproteinization," *Process Biochem*, vol. 47, no. 12, pp. 2032–2039, 2012, doi: 10.1016/j.procbio.2012.07.017.

17. Y. Yang, S. Wang, Y. Wang, X. Wang, Q. Wang, and M. Chen, "Advances in self-assembled chitosan nanomaterials for drug delivery," *Biotechnol Adv*, vol. 32, no. 7, pp. 1301–1316, 2014, doi: 10.1016/j.biotechadv.2014.07.007.
18. P. J. Rossky, "Exploring nanoscale hydrophobic hydration," *Faraday Discuss*, vol. 146, pp. 13–18, 2010, doi: 10.1039/c005270c.
19. H. S. Mansur and H. S. Costa, "Nanostructured poly(vinyl alcohol)/bioactive glass and poly(vinyl alcohol)/chitosan/bioactive glass hybrid scaffolds for biomedical applications," *Chem Eng J*, vol. 137, no. 1, pp. 72–83, 2008, doi: 10.1016/j.cej.2007.09.036.
20. F. Mirahmadi, M. Tafazzoli-Shadpour, M. A. Shokrgozar, and S. Bonakdar, "Enhanced mechanical properties of thermosensitive chitosan hydrogel by silk fibers for cartilage tissue engineering," *Mater Sci Eng C*, vol. 33, no. 8, pp. 4786–4794, 2013, doi: 10.1016/j.msec.2013.07.043.
21. T. M. Budnyak, I. V. Pylypchuk, V. A. Tertykh, E. S. Yanovska, and D. Kolodynska, "Synthesis and adsorption properties of chitosan-silica nanocomposite prepared by sol-gel method," *Nanoscale Res Lett*, vol. 10, no. 1, pp. 1–10, 2015, doi: 10.1186/s11671-014-0722-1.
22. F. L. Mi, S. S. Shyu, Y. B. Wu, S. T. Lee, J. Y. Shyong, and R. N. Huang, "Fabrication and characterization of a sponge-like asymmetric chitosan membrane as a wound dressing," *Biomaterials*, vol. 22, no. 2, pp. 165–173, 2001, doi: 10.1016/S0142-9612(00)00167-8.
23. A. P. Quist, E. Pavlovic, and S. Oscarsson, "Recent advances in microcontact printing," *Anal Bioanal Chem*, vol. 381, no. 3, pp. 591–600, 2005, doi: 10.1007/s00216-004-2847-z.
24. S. Ifuku and H. Saimoto, "Chitin nanofibers: Preparations, modifications, and applications," *Nanoscale*, vol. 4, no. 11, pp. 3308–3318, 2012, doi: 10.1039/c2nr30383c.
25. V. G. L. Souza, J. R. A. Pires, C. Rodrigues, I. M. Coelhoso, and A. L. Fernando, "Chitosan composites in packaging industry-current trends and future challenges," *Polymers (Basel)*, vol. 12, no. 2, pp. 1–16, 2020, doi: 10.3390/polym12020417.
26. S. Sahraee and J. M. Milani, *Chitin and Chitosan-Based Blends, Composites, and Nanocomposites for Packaging Applications*. INC, 2020. doi: 10.1016/B978-0-12-817968-0.00008-1.
27. Y. Luo and Q. Wang, "Recent development of chitosan-based polyelectrolyte complexes with natural polysaccharides for drug delivery," *Int J Biol Macromol*, vol. 64, pp. 353–367, 2014, doi: 10.1016/j.ijbiomac.2013.12.017.
28. Y. X. Xu, K. M. Kim, M. A. Hanna, and D. Nag, "Chitosan-starch composite film: Preparation and characterization," *Ind Crops Prod*, vol. 21, no. 2, pp. 185–192, 2005, doi: 10.1016/j.indcrop.2004.03.002.
29. M. Jindal, V. Kumar, V. Rana, and A. K. Tiwary, "An insight into the properties of Aegle marmelos pectin-chitosan cross-linked films," *Int J Biol Macromol*, vol. 52, no. 1, pp. 77–84, 2013, doi: 10.1016/j.ijbiomac.2012.10.020.
30. F. Freitas, V. D. Alves, M. A. Reis, J. G. Crespo, and I. M. Coelhoso, "Microbial polysaccharide-based membranes: Current and future applications," *J Appl Polym Sci*, vol. 131, no. 6, pp. 1–11, 2014, doi: 10.1002/app.40047.
31. S. Rivero, M. A. García, and A. Pinotti, "Composite and bi-layer films based on gelatin and chitosan," *J Food Eng*, vol. 90, no. 4, pp. 531–539, 2009, doi: 10.1016/j.jfoodeng.2008.07.021.
32. C. O. Ferreira, C. A. Nunes, I. Delgadillo, and J. A. Lopes-da-Silva, "Characterization of chitosan-whey protein films at acid pH," *Food Res Int*, vol. 42, no. 7, pp. 807–813, 2009, doi: 10.1016/j.foodres.2009.03.005.
33. M. Z. Elsabee and E. S. Abdou, "Chitosan based edible films and coatings: A review," *Mater Sci Eng C*, vol. 33, no. 4, pp. 1819–1841, 2013, doi: 10.1016/j.msec.2013.01.010.
34. M. Kurek, S. Galus, and F. Debeaufort, "Surface, mechanical and barrier properties of bio-based composite films based on chitosan and whey protein," *Food Packag Shelf Life*, vol. 1, no. 1, pp. 56–67, 2014, doi: 10.1016/j.fpsl.2014.01.001.
35. V. G. L. Souza and A. L. Fernando, "Nanoparticles in food packaging: Biodegradability and potential migration to food-A review," *Food Packag Shelf Life*, vol. 8, pp. 63–70, 2016, doi: 10.1016/j.fpsl.2016.04.001.
36. J. R. A. Pires, V. G. L. de Souza, and A. L. Fernando, "Chitosan/montmorillonite bionanocomposites incorporated with rosemary and ginger essential oil as packaging for fresh poultry meat," *Food Packag Shelf Life*, vol. 17, no. December 2017, pp. 142–149, 2018, doi: 10.1016/j.fpsl.2018.06.011.
37. H. Celebi and A. Kurt, "Effects of processing on the properties of chitosan/cellulose nanocrystal films," *Carbohydr Polym*, vol. 133, pp. 284–293, 2015, doi: 10.1016/j.carbpol.2015.07.007.
38. H. Mao, C. Wei, Y. Gong, S. Wang, and W. Ding, "Mechanical and water-resistant properties of eco-friendly chitosan membrane reinforced with cellulose nanocrystals," *Polymers (Basel)*, vol. 11, no. 1, p. 166, 2019, doi: 10.3390/polym11010166.

39. H. P. S. Abdul Khalil *et al.*, "A review on chitosan-cellulose blends and nanocellulose reinforced chitosan biocomposites: Properties and their applications," *Carbohydr Polym*, vol. 150, pp. 216–226, 2016, doi: 10.1016/j.carbpol.2016.05.028.

40. J. R. A. Pires, V. G. L. Souza, and A. L. Fernando, "Valorization of energy crops as a source for nanocellulose production – Current knowledge and future prospects," *Ind Crops Prod*, vol. 140, no. August, p. 111642, 2019, doi: 10.1016/j.indcrop.2019.111642.

41. Y. Tang, X. Zhang, R. Zhao, D. Guo, and J. Zhang, "Preparation and properties of chitosan/guar gum/nanocrystalline cellulose nanocomposite films," *Carbohydr Polym*, vol. 197, no. May, pp. 128–136, 2018, doi: 10.1016/j.carbpol.2018.05.073.

42. S. M. Noorbakhsh-Soltani, M. M. Zerafat, and S. Sabbaghi, "A comparative study of gelatin and starch-based nano-composite films modified by nano-cellulose and chitosan for food packaging applications," *Carbohydr Polym*, vol. 189, no. September 2017, pp. 48–55, 2018, doi: 10.1016/j.carbpol.2018.02.012.

43. V. G. L. Souza, P. F. Rodrigues, M. P. Duarte, and A. L. Fernando, "Antioxidant migration studies in chitosan films incorporated with plant extracts," *J Renew Mater*, vol. 6, no. 5, pp. 548–558, 2018, doi: 10.7569/JRM.2018.634104.

44. J. Hafsa *et al.*, "Physical, antioxidant and antimicrobial properties of chitosan films containing Eucalyptus globulus essential oil," *LWT Food Sci Technol*, vol. 68, pp. 356–364, 2016, doi: 10.1016/j.lwt.2015.12.050.

45. G. Yuan, H. Lv, B. Yang, X. Chen, and H. Sun, "Physical properties, antioxidant and antimicrobial activity of chitosan films containing carvacrol and pomegranate peel extract," *Molecules*, vol. 20, no. 6, pp. 11034–11045, 2015, doi: 10.3390/molecules200611034.

46. Z. Kalaycıoğlu, E. Torlak, G. Akın-Evingür, İ. Özen, and F. B. Erim, "Antimicrobial and physical properties of chitosan films incorporated with turmeric extract," *Int J Biol Macromol*, vol. 101, pp. 882–888, 2017, doi: 10.1016/j.ijbiomac.2017.03.174.

47. V. G. L. Souza, A. L. Fernando, J. R. A. Pires, P. F. Rodrigues, A. A. S. Lopes, and F. M. B. Fernandes, "Physical properties of chitosan films incorporated with natural antioxidants," *Ind Crops Prod*, vol. 107, no. February, pp. 565–572, 2017, doi: 10.1016/j.indcrop.2017.04.056.

48. M. Raji, A. Halloub, A. el K. Qaiss, and R. Bouhfid, "Bioplastic-Based Nanocomposites for Smart Materials," *Handbook of Bioplastics and Biocomposites Engineering Applications*, Inamuddin, Tariq Altalhi (ed.), pp. 457–470, Wiley, Jan. 2023, doi: 10.1002/9781119160182.CH21.

49. V. G. L. Souza, J. R. A. Pires, É. T. Vieira, I. M. Coelhoso, M. P. Duarte, and A. L. Fernando, "Activity of chitosan-montmorillonite bionanocomposites incorporated with rosemary essential oil: From in vitro assays to application in fresh poultry meat," *Food Hydrocoll*, vol. 89, no. May 2018, pp. 241–252, 2019, doi: 10.1016/j.foodhyd.2018.10.049.

50. J. Quesada, E. Sendra, C. Navarro, and E. Sayas-Barberá, "Antimicrobial active packaging including chitosan films with thymus vulgaris l. Essential oil for ready-to-eat meat," *Foods*, vol. 5, no. 3, pp. 1–13, 2016, doi: 10.3390/foods5030057.

51. S. Lekjing, "A chitosan-based coating with or without clove oil extends the shelf life of cooked pork sausages in refrigerated storage," *Meat Sci*, vol. 111, pp. 192–197, 2016, doi: 10.1016/j.meatsci.2015.10.003.

52. W. Xu *et al.*, "Carboxymethyl chitosan/gelatin/hyaluronic acid blended-membranes as epithelia transplanting scaffold for corneal wound healing," *Carbohydr Polym*, vol. 192, no. September 2017, pp. 240–250, 2018, doi: 10.1016/j.carbpol.2018.03.033.

53. N. Charoenthai *et al.*, "Heparin-chitosan complexes stimulate wound healing in human skin," *J Sci Ind Res (India)*, vol. 4, no. 2, pp. 27–52, 2012.

54. M. Ishihara *et al.*, "Photocrosslinkable chitosan as a dressing for wound occlusion and accelerator in healing process," *Biomaterials*, vol. 23, no. 3, pp. 833–840, 2002, doi: 10.1016/S0142-9612(01)00189-2.

55. K. Obara *et al.*, "Photocrosslinkable chitosan hydrogel containing fibroblast growth factor-2 stimulates wound healing in healing-impaired db/db mice," *Biomaterials*, vol. 24, no. 20, pp. 3437–3444, 2003, doi: 10.1016/S0142-9612(03)00220-5.

56. H. Sosiati, D. A. Wijayanti, K. Triyana, and B. Kamiel, "Morphology and crystallinity of sisal nanocellulose after sonication," *AIP Conf Proc*, vol. 1877, no. September 2017, pp. 030003-1–030003-7, 2017, doi: 10.1063/1.4999859.

57. H. K. No and S. P. Meyers, "Application of chitosan for treatment of wastewaters," *Rev Environ Contam Toxicol*, vol. 163, pp. 1–27, 2000, doi: 10.1007/978-1-4757-6429-1_1.

58. R. R. L. Vidal and J. S. Moraes, "Removal of organic pollutants from wastewater using chitosan: A literature review," *Int J Environ Sci Technol*, vol. 16, no. 3, pp. 1741–1754, 2019, doi: 10.1007/s13762-018-2061-8.

59. G. O. Odu and O. E. Charles-Owaba, "Review of multi-criteria optimization methods – Theory and applications," *IOSR J Eng (IOSRJEN)*, vol. 3, no. 10 October, pp. 1–14, 2012.
60. L. Rahmi and R. Nurfatimah, "Preparation of polyethylene glycol diglycidyl ether (PEDGE) cross-linked chitosan/activated carbon composite film for Cd^{2+} removal," *Carbohydr Polym*, vol. 199, pp. 499–505, 2018, doi: 10.1016/j.carbpol.2018.07.051.
61. E. A. Abdelrahman and R. M. Hegazey, "Utilization of waste aluminum cans in the fabrication of hydroxysodalite nanoparticles and their chitosan biopolymer composites for the removal of Ni(II) and Pb(II) ions from aqueous solutions: Kinetic, equilibrium, and reusability studies," *Microchem J*, vol. 145, no. October 2018, pp. 18–25, 2019, doi: 10.1016/j.microc.2018.10.016.
62. V. Jahed, E. Vasheghani-Farahani, F. Bagheri, A. Zarrabi, H. H. Jensen, and K. L. Larsen, "Quantum dots-βcyclodextrin-histidine labeled human adipose stem cells-laden chitosan hydrogel for bone tissue engineering," *Nanomedicine*, vol. 27, p. 102217, 2020, doi: 10.1016/j.nano.2020.102217.
63. R. Niranjan, C. Koushik, S. Saravanan, A. Moorthi, M. Vairamani, and N. Selvamurugan, "A novel injectable temperature-sensitive zinc doped chitosan/β-glycerophosphate hydrogel for bone tissue engineering," *Int J Biol Macromol*, vol. 54, no. 1, pp. 24–29, 2013, doi: 10.1016/j.ijbiomac.2012.11.026.
64. T. K. Giri, A. Thakur, A. Alexander, Ajazuddin, H. Badwaik, and D. K. Tripathi, "Modified chitosan hydrogels as drug delivery and tissue engineering systems: Present status and applications," *Acta Pharm Sin B*, vol. 2, no. 5, pp. 439–449, 2012, doi: 10.1016/j.apsb.2012.07.004.
65. S. Peers, A. Montembault, and C. Ladavière, "Chitosan hydrogels for sustained drug delivery," *J Control Release*, vol. 326, no. February, pp. 150–163, 2020, doi: 10.1016/j.jconrel.2020.06.012.
66. A. M. Craciun, L. Mititelu Tartau, M. Pinteala, and L. Marin, "Nitrosalicyl-imine-chitosan hydrogels based drug delivery systems for long term sustained release in local therapy," *J Colloid Interface Sci*, vol. 536, pp. 196–207, 2019, doi: 10.1016/j.jcis.2018.10.048.
67. L. Das, P. Das, A. Bhowal, and C. Bhattacharjee, "Synthesis of hybrid hydrogel nano-polymer composite using graphene oxide, chitosan and PVA and its application in waste water treatment," *Environ Technol Innov*, vol. 18, p. 100664, 2020, doi: 10.1016/j.eti.2020.100664.
68. P. Mohammadzadeh Pakdel and S. J. Peighambardoust, "Review on recent progress in chitosan-based hydrogels for wastewater treatment application," *Carbohydr Polym*, vol. 201, no. August, pp. 264–279, 2018, doi: 10.1016/j.carbpol.2018.08.070.
69. G. Crini, G. Torri, E. Lichtfouse, G. Z. Kyzas, L. D. Wilson, and N. Morin-Crini, "Dye removal by biosorption using cross-linked chitosan-based hydrogels," *Environ Chem Lett*, vol. 17, no. 4, pp. 1645–1666, 2019, doi: 10.1007/s10311-019-00903-y.
70. S. Zhao, W. C. Tsen, and C. Gong, "3D nanoflower-like layered double hydroxide modified quaternized chitosan/polyvinyl alcohol composite anion conductive membranes for fuel cells," *Carbohydr Polym*, vol. 256, no. October 2020, p. 117439, 2021, doi: 10.1016/j.carbpol.2020.117439.
71. F. Hu *et al.*, "Preparation and properties of chitosan/acidified attapulgite composite proton exchange membranes for fuel cell applications," *J Appl Polym Sci*, vol. 137, no. 36, pp. 1–9, 2020, doi: 10.1002/app.49079.
72. J. Wang *et al.*, "Proton exchange membrane based on chitosan and solvent-free carbon nanotube fluids for fuel cells applications," *Carbohydr Polym*, vol. 186, no. September 2017, pp. 200–207, 2018, doi: 10.1016/j.carbpol.2018.01.032.
73. L. Wang *et al.*, "Chitosan for constructing stable polymer-inorganic suspensions and multifunctional membranes for wound healing," *Carbohydr Polym*, vol. 285, no. January, p. 119209, 2022, doi: 10.1016/j.carbpol.2022.119209.
74. T. M. Tamer *et al.*, "MitoQ loaded chitosan-hyaluronan composite membranes for wound healing," *Materials*, vol. 11, no. 4, pp. 1–14, 2018, doi: 10.3390/ma11040569.
75. A. Enumo, D. F. Argenta, G. C. Bazzo, T. Caon, H. K. Stulzer, and A. L. Parize, "Development of curcumin-loaded chitosan/pluronic membranes for wound healing applications," *Int J Biol Macromol*, vol. 163, pp. 167–179, 2020, doi: 10.1016/j.ijbiomac.2020.06.253.
76. Z. Zhou *et al.*, "Biomaterials based on *N,N,N*-trimethyl chitosan fibers in wound dressing applications," *Int J Biol Macromol*, vol. 89, pp. 471–476, 2016, doi: 10.1016/j.ijbiomac.2016.02.036.
77. M. A. Matica, F. L. Aachmann, A. Tøndervik, H. Sletta, and V. Ostafe, "Chitosan as a wound dressing starting material: Antimicrobial properties and mode of action," *Int J Mol Sci*, vol. 20, no. 23, pp. 1–33, 2019, doi: 10.3390/ijms20235889.
78. S. Ahmadi Majd, M. Rabbani Khorasgani, S. J. Moshtaghian, A. Talebi, and M. Khezri, "Application of chitosan/PVA nano fiber as a potential wound dressing for streptozotocin-induced diabetic rats," *Int J Biol Macromol*, vol. 92, pp. 1162–1168, 2016, doi: 10.1016/j.ijbiomac.2016.06.035.

79. M. Z. Albanna, T. H. Bou-Akl, O. Blowytsky, H. L. Walters, and H. W. T. Matthew, "Chitosan fibers with improved biological and mechanical properties for tissue engineering applications," *J Mech Behav Biomed Mater*, vol. 20, pp. 217–226, 2013, doi: 10.1016/j.jmbbm.2012.09.012.

80. S. Ranganathan, K. Balagangadharan, and N. Selvamurugan, "Chitosan and gelatin-based electrospun fibers for bone tissue engineering," *Int J Biol Macromol*, vol. 133, pp. 354–364, 2019, doi: 10.1016/j.ijbiomac.2019.04.115.

81. S. Yang *et al.*, "*N*-Carboxyethyl chitosan fibers prepared as potential use in tissue engineering," *Int J Biol Macromol*, vol. 82, pp. 1018–1022, 2016, doi: 10.1016/j.ijbiomac.2015.10.078.

82. F. Liang, T. Hou, S. Li, L. Liao, P. Li, and C. Li, "Elastic, super-hydrophobic and biodegradable chitosan sponges fabricated for oil/water separation," *J Environ Chem Eng*, vol. 9, no. 5, p. 106027, 2021, doi: 10.1016/j.jece.2021.106027.

83. Z. Yin, X. Sun, M. Bao, and Y. Li, "Construction of a hydrophobic magnetic aerogel based on chitosan for oil/water separation applications," *Int J Biol Macromol*, vol. 165, pp. 1869–1880, 2020, doi: 10.1016/j.ijbiomac.2020.10.068.

84. N. Cao *et al.*, "Facile synthesis of fluorinated polydopamine/chitosan/reduced graphene oxide composite aerogel for efficient oil/water separation," *Chem Eng J*, vol. 326, pp. 17–28, 2017, doi: 10.1016/j.cej.2017.05.117.

85. M. Raji *et al.*, "pH-indicative films based on chitosan-PVA/sepiolite and anthocyanin from red cabbage: Application in milk packaging," *J Bionic Eng*, vol. 19, 837–851, Feb. 2022, doi: 10.1007/s42235-022-00161-9.

9 Glucans
Safe-by-Design and Applications

Nishat Khan and Seema Garg
Amity University

9.1 INTRODUCTION

Glucans are the most common polysaccharides found in nature. There is a wide range of molecular weight and configuration depending on the source, which is made up of more than ten distinct forms of monosaccharides linked by glycosidic bonds and is found in many organisms [1–4]. It is known as a bioactive glucan or biological response modifier because it can promote body health, control cell differentiation, regulate cell growth and senescence, and participate in cell recognition, cell metabolism, embryonic development, viral infection, immune response, and other life activities (BRM) The chemical structure of glucan determines its biological activity; glucans are identified by the presence of side chains attached to the backbone and emanating outward like tree branches [5]. The side chains' frequency and nature strongly influence the glucan's ability to help facilitate binding to surface receptors on target cells, thereby influencing the glucan's effectiveness as an immunostimulant. As a result, research into the ultrastructure-activity relationship of glucan can provide theoretical guidance for screening biological activities. Active glucans have a high safety profile, few side effects, a good curative effect, and a diverse source. They function in immune regulation, antitumor, antivirus, antioxidation, and other ways. Natural glucans and synthetic glucans are the two types of general glucans, whereas natural active glucans are found in animals, plants, and microorganisms. Lentinan and *Ganoderma lucidum* glucan, for example, were frequently used for antineoplastic treatment [5]. β-Glucans are naturally occurring polysaccharides. These glucose polymers are constituents of the cell wall of certain pathogenic bacteria and fungi. The healing and immune-stimulating properties of mushrooms have been known for thousands of years in Eastern countries. These mushrooms contain biologically active polysaccharides that mostly belong to the group of β-glucans. These substances increase host immune defense by activating the complement system, enhancing macrophages and natural killer cell function [6].

9.2 SOURCES OF ACTIVE GLUCAN

Active glucans come from a wide range of sources, mainly including fungal glucans, marine biological glucans, and plant glucans. Fungal glucans are active glucans isolated from mycelia, fruiting bodies, and fermentation broth of fungi. According to different sources, marine biological glucans can be divided into seaweed glucans, marine animal glucans, and marine microbial glucans. Plant glucans can be divided into intracellular glucans, cell wall glucans, and extracellular glucans according to their existing positions. Lentinan, *Coriolus versicolor* glucan, schizophyllan, and polyporus glucan have been used clinically, and fungal glucan health products such as lentinan tablets, *Hericium erinaceus* glucan tablets, and *Ganoderma lucidum* glucan tablets have also been put on the market. Common main active glucans and sources were shown in Table 9.1 [7].

DOI: 10.1201/9781003265054-9

TABLE 9.1

Common Main Active Glucans and Sources

Active Glucan	Source
Fungal glucan	Mushrooms, black fungus, marine shiitake mushrooms, etc.
Plant glucan	Chinese wolfberry, pumpkin, Astragalus, Aloe vera, jujube, tea, achyranthes root, kiwi fruit, etc.
Marine glucan	Spirulina, brown algae, marine algae, such as *Porphyridium*, crustaceans, fish, shellfish, etc.

9.3   STRUCTURAL ACTIVITY OF GLUCAN

Different types of glucans have different structural units, and their biological activities are very different. This is because the structural units largely determine the biological activities of glucans. The first one is glucan, which is the basic structural unit of many biological glucans in nature, including animals, plants, and microorganisms. Among glucans, glucan is the most important structural unit, for example, lentinan glucan, schizophyllum glucan, *Grifola frondosa* mycelium glucan, *Ganoderma lucidum* glucan, *Poria* glucan, etc [8]. The relationship between the primary structure and higher structure of glucan and its biological activity is called the structure-activity relationship of glucan [9]. It shows that the structure of glucan is an important factor that affects the expression of its biological activities. At present, the most important thing is to study the structure-activity relationships of glucans from the following aspects: monosaccharide composition, category of glycosidic bond, branch chain position, spatial configuration, substituent type, number, etc. In addition, glucans also have physical and chemical properties, such as relative molecular weight, viscosity, solubility, and so on, which can also affect the biological activities of glucans. The primary structure of glucan includes monosaccharide composition, connection mode, category of glycosidic bond, branching degree, etc. Various factors have different effects on the activities of glucans according to the influence of various factors on biological activities, the order is as follows: main chain structure (category of glycosidic bond, monosaccharide connection mode)>branching degree (presence or absence of branching)>monosaccharide composition (types of monosaccharide in homoglucan or heteroglucan) [10]. Glucans composed of other kinds of sugar units also have effects on their biological activities. For example, the main chain of tremella glucan is to use mannan as a sugar unit linked by α-(13)-glycosidic bond. Studies have found that tremella glucan has immunoregulatory, antithrombotic, anticoagulant, antitumor, and other functions. Mannan obtained from yeast cell walls could inhibit mutation of human cells and had an antioxidant effect. The galactomannan and arabinogalactan were the main insoluble glucans in waste coffee grounds, accounting for 24%–39% of their dry weight. The glucan consists of a linear β-(1–4)-linked mannan backbone connected to a galactose at O-6 by a (1–6)-glycosidic bond [11]. Based on these characteristics, galactomannan could be used as an additive in foods, infant milk powder, etc. The chemical structure of active glucan is the basis of its biological activity. The structure-activity relationships of glucans refer to the relationship between the primary and advanced structures of glucans as shown in the Figure 9.1.

9.4   EFFECT OF GLYCOSIDIC BONDING

Glycosidic bond refers to the acetal or ketal bond formed between the semiacetal group of one sugar and the hydroxyl group, amine, or mercapto group (such as alcohol, sugar, purine, or pyrimidine) of another molecule. Common glycosidic bonds are the N-glycosidic bond and the O-glycosidic bond. However, the connection between the glycosyl group and glycosyl group in glucan is called the glycosidic bond. Different types of glycosidic bonds have different biological activities. For example, the main chain sugar unit of lentinan and amylose is glucan, but their biological activities are quite different. The reason is that the former has a (1→3)-glycosidic bond, which has strong antitumor activity, while the latter is linked with a (1→4)-glycosidic bond and

**THE STRUCTURAL ACTIVITY
RELATIONSHIP OF GLUCAN**

PRIMARY
STRUCTURE

ADVANCED
STRUCTURE

Monosaccharide composition type of glycosidic
bond order of connection configuration of
anomeric carbon, molecular weight position, and
branch length.

Three-dimensional structure, tertiary
structure, etc.

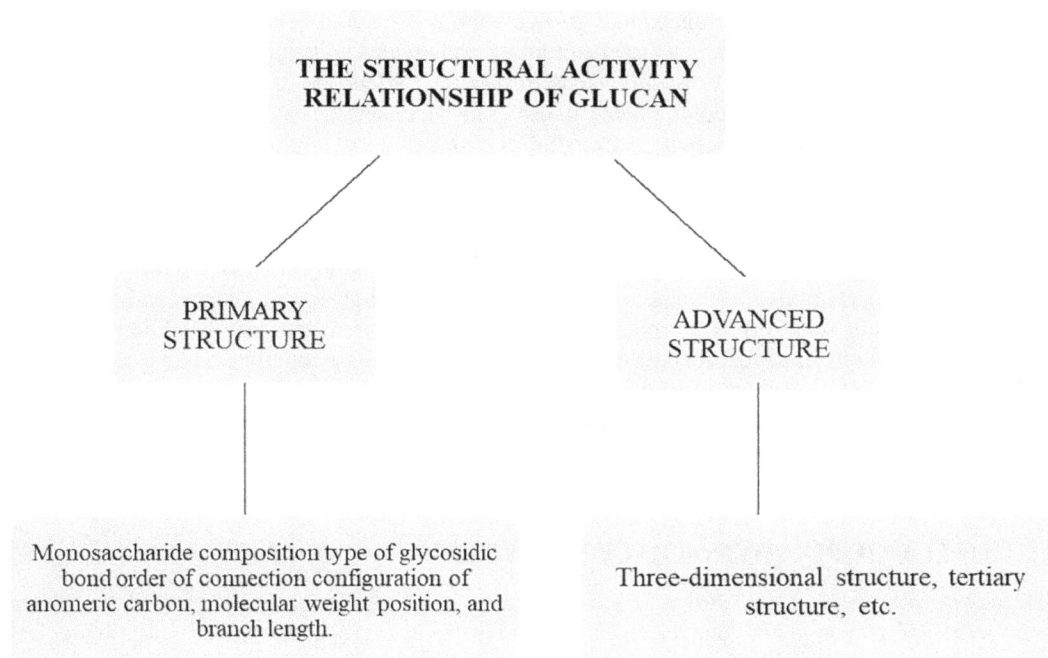

FIGURE 9.1 Structure-activity relationships of glucans.

has no biological activity, which indicates that the type of glycosidic bond also affects the biological activity of glucan [12].

The glycosidic bond type of glucan with strong biological activity is (1→3)-glycosidic bond, such as lentinan, *Ganoderma* glucan of Polygonatum, etc. β-(1→3)-Linked glucan had a strong inhibitory effect on S180 and a strong antitumor effect. For example, anticoagulant glucan is a homogeneous glucan with (1→3)-glucan as the main chain; lichen glucan is a mixed glucan composed of 66% (1→4)-glucan and 33% (1→3)-glucan. Under the same substitution conditions, the inhibition rate of anticoagulant glucan derivatives to tumor cells was much higher than that of lichen glucan derivatives [13].

9.5 EFFECT OF MAIN CHAIN CONFIGURATION

The activity of glucan is not only affected by the sugar unit and glycosidic bond but also affected by the configuration of the main chain. If the configuration was different, its biological activity might be completely different. Most glucans with antitumor activity have the main chain structure of β-(1→3)-D-glucan, while glucans with α-glucan as the main chain structure generally have no antitumor activity, such as active glucans in edible and medicinal bacteria, lentinan, polyporus glucan, and so on. The main chain structure of the antitumor active component in the glucan was β-(1→3)-D-glucan [2].

9.6 EFFECT OF SUBSTITUTION DEGREE, LENGTH, AND POSITION OF BRANCH

Through the analysis of these glucans with antitumor activity, it is found that their structures contain water-soluble D-glucan, and most of them are straight-chain glucans without extra-long branch chains. Therefore, the length and shape of the branch chain may affect the antitumor activity of glucans. For example, although the main chain structure of Bachman is β-(1→3)-D-glucan, each β-(1→5)-bonded glucosyl branch chain of Bachman is separated from 1 to 2 β-(1→6)-bonded glucosyl groups. Therefore, Bachman has no antitumor activity due to its long branched chain, but its main chain structure indicates that Bachman has an antitumor effect and immunomodulatory ability. So, we need to modify its structure to make full use of its characteristics and reduce its branch chain length, so it can have antitumor activity. Some studies have found that controlled oxidative hydrolysis can effectively reduce the branch chain length of Bachman, thus making it biologically active. The antitumor activity of branched glucan is negatively correlated with the degree of branching, that is to say, the biological activity of glucan decreases with the increase of the degree of branching. When the degree of branching exceeded four sugar units, the glucan lost its biological activity [14].

The excessive substitution will lead to a reduction of hydroxyl groups in glucans and poor water solubility, but water solubility is a prerequisite for glucans to exert their biological activities. For example, the level of sulfate affects the activity. Heparin had the best activity when each sugar residue contained 1.5–2.0 sulfates. When the sulfated glucan contained one sulfate group per hexose unit, it had the strongest inhibitory effect on the herpes simplex virus. The fungal pestalotan, which was a highly branched glucan, had a DS as high as 2.8, but the biological activity at this time was relatively low. The tumor inhibition rate in rats was only 57% [15].

9.7   EFFECT OF RELATIVE MOLECULAR WEIGHT

The biological activity of glucan is affected by molecular weight. If the relative molecular weight is too large, the glucan is not conducive to cross-membrane transport, thus affecting the biological activity of glucan. However, the relative molecular weight is too small to form a bioactive glucan structure. Glucan with a relative molecular weight of 90,000 had certain biological activity, and its activity decreased quickly when the relative molecular weight was greater than or less than 90,000 [16]. Heparin had the best anticoagulant activity when its relative molecular weight was from 4,000 to 12,000. Only glucans with a relative molecular weight of over 50,000 and a triple structure had antitumor activity. For example, the anti-HIV activity of glucan sulfate increased with the increase of the relative molecular weight, and the maximum activity was maintained when the molecular weight was between 10,000 and 500,000. However, when the molecular weight continued to increase, its antiviral activity gradually weakened. Generally speaking, glucans with molecular weights between 100 and 200 kD have higher biological activities, while glucans with molecular weights between 5 and 10 kD from the same source have no biological activity. For example, *Schizophyllum commune* glucan with a molecular weight greater than 100 kD had strong antitumor activity, while *Schizophyllum commune* glucan with a molecular weight less than 50 kD had no biological activity [16].

9.8   EFFECT OF VISCOSITY

The viscosity of glucan is mainly due to the hydrogen bond interaction between glucan molecules. It is mainly the comprehensive expression of internal friction between the sugar chain and solvent. Its size is related to the structure and conformation of glucans, charged properties, etc. It is not only positively correlated with solubility to some extent but also one of the key control factors of clinical efficacy. If the viscosity is too high, it is not conducive to the diffusion and absorption of glucan drugs. The viscosity of glucan can be reduced, and its activity can be improved by introducing a branched chain to destroy its hydrogen bond and degrading the main chain. Among them, controlling the main chain by ultrasound therapy is an effective method. For example, the schizophyllan has great application prospects. It has obvious effects in regulating immunity, antitumor, antiradiation, etc., but it has not been degraded. Due to the high molecular weight of schizophyllan, its aqueous solution has a high viscosity; it can be precipitated with ethanol and freeze-dried to form a crystalline structure, making it difficult to redissolve in water. After controlled ultrasound, it was partially depolymerized to reduce molecular weight and viscosity, but its antitumor activity remained unchanged due to its repetitive structure [17].

9.9 CLASSIFICATION OF GLUCANS

Glucans are complex polysaccharides consisting of repeated units of d-glucose linked by glycosidic bonds. Generally, glucans are divided into classes as follows:

  i. α-Glucan
  ii. β-Glucan

9.9.1 α-GLUCANS

α-Glucans are polysaccharides comprising polymerized glucose joined by α-glycosidic linkages. The most commonly studied α-glucans concerning CBM research are starch, glycogen, and pullulan. The glucan polymers consist of α-D-glucosyl residues, connected via α-1,4- and α-1,6-glycosidic bonds. α-1,4-Glucan chains are connected via α-1,6-linkages occurring approximately every 24–30 glucose residues. While both, starch and glycogen, are chemically identical, major differences in their physico-chemical properties are related to the molecular organization of glucan chains within the molecules. In starch, branching points are clustered, in contrast to glycogen, resulting in longer linear glucan chains, that can form double helices, and water is excluded. Pullulan is an imperative microbial exo-polymer commercially produced by the yeast-like fungus *Aureobasidium pullulans*. Its structure contains malt-ose repeating units which comprise two α-(1→4) linked glucopyranose rings attached to one glucopy-ranose ring through an α-(1→6) glycosidic bond. The co-existence of α-(1→6) and α-(1→4) glycosidic linkages endows distinctive physicochemical properties to pullulan. It is highly biocompatible, non-toxic, and non-carcinogenic [18]. The α-D-glucans have been widely distributed among the microbial world concerning plants and animals [19] and differ in their structure and function based on the source. Glucans associated with α-D-(1→3) linkages play a pivotal role in the pharmaceutical industry as a prebiotic [5] and as a biodegradable polymer in the plastic industry [20]. Nevertheless, the applica-tion potential of a majority of the α-D-(1→3) glucans in the health sector are limited because of their inherent insolubility. The solubility of the glucans majorly depends upon molecular weight, glycosidic linkage pattern, configuration (α or β), surface ionic properties, and degree of branching. These proper-ties in turn alter the application potential of glucans in food and non-food industries [5]. Some of the glucans are amorphous and soluble in hot water, because of their homogenous nature, and branching structure which can reduce the inter and intra-molecular interactions between polymer segments [5]. For example, glucans with (1→6) branching lead to easier solubility because of a largely favorable entropy of the solution. However, high molecular weight glucans exhibiting ordered structures with lin-ear chains (especially linear 1→3 linkages) are water insoluble and microfibrillar in nature. These can act as a structural element of fungal cell walls [20]. Therefore, to improve the solubility and the appli-cation potential of the glucans, it is essential to improve their hydrophilicity. In this regard, molecular weight reduction by specific techniques such as chemical derivatization and physical irradiation plays a significant role and has also been reported in earlier studies. The majority of the α-D-(1→3) glucans upon derivatization have shown improved water solubility as well as bioactivity profile [21].

α-Glucan-coated starch (α-GCS) is produced at temperatures below the general gelatinization temperature of raw NCS; thus, it is assumed that α-GCS retains the structural characteristics and physicochemical properties of raw starch. Here, we hypothesize that the degree of recrystallization can be reduced at the initial stage of retrogradation because it is possible to control premature ret-rogradation by inhibiting the loss of the amylose sheet layer by α-glucan coating, which affects the early stage of retrogradation and increases the degree of amylose aggregation.

9.9.2 SOURCES OF α-GLUCAN

The structure of α-D-glucans differs mainly in their α-glycosidic linkages, arrangement of glu-cose units in the chain (linear or branched) as well as source. Glycogen or other similar α-D-(1→4) glucans have been reported in more than 50 different bacterial species. Many Gram-negative and Gram-positive bacteria as well as archaea have been reported to accumulate glycogen (containing branched α-D-(1→4)(1→6) linkages) as a storage material used under starvation conditions. However, lactic acid bacteria, belonging to the genera *Leuconostoc*, *Streptococcus*, and *Lactobacillus*, produce α-D-(1→3) glucan as a capsule on the outside of the cell wall. A variety of α-D-glucans such as dextran, mutant, alternan, and veteran are also produced by different bacterial species using sucrose as a substrate material. Source and linkage of different bacterial α-D-glucans are shown in Table 9.2.

α-D-glucans have been synthesized by a variety of natural sources such as bacteria, fungi, plants, algae, lichens, insects, and mammals. Their characteristics such as molecular weight, linkage pattern, and branching as well as function differs with the source (see Figure 9.1). Mostly α-D-(1→4) glucans function as an energy source in plants (starch), animals (glycogen), bacteria (glycogen), and fungi (glycogen, amylose). However, α-D-(1→3) glucans majorly existed as a cell wall element in fungi [13] and as an exopolysaccharide (EPS) in bacteria [2] (see Figure 9.1). Apart from natural sources, α-D-(1→3) glucans were also synthesized by in vitro methods using rDNA technology and recombinant microbes.

The lichen thalli are reported to be comprised of majorly three different types of polysaccharides such as α-D-glucans, β-D-glucans, galactomannans, and a few complex heteroglucans such as rhamnopyranosyl galactofuranan galactomannoglucans, and thamnolan. The major lichen polysaccharides of higher percentage (glucans, galactomannans) are part of a mycobiont showing primarily linear or little substitution in α, β-D-glucans and slight branching in galactomannans structure. The polysaccharides extracted from the part of photobiont in lichen symbiont are little in percentage (e.g., thamnolan). However, the localization of polysaccharides in the mycobiont has not been explored. It could be either a fungal cell wall structural component or a reserve polymer and could also be a fungal intracellular or part of intracellular material, which surrounds both fungal and algal cells.

Organisms that possess glucans with α-D-(1→4) and α-D-(1→6) linkages, primarily functioning as storage polymers. They are intracellular in nature and resemble starch. In contrast to glycogen granules observed in the prokaryotic cyanobacteria and distinct α-polyglucan from *Cyanobacterium* sp. MBIC10216, *Myxosarcina burdens*, and *Synechococcus* sp. BG043511, a new water extractable α-D-glucan with α-D-(1→6) linkages connected through (1→3) branches were reported from *Chlorella vulgaris*, reported to exhibit immune-enhancing functionality. Plants also synthesize α-D-glucans, popularly known as starch molecules and utilized mainly as a storage material for energy. However, apart from the storage polysaccharide (starch), some of the plants also synthesize glycogen like α-D-(1→4) glucan with α-D-(1→4) linked branch attached to the O-6 position. The branching differs from source to source and exhibits pharmacological activity [22] (Figure 9.2).

TABLE 9.2

Source and Linkage of Different Bacterial α-D-Glucans

Species	Trivial Name	Type of α-Linkage	Solubility	Uses
Lactobacillus species	Dextran	A linear chain of (1→6) with the side chain of (1→3);(1→2)or (1→4)	Water soluble but insoluble if (1→3) units are >43%	As a capsule or slime
Lactobacillus reuteri	Mutant	A linear chain of (1→3) with the side chain of (1→6)	Water insoluble	As a capsule or slime
Leuconostoc mesenteroides	Alternan	Alternate (1→3)and (1→6) linkage with some degree of (1→3) branching	Water soluble	As a capsule or storage polymer

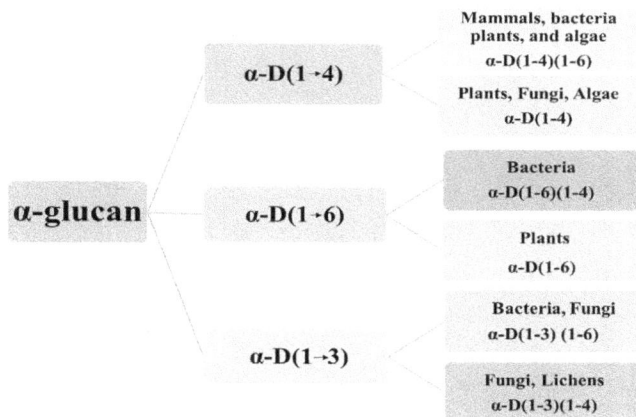

FIGURE 9.2   Some sources of α-glucan.

9.9.3   Stereoselective Synthesis of α-Glucans

Even though great progress in the synthesis of (complex) oligosaccharides, no general solution for the stereoselective development of difficult glycosidic bonds, such as 1,2-cis and 2-deoxy linkages, exists. The immense variation in carbohydrate building blocks and the various mechanism of action pathways that can be followed in the union of these are at the root of this persistent problem. The majority of glycosylation reactions rely on the activation of a glycosyl donor with a (Lewis) acid catalyst to produce a strong electrophile, which can be a covalent species, a close ion pair, or a solvent-separated ion pair in which the glycosyl oxocarbenium ion and the counterion are fully dissociated (as shown in Figure 9.3). Triflate-based activators are most commonly used, and a plethora of π-conjugated anomeric triflates have been characterized over the last two decades. These triflates may participate in an SN_2-type substitution reaction, but they are more commonly used as a reservoir to hold the more reactive glycosyl cation-triflate ion pair, resulting in SN_1-type reactions. The equilibrium of the covalent species and ion pairs, combined with the reactivity of the incoming nucleophile the acceptor, determines which pathway(s) will be followed. The activity of the donor-building block is determined by the natural position of the functional groups on the carbohydrate ring, and the different reactivity of donor glycosides has been used in one-pot chemo-selective glycosylation sequences based on reactivity. It is also well appreciated but less well studied that the reactivity of the acceptor alcohol can differ as a result of the protecting/functional group pattern on the ring, and the intrinsic reactivity difference between primary and secondary alcohols often leads to a different stereochemical outcome when glycosylating these acceptors. It is a tremendous challenge to design a general glycosylation strategy that accommodates the varying reactivity of different donor–acceptor glycoside combinations and ensures a fully stereoselective glycosylation process. An attractive way to modulate the reactivity of a glycosyl donor is through the use of an exogenous nucleophile that can be added to the coupling reaction. These nucleophilic additives or reactivity modulators react with the activated donor to form new covalent species as shown in the figure. Various additives have been probed over the years, including sulfides, sulfoxides/sulfinamides, phosphine oxides, amides and formamides, and iodide-based reagents, and stereoselective 1,2-cis-glycosylation procedures have been reported based on their use. The most often invoked mechanistic rationale to account for the observed stereoselectivity involves the generation of a stable α-covalent species (often identified and characterized by NMR spectroscopy), which is in equilibrium with its less stable and more reactive β-counterpart (often not detected by NMR), following an in situ isomerization kinetic scenarios as first introduced by Lemieux and coworkers. We reasoned that modulation of donor reactivity through external nucleophiles would

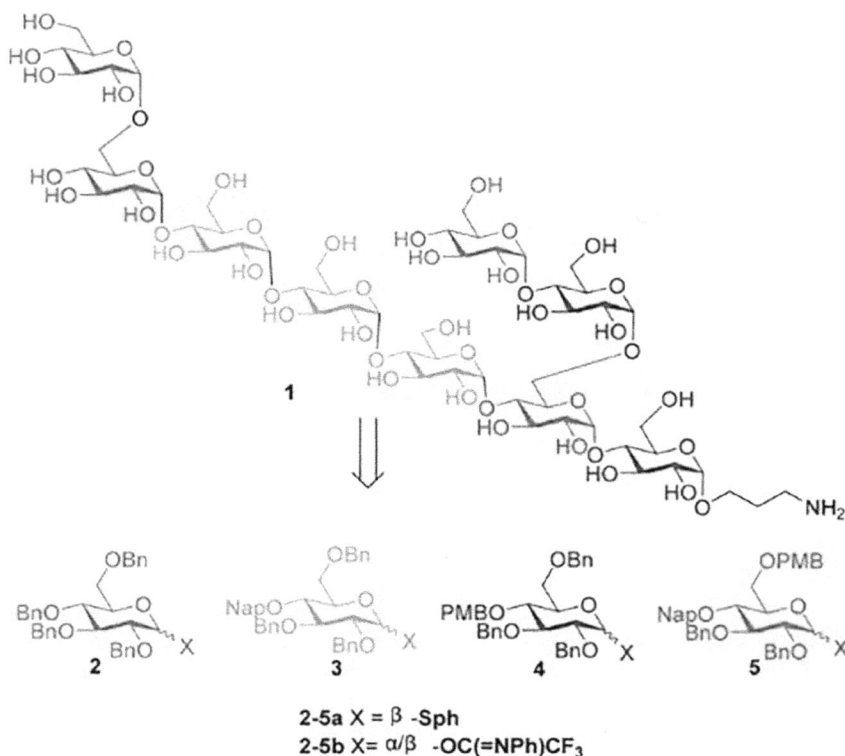

FIGURE 9.3   Synthesis of α-glucans.

be very attractive to match the reactivity of acceptoralcohols of different nucleophilicity to achieve fully stereoselective glycosylation reactions with both partners. We hear a re-report of how a single type of donor glycoside can be used for the fully stereoselective glycosylation of both secondary alcohol acceptors. Different additives have been used to accommodate the intrinsic reactivity that differs between these two types of alcohol. Key to the success of the strategy is protecting the group strategy that ensures identical reactivity of the parent donor-building blocks used so that the reactivity of the system is under the direct control of the active additive used. We show the applicability of this approach in the assembly of *Mycobacteriumium tuberculosis* (Mtb) derived branched α-glucans. Mtbα-Glucans play an important role in allowing the bacterium to evade the human immune system, and the molecular details behind this process remain obscure. To unravel how α-glucans interact with our immune system, well-defined α-glucans fragments will be valuable tools. The structures represent excellent target molecules to test the proposed synthetic strategy, as they only contain 1,2-cislinkages and carry different branches, necessitating flexible building blocks and stereoselective glycosylation methodology for the construction of glycosidic linkages to both primary and secondary alcohol functions [23].

All protecting groups on the building blocks are benzyl ethers, and the reactivity of the building blocks is as similar as possible. The only factors influencing the relative reactivity of the acceptors are the intrinsic difference between the primary and secondary alcohols and the effect of the growing chain length on the reactivity of the acceptor. Second, global protection of the donor glycoside with benzyl ethers leads to a donor that is as reactive as possible. Previous reports employing nucleophilic additives glycosylations have shown that this type of glycosylation is generally very slow. The reactive intermediates that are generated are relatively stable necessitating long reaction times. The use of acyl type protecting groups would make the system less reactive leading to even longer reaction times.

9.9.4   APPLICATIONS OF α-GLUCAN

- α-D-glucans have become the choice of natural biopolymer owing to their unique structural and functional properties with potential applications in the food, medical, pharmaceutical, cosmetic, metallic industries, etc.
- The α-D-(1→3) glucans have gained much attention in the medical field due to their potential applications as prebiotics and therapeutic agents in the prevention and treatment of various diseases such as cancer in mouse models, dengue fever, schistosomiasis cytomegalovirus infections, infectious illnesses, and various autoimmune diseases. However, the biological applications of α-D-glucans depend mainly on the solubility property which in turn is influenced by the length, branching, linkage, and molecular weight of polysaccharide.
- The extracellular glucans (EPS) play a vital role in pharmaceutical applications as excipients in formulation, carriers in the delivery of drugs, diagnostic agents in disease diagnosis, as well as theranostics in the treatment and diagnosis of disease, and as vaccines to provide artificial immunization. Dextran, the bacterial EPS, has potential pharmaceutical applications as a plasma volume expander.
- The cholesteryl-pullulan nanoparticle composites act as drug-targeting carriers with the selective binding ability to ß-amyloid receptors and reduce the toxicity of neurological disorders, e.g., Alzheimer's disease.
- The immune stimulation properties of α-D-glucans consist of (1→6) linkage extracted from the roots of *Ipomea batatas* had shown by increasing the phagocytic function, hemolytic activity, and concentration of IgG antibodies in a dose-dependent manner.
- The α-D-(1→3) glucans had better prebiotic potential and can stimulate the growth of *Lactobacillus* and *Bifidobacterium* bacteria, but therapeutic properties such as immunostimulation and antitumor potential are limited in the native form. However, the enhanced immunomodulatory and antiproliferative potential along with less systemic toxicity than marketed anticancer drugs was shown by derivatized α-D-(1→3) glucans. This has been achieved due to a reduction in molecular weight of glucans after derivatization, which leads to increased solubility as well as bioactivity.
- The dextrans are extensively used as a column material in size-exclusion chromatography owing to their gel-forming capacity.
- The structure and concentration of the cell wall glucan are vital for the maintenance of the fungal cell morphology and integrity. Hochstenbach and coworkers reported that a decrease in the concentration of either α-D-(1→3) or β-D-(1→3) glucan results in an abnormality of cell shape in fission yeast.
- In some fungi such as *A. nidulans* and *S. commune*, the cell wall α-D-(1→3) glucan functions as a reserve carbohydrate under nutrient-deprived conditions and is involved in the aggregation of hyphae and conidia in *A. nidulans*, which has been exploited in the development of α-D-(1→3) glucan-based polymer development.
- The film-forming capacity of starch could be used to replace the plastics in food packaging applications Similarly, the bacterial α-D-glucans, viz. dextran, are known to act as a gelling, viscosity modifier, and emulsifying agent in food.
- The α-D-(1→3) glucans act as mutants inducers [α-(1→3) glucanases] from microorganisms that can be used to treat dental caries and as a biocontrol agent for pest control.
- α-Glucan phosphorylases (α-GPs) act as a catalyst for reversible phosphorolysis of α-1,4-linked polysaccharides such as glycogen, starch, and maltodextrins, and thus play an important role in the utilization of storage polysaccharides. Because of their high operating temperatures, GPs provide significant benefits and facilitate bioprocess design as shown in Figure 9.4 [22].
- 1,4-α-Glucan branching enzymes (GBEs) are found throughout all animals, microorganisms, and plants. These enzymes transform the structure of both starch and glycogen,

Inorganic phosphate

α-glucan phosphorylase

α1,4-linked glucan (DP$_{n+1}$)

α-D-glucose 1-phosphate

α1,4-linked glucan (DP$_n$)

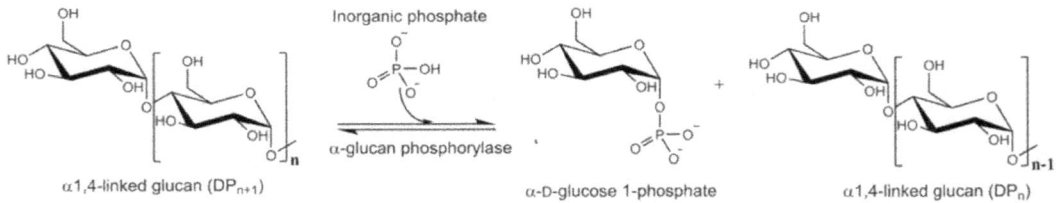

FIGURE 9.4 Bioprocess design of GP.

and they play a critical role in regulating the frequency and location of branch points in glucan chains. Thus, GBEs can change the structure of starches, thereby influencing their functional properties. The action patterns of 1,4-α-glucan branching enzymes allow several crucial downstream uses. For instance, modification of the digestibility and retrogradation characteristics of starch can be achieved by adjusting the points and frequency of branch points in glucan chains.

- α-1,6 Linkages in starch chains produce a large amount of α-limit dextrins (α-LDx) during α-amylolysis, decelerating glucose release at the intestinal α-glucosidase level.
- Macro-sized branched-glucans with high α-LDx has the potential to be used as a slowly digestible material to reduce postprandial glycemic response.

9.9.5 β-GLUCAN

β-Glucan is a polysaccharide that is found in yeast, mushrooms, bacteria, algae, barley, and oats. Because of its numerous health benefits, it is regarded as a functional food ingredient. The hypocholesterolemic and hypoglycemic properties of β-glucan are due to its high molecular weight (Mw) and viscosity. Lentinan, a β-glucan derived from the fungus *Lentinus edodes*, exists as triple helical structures at room temperature, resulting in high viscosity and excellent tolerance to a wide range of pH, temperature, and salt concentrations in aqueous solution. The structural component of fungal cell walls consists of chitin covalently bonded to glucan, forming a native composite material (chitin-glucan). The foaming capability and emulsifying properties of β-glucan products were better in neutral solutions (pH=7) than in the acidic (pH=4) and alkaline (pH=9) [24]. β-Glucan is a non-starch soluble polysaccharide widely present in yeast, mushrooms, bacteria, algae, barley, and oat. β-Glucan is regarded as a functional food ingredient due to its various health benefits. The high molecular weight (Mw) and high viscosity of β-glucan are responsible for its hypocholesterolemic and hypoglycemic properties. β-Glucan are also used in the food industry for the production of functional food products. The inherent gel-forming property and high viscosity of β-glucan lead to the production of low-fat foods with improved textural properties [25].

β-Glucans are naturally occurring glucose polymers that are present in abundance in plants, bacteria, and fungi. For centuries, traditional Chinese medicine uses fungi for healing, and currently, interests have focused on polysaccharides that are a crucial component of fungi cell walls. Within the multitude of polysaccharides present, β-glucans are a key reason in why fungi are used in cosmetics, as food additives, or for medicinal purposes. β-Glucans share a common structure consisting of a backbone of β(1,3)-linked β-d-glucopyranosyl units. However, they can strongly differ by their length and branching structure. Fungal β-glucans, which represent the most abundant polysaccharides found in the cell wall of fungi, are mainly characterized by the presence of β(1,6)-linked branches coming off of the β(1,3) backbone. The structural diversity also depends on the fungal source. For example, β-glucans of mushrooms have short β(1,6)-linked branches whereas those of yeast have β(1,6)-side branches with additional β(1,3) regions [26]. Of note, these structural differences may influence the immunogenic properties of β-glucans, and many studies have suggested that a higher degree of structural complexity is associated with enhanced β-glucans-induced antimicrobial and anticancer activity. Mushrooms have unique sensory properties and nutritional values as well as health benefits due to their bioactive compounds, especially β-glucans. Well-known edible and medicinal mushroom species, as well as uncommon or unknown species representing interesting sources of bioactive β-glucans, have been widely studied [6].

The major structural feature of mushroom β-glucans is a β-1,3-D-glucan main chain with single D-glucosyl residues linked to β-1,3 along this main chain. Some of this glucan can be extracted from the fruiting body of the mushroom, and soluble β-glucans are also produced by cultured mycelia. Because β-glucans are not synthesized by the human body, they are recognized by the immune system and induce both adaptive and innate immune responses. In this context, the use of mushroom extracts with soluble β-glucans vs the consumption of the whole fruiting body is discussed for digestibility and bioactivity In addition, chitin and α-glucans are present in mushrooms; the total polysaccharide contents of mushrooms range between 50% and 90%.

9.9.6 Sources of β-Glucan

β-Glucans can be divided into two subgroups: cereal and non-cereal. Cereal or grain-derived β-glucans usually have 1,3 1,4 glycosidic linkages without any 1,6-bonds or branching. They are fibrous structures found in aleurone (proteins stored as granules in the cells of plant seeds), in the sub-aleurone layer, and in the cell wall of endospores. Cereals include oat, barley, wheat, and rice β-glucans share similar structures; some differences include variation in 1,3 and 1,4 linkage ratio and molecular size; and some have large cellulose structures. β-Glucan content also varies among cereal sources—there is higher glucan content in barley than oats, and the least is found in rice and wheat. Mushroom is one of the major sources of β-glucan used in medical applications and traditional therapies. Thus, structure analysis and quantification of β-glucan content are crucial to evaluating medicinal mushrooms [19].

Non-cereal β-glucans are fibrous structures found in yeast, fungi, bacteria, and algae. β-Glucans originating from yeast have linear (1,3) backbones with long chains of 1,6 branching. Unlike grain β-glucans, fungal β-glucans differ between species concerning the degree of branching and distribution. Curdlan, a glucan isolated from *Agrobacterium*, contains no side branching, just a β-D glucan backbone. The solubility of the molecule is reliant on 1,3 linkages. β-Glucans are classified as soluble dietary fiber as the beta configuration is not digestible by enzymes in the human gastrointestinal (GI) tract. They are further classified pharmacologically as biological response modifiers (BRMs) as they influence their immune system counterparts. For cereal-based β-glucan at least, higher molecular weight β-glucans appear to be more effective than lower weight molecules. With this level of variance, it is not surprising that there is a range of diverse applications of β-glucans in clinical trials ranging from the alleviation of respiratory illnesses to improving fatigue and weight loss (Figure 9.5).

9.9.7 Stereoselective Solid-Phase Synthesis of β-Glucans

The development of an iterative automated solid-phase synthesis of linear β-(1,3)-glucans that can produce oligosaccharides of any chain length. The power of that approach is highlighted by the synthesis of the linker-functionalized β-(1,3)-glucan dodecasaccharide. High coupling efficiencies for each glycosylation step are required to synthesize β-(1,3)-glucan oligosaccharides such as dodecasaccharide on solid support (Figure 9.1). Control of the anomeric configuration of the newly formed glycosidic bonds is imperative to avoid the formation of complex mixtures of products. The design of the building block including protecting group pattern and anomeric leaving group, the choice of the linker, and the type of solid support were made mindful of these requirements (Scheme 9.1). Thioglucoside was envisioned as a reliable building block to produce β-configured glucans. The anomeric leaving group was chosen since it has already been used for the introduction of β-glucosidic linkages, while the pivaloyl ester in the C_2 position was introduced to ensure the selective formation

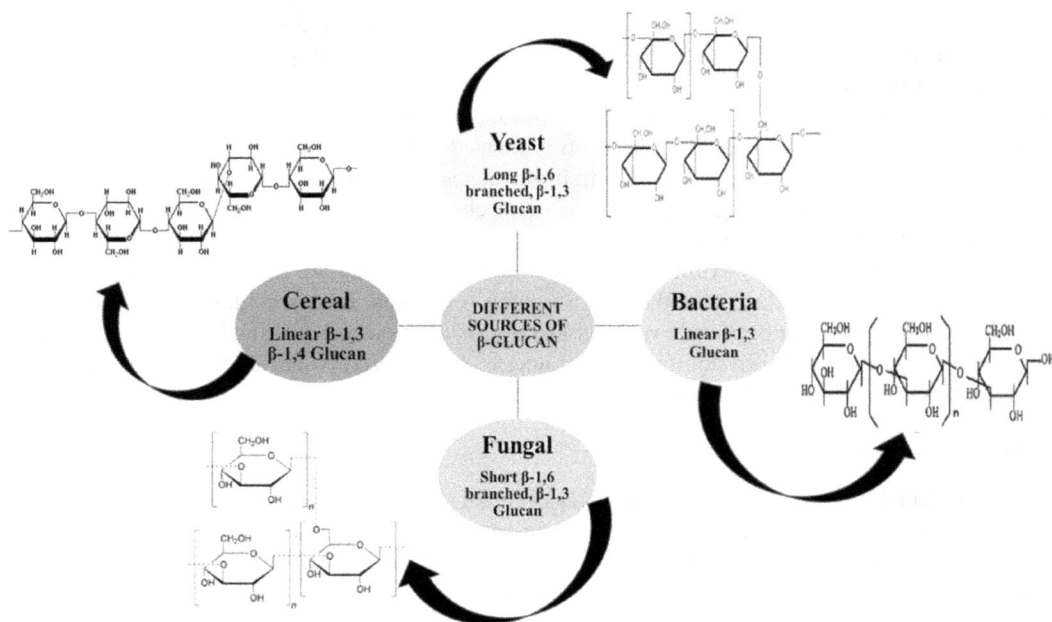

FIGURE 9.5 Some sources of β-glucan.

of β-glycosidic link. The 4,6-O-benzylidene acetal was seen as suitable protection for the C_4 and C_6 hydroxyl groups as it has already been successfully employed in several syntheses of β-(1,3)-glucans. The Fmoc group was chosen for temporary protection of the C_3 hydroxyl for its stability to acids, the ease of assessing the efficiency of each glycosylation, and the mild conditions required for its removal after each coupling step. Merrifield resin was selected as solid support since it has been employed successfully in the solid-phase synthesis of carbohydrates. The versatility of functionalized Merrifield resin in the synthesis of various oligosaccharides has recently been demonstrated. The base-labile linker allows for the simultaneous cleavage of the crude products from the solid support and the removal of base-labile protecting groups present on the oligosaccharide. Furthermore, it is stable to the most common activation conditions for thioglycosides. Building block was synthesized in seven steps starting from β-d-glucose pentaacetate (Scheme 9.1). Intermediate was prepared according to previously reported procedures. Esterification of the C_2 hydroxyl group with pivaloyl chloride afforded thioglycoside. Removal of the TBS silyl ether using a buffered solution of TBAF·3H$_2$O and glacial acetic acid gave alcohol that was treated to afford thioglycoside building block. Functionalized Merrifield resin was prepared according to previously described procedures. The solid-phase assembly of β-(1,3)-glucans on functionalized solid support was performed using three equivalents of thioglycoside in three repetitions per glycosylation cycle (Scheme 9.2). The glycosylations were carried out on the automated synthesizer by suspending functionalized resin, and three equivalents of the glycosylating agent in dichloromethane in dichloromethane/acetonitrile (1:1) were added dropwise via a syringe pump. The reaction mixture was agitated using an argon flow while maintaining the temperature for 5 minutes. Then the temperature was increased to 108°C, and the mixture was agitated for another 40 minutes before draining the reaction vessel using argon pressure. This procedure was executed three times, followed by washing steps using different solvents. Following each glycosylation cycle, the temporary Fmoc protecting group was removed using a 20% (v/v) solution of piperidine in DMF. The reaction vessel was drained, and the resin was washed. These steps were repeated until oligosaccharides of the desired chain lengths were assembled. The crude product was cleaved from the solid support by saponification with sodium methoxide in a mixture of methanol and dichloromethane.

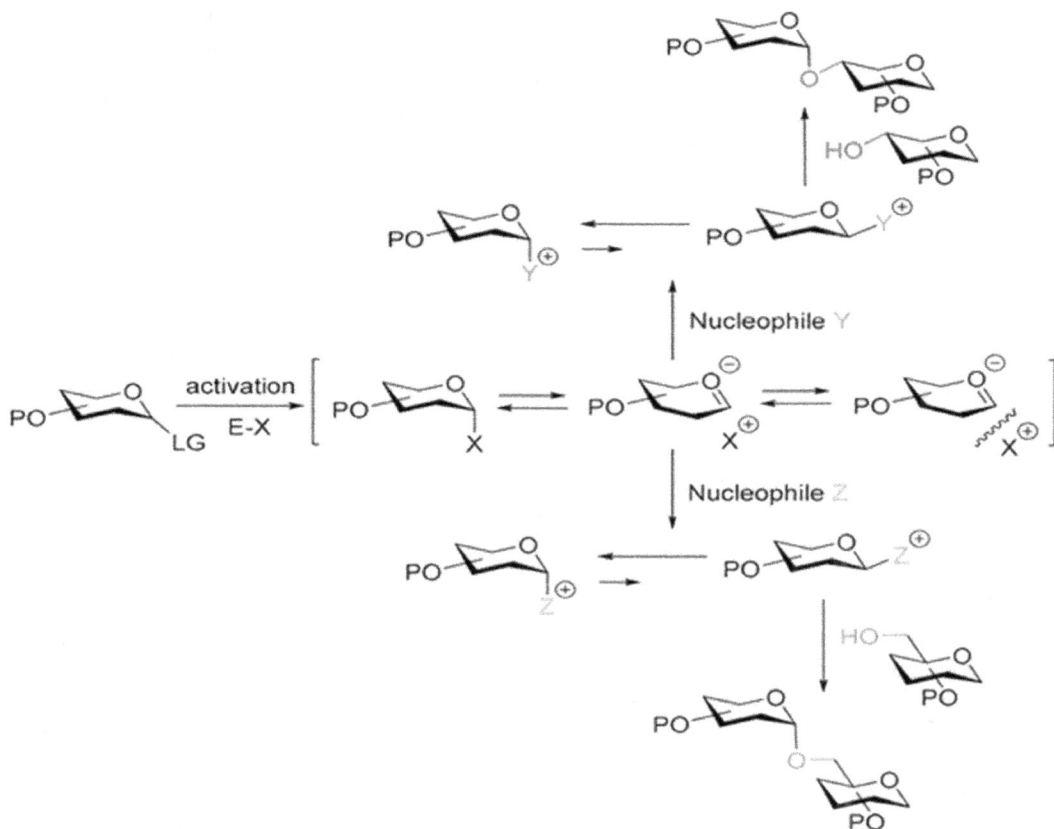

SCHEME 9.1 The target α-glucans of this study and the employed building blocks are depicted.

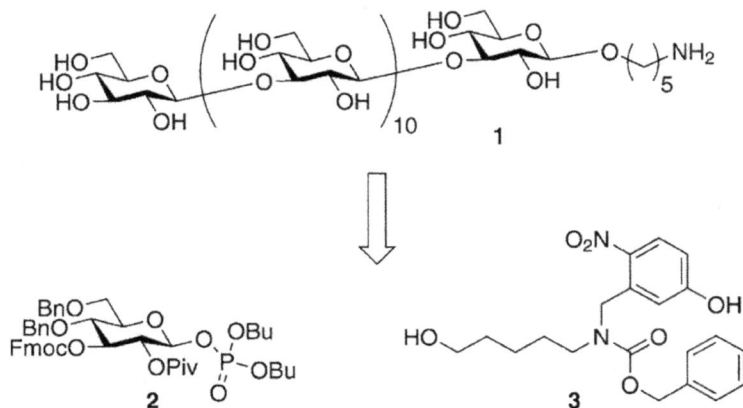

SCHEME 9.2 The solid-phase assembly of β-(1,3)-glucans on functionalized solid support was performed using three equivalents of thioglycoside in three repetitions per glycosylation cycle.

9.9.8 APPLICATIONS OF β-GLUCAN

- The mushrooms contain bioactive polysaccharides, the majority of which are β-glucans.
- These substances boost the host immune defense system by activating the humoral immune system and boosting the function of macrophages and natural killer cells. The induction

of cellular responses by mushrooms and other β-glucans is most likely because of their unique interaction with several cell-specific receptors, such as complement receptors.

- β-Glucans may inhibit the body's absorption of cholesterol from food. They may also stimulate the immune system by increasing anti-infection chemicals.
- β-Glucans are powerfully bioactive compounds with anticancer, anti-inflammatory, and immune-modulating properties. β-Glucan is increasingly being used in food and other industries due to its unique physical properties such as water solubility, viscosity, and gelation. Fungal chitin nanofibrils can be used to make nano papers (FChNFs). Because of their distinct surface properties inherited from the chitin and glucan composition, FChNF nano papers have the potential to be used in packaging films, composites, or membranes for water treatment.
- Several fungal pathogens by binding to β-glucans, and it is important in innate immune responses.
- β-Glucan is also extensively used in the food industry for its ability to form a gel and enhance the viscosity of aqueous solutions. It is used for enhancing the texture and appearance of salad dressings and gravies.
- β-Glucan is also used as a fat mimetic to develop calorie-reduced food production.
- Bile acid is produced in the liver from cholesterol. The binding of β-glucan with bile acid and its fecal excretion lead to a decrease in the cholesterol levels in the body.
- β-Glucan has various physical properties such as thickening, stabilizing, emulsification, and gelation. β-Glucan has the potential to be used in acceptable health products that offer a wide range of added health benefits.
- A high levels of β-glucan were obtained from oat or barley grain. This product can be produced as an agglomerated food additive having at least about 18% β-glucan by dry weight. Methods are providing for enriching a product with the β-glucan agglomerated food additive.
- β-Glucans have a potential applications in medicine and pharmacy, food, cosmetic and chemical industries, as well as in veterinary medicine and feed production.

9.10 CONCLUSION

Glucan is a key food component in the treatment of the physiologic dysregulations-related metabolic syndrome. Although the physiological effects of ingested glucan are similar to those of other soluble fibers, its availability and ease of handling led to its increasing incorporation into foods, with the goal of increasing daily fiber consumption. The chapter has also explored the role of glucan in fungal pathogenesis and how it promotes phagocytic-mediated immune responses. Similarly, fungal glucan administration is well known to enhance the immune system and boost resistance to numerous infectious diseases and cancers. Glucans with different structures and molecular weight interact with each receptor, and specific signaling pathways are triggered by glucan molecules with described biochemical properties. With such a common control, we may attempt an intellectually honest use of this promising molecule as an adjuvant or therapeutic age in the future.

REFERENCES

1. Sima P, Richter J, Vetvicka V. Glucans as new anticancer agents. *Anticancer Res.* 2019 Jul;39(7):3373–3378. DOI: 10.21873/anticanres.13480.
2. Huang G, Huang S. The structure-activity relationships of natural glucans. *Phytother Res.* 2021 Jun;35(6):2890–2901. DOI: 10.1002/ptr.6995.
3. Alonso N, Goldman GH, Gern RM. β-(1→3),(1→6)-glucans: medicinal activities, characterization, biosynthesis, and new horizons. *Appl Microbiol Biotechnol.* 2015 Oct;99(19):7893–7906. DOI: 10.1007/s00253-015-6849-x.

4. Singh RS, Kaur N, Hassan M, Kennedy JF. Pullulan in biomedical research and development—a review. *Int J Biol Macromol*. 2021 Jan 1;166:694–706. DOI: 10.1016/j.ijbiomac.2020.10.227.

5. Ubiparip Z, Beerens K, Frances J, Vercauteren R, Desmet T. Thermostable alpha-glucan phosphorylases: characteristics and industrial applications. *Appl Microbiol Biotechnol*. 2018 Oct;102(19):8187–8202. DOI: 10.1007/s00253-018-9233-9.

6. Akramiene D, Kondrotas A, Didziapetriene J, Kevelaitis E. Effects of beta-glucans on the immune system. *Medicine (Kaunas)*. 2007;43(8):597–606.

7. Chen L, Huang G. The antiviral activity of glucans and their derivatives. *Int J Biol Macromol*. 2018;115:77–82. DOI: 10.1016/j.ijbiomac.2018.04.056

8. Cheng H, Huang G. The antioxidant activities of carboxymethylated garlic glucan and its derivatives. *Int J Biol Macromol*. 2019;140:1054–1063. DOI: 10.1016/j.ijbiomac.2019.08.204

9. Chen F, Huang G. Extraction, derivatization, and antioxidant activity of bitter gourd glucan. *Int J Biol Macromol*. 2019;141:14–20.

10. Chen L, Long R, Huang G, Huang H. Extraction and antioxidant activities in vivo of pumpkin glucan. *Ind Crops Prod*. 2020;146:112199.

11. Chen F, Huang G, Yang Z, Hou Y. Antioxidant activity of Momordica charantia glucan and its derivatives. *Int J Biol Macromol*. 2019;138:673–680.

12. Liu HY, Geng MY, Xin XL. Multiple and multivalent interactions of novel anti-AIDS drug candidates, sulfated polymannuronate (SPMG)-derived oligosaccharides with gp120 and their anti-HIV activities. *Glycobiology*. 2005;15:501–510.

13. Liu X, Xie JH, Jia S. Immunomodulatory effects of an acetylated Cyclocarya paliurus glucan on murine macrophages RAW264.7. *Int J Biol Macromol*. 2017;98:576–581. DOI: 10.1016/j.ijbiomac.2017.02.028

14. Ishu RD, Ken JF. The anticancer activity of polysaccharide prepared from Li by a dates Phoenix dactylifera L. *Carbohydr Polym*. 2005;59, 531–535. DOI: 10.1016/j.carbpol.2004.11.004

15. Sun Z, He Y, Liang Z. Sulfation of (1! 3)-β-D-glucan from the fruiting bodies of Russula virescens and antitumor activities of the modifiers. *Carbohydr Polym*. 2009;77(3):628–633. DOI: 10.1016/j.carbpol.2009.02.001

16. Ohnon N, Miura T, Miura NN, Adachi Y, Yadomae T. Structure and biological activities of hypochlorite oxidized zymosan. *Carbohydr Polym*. 2001;44:339–349. DOI: 10.1016/S0144-8617(00)00250-2

17. Nie Y, Luo FJ, Zeng XN. Research progress on physiological functions and application of rice bran glucans. *Food Oils*. 2015;28(11):10–13.

18. Ban X, Dhoble AS, Li C, Gu Z, Hong Y, Cheng L, Li Z. Bacterial 1,4-α-glucan branching enzymes: characteristics, preparation, and commercial applications. *Crit Rev Biotechnol*. 2020;40(3):380–396. DOI: 10.1080/07388551.2020.1713720

19. Du B, Meenu M, Liu H, Xu B. A concise review on the molecular structure and function relationship of β-glucan. *Int J Mol Sci*. 2019 Aug 18;20(16):4032. DOI: 10.3390/ijms20164032.

20. Liu H, Li Y, You M, Liu X. Comparison of physicochemical properties of β-glucans extracted from hull-less barley bran by different methods. *Int J Biol Macromol*. 2021 Jul 1;182:1192–1199. DOI: 10.1016/j.ijbiomac.2021.05.043.

21. Zhu F, Du B, Xu B. A critical review on production and industrial applications of beta-glucans. *Food Hydrocoll*. 2016;52:275–288. DOI: 10.1016/j.foodhyd.2015.07.003

22. Reddy Shetty P, Batchu UR, Buddana SK, Sambasiva Rao K, Penna S. A comprehensive review on α-D-Glucans: structural and functional diversity, derivatization and bio applications. *Carbohydr Res*. 2021;503:108297. DOI: 10.1016/j.carres.2021.108297

23. Wang L, Overkleeft HS, van der Marel GA, Codée JDC. Reagent controlled stereoselective synthesis of α-glucans. *J Am Chem Soc*. 2018 Apr 4;140(13):4632–4638. DOI: 10.1021/jacs.8b00669.

24. Nawawi WMFW, Lee KY, Kontturi E, Bismarck A, Mautner A. Surface properties of chitin-glucan nano papers from Agaricus bisporus. *Int J Biol Macromol*. 2020 Apr 1;148:677–687. DOI: 10.1016/j.ijbiomac.2020.01.141.

25. Ariyoshi W, Hara S, Koga A, Nagai-Yoshioka Y, Yamasaki R. Biological effects of β-glucans on osteoclastogenesis. *Molecules*. 2021 Apr 1;26(7):1982. DOI: 10.3390/molecules26071982.

26. Camilli G, Tabouret G, Quintin J. The complexity of fungal β-glucan in health and disease: effects on the mononuclear phagocyte system. *Front Immunol*. 2018 Apr 16;9:673. DOI: 10.3389/fimmu.2018.00673

10 Polysaccharides in Sensors and Actuators

Richika Ganjoo, Shveta Sharma,
Humira Assad, Abhinay Thakur
Lovely Professional University

Ashish Kumar
Bihar Engineering University

10.1 INTRODUCTION

Polysaccharides are the most common supramolecular polymers found naturally, and they can be identified in phytoplankton, kingdom plantae, and microorganisms including fungi and bacteria [1,2]. Different investigations have been conducted to date on the numerous kinds of polysaccharides found in sensors and actuators, including cellulose, starch, chitin, and chitosan [3]. A sensor is a sort of electrical equipment that is hypothetically characterized as a transducer that changes one type of energy to another that may be processed further [4]. Sensors convert physical inputs to electrical outputs, while actuators do the reverse [5]. They convert the electrical signals generated by control modules to physical outputs [6,7]. Polysaccharides are the most prevalent class of materials derived from reusable and maintainable sources. They are decomposable, carbon (C) neutral; pose little dangers to the environment, human health, or safety; and act as a structural component of plant cell walls [8–10]. The production of synthetic textile fibers (e.g., organic and inorganic esters, alkyl, hydroxy alkyl, and carboxyl alkyl ethers), as well as coating materials, optical films, food products, medical devices, and cosmetics, all relies on cellulose and hemicelluloses, which have been used as structural materials in a broad range of applications for many years [11]. Additionally, because of the characteristics, usefulness, durability, and homogeneity of polysaccharides, their usage in sensors and actuators has increased in recent years [9]. This chapter will address the usage of several kinds of polysaccharides in sensors and actuators [12].

10.2 POLYSACCHARIDES IN SENSORS AND ACTUATORS

10.2.1 CELLULOSE IN SENSORS AND ACTUATORS

Cellulose is the biosphere's most plentiful inexhaustible organic element, with a year-round production of around 5×10^{11} metric tons [13]. Cellulose is a syndiotactic homopolymer made up of D-anhydroglucopyranose units connected by β-(1→4)-glycosidic bonds as shown in Figure 10.1 [14].

Cellulose, a bio-based product, may be efficiently generated either top-down, by subjecting biomass to strong shear pressures to reduce the dimension of cellulose molecule in suspension, or bottom-up, by employing bacterial cellulose biosynthesis, with *Acetobacter xylinam* being the most effective bacterial specie [3]. Due of its cost efficiency, cellulose derived from plants was favored for mass manufacture of engineering properties. The notion of a renewable and sustainable product derived from cellulose was deeply investigated. As a consequence of their behavior changes in response to stimuli, "smart" materials based on cellulose provide a diverse set of applications in the sensing domain [14]. Due to cellulose's biodegradability and biocompatibility, it has recently been used to new sectors. One of the emerging disciplines is the unearthing of cellulose paper as

DOI: 10.1201/9781003265054-10

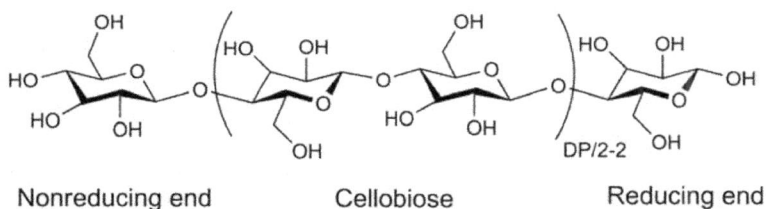

FIGURE 10.1   Cellulose. Adapted from ref. [11].

a smart substance capable of being employed as sensors and actuators. The name "electro-active paper" (EAPap) refers to this kind of smart material. Yun et al. combined F-MWNTs (functionalized multiwall carbon nanotubes) with cellulose mixture to form F-MWNTs/cellulose electroactive paper (EAPap) actuators. Cellulose was liquefied in a LiCl/N, DMAc (N-dimethylacetamide) solution and combined with F-MWNTs to form F-MWNTs/cellulose EAPap actuators. The cellulose EAPap's actuation notion was most likely inspired by the ion transfer mechanism, which occurs frequently in humid environments. Hydrophilic moieties on MWNTs enhanced cellulose's moisture retention in a somewhat humid conditions, while enhancing Cl– ion transportation in F-MWNT/cellulose EAPap. Orelma et al. have developed an optical cellulose fiber for water sensing utilizing a progressive preparation technique [15]. The fiber's core was generated by dry–wet spinning liquefied cellulose in [EMIM]OAc into H_2O. Cladding layer was developed by covering a sheet of $C_{164}H_{174}O_{111}$ (cellulose acetate) liquefied in acetone onto the cellulose core using a filament coater. Employing UV and FTIR spectroscopy, cast films were used to test the chemical and optical properties of regenerated cellulose and cellulose acetate. It was discovered that regenerated cellulose sheet absorbs UV rays while fleeting visible light wavelengths. $C_{164}H_{174}O_{111}$ sheet was revealed to pass the entire visible light spectrum. The manufactured optical fiber directed light in the 500–1,400 nm wavelength range. At 1,300 nm, the attenuation constant of the $(C_6H_{10}O_5)n$ fiber was determined verified to be $63 \times 10^{-1} dB\ cm^{-1}$. When the fiber was immersed in H_2O, a significant decrease in light intensity was noticed. Table 10.1 summarizes the various types of cellulose used in sensors and actuators [4,8,16–21].

Cellulose paper, a naturally existing material that is cost-effective, compact, harmless to environment, and physically bendable, is crucial in the creation of next-generation wearable electronics. Soft actuators constructed of cellulose paper have attracted a lot of attention since

TABLE 10.1

Types of Cellulose and Its Applications in Sensors and Actuators

Type of Sensor/Actuator	Type of Cellulose	Application
Solvent sensor	Cellulose nanocrystal	For gravimetric detection of water molecules
Solvent sensor	Graphene/cellulose nanocomposite	To differentiate between different organic solvents relying on their diffusion mechanisms
Humidity sensor	Cellulose acetate butyrate	To measure humidity
Vibration sensor	Carboxymethyl cellulose (G-CMC)	As a sensor, the capability to provide reliable signal output for both low and high frequency purposes
Humidity sensor	Nanocrystalline cellulose (NCC)	Structured color humidity indicator
pH sensor	Cellulose nanocrystal (CNC) matrix	To study the spoiling process of milk
Tactile actuators	Cellulose acetate composite	For preferable execution of a braille tactile display
Gel electrode actuators	Oxidized cellulose nanofiber	Utilized in electronic and energy conversion systems that are accessible and transparent
pH-sensor	Cotton cellulose	A color change is used to evaluate pH fluctuations

the boom in the development of functional electronics [22]. Soft actuators based on cellulose paper provide a legitimate response to the environmental concerns associated with the extensive use of flexible actuators. While far from flawless, advances in cellulose paper actuators have enabled simple and ecologically friendly fabrication of beneficial operating gadgets, demonstrating their potential uses in paper robots, do-it-yourself projects, inquisitive education, and smart home gadgets as well. As materials science, fabrication methods, and construction technology develop, CPap actuators will emerge into elite device platforms competent of implementing breakthrough flexible paper machineries.

10.2.1.1 Cellulose in Humidity Sensors

The presence of water vapor in the air is referred to as humidity. Humidity, as a fundamental yet critical environmental component, has a significant impact on manufacturing, assessment, technical performance, convenience, and healthcare [17]. With the advancement of sophisticated advanced urban areas and manufacturing and agribusiness activities, it is becoming more necessary to monitor and adjust the humidity level. The demand for very precise humidity sensors has sped up the progression of various transducer innovations and material science. Capacitive sensors now monopolize the humidity sensor market; however, investigation into other types of sensors, including resistive, optical, and gravimetric sensors, is still ongoing [16]. It is widely accepted that inorganic materials or ceramics have effectively developed and improved the technical qualities of a humidity sensor in the past [16].

Guan et al. have developed humidity sensors through a straightforward solution approach using EPTAC (glycidyl trimethyl ammonium chloride) improved cellulose paper, with the paper serving as both the humidity detecting system and the sensor material [23]. By modifying EPTAC, the resultant sensor's sensitivity was increased and the reaction time (R_T) was lowered to 25 s, which is equivalent to the performance of high-quality paper-based humidity sensing devices. The paper-based moisture sensor is also exceedingly elastic and biocompatible, rendering it ideal for a wide range of functions such as respiration surveillance, non-contact shifting, and cutaneous moisture surveillance. The cheap and straightforward fabrication approach described in this paper may be an advantageous option for producing multifunctional humidity sensors.

Cellulose, termed electro-active paper, has been reinvented as a functional materials suitable of being used as sensors and actuators. Mahadeva and coworkers demonstrate the versatility of cellulose as a humidity and temperature sensor. Nano-scaled polypyrrole (PPy) was put onto the cellulose interface as a temperature and relative humidity sensitive layer without changing the cellulose architecture using an in-situ polymerization procedure. The successful application of a polypyrrole nanolayer onto the interface of cellulose, progress in the production of a cellulose–PPy nanocomposite, was confirmed using atomic force microscopy, ultraviolet–visible spectroscopy, transmission electron microscopy, and secondary ion mass spectrometry. The influence of polymerization time on cellulose–PPy nanocomposite sensing performance was investigated; experimental outcomes revealed that cellulose–PPy nanocomposite with a 16-hour polymerization time is well suited for temperature and humidity detectors [24].

For enhanced human-machine interactions, the development of non-contact bendable electronic sensors with elevated sensitivity and robust association is critical. Meng and coworkers designed a novel cellulose acetate (CA)/aluminum (Al) flexible humidity sensor using theoretical simulations [25]. Aligned CA fiber mats with groove architectures demonstrated a high sensitivity to humidity and a programmable deformation capacity. The conductive Al layer had an organized wrinkle structure that was ideal for balancing high deformability and reliable electrical connection. On the basis of these characteristics, the non-contact CA/Al sensor can detect changes in sweating on a human finger and produce consistent on–off signals under ongoing deformations, making it ideal for wearable human-machine interface equipment. This research paves the way for the creation of high-performance non-contact bendable electronic sensors that are directed by artificial intelligence in the future.

10.2.1.2 Cellulose in pH Sensors

Halochromic or pH-sensitive textiles are constituents that alter color in response to changes in pH and may be utilized for injury treatment, defensive apparel, and purification, among other purposes. The majority of synthetic dyes are halochromic, meaning they change color depending on the pH. They do, however, have a negative impact on the atmosphere and are related with allergic, noxious, and other adverse reactions [26]. Because the process of rotting is generally accompanied by a pH shift, pH indicators have been used to monitor and signal the freshness of food in storage due to their efficiency and simplicity [19]. Ma and Wang synthesized a colorimetric pH-sensing coating from grape skin (EGS) extracts and embedded it in a tara gum (TG)/cellulose nanocrystal (CNC) matrix [19]. The EGS's UV-vis spectra were examined throughout a pH range of 1–10, and the color varied dramatically from brilliant red to dark green. Ji et al. developed color-tunable luminous macrofibers centered on CdTe quantum dots (QDs) utilizing bacterial cellulose nanofibers and wet spinning [27]. The luminous macrofiber's green, yellow, and orange fluorescence can be readily modified by adjusting the dimension of the CdTe QDs. The luminous macrofiber exhibited a sigmoidal pH dependency, and this pH-dependent activity varied with the size of CdTe QDs.

Devarayan and Kim discuss the use of electrospun cellulose nanofibers chemically modified with a natural color to create an ecologically responsible, bidirectional, and global pH sensor [26]. A natural color originating from red cabbage was incorporated into electrospun non-woven cellulose fibers using adsorption and chemical crosslinking processes. According to the observations, the biocomposite produced is a ubiquitous pH sensor that can recognize pH values between 1 and 14 by presenting a different color code for each pH value. Additionally, it was discovered that the pH detecting was even across a wide range of temperatures and time periods. Additionally, the colors could be reversed and the pH sensor could be recycled.

10.2.1.3 Cellulose in Gas Detection Sensors

Gas sensors (also known as gas detectors) are electronic equipment that recognize and classify a wide range of gases. They're frequently used to identify and analyze potentially hazardous or combustible gases. Gas sensors are used to monitor gas leaks in factories and industrial complexes, as well as smoke and carbon monoxide in residential settings. Gas sensors come in a wide range of dimensions (transportable and stationary), sensing capabilities, and ranges. They're usually part of a larger embedded device, like hazmat or security, and they're usually connected to an audible alarm or user interface. Gas sensors require calibration more frequently than other types of sensors due to their constant interaction with air and other gases. Mun et al. created a titanium dioxide MWCNT composite in nanoscale using a hydrothermal technique, and further composite formation was verified by utilizing X-ray diffractometry and transmission electron microscopy. After that for using in ammonia gas sensor, a cellulose–TiO_2–MWCNT composite was created. The cellulose–TiO_2–MWCNT hybrid nanocomposite displayed substantially better responsiveness and faster recuperation when the quantity of NH_3 gas was elevated from 50 to 500 ppm. This versatile and low-cost hybrid gas sensor provides a high level of sensitivity and repeatability [28].

Hittini et al. discovered the invention of a new class of hydrogen sulfide gas sensors manufactured by using metal-oxide and polymeric matrix, which are responsive, specific, and low temperature. Copper oxide nanoparticles were used for precise nanoparticle size control. To make elastic and semi-conductive polymeric matrix membranes, they mixed sodium carboxymethyl cellulose dust with 5% glycerol and varied amounts of copper oxide nanoparticles. An H_2S gas sensor was created by sandwiching each membrane between two electrodes. Investigation of the sensor's temperature-dependent gas detection capabilities was carried out from 40°C to 80°C. At a low processing temperature and a gas concentration of 15 ppm, the sensors demonstrated good sensitivity and rapid response times to H_2S gas. The sensors were also extremely specific for H_2S gas and had a low humidity reliance, demonstrating that they will work dependably in humid conditions. This organic-inorganic hybrid materials gas sensor is versatile, sensitive, and energy efficient, making it ideal for use in harsh environments [29].

10.2.1.4 Cellulose in Strain Sensor

Strain sensors are an example of a highly effective electronic sensor. They're also known as electro-mechanical sensors or sensor and actuator. This sensor has a wide range of industrial applications. Construction, bridges, and other critical infrastructure, for example, are occasionally vulnerable to major natural disasters like tsunamis, etc. It is vital to use a strain sensor to continuously check the condition of these infrastructures in order to avoid a catastrophic failure. Fundamentally, strain sensors with high flexibility and stretch ability play a crucial part in personal health monitoring, human-friendly gadgets, and extremely sensitive devices, or what are commonly referred to as smart devices, due to their widespread use in outdoor applications [3]. From a structural aspect, this type of sensor is frequently made as a thin sheet. To promote its performance, this sheet is frequently made as an ultra-thin layer. Strain sensor, also referred as electromechanical sensor, uses highly polarized nanoparticles as an active medium to generate an electrical charge in reaction to an external electric field. In most cellulose composites, cellulose serves as a matrix for nanoparticles. As a result, considerable polarization occurs when an external electric field is applied. Wu et al. have presented a novel way for constructing stretchable strain sensors by using cellulose [30]. The results demonstrate that combining reduced graphene oxide with a tiny amount of carbon nanofibers increases the electrical conductivity and piezoresistive sensitivity of the composite sensor. Wu et al. also proposed polymer which is a hybrid made up of crosslinking hydroxyethyl cotton cellulose in the form of nanofibers and carbon nanotubes with polyurethane and investigated response of amount of CNTs on the electrical conductance nanocomposite formed to see if it could be used as a strain [31]. The structural and water-responsive shape memory characteristics were thoroughly examined to assess the substance's durability. The fully complexed TPU matrix was shown to have outstanding mechanical attributes and sensing capability. Figure 10.2 depicts a mechanism for interfacial interactions based on hydrogen bonding [31].

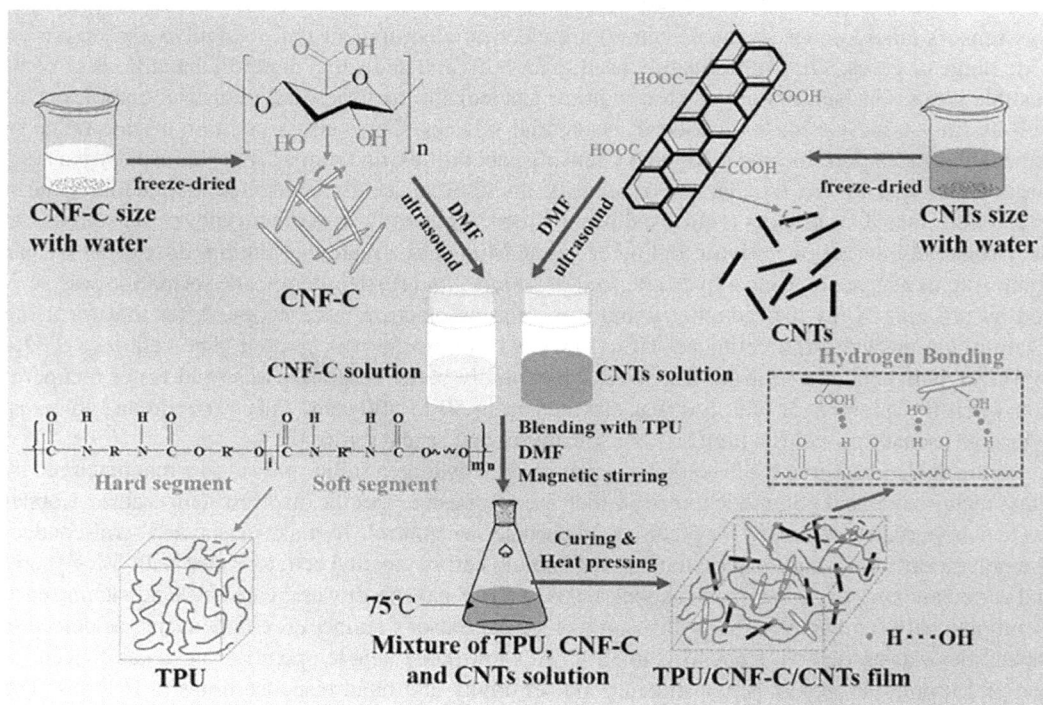

FIGURE 10.2 Schematic illustration of the hydrogen bonding interaction mechanism among molecules of TPU, CNF-C, and CNTs, and the film formation of TPU/CNF-C/CNTs. Adapted from ref. [32].

The water-induced shape fixity ratio (Rf) and shape recovery ratio (Rr) demonstrating that the alteration in the composite could be returned back to its original shape in response to a stimulus. The sensing features of the resulted composite was also studied at different conditions. The hybrid composite can detect huge stresses correctly for more than 103 times, according to the findings, and water-induced shape regain capacity may help sustain sensing accuracy after material fatigue to some degree. It is predicted that future water-responsive sensors and actuators could benefit greatly from using this type of composite material.

10.2.1.5 Cellulose in Ultraviolet Sensor

UV sensors detect the amount of ultraviolet (UV) light that is emitted. Although the wavelengths of this kind of electromagnetic radiation are less than those of visible light, they are still longer than those of x-rays. UV sensors are used to determine how much ultraviolet light has been exposed to in a laboratory or in the environment. Cellulose was employed as the matrix and organic polymer which are conductive in nature was used to enhance the bio-based composite for the UV sensor, as was the case with gas detection and humidity sensor. Gimenez et al. suggested a unique technique for fabricating a composite material consisting of zinc oxide (ZnO) crystals and cellulose fibers by compressing them together. When subjected to ultraviolet (UV) light, the manufactured material's electrical photoconductivity changed. When cellulose fibers are mixed with ZnO, the photosensitivity is greater than when the device is built entirely of ZnO [32].

Komatsu et al. used a simple spray coating approach to create transparent films made of cellulose and ZnO, in the form of nanofibers and nanoparticles, respectively [33]. The nanocomposite films that were generated were thin (~10 m) and bendable. The formation of film was verified by by scanning electron microscopy and atomic force microscopy, and it was found that film consisted of mainly cellulose nanofibers containing ZnO nanoparticles in it and sections of agglomerated ZnO nanoparticles, also electrical conductance of the resulted film rose significantly above 40 wt.% ZnO, reaching >50 nA at 60 wt.% ZnO. When a particular point, the ZnO NPs present in the film produced an increasing number of conductive pathways, according to the researchers. The nanocomposite film displayed a consistent reactivity to UV light exposure during on/off cycles. Even if the device was bent, the responsiveness and sensitivity remained robust (radius of curvature equals to 3 mm). This new nanocomposite film could be utilized to make UV sensors that are environmentally friendly and versatile.

10.2.2 Starch in Sensors and Actuators

Starch is a significant carbohydrate found in plants. Amylopectin and amylose make up this carbohydrate's structure [34]. Amylose is a linear polymer made up of α-D-glucopyranose units connected by a α-(1–4) linkage; amylopectin is comparatively large with many branches composed of linear α-(1–4) polymer attached with side chains via a single (1–6) junction [35].

Garcia et al. demonstrated the use of screen-printed carbon electrodes (SPCE) to detect salivary α-amylase (sAA) in human saliva samples (SPCE). The suggested approach included the indirect estimation of sAA via the use of a series of two chemical processes. Essentially, the first reaction occurs when sAA hydrolyzes starch to form maltose. The resulting reducing sugar then aids in converting the compound $[Fe\,(CN)_6]^{3-}$ to $[Fe\,(CN)_6]^{4-}$ in a subsequent reaction. Numerous variables, including pH, time of reaction, sAA volume, and amount of starch, were optimized exhaustively. A better electrochemical results were obtained with optimized conditions of 5 milli mol L^{-1} sodium hydroxide at 12 pH, for 20 minutes, 15 μL of sAA quantity, and a starch concentration of 0.5% (w/v). The results demonstrated a strong association between sAA concentrations of 100 and 1,200 U mL^{-1}, which was linear. Limits of detection (LOD) and sensitivity were determined to be 1.1 U mL^{-1} and 10.7 μA/(log U mL^{-1})10.7 A/(log U mL^{-1}), respectively. The electrochemical sensor presented here displayed a high degree of selectivity [36].

The influence of porous structure (larger in size) on the electrical characteristics of SnO_2 gas sensors was investigated by Lee and Kang. For this, starch powder of varied concentration (0%–15% by weight) was added in SnO_2 gas sensors resulted in large holes at intergranular level. Except for the creation of big holes, the addition of starch powder had no effect on the sintering behavior or microstructure. The inclusion of starch enhanced the gas sensitivity [37].

Choi et al. produced a novel film potato starch, utilizing agar, and natural colors taken from purple sweet potato, *Ipomoea batatas* [38]. A colorimeter was used to determine the color changes of pH indicator films following immersion in buffer solution with different pH. A feasibility study was done to determine the sensor's suitability for usage as a meat-rotting sensor. These resulted films demonstrated how pork sample's pH and spoilage point changed from red to green. As a result, these films generated might be utilized for detecting food deterioration.

Elgamouz et al. produced silver nanoparticles using starch, AgNO3, sodium borohydride as a capping shell, silver source, and reducing agent, respectively [39]. The detection of hydrogen peroxide in water taken as solvent and samples of urine were explored using starch-capped silver nanoparticles. It was discovered that the decomposition of H_2O_2 on the catalytic surface is pH, temperature, and time dependent [39].

Additionally, starch-based nanocomposites are employed as a sensor for heavy metals. Khachatryan et al. demonstrated the fabrication of new starch-based nanocomposites comprising zinc sulfide QDs capped with L-cysteine [40]. These nanoscale composites were made using potato starch gels and foils embedded with spherical QDs ranging in size from 10 to 20 nm and were verified by using photoluminescence, infrared and ultraviolet spectra, and transmission electron microscopy/scanning electron microscopy. Pb^{2+} and Cu^{2+} ions lowered the photoluminescent spectral band's emission intensities.

10.2.3 CHITIN AND CHITOSAN IN SENSORS AND ACTUATORS

Chitin and chitosan have gained considerable interest in various fields of sciences because of their outstanding structural and functional features. Chitin is a polymer consisted of long-chain N-acetyl glucosamine that serves as the structural component of a variety of organisms, including fungal cell walls, insect and crustacean exoskeletons, mollusk radulas, and squid beaks [41]. Chitin deacetylation ultimately resulted in to chitosan and may be tailored to have a wide variety of chemical and physical properties by altering the chain length of the starting constituent or varying the extent of changing the polymers back into monomers and by removing the acetyl groups. Chitin and, in particular, chitosan have been widely used in advanced biofabrication. There are various functional groups having internal oxygen and nitrogen in its chemical structure that can be employed as starting points for covalent modification of the chitosugar chain, which is another favorable attribute. Chitosan has been more useful as compared to chitin because it can be easily dissolved into dilute acids, they can be easily deposited like hydrogel and have the ability to electrodeposit a thin film on electrode surfaces, which is very important in nanoscale applications like micro to nanobiosensor formation. It's not unexpected that chitosan is more often utilized in sensor technology than chitin because of these additional benefits. Polymeric matrix composites that are ecofriendly and long lasting have recently piqued the curiosity of scientists and industry alike, and because of vast availability and environmental appeal, naturally produced materials have been intensively investigated.

Chen et al. proposed a unique small chitosan diaphragm-based extrinsic Fabry–Perot sensors acoustic pressure sensor that retains the benefits of traditional all-silica micro EFPI sensors while increasing sensitivity [42]. The sensor is manufactured without the use of chemical processes, which results in cost efficiency. The studies were conducted at a range of frequencies (20 Hz to 20 kHz), and the sensor demonstrated good accuracy and consistency in sensing acoustic pressure. By removing retained expanding air within the space, the chitosan diaphragm solves the issue of background pressure shift and ensures temperature stability. The sensor's signal-to-noise ratio may be increased by utilizing a narrowband filter and amplifier. Additionally, the fiber tip is coated with chitosan diaphragm material, which has a good sensitivity of 6 dB and significant promise in biological applications owing to its similar properties. In a separate investigation, Ezati et al. create a colorimetric indicator for evaluating the freshness of minced beef in real time [43]. The indicator was synthesized by incorporating alizarin into a cellulose-chitosan film. The addition of alizarin had no effect on the swelling index but lowered the water solubility. At pH 2–11, the indicator demonstrated a distinct transition from yellow to brown to purple and exhibited greater color stability at 4°C than at 25°C after 2 months of storage. Notably, the indicator changed color from brown to purple when the critical level of total volatile basic nitrogen reached 20.53 mg N/100 g.

Khatri et al. established that chitosan films function polysaccharide bioreceptor for the detecting α-synuclein, and the receptor was very sensitive in their investigation [44]. As a sensing platform, a U-shaped fiber optic probe whose movement was restricted with gold nanoparticles and then chitosan was employed to functionalized. The interaction of amyloid proteins with this film results in a localized change in the dielectric constant around the nanoparticles, resulting in a shift in the spectrum at the fiber's output end. To establish a robust sensing platform, the binding rate kinetics and intensity change were combined.

Nguyen et al. investigated the use of a new polyvinyl alcohol/chitosan-thermally reduced graphene modified glassy carbon electrode for the measurement of lead in aqueous samples using square-wave anodic stripping voltammetry [45]. Graphene was synthesized by an environmentally benign process of in a nitrogen atmosphere. Polyvinyl alcohol was added to chitosan to increase its electrical and physical properties. Subsequently, FESEM images were employed to the surface of the PVA/chitosan-TRG composite to confirm its homogenous and well-dispersed nature. Electrochemical impedance spectroscopy and SWASV were employed for analyzing the electrode's electrochemical properties before and after modification. The impacts of various variables like pH, time and voltage, and ratio of various constituents were examined in a systematic manner. The redesigned electrode considerably boosted the sensitivity and selectivity for Pb detection. The selectivity analysis demonstrated that the majority of common foreign ions had no substantial interference with the detection of Pb. Additionally, the electrode has been successfully used to detect lead in real-world water samples. The suggested technique enables on-site detection of Pb at trace levels in aqueous samples utilizing a simple, cost-effective, and environmentally friendly process that does not need the use of sophisticated gear.

Glucose is necessary for the human body to function properly. As a result, it is critical to maintain control over its level. Lipińska et al. reported a research on a new electrode made up of titanium foil with Au coating nanoparticles, and further modifications were considered by using glucose oxidase with enclosed chitosan [46]. Initially, the Ti foil was anodized, then chemical etching is performed, which then resulted in the development of uniformly distributed inverted caps, after that Au nanoparticles with a size restricted by the cavity dimensions are generated and deposited. Finally, chitosan and enzymes are used to alter the surface of the material, and immobilization of glucose oxidase was confirmed with FTIR. The electrode's response to glucose was determined in 0.1 M PBS containing a variety of interfering agents and biological fluids. The high selectivity and stability of the material encourage its use as the primary biosensor component. Additionally, no extra mediators were given to increase the activity of the enzyme. By and large, the suggested electrode material, owing to its superior performance characteristics, biocompatibility, and relatively easy manufacturing technique, shows promise as a candidate for glucose monitoring, particularly non-invasive and painless monitoring.

Developing an electroconductive and resilient hydrogel that can self-heal and recover from wearable strain sensors is an ongoing issue. Presently hydrogels exhibit a trade-off between static crosslinks and dynamic crosslinks for SELF characteristics. Heidarian et al. worked on developing a simple procedure for creating a dynamic hydrogel, by using starch/polyacrylic acid, ferric ions, and tannic acid-coated chitin nanofibers. According to our results, the toughest hydrogel (1.43 MJ m^{-3}) with a 1 wt.% TA-ChNF reinforcement has a toughness that is 10.5 times that of the unreinforced equivalent. The improved hydrogel was subjected to morphological and FTIR analyses [47]. The fracture surface of the pristine and TA-ChNF-reinforced hydrogels is shown in Figure 10.3. Additionally, the TA-ChNF-reinforced hydrogel at 1 wt.% demonstrated the strongest resistance to fracture propagation and a 96.5% healing efficiency after 40 minutes. As a result, it was selected as

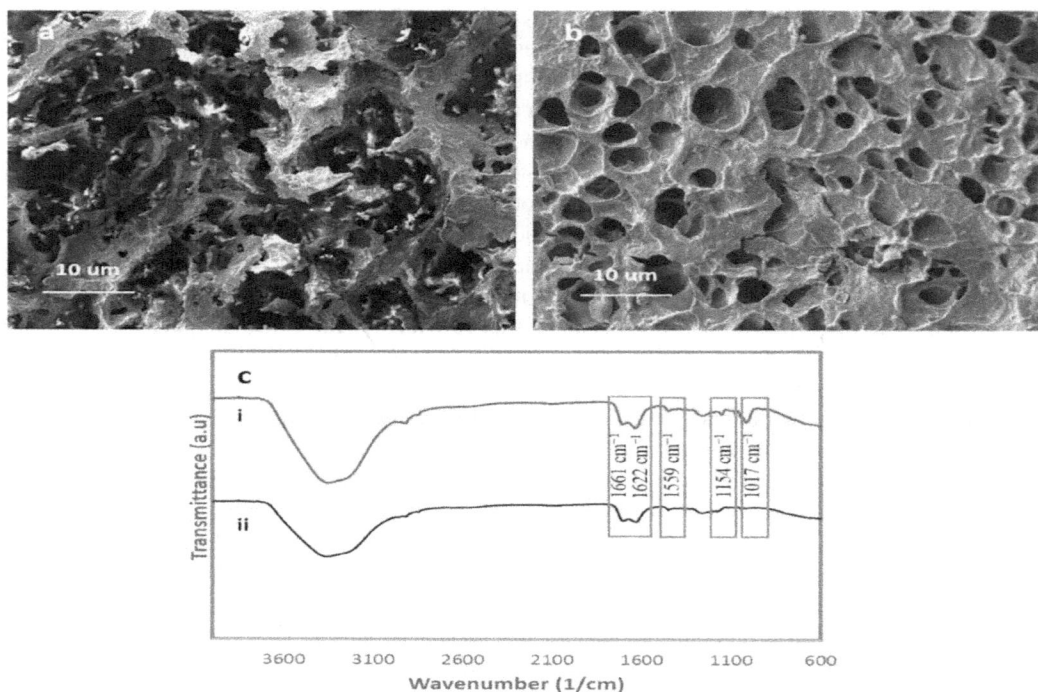

FIGURE 10.3 SEM pictures of (a) unreinforced hydrogels and (b) reinforced hydrogels containing 1% TA-ChNFs; FTIR findings of (ci) TA-ChNF-reinforced hydrogels containing 1% TA-ChNFs and (cii) unreinforced hydrogels. Adapted from ref. [48].

the optimized hydrogel on which the subsequent tests would be conducted. Self-adhesive hydrogels could be used in wearable strain sensors because of their unique self-recovery ability, network resilience, and machinability [47].

Brondani et al. developed ionic liquid 1-butyl-3-methylimidazolium hexafluorophosphate-based biosensor for the purpose of determining adrenaline in medications using square-wave voltammetry [48]. Nujol was utilized as a binder in these biosensors, and a maize homogenate was chemically attached (provides peroxidase, which catalyzed the reaction of adrenaline to the corresponding O-quinone), using carbodiimide and glyoxal. The best results were obtained when peroxidase (7.5 unitsmg^{-1} carbon paste), BMIPF$_6$: Nujol in equal ratio by weight and 0.1 mol L^{-1} phosphate buffer at pH 7.0. Linear curve was there for adrenaline concentrations ranging from 9.89×10^{-7} to 1.22×10^{-4} mol L^{-1}. These biosensors demonstrated stability and reproducibility and tests with the suggested biosensors recovered adrenaline in pharmaceutical samples at a rate ranging from 97.8% to 100.8%, and the results were consistent with those obtained using the standard method.

Chen et al. have developed an ionic liquid-assisted approach for producing chitin films with ordered microstructures [49]. Chitinvitrimer composites had an unusual two-stage stress-strain behavior, and their failure mechanisms were explored in terms of fracture mechanics. The composites were found to be weldable and recyclable after the hot-pressing procedure. Additionally, because to the shape memory effect, the welded bilayer structure displayed a smart actuator function. A green and sustainable industry could benefit from the employment of dynamic chemistry and natural materials, thanks to this simple yet successful production approach. While chitin films cannot compete with synthetic fibers in some technical applications, the layer-by-layer construction technique is very efficient and avoids the problem of dispersion, which may be used to a wide variety of different polymeric composites. Additionally, this study demonstrates how dynamic chemistry and natural materials may be used in a green and sustainable sector.

10.3 CONCLUSION

There has been substantial advancement and development in cellulose and cellulose-based composites during the last 30 years in a wide range of application areas. Cellulose as a sensor material was the subject of this review's discussion. There were numerous sensors discussed in the article ranging from gas detectors to capacitance detectors to UV detectors to humidity and strain sensors. We were encouraged by the value-added potential of cellulose in any application, which is sustainable and regenerative. Sensors for energy research and technology have been proposed in a variety of ways. However, the technical features and performance of the sensors are still being improved. To fulfill the requirement of employing cellulose for more efficient sensors, more improvement of cellulose's engineering qualities is necessary in order to increase cellulose's usage as a sensor in energy science and technology. Because they are abundant polysaccharides, chitin and chitosan are readily available and biocompatible. In addition to their thin-layer formation capacity and abundance of oxygen and nitrogen, which can be used to chemically change the various properties to match the requirements, the two materials have several medical and technical applications. An issue with chitin as a biosensor component is that it cannot be dissolved in aqueous fluids at temperatures low enough to prevent the inactivation of biosensor recognition elements, and one possible solution to this problem would be to suspend colloidal chitin in the preferred buffers for biomatrix formation on sensor surfaces. For chitosan, the chemical reversibility of hydrogel film production may provide difficulties when exposed to fluids with a low enough pH to induce substantial reprotonation of the amine groups, resulting in unfavorable internal matrix conversion and/or loss of surface adherence and exfoliation.

REFERENCES

1. Mohammed ASA, Naveed M, Polysaccharides JN. Classification, chemical properties, and future perspective applications in fields of pharmacology and biological medicine (a review of current applications and upcoming potentialities). *J Polym Environ [Internet]*. 2021;29(8):2359–71. Available from: https://doi.org/10.1007/s10924-021-02052-2
2. Yun S, Chen Y, Nayak JN, Kim J. Effect of solvent mixture on properties and performance of electroactive paper made with regenerated cellulose. *Sensors Actuators, B Chem*. 2008;129(2):652–8.
3. Ummartyotin S, Manuspiya H. A critical review on cellulose: From fundamental to an approach on sensor technology. *Renew Sustain Energy Rev*. 2015;41:402–12.
4. Kafy A, Sadasivuni KK, Akther A, Min SK, Kim J. Cellulose/graphene nanocomposite as multifunctional electronic and solvent sensor material. *Mater Lett [Internet]*. 2015;159:20–3. Available from: https://doi.org/10.1016/j.matlet.2015.05.102
5. Duy LT, Noh YG, Seo H. Improving graphene gas sensors via a synergistic effect of top nanocatalysts and bottom cellulose assembled using a modified filtration technique. *Sensors Actuators, B Chem [Internet]*. 2021;334(November 2020):129676. Available from: https://doi.org/10.1016/j.snb.2021.129676
6. Kanaparthi S, Badhulika S. Low cost, flexible and biodegradable touch sensor fabricated by solvent-free processing of graphite on cellulose paper. *Sensors Actuators, B Chem [Internet]*. 2017;242:857–64. Available from: https://doi.org/10.1016/j.snb.2016.09.172
7. Yao Y, Huang XH, Zhang BY, Zhang Z, Hou D, Zhou ZK. Facile fabrication of high sensitivity cellulose nanocrystals based QCM humidity sensors with asymmetric electrode structure. *Sensors Actuators, B Chem [Internet]*. 2020;302(August 2019):127192. Available from: https://doi.org/10.1016/j.snb.2019.127192
8. Zhanga YP, Chodavarapua VP, Kirka AG, Andrews MP. Structured color humidity indicator from reversible pitch tuning in self-assembled nanocrystalline cellulose films. *Sensors Actuators, B Chem [Internet]*. 2013;176:692–7. Available from: https://doi.org/10.1016/j.snb.2012.09.100
9. Lee SW, Kim JH, Kim J, Kim HS. Characterization and sensor application of cellulose electro-active paper (EAPap). *Chinese Sci Bull*. 2009;54(15):2703–7.
10. Rajala S, Siponkoski T, Sarlin E, Mettänen M, Vuoriluoto M, Pammo A, et al. Cellulose nanofibril film as a piezoelectric sensor material. *ACS Appl Mater Interfaces*. 2016;8(24):15607–14.
11. Zhu Q, Liu S, Sun J, Liu J, Kirubaharan CJ, Chen H, et al. Stimuli-responsive cellulose nanomaterials for smart applications. *Carbohydr Polym [Internet]*. 2020;235:115933. Available from: https://doi.org/10.1016/j.carbpol.2020.115933

12. Guido E, Colleoni C, De Clerck K, Plutino MR, Rosace G. Influence of catalyst in the synthesis of a cellulose-based sensor: Kinetic study of 3-glycidoxypropyltrimethoxysilane epoxy ring opening by Lewis acid. *Sensors Actuators, B Chem [Internet]*. 2014;203:213–22. Available from: https://doi.org/10.1016/j.snb.2014.06.126

13. Teodoro KBR, Sanfelice RC, Migliorini FL, Pavinatto A, Facure MHM, Correa DS. A Review on the role and performance of cellulose nanomaterials in sensors. *ACS Sensors*. 2021;6(7):2473–96.

14. Qiu X, Hu S. "Smart" materials based on cellulose: A review of the preparations, properties, and applications. *Materials (Basel)*. 2013;6(3):738–81.

15. Orelma H, Hokkanen A, Leppänen I, Kammiovirta K, Kapulainen M, Harlin A. Optical cellulose fiber made from regenerated cellulose and cellulose acetate for water sensor applications. *Cellulose*. 2020;27(3):1543–53.

16. Zheng Z, Tang C, Yeow JTW. A high-performance CMUT humidity sensor based on cellulose nanocrystal sensing film. *Sensors Actuators, B Chem [Internet]*. 2020;320(May):128596. Available from: https://doi.org/10.1016/j.snb.2020.128596

17. Zhou R, Li J, Jiang H, Li H, Wang Y, Briand D, et al. Highly transparent humidity sensor with thin cellulose acetate butyrate and hydrophobic AF1600X vapor permeating layers fabricated by screen printing. *Sensors Actuators, B Chem [Internet]*. 2019;281:212–20. Available from: https://doi.org/10.1016/j.snb.2018.10.061

18. Sinar D, Knopf GK. Disposable piezoelectric vibration sensors with PDMS/ZnO transducers on printed graphene-cellulose electrodes. *Sensors Actuators, A Phys [Internet]*. 2020;302:111800. Available from: https://doi.org/10.1016/j.sna.2019.111800

19. Ma Q, Wang L. Preparation of a visual pH-sensing film based on tara gum incorporating cellulose and extracts from grape skins. *Sensors Actuators, B Chem [Internet]*. 2016;235:401–7. Available from: https://doi.org/10.1016/j.snb.2016.05.107

20. Terasawa N. High-performance TEMPO-oxidised cellulose nanofibre/PEDOT: PSS/ionic liquid gel actuators. *Sensors Actuators, B Chem [Internet]*. 2021;343(May):130105. Available from: https://doi.org/10.1016/j.snb.2021.130105

21. Van Der Schueren L, De Clerck K, Brancatelli G, Rosace G, Van Damme E, De Vos W. Novel cellulose and polyamide halochromic textile sensors based on the encapsulation of Methyl Red into a sol-gel matrix. *Sensors Actuators, B Chem [Internet]*. 2012;162(1):27–34. Available from: https://dx.doi.org/10.1016/j.snb.2011.11.077

22. Liu Y, Shang S, Mo S, Wang P, Yin B, Wei J. Soft actuators built from cellulose paper: A review on actuation, material, fabrication, and applications. *J Sci Adv Mater Devices [Internet]*. 2021;6(3):321–37. Available from: https://doi.org/10.1016/j.jsamd.2021.06.004

23. Guan X, Hou Z, Wu K, Zhao H, Liu S, Fei T, et al. Flexible humidity sensor based on modified cellulose paper. *Sensors Actuators, B Chem*. 2021;339(January):129879.

24. Mahadeva SK, Yun S, Kim J. Flexible humidity and temperature sensor based on cellulose-polypyrrole nanocomposite. *Sensors Actuators, A Phys [Internet]*. 2011;165(2):194–9. Available from: https://dx.doi.org/10.1016/j.sna.2010.10.018

25. Meng X, Yang J, Liu Z, Lu W, Sun Y, Dai Y. Non-contact, fibrous cellulose acetate/aluminum flexible electronic-sensor for humidity detecting. *Compos Commun [Internet]*. 2020;20:100347. Available from: https://doi.org/10.1016/j.coco.2020.04.013

26. Devarayan K, Kim BS. Reversible and universal pH sensing cellulose nanofibers for health monitor. *Sensors Actuators, B Chem [Internet]*. 2015;209:281–6. Available from: https://dx.doi.org/10.1016/j.snb.2014.11.120

27. Yao J, Ji P, Wang B, Wang H, Chen S. Color-tunable luminescent macrofibers based on CdTe QDs-loaded bacterial cellulose nanofibers for pH and glucose sensing. *Sensors Actuators, B Chem [Internet]*. 2018;254:110–9. Available from: https://doi.org/10.1016/j.snb.2017.07.071

28. Mun S, Chen Y, Kim J. Cellulose-titanium dioxide-multiwalled carbon nanotube hybrid nanocomposite and its ammonia gas sensing properties at room temperature. *Sensors Actuators, B Chem [Internet]*. 2012;171–172:1186–91. Available from: https://doi.org/10.1016/j.snb.2012.06.066

29. Hittini W, Abu-Hani AF, Reddy N, Mahmoud ST. Cellulose-copper oxide hybrid nanocomposites membranes for H2S gas detection at low temperatures. *Sci Rep*. 2020;10(1):1–9.

30. Wu S, Peng S, Wang CH. Stretchable strain sensors based on PDMS composites with cellulose sponges containing one- and two-dimensional nanocarbons. *Sensors Actuators, A Phys [Internet]*. 2018;279:90–100. Available from: https://doi.org/10.1016/j.sna.2018.06.002

31. Wu G, Gu Y, Hou X, Li R, Ke H, Xiao X. Hybrid nanocomposites of cellulose/carbon-nanotubes/polyurethane with rapidly water sensitive shape memory effect and strain sensing performance. *Polymers*. 2019;11(10):1586.

32. Gimenez AJ, Yáñez-Limón JM, Seminario JM. ZnO-cellulose composite for UV sensing. *IEEE Sens J.* 2013;13(4):1301–6.

33. Komatsu H, Kawamoto Y, Ikuno T. Freestanding translucent ZnO-cellulose nanocomposite films for ultraviolet sensor applications. *Nanomaterials.* 2022 Mar 12;12(6):940.

34. Janani B, Syed A, Raju LL, Marraiki N, Elgorban AM, Zaghloul NSS, et al. Highly selective and effective environmental mercuric ion detection method based on starch modified Ag NPs in presence of glycine. *Opt Commun.* 2020;465:125564. Available from: https://doi.org/10.1016/j.optcom.2020.125564

35. Chough SH, Kim KS, Hyung KW, Cho HI, Park HR, Choi OJ, et al. Identification of botanical origins of starches using a glucose biosensor and amyloglucosidase. *Sensors Actuators, B Chem.* 2006;114(2):573–7.

36. Garcia PT, Guimarães LN, Dias AA, Ulhoa CJ, Coltro WKT. Amperometric detection of salivary A-amylase on screen-printed carbon electrodes as a simple and inexpensive alternative for point-of-care testing. *Sensors Actuators, B Chem [Internet].* 2018;258:342–8. Available from: https://doi.org/10.1016/j.snb.2017.11.068

37. Lee GG, Kang SJL. Formation of large pores and their effect on electrical properties of SnO2 gas sensors. *Sensors Actuators, B Chem.* 2005;107(1):392–6.

38. Choi I, Lee JY, Lacroix M, Han J. Intelligent pH indicator film composed of agar/potato starch and anthocyanin extracts from purple sweet potato. *Food Chem.* 2017;218:122–8. Available from: https://dx.doi.org/10.1016/j.foodchem.2016.09.050

39. Elgamouz A, Bajou K, Hafez B, Nassab C, Behi A, Haija MA, et al. Optical sensing of hydrogen peroxide using starch capped silver nanoparticles, synthesis, optimization and detection in urine. *Sensors Actuators Rep.* 2020;2(1):100014. Available from: https://doi.org/10.1016/j.snr.2020.100014

40. Khachatryan G, Khachatryan K. Starch based nanocomposites as sensors for heavy metals - Detection of Cu^{2+} and Pb^{2+} ions. *Int Agrophysics.* 2019;33(1):121–6.

41. Suginta W, Khunkaewla P, Schulte A. Electrochemical biosensor applications of polysaccharides chitin and chitosan. *Chem Rev.* 2013;113(7):5458–79.

42. Chen LH, Chan CC, Yuan W, Goh SK, Sun J. High performance chitosan diaphragm-based fiber-optic acoustic sensor. *Sensors Actuators, A Phys [Internet].* 2010;163(1):42–7. Available from: https://doi.org/10.1016/j.sna.2010.06.023

43. Ezati P, Tajik H, Moradi M. Fabrication and characterization of alizarin colorimetric indicator based on cellulose-chitosan to monitor the freshness of minced beef. *Sensors Actuators, B Chem.* 2019;285:519–28. Available from: https://doi.org/10.1016/j.snb.2019.01.089

44. Khatri A, Punjabi N, Ghosh D, Maji SK, Mukherji S. Detection and differentiation of A-Synuclein monomer and fibril by chitosan film coated nanogold array on optical sensor platform. *Sensors Actuators, B Chem.* 2018;255:692–700. Available from: https://dx.doi.org/10.1016/j.snb.2017.08.051

45. Nguyen LD, Doan TCD, Huynh TM, Nguyen VNP, Dinh HH, Dang DMT, et al. An electrochemical sensor based on polyvinyl alcohol/chitosan-thermally reduced graphene composite modified glassy carbon electrode for sensitive voltammetric detection of lead. *Sensors Actuators, B Chem.* 2021;345(July):130443. Available from: https://doi.org/10.1016/j.snb.2021.130443

46. Lipińska W, Siuzdak K, Karczewski J, Dołęga A, Grochowska K. Electrochemical glucose sensor based on the glucose oxidase entrapped in chitosan immobilized onto laser-processed Au-Ti electrode. *Sensors Actuators, B Chem.* 2021;330(December 2020):129409.

47. Heidarian P, Kouzani AZ, Kaynak A, Zolfagharian A, Yousefi H. Dynamic mussel-inspired chitin nanocomposite hydrogels for wearable strain sensors. *Polymers (Basel).* 2020;12(6):1–15.

48. Brondani D, Dupont J, Spinelli A, Vieira IC. Development of biosensor based on ionic liquid and corn peroxidase immobilized on chemically crosslinked chitin. *Sensors Actuators, B Chem.* 2009;138(1):236–43.

49. Chen Z, Wang J, Qi HJ, Wang T, Naguib HE. Green and sustainable layered chitin-vitrimer composite with enhanced modulus, reprocessability, and smart actuator function. *ACS Sustain Chem Eng.* 2020;8(40):15168–78.

11 Polysaccharide Applications in Functional Textiles and Textile Wastewater Treatment

Tanvir Mahady Dip
Bangladesh University of Textiles

Md Humayun Kabir and Muhammet Uzun
Marmara University

11.1 INTRODUCTION

Biopolymers are interconnected chains of biodegradable monomers primarily produced by living entities. The lion's share of these producers are microorganisms, plant biomass, or agro-wastes. Chemically synthesized substrates from biological sources like vegetable oil, sugar, fat, protein, etc., also fall under this category. Major components of most biopolymers include proteins, fats, nucleotides, and polysaccharides [1,2]. History suggests mankind have been using biopolymers since ancient time. However, the last few decades have seen a decline in their applications due to the emergence of less expensive and tailorable synthetic polymers. Nevertheless, in recent times, the realization of environmental costs and concern for the future has led to the re-insurgence of biodegradable polymers from renewable sources [3]. Moreover, renewability, availability, biodegradability, high absorption, and ease of functionalization have instigated interest from both academia and different industries. Compared with their synthetic counterparts, biopolymers display a more complex molecular arrangement with a specified 3D shape [4]. The nature-originated biopolymers are categorized as proteins, esters, and PSs biopolymers. Biopolymers can further be classified into three other broad categories based on occurrence and abundance. They are polynucleotides, polypeptides, and PSs. The monosaccharides are linked via glycoside bonds to form PSs [2]. Variations in physicochemical and mechanical properties such as viscosity, gelling properties, interfacial characteristics, chain length, linkage variety, etc., allow PS to derive novel materials for a wide range of functionalities [4].

A powerful multifunctional biopolymer like PS can create high performance, value-added textile products. For instance, PS in the gel form can enhance the thickening quality of printing paste during textile printing. Along with an improvement in the rheological properties of the paste, PSs like alginate and carboxymethyl cellulose (CMC) accounted for better color fastness and hand feel than traditional thickeners [5]. Thus, PSs can be considered a suitable, environmentally friendly alternative to conventional thickeners. Textile substrate finished with PS will add functionalities to the textile itself. UV ray protection, contact angle and air permeability improvement, aroma generation, insect repellent, and microbial growth-inhibiting capabilities can be introduced [6]. Adding to that, PS's ability to reduce, stabilize, and perform adhesion of metal nanoparticles (MNPs) with multifunctionalities can enhance textiles' performance. PSs like chitosan, alginate, starch, cellulose, and cyclodextrins can interact with silver, zinc, copper, titanium, etc., to introduce antimicrobial, UV protection, conductivity, etc. [7]. With the increasing health concern, textiles with the ability to repel micro-organism accommodation are becoming a popular area of research [8]. Biopolymers like chitosan, starch, and alginate have become crucial for many biomedical operations in the form of textile substrates, flexible films, or an antibacterial finish. Ease of modifications, controlled release

DOI: 10.1201/9781003265054-11

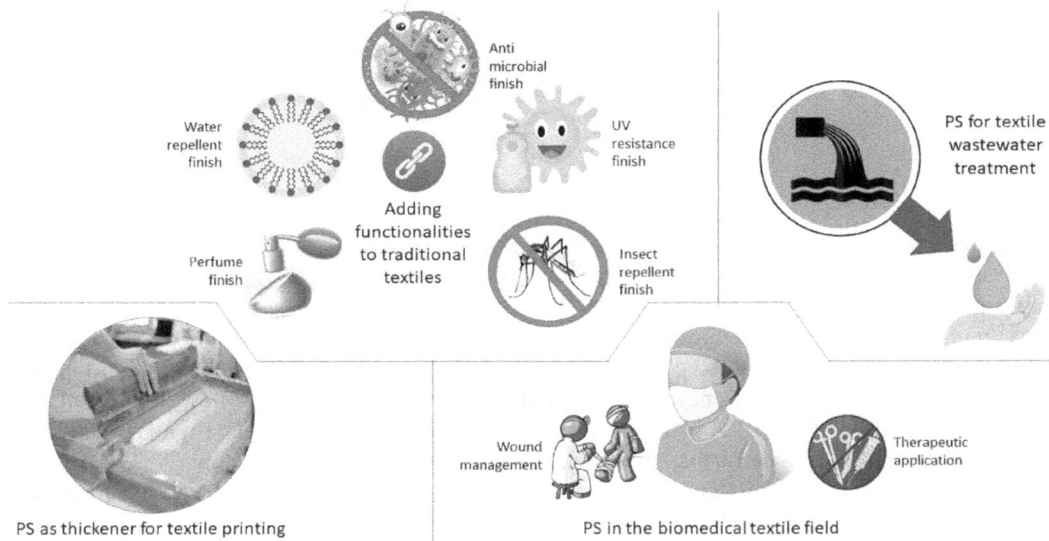

FIGURE 11.1 Multilayers of textile application areas for PS.

of drugs, and enriched biocompatibility have contributed to enhanced pharmaceutical applications of these PSs [1]. Furthermore, the synergistic effects of various complex structures can account for control, target, and delivery actions in pharmaceutical areas [9]. In wound dressings, bioactive polymers containing functional components can show superior performance compared to traditional cotton gauzes [10]. The main advantage lies with PS being a versatile biopolymer that can introduce functionalities at multiple levels, making them unique compared to most other compounds. On another note, the textile effluent is mostly composed of unused dyes, auxiliaries, and other chemicals. PSs, their derivatives, or modified PSs [11] as absorbent [12], flocculant [13], or coagulant can effectively remove contaminants from textile discharge.

Apart from being used as traditional textile raw material, PSs can be applied to textile industries in multilayers of application areas (as shown in Figure 11.1). This chapter is dedicated to natural PSs being used in various platforms to assist in the textile operation, adding functionalities and reducing textile waste. Starting with introducing the specific PSs involved with the targeted area, the chapter further represents a detailed overview of those PSs in action to improve textile printing. It further sheds light on adding functionalities to conventional textile materials. It also covers various textile-based operations in the biomedical field. The chapter also includes a brief discussion on the PS-associated textile effluent treatment mechanism and a review of recent research works. Finally, the chapter concludes with accumulating challenges and limitations with the existing setup insinuating some directions for future researchers.

11.2 POLYSACCHARIDES FOR TEXTILES APPLICATIONS

An uncountable number of PSs can be found in the world. A few of them have found their worth in the textile application arena. This section talks about the specific PSs that are being used in the textile industry briefly.

11.2.1 ALGINATE

Alginate is one of the most commonly used PSs in the textile industry. It is a natural PS that can be found in the embankment of brown algae [14]. The most commonly used species are *Laminaria*

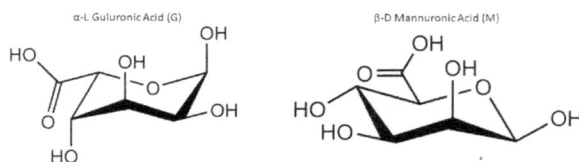

FIGURE 11.2 Chemical structure of alginate.

hyperborea, *Macrocystis pyrifera*, and *Ascophyllum nodosum*. E.C.C. Stanford discovered this PS in 1881 while he was looking for beneficial products in the kelp [2]. Chemically, alginate is a polymeric acid that is constructed from two units of monomers (as shown in Figure A1). The two monomeric units are (G) and D-mannuronic acid (M) (as shown in Figure 11.2) [15].

Alginate is a naturally occurring PS, and it has some special features. The most important and notable features are high absorbency, biocompatibility, good mechanical properties, good breathability, and nontoxicity [1617]. It has better properties for gelling and exceptional ease for modification and processing [18]. Alginate further possesses fundamental flame-retardant property [19]. Alginate as a textile filament is generally produced by using the wet spinning technique, where it is forced out in the form of a solution (sodium alginate) into a bath of aqueous calcium chloride [16,17,20].

Alginate in textile is used mainly for medical textile-related applications. It can immobilize enzymes or act as an assistance for bioactive molecules [16]. Numerous studies report its application in medical wound dressing [16–18]. Alginate can also be used to manufacture tissue engineering scaffolds and facilitate drug delivery systems. Textile finishing is a popular technique to enhance functionalities. In this case, various textile products' functional and antibacterial finishing can also be done using alginate-based PSs [18]. Alginate can be incorporated into the textile industries in various forms, like, sliver, yarn, woven and nonwoven fabric, knitted fabric, crochet, braided, composite materials, nanocomposite coating, finishing agents, etc. Good quality fibers are also manufactured industrially in mixed ionic form of sodium/calcium [16,18].

Scientists believe more potential applications of alginate-based PS are yet to be discovered. Among them, hemostasis and tissue repair is an important domain under development. Moreover, blended yarns made out of alginate have provided better strength and hand feeling [16]. It was also seen from a previous study that an aqueous solution of sodium alginate and CMC could produce a composite fiber with improved features [17].

11.2.2 Chitin and Chitosan

Chitin and chitosan are exceptionally convertible biomaterials having unique characteristics like recyclability, biodegradability, low price, etc. Chitin is ranked as the second most available polymer present on earth after cellulose. Biodegradability is one of the most useful features of polymers, and chitin is one of such polymers which will decompose after usage. It is attained from the shell of crabs, shrimps, crustaceans, and cell walls of fungi [14]. An amended biopolymer attained from chitin is chitosan. Chitosan is basically a group of polymers extracted after the deacetylation of chitin to different degrees [21–24]. Deacetylation means eliminating acetyl groups of chitin and keeping only the amino group (–NH2). Hence, the prime difference between chitin and chitosan (as shown in Figure 11.3) is the acetyl content of the polymer [21,24].

Bioactive and natural PS chitin and chitosan are used in several textile fields, especially in medical textile and functional finishing. It is used in biomedical applications like tissue repairing, healing wounds, and producing scaffolds [14]. Different functional finishing and surface modification is done using chitin and chitosan. Antimicrobial finishing can be achieved by either attaching chitosan to the fabric or by preparing antibacterial fibers. Similarly, antiwrinkle finishing can be done by dipping the fabric in chitosan solution by regular dipping-rolling-baking process, which reinforces the amorphous regions of the fibers and reduces the mobility [14,25]. Another special finish is antistatic

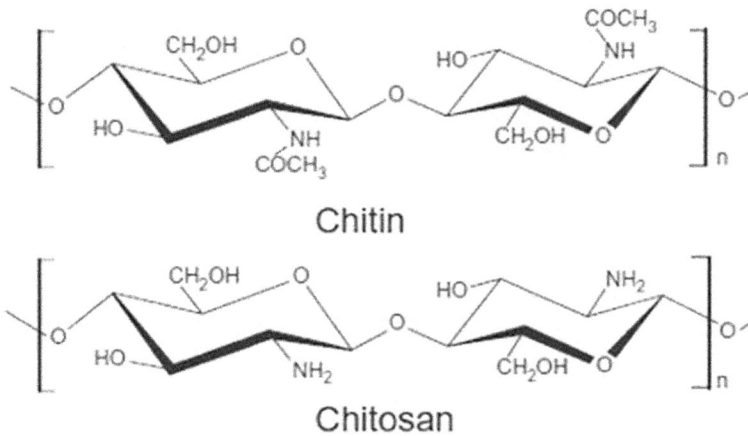

FIGURE 11.3   Chitin and chitosan chemical structure.

finishing to produce antistatic workwear for the employees working in industrial fields and using sophisticated electric instruments. Dyeing is one of the most crucial processes in the textile industry. Chitosan is used for the degeneration of the dye effluents. It is also a flawless agent for dye fixation for anionic dyes as it minimizes the Coulomb repulsion between the anionic dyes and the fiber [25,26].

11.2.3   STARCH

Starch is a well known polysaccharide consisting of monomers of glucose joined in α-1,4-linkages. It is an organic chemical which is white in color and generated by all types of green plants. It is insoluble in alcohol and cold water. The general chemical formula of starch is $(C_6H_{10}O_5)_n$ [27]. The major sources of starch are cassava, corn, wheat, sweet potato, and potato. Minor sources of starch are rice, barley, sorghum, etc. Raw starch generally comes up in the form of grains. The dimension and the arrangement of molecules in the grains depend on the species of plants and the environmental interactions [28]. Egyptians developed the main use of starch-based products and starch during the pre-dynastic era. Starch industry expanded extensively after the 19th century because of the growing needs from the textile and clothing industry, paper industry, and color printing. The ability of starch to be converted into a gum-like material, dextrin, accelerated its demand [29]. Since 1930, different scientists have developed other starch-based products like waxy corn starch, corn starch with high-amylose, sweeteners, ethanol, amino acid, etc. [29]. Starch is composed of two polymers of D-glucose. Amylose consists of linear α-1,4-linked glucose, whereas amylopectin is composed of α-1,4 and α-1,6 links which are branched. Amylose and amylopectin (as shown in Figure 11.4) act for around 98%–99% of the entire dry weight of starch, while the other 1%–2% represent phosphorus, mineral, protein, and lipid [28].

Starch is widely used in the textile industry for sizing, finish, and print purposes. It is used to size warp yarn to improve the yarn strength and the abrasion resistance during the weaving process. It should have the ability to form strong films to provide a protective coating on the yarn surface, flexibility, and consistent viscosity during applications. Modified tapioca starch is suitable for these applications [29,30]. Starch produced from wheat is used to reinforce textile yarn [29]. Rye starch can be used as a thickening agent instead of potato starch. Pure starch fibers can be produced by a modified electrospinning technique, i.e., "electro-wet spinning" [31].

In the last 10 years, fibers produced from starch-based materials by electrospinning process have exhibited prospects in different important fields like tissue engineering, wound dressing, and drug delivery. As the electrospinning of native starch is difficult as they have weak mechanical properties, researchers have developed various modified starch suitable for electrospinning [32]. Modified starch

Amylose:

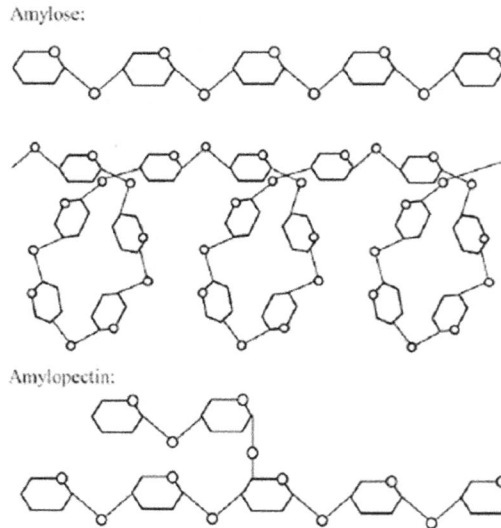

Amylopectin:

FIGURE 11.4   Amylose (linear, helical) and amylopectin.

fibers produced by the electrospinning process can be used in diaper, paper towels, sanitary tissue, biomedical products such as drug delivery, sustainable wound dressing, and scaffold of tissue [31]. Starch graft copolymerization is one of the most efficient methods to introduce desirable sizing properties in starch. This method can be improved to further enhance the sizing quality of the yarns [33].

11.3   POLYSACCHARIDE COMPOSITES

Two or more PSs together can enhance their performance significantly. PS can form various composites with other materials (for instance, cellulose polysaccharide complex, protein-polysaccharide complex, nanocomposites, etc.) also. PS-based composites have become a recent research attraction for biodegradability, renewability, and good sustainability. Some of the composites are listed as follows [34–37]:

   i. Fish gelatin-polysaccharide composites: This composite has a wide range of applications specially in wound dressing. The foremost protein gelatin is produced by degrading collagen. Nanofibrous structure produced from this composite is used in tissue engineering and wound healing applications.
   ii. CuO-Chitosan nanocomposite: This composite cotton fabric is treated in a pad-dry-thermofixation technique. The final fabric shows remarkable antibacterial reduction, which withstands 30 laundering cycles.
   iii. Neem/Cs nanocomposite: This composite is used to treat cotton fabric to impart antibacterial activity.
   iv. 1-Hydroxymethyl-5,5-dimethylhydantoin chitosan: Cotton is treated with improved chitosan in BTCA/sodium hypophosphite using the pad-dry-cure technique. This improves the antibacterial functionality of treated cotton against bacteria such as *S. aureus* and *E. coli*.
   v. Cs/Ag/ZnO nanocomposite: Pad-dry technique is used to treat cotton gauze with chitosan/Ag, or chitosan/ZnO, or chitosan/Ag/ZnO. Fabric treated with chitosan/Ag/ZnO nanocomposite exhibited improvement in different characteristics of wound dressing, such as water absorbency, time of drying, or wicking ability.

vi. Cs/ZnO nanocomposites (CZNCs): Using pad-dry-thermofixation, cotton fabric was treated. The treated fabric with nanocomposite showed antibacterial behavior and UV protection without hampering other physical properties.

vii. Cs/ ZnONPs, Cs/TiO^{2}NPs, Cs/SiO^{2}NPs: Cotton/PET blended fabric of two different fiber ratio (50/50) and (65/35) was treated using pad-dry-cure method. This treatment improved the antibacterial activity, self-cleaning resiliency, and UV protection properties of the fabric.

viii. Ag/TiO^2/βCD nanocomposites: Using the pad-dry-cure technique, cotton fabric was treated, and it enhanced the self-cleaning characteristics. The treated fabric also exhibited antibacterial activity against *Staphylococcus aureus* and improved crease recovery angle and tensile strength.

ix. Alginate microcapsules containing herbal extract: Cotton denim was treated using microcapsules and citric acid by drying and curing technique. Herbal extract and sodium alginate were mixed, and CaCl$_2$ solution was sprayed to produce the microcapsules. It was visible that microcapsule enhanced the durability of the antimicrobial functionality even after 15 washes.

x. CuNPs/Ca-alginate: CuNPs/Ca-alginate treated cotton exhibited superior antibacterial effect, especially against *E. coli* bacteria, without impacting the mechanical properties of the fabric.

xi. AgNPs–alginate composite: The cotton fabric was treated using AgNPs–alginate composite using the pad-dry cure method in the presence of a binder. The usage of a binder enhanced the wash durability of the treated sample.

xii. Carboxymethyl starch/polyvinyl-alcohol electrospun composite nanofibers: PVA was blended with carboxymethyl starch due to its nontoxicity, water solubility, and biocompatibility, which enhanced the spinnability of the solutions blended. Moreover, this solution gives continuous, smooth, and better-quality fibers. Starch-based electrospun fibers have potential applications in medical textiles, which are currently being researched.

xiii. PAFC-Starch-gp (AM-DMDAAC) composite: To be used in textile wastewater treatment, a polymer composite which is inorganic/natural named PAFC-starch-gp(AM-DMDAAC) (polyaluminum ferric chloride-starch graft copolymer with acrylamide and dimethyl diallyl ammonium chloride) was produced. It was seen that PAFC-Starch-gp (AM-DMDAAC) gives better and steady dye removal efficiency. PAFC-Starch-g-p(AM-DMDAAC) can deduct by more than 50% compared to the conventional technique of coagulation-flocculation for wastewater treatment of synthetic textile.

11.4   OTHER POLYSACCHARIDES

Other than the usual PSs, some other important PSs that have found their worth in various textile applications are given as follows:

i. *Zizylphus vulgaris* PS: *Zizylphus vulgaris* plant has been cultivated for almost 4,000 years. Polysaccharide, a necessary bioactive material, is present in *Zizylphus vulgaris*. It is one kind of complex heteropolysaccharide. Its unique biomedical benefits are protecting the liver, nourishing blood functionality, antioxidants, etc. A proverb says, "three Zizylphus vulgaris in solar eclipse will not show aging in one's life." There is a scope for producing sustainable wound dressings from this polysaccharide, which will improve the existing application of technical textile of polysaccharides in meditex [38].

ii. Psyllium PS: For a long time, psyllium has been used for curing various health-related problems like diarrhea, constipation, high blood pressure, and hemorrhoids. The principal element of the seeds and husk of psyllium is a mucilaginous polysaccharide [39,40]. Psyllium polysaccharide-based hydrogels have attracted a lot of attention for their distinctive characteristics like sustainability and biodegradability. This hydrogel can be used in various meditex applications like scaffolds for tissue engineering and cell encapsulating biomaterials.

It is used widely as biopolymer-based wound dressings because of its outstanding water absorbency, biodegradability, and biocompatibility. Hemicellulose extracted from psyllium seed husk (PSH) is Arabinoxylan, and it is also used in biomedical applications. The healing property of these gels was improved by mixing gelatin with Arabinoxylan. Arabinoxylan, gelatin, glycerol, and gentamicin antibiotics were blended to form AX-gelatin (GL) film [14]. Glycerol as a plasticizer facilitated the localized delivery of antibiotic gentamicin (GM) at the infection location. A high release of 89% of the GM in 24 hours was recorded, and thus, it exhibited superior antibacterial performance to standard GM solution [39]

iii. Carboxymethylcellulose (CMC) PS: CMC is one of the most widely used cellulose derivatives. It is nontoxic and obtained generally from softwood pulp or cotton linter. Cotton fabric is treated using carboxymethylcellulose (CMC) polysaccharide nanocomposites to impart antibacterial properties. Fumaric acid crosslinked CMC hydrogel was dispersed with silver ions and then applied on the surface of the cotton fabric. This FA is a biodegradable and unsaturated dicarboxylic acid. The silver particles are distributed homogenously on the surface of the cotton in the size of nanoparticles. The resulting fabric shows excellent antibacterial properties for using this in medical textile as wound dressing [41]. Apart from the PSs mentioned above, some other PSs are found in nature on small scales. Table 11.1 enlists some of the other important natural PSs with respective sources [42–47].

TABLE 11.1
List of PSs with Respective Sources That Are Produced in the Nature in a Small Scale

Polysaccharide Name	Source	Polysaccharide Name	Source
Polysaccharide extracted from leaves of *Hammada scoparia*	Leaves	*Anadenanthera colubrine*	Barks
Polysaccharide extracted from *Trigonella foenum-graecum*	Leaves of *Hammada scoparia*	*Thymus quinquecostatus* Celak.	Leaves
Sanguisorba officinalis L. polysaccharide	Roots of the plant	*Ginkgo biloba* L.	Leaf
P. anisum seed polysaccharide	Polysaccharide from the seeds of *Pimpinella anisum*	*Prosopis juliflora*	Fruits
Linum usitatissimum L.	Inner bark of *Grewia mollis* and from the leaves of *Hoheria populnea*	PGB (Polysaccharide of *Gastrodia elata blume*)	Rhizome
Daucus carota polysaccharides	Roots	*Lycium barbarum* polysaccharides (LBPs)	Fruit
Crataegus azarolus L. var. aronia polysaccharides	Pulps and seeds	MCP (*Momordica charantia* polysaccharide)	Fruit
Annona muricata leaf polysaccharide	Leaf	WMFP (White mulberry fruit polysaccharides)	Fruits
Salicornia arabica	Tunisian halophyte (SA)	CSP (*Cordyceps sinensis* polysaccharide)	Mycelia
Phyllostachys pubescens	Bamboo leaves	Fucoidan polysaccharide	Cell walls of seaweeds
Trigonella foenumgraecum	Fenugreek seeds	P. ginseng	Roots
Sorghum bicolor L.	Seeds	CGP (*Crassostrea gigas* polysaccharide)	Oyster meat
Caesalpinia ferrea	Stem barks	DOP (*Dendrobium officinale* polysaccharide)	Stem
Lycium barbarum	Fruits of Guoqi	*Prunus amygdalus* polysaccharide	Seeds

(Continued)

TABLE 11.1 (*Continued*)
List of PSs with Respective Sources That Are Produced in the Nature in a Small Scale

Polysaccharide Name	Source	Polysaccharide Name	Source
Sophora japonica L.	Flower buds	*Aesculus chinensis* Bunge polysaccharide	Leaf
Aegle marmelos L.	Pulp	WMRP (*Citrullus lanatus* polysaccharide)	Fruit
Agave sisalana leaf polysaccharide	Leaf		

11.5 POLYSACCHARIDE APPLICATIONS IN MULTILAYERS OF TEXTILE ACTIVITIES

Other than its primary use as textile fibers in the textile industry, PSs have some other important roles to play. Specially for enhancing performance in textile operations like printing, adding mono and multifunctionalities (for instance, antimicrobial, water repellent, UV resistant, insect repellent, aroma finishing, etc.) to the existing textile substrates. The following section is dedicated to introducing various PSs for assisting in textile printing stage and adding functionalities to functional textiles in day-to-day lives and biomedical field.

11.5.1 THICKENER FOR TEXTILE PRINTING

Printing is an important textile operation that is applied for localized color placement on the textile substrate. Print paste is semisolid and thickener is an inevitable ingredient of printing paste that maintains uniform and even flow of the paste during printing [48–50]. The importance is reflected as they provide necessary plasticity and adhesion to the paste as it delivers design without color bleeding [48]. Rheological behavior is an important parameter of printing pastes for determining printing quality as it refers to paste liquidity, viscoelasticity, and thixotropic behavior [51]. The shear thinning during print paste application is observed as the applied force disrupts the paste structure allowing the paste to be pushed through and spread over the surface as intended design. As soon as the printing action is withdrawn, new entanglement forms and resists unwanted flow of paste [52]. Therefore, thickeners with good viscoelasticity property reflect good printing quality. The nontoxic and nonallergic nature of ecofriendly natural thickeners gets preference over their synthetic counterparts. PS macromolecules have demonstrated high viscosity and good film-forming ability [53]. Most of the organic thickeners like starch, gum, or gelatins are polysaccharides [54]. Some of the other mentionable polysaccharides are carrageenan, methylocellulose, hydromethylocellulose, agar, etc. have excellent gelling properties. Sodium alginate (SA) was a widely accepted natural thickener as the containing -COO- group could efficiently migrate the dye particles from paste to fabric [55]. Carboxymethyl cellulose (CMC), a modified PS, has also gained popularity because of solubility, film-forming character, and activity over a wide pH range [50]. Degree of substitution (DS) refers the number of functional groups [–OH] substituted by other groups in a PS derivative. Lower DS in many PS like CMC impacts the handle after printing [50].

Obele et al. reported CMC and cassanova stem waste cellulose nanocrystals (CNC) as thickeners for printing reactive dyes on cotton. The amount of CMC and CNC controlled the prepared printing paste viscosity. The color fastness (change and stain) and handle comparison between the prepared paste and commercial paste revealed almost similar results (change: 3 and 4, stain: 4, handle: soft) [50]. Li et al. compared various intermixtures of industrially used polysaccharide thickeners (hydroxypropyl cellulose (H-HPC)/SA; H-HPC/CMC; H-HPC/hydroxyethyl cellulose (HEC);

TABLE 11.2
PSs Thickeners with Relevant Parameters

Polysaccharide Thickener	Viscosity	Unit	PVI*	Substitution Level/Degree of Substitution	Reference
Hydroxypropyl cellulose (HPC)	0.74	Pa.s	0.33	2.5	[52]
Sodium alginate (SA)	1.02	Pa.s	0.44	0	
Carboxymethyl cellulose (CMC)	3.35	Pa.s	0.23	1.05	
Hydroxyethyl cellulose (HMC)	0.66	Pa.s	0.66	0.2	
Guar gum (GG)	0.11	Pa.s	0.73	0	
Carboxymethyl starch (CMS)	1.66	Pa.s	0.27	0.9	
Carboxymethyl cellulose (CMC)	39.5777	cP	–	1.09	[50]

H-HPC/Guar Gum (SG-9) and H-HPC/carboxymethyl starch (SG-24)) to determine their transfer printing quality with reactive dyes on cotton and silk fabrics. The mixed thickener was prepared by generally mixing a second thickener with H-HPC in the deionized water followed by electric stirring and refrigeration. Reactive dye was printed after coating and baking the printing paper surface with this prepared paste. The result concluded H-HPC/SA showed the highest and H-HPC/ HEC showed the lowest result in terms of color depth; combinations of HHPC/CMC, H-HPC/SG-9, and H-HPC/SG-24 accounted for the clearest print sharpness; no thickener combination had any significant impact in declining fabric handle and color fastness [52]. An et al. further evaluated the rheological behavior of three types of PS thickeners (carboxymethyl hydroxypropyl cellulose (CMHPC), CMC, and SA) on printing performance with reactive dye. Wool fabric was used as base material for printing on. Separate pretreatment solutions of CMHPC, CMC, and SA were prepared and padded on the wool fabric surface before printing. The CMHPC sample showed the highest color strength (K/S value 28.4), lowest permeability (2.4%), lowest color bleeding, and therefore the highest color sharpness [53]. Parameters relevant to thickening are important to determine the thickening performance of a PS. Table 11.2 represents some PS in application as a thickening agent for textile printing along with relevant parameters.

Reusability is of great demand in this era. PS has the ability to reproduce thickening agent from the textile wastewater. Fijan et al. reported alginate, CMC, and carboxymethylated guar gum (CGG) PS-based thickeners recycled from concentrated wastewater. However, the performance in terms of viscosity, paste adhesion, and penetration was not as good as alginate or CMC-based thickeners [56].

11.6 ANTIMICROBIAL FINISHING

Textiles made from natural substances have the tendency to allow microorganisms and pathogens to grow because of their hydrophilic, nutrients containing nature, porous structure, and moisture attainment ability. Contact with human body while wearing those textiles provide the ideal scenario for those organisms to grow as they can draw the required nutrients ceaselessly [8]. Other organisms like molds and fungi causes damage, strength loss, staining, and discoloration of textile matters [6]. Attempts to hinder those growths with synthetic agents is limited by the non-biodegradability and polluting nature of the acting substances [57,58]. Hence, various polysaccharides like chitosan [57], their derivatives [58], alginate [59], and starch [60] have found their application in the antimicrobial textile furnishing. The anti-microorganism mechanism involves electrostatic interaction that causes alteration of the subject's cell permeability. Consequently, the intracellular matrix disrupts to leak cytoplasmic constituents and hinder the normal metabolic operations ultimately resulting in cell death. The electrostatic interaction between the positively charged amine (NH^{3+}) groups in cationic polysaccharides (such as chitosan) and negatively charged cell surfaces of bacteria, fungi, etc. is responsible for the effective antimicrobial performance (as shown in Figure 11.5) [61]. Thus, increasing the density of positive charge will enhance antimicrobial mechanism in case of cationic

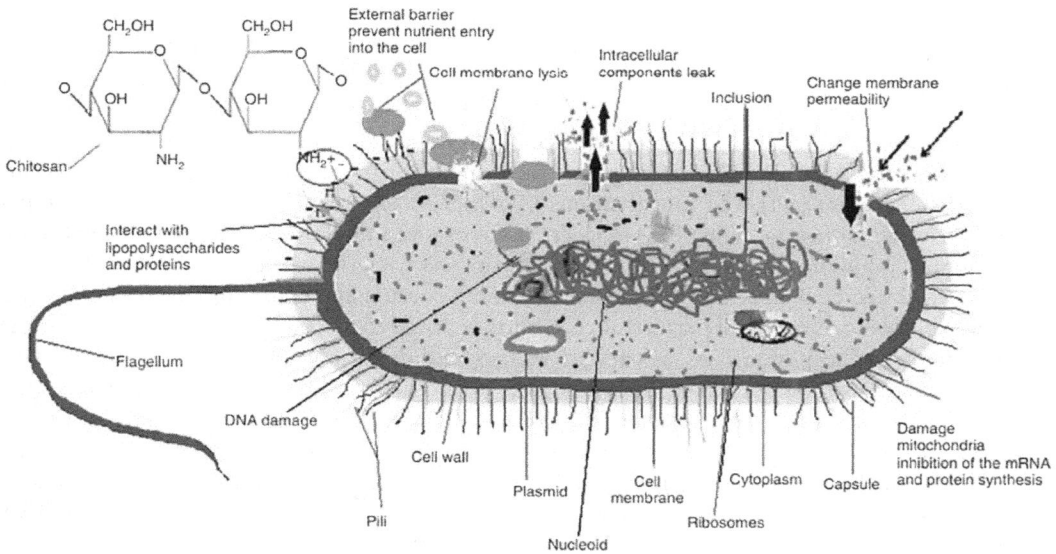

FIGURE 11.5 Antimicrobial mechanism of a polysaccharide (chitosan), reused with permission from [65] © De Gruyter, 2018.

PSs [35]. The polycationic nature of these PSs attaches efficiently to anionic microbe proteins [62]. Among other antibacterial mechanisms, protein synthesis, lipid peroxidation, degradation of endotoxin, and nucleic acid synthesis are worth mentioning [59,63]. Furthermore, PSs (SA) can also act as the anchoring polymer for antibacterial agents like quaternary ammonium complex (QAC) [8], TiO2 [59], micro/nanocapsules containing antimicrobial extracts [64], etc.

The antimicrobial activities of PSs can be regulated via alteration of parameters. For example, molecular weight, degree of acetylation, and pH can regulate the antimicrobial performance of chitosan [66]. Table 11.3 accumulates the parameters relevant for antimicrobial performance mentioned in some studies. Apart from providing protection against micro-organism, antimicrobial agent integrated textiles can exhibit rot proof, stain proof, odor proof functionality, etc. additionally [67].

Chitosan is a natural antibacterial component [57]. Both chitin and chitosan PSs are well known for bacteriostatic and bactericidal in vitro performances, while maintaining good in vivo protection against infections [68]. Alginate is another popular bioPS. It does not have antibacterial property of its own but can demonstrate firm protection against bacterial growth by forming a barrier moisture transfer, oxidation, loss of flavors, etc. [69]. It has gained importance in providing a stable

TABLE 11.3

Parameters Relevant for Antimicrobial Performance of PS

Polysaccharide	Molecular Weight	Unit	Degree of Acetylation	Micro-organism Type	pH	Reference
Chitosan	440	kDa	82%	Bacteria	6.0	[67]
Chitosan	100–300	kDa	85%	Bacteria	-	[73]
Chitosan	2.1×10^5	kD	82%	Bacteria	5.0	[72]
Chitosan	50	kDa	90.32%	Bacteria	7.0	[74]
Chitosan	200	kDa	95%	Bacteria	7.0	[58]
Chitosan	10^6	Da	85%	Bacteria	7.5	[62]
Chitosan	80	Mn	90%	Bacteria, microfungi	7.0	[66]
Sodium alginate	2.5×10^5	kD	–	Bacteria	4.0	[72]
Sodium alginate	8,000	–	–	Bacteria	–	[71]

deposition platform on textiles for other antimicrobial agents [70]. Starch has also shown similar activity of improving textile fibers adhesion toward antibacterial NPs and consequently enhancing durability of such performance [60]. Kudzin and Giełdowska recently developed antimicrobial properties on melt-blown PLA nonwoven textile by coating it with alginate and zinc solutions. Dip coating was conducted in two steps with alginic sodium salt solution followed by immersing in ZnCl$_2$ solution. Antimicrobial tests against the gram-positive (*Staphylococcus aureus*), gram-negative (*Escherichia coli*), and fungal strains (*Aspergillus niger* van Tieghem and *Chaetomium globosum*) were carried out. The PLA-Alg-Zn$^{(2+)}$ composite sample with 39 g/Kg concentration of Zn showed inhibition zones of >1 mm area for both the bacteria and <1 mm area for the fungi strains [70]. Grgac et al. applied chitosan solution on the cotton and cotton/polyester blend fabric to develop durable antibacterial performance. Chitosan-powered dissolved in carboxylic acid (CA) and sodium hypophosphite (SHP) as catalyst was applied to the open structure of both the fabric during mercerization operation. Experimenting with gram-positive bacteria (*Staphylococcus aureus*), gram-negative bacteria (*Escherichia coli*), and microfungi (*Candida albicans*), it can be concluded that cotton/chitosan and cotton-polyester/chitosan showed good antibacterial activities for both the bacteria. In terms of microfungi, no activity was recorded in for cotton/chitosan, but cotton/polyester/chitosan showed good results [66]. Kim et al. reported a PS-based SA-TSA (trimethoxysilyl propyl-octadecyl-dimethylammonium chloride) complex as an antibacterial polymer. The colloidal solution was applied on a preprocessed cotton fabric surface via pad-dry-cure mechanism. The concentration of the applied SA-TSA NPs showed a positive correlation with bacterial reduction. An exposure of 8 hours at 100 ppm concentration allowed 75% and 90% reduction in the cases of gram-negative and positive bacteria, respectively [8]. In another study, El-Nahhal et al. functionalized cotton fiber with corn starch and ZnO NPs to impart antimicrobial ability. Starch was coated on purified cotton fibers in three separate concentrations and dried in vacuum condition. The starched cotton was uniformly coated with ZnO NPs using ultrasonic irradiation technique. Testing against *Staphylococcus aureus* and *Escherichia coli* revealed a significant improvement in antibacterial activities, 50% and 21.5%, respectively [60]. Furthermore, sometimes, one antimicrobial agent cannot perform satisfactorily, in which case, combining other types of antimicrobial agent can provide better result. For example, silver (Ag) NP is a popular antibacterial agent which is difficult to be integrated with textile substrate due to their poor charge density. To resolve such situation, PSs like chitosan and alginate can be equipped noble metals like Ag, Cu, gold, etc. to impart a more effective performance [67,71]. This provides with the additional facility of immobilizing the metallic NPs along with enhanced antimicrobial performance [69]. Hence, composite antibacterial materials have been reported to show a better microbe cleansing [72]. Bajpai et al. reported a Cu (II) loaded alginate impregnated textile fabric with antimicrobial ability. The subjected cotton fabric was dipped into the SA aqueous solution followed by CaCl$_2$ crosslinking and Cu (II) loading by immersing into the Cu (II) solution. The trapped SA on the fabric surface readily captured Cu (II) parts via ion-exchange mechanics. Higher concentration of Cu (II) solution resulted in greater captivation of metallic particle and therefore, greater area of bacterial inhibition zone [71]. In a similar study, Zahran et al. (2014) fabricated an AgNPs–alginate composite coated cotton fabric. Pad-dry-cure method was used to deposit and fixate the colloidal composite solution of the fabric surface. A similar result is observed as samples with higher concentration accounted for greater inhibition zones [69]. According to Mihailović et al., TiO$_2$ was loaded on polyester fabric using alginate which enhanced the antibacterial performance remarkably to account for 99.9% reduction in the *E. coli* bacterial colony [59].

11.7 SURFACE MODIFICATION AND MULTIFUNCTIONAL TEXTILES

Functional finishing by coating, deposition, or spraying are popular surface modification principles for textile substrates. A variety of surface modification techniques are available. They can be categorized as physical (corona discharge, dielectric barrier discharge, neutron activation, laser,

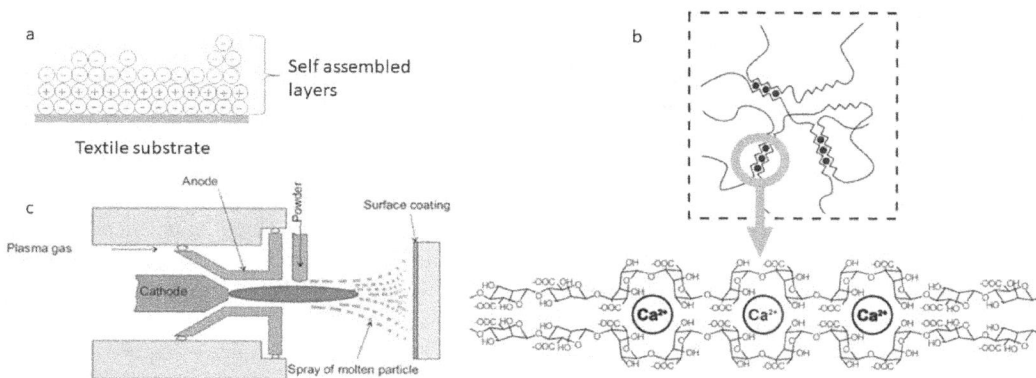

FIGURE 11.6 (a) Schematic of surface modification technique by LBL self-assembly mechanism with polysaccharide, (b) Ionotropic crosslinking of a polysaccharide, reused with permission from [75] (c) 2016, Elsevier, (c) Plasma spray surface modification schematics, reused with permission from [76] (c) 2021, Elsevier.

ion beam and electron beam), chemical (enzymatic approach, sol–gel technique, chemical reagents treatment, surface grafting/polymerization), and microencapsulation [63]. Introduction of nanotechnology has introduced more sophisticated coating (aka nanocoating with thickness <1 μm) methods like plasma-assisted polymer coating (as shown in Figure 11.6c), layer-by-layer (LBL) self-assembly (as shown in Figure 11.6a), etc. [73]. Adsorption between oppositely charged cationic and anionic matters allows formation and adhesion of thin films or coatings on textile surfaces [74]. PSs being positive or negatively charged can easily form electrostatic interaction with oppositely charged surface [72]. Chemical crosslinking (as shown in Figure 11.6b) is another technique that can be used efficiently to impart PS-based surface coating on textile substrates [58].

While being used for finishing textile substrates, PSs provide the opportunities to introduce additional functionalities like UV resistance, water resistance, insect repellant, deodorant, etc. Various chemical treatments to improve one property may instigate potentiality for multifunctionality. For instance, Cheng et al. performed a chemical modification to synthesize halamine derivative of chitosan. Although the primary target was to enhance antimicrobial ability, the halamine structure apparently introduced and enhanced durable press finishing functionality [58]. Furthermore, nanostructured biopolymer surface coating can introduce functionalities to textile surface [77]. PSs like alginate has abundance of carboxylic groups in its chemical structure, and this makes it a good modifier of textile surface by allowing enhanced bindings of NPs [59].

UV radiation is a major concern as it is highly detrimental for animal health apart from causing severe damages like discoloration, chalking, and mechanical degradation to textiles [77]. UV can cause skin damages such as sunburn, skin aging, allergies, cancer, etc. [78]. UV-resistant textiles can protect the human body from this bad radiation. UV resistance in textiles can be developed either by selecting the right textile raw materials or by treating textile surface with additives like titanium oxide, zinc oxide, carbon black, etc. This approach impedes the UV ray penetration by forming a layer on the subjected material. Reflection, absorption, or scattering are the typical reasons for showcasing UV protection behavior [6]. Dyeing of textile substrates further acts as reinforcements for improving UV blocking [79]. In order to protect textiles and consequently protecting the living beings from UV radiation, UV absorber biopolymers like PSs have shown great potential. Muzaffar et al. reported nanochitosan-PU (NCS-PU) based finishing of polyester-cotton fabric for introducing UV resistance performance. The PU extended with NCS was applied on the pre-dyed and printed fabric surface by pad-dry-cure method. The ultra-violation protection factor (UPF) readings for the applied samples were between 37.80 and 54.30 [63]. Ramadan et al. conducted a study on cotton fabric with chitosan finishing to impart UV resistance behavior. Cotton fabric of multiple constructions (shibeka, honeycomb, and crepe) was dipped into chitosan solution prepared with 5% acetic acid. The UPF characterization revealed that the UPF demonstrated a positive correlation with chitosan concentration regardless of the construction type. The maximum UPF reading was recorded 47 at $6 \, g \, L^{-1}$ chitosan concentration for crepe construction [6]. Mihailović et al carried out a research work based on polyester fabric functionalized with alginate and TiO_2 NPs for UV protection. Desized and bleached polyester fabric was dipped and cured in 0.1% SA solution before immersion in 0.1 M TiO_2 solution. Primarily the UPF value and UPF rating of the fabric was 43 and 40 which excelled to 119.8 and 50+ after the SA-TiO_2 NPs treatment accounting for an excellent UV blocking performance [59].

Water repellency can be achieved either by minimizing the subject's surface energy to the minimum or by changing the surface topography to multiscaled textured roughness. Nanoscaled formation on the surface also can enhance the contact angle to make it hydrophobic [80]. Polysaccharides have found their worth in this regard as they can be chemically modified to provide adequate sizes and rough surface topography. Additionally, the mechanical stability can also be regulated in terms of dispersion and wettability. Ivanova and Philipchenko reported a superhydrophobic coating of chitosan on cotton fabric. Multiscaled layers of texturization were deposited on cotton fabric via spraying chitosan in nanoparticle form. The nanoparticles were formed by hydrolysis in lactic acid. The maximum contact angle was recorded $157.2 \pm 2°$ for particle size of 63 nm and rolling angle $14 \pm 5°$. Even after, 10 cycles of washing for 30 minutes, the samples demonstrated significant amount of water repellency [80].

Microencapsulation is a promising technique for incorporating and protective sensitive components in textile finishing. PSs like chitosan 81], CMC [82], etc. have shown microencapsulation performance. Most textile fibers are anionic facilitating the ionic affinity with cationic PSs [81]. Perfume and other aroma components can be incorporated more efficiently on textile surface by solid nanocarriers. The major limitations with aromatherapeutic textiles are concerned with controlled release, losses by evaporation, and durability of the fragrance. The microencapsulation technique can provide sustainable solution [83]. PSs as the surfactant to produce stable microdroplets can very easily be deposited on textile surface via immersion, curing, grafting, spraying, printing, etc. [84].

CMC PS film encapsulating *Citrus grandis* peel oil (CGPO) and *N,N*-diethyl-m-toluamide (DEET) was studied by Misni et al. for insect repellency application. Interfacial precipitation technique was used to carry out the microcapsule formation. The flexible film encapsulated microcapsules were dispersed on a lotion base to apply on human skin. The experiment conducted on *Ae. Aegypti* females revealed the results that the formulation provided 100% repellency for 2 hours, a high 98.82% repellency for the next 4 hours, and a reduced 75.86% repellency after 8 hours [82]. Singh et al. studied fragrance and antimicrobial multifunctional textile finishing using chitosan citrate, β-cyclodextrin, and β-cyclodextrin/grafted chitosan. 1%–10% lavender oil solution with β-cyclodextrin, chitosan citrate, and β-cyclodextrin/grafted chitosan (β-CD-CS-L) was applied on cotton fabric via pad-dry technique. Six percent of lavender-associated β-CD-CS-L sample accounted for the maximum fragrance release rating after 72 hours [85]. In a similar attempt to fabric multifunctional textiles with antimicrobial and aroma functionality, Cerempei et al. studied chitosan emulsion-encapsulated geranium oil. The chitosan dissolved in acetic acid solution was mixed with oil/glycerin emulsion before padding on cotton knit fabric with a 100% wet pick up. Controlled release of the core oil can be achieved as increased release over time with incrementing oil concentration can be observed. Contrarily, with the increasing chitosan amount, the oil chamber dimension declines, and therefore, release becomes more difficult [86]. Li et al. reported orange oil as fragrance particle encapsulated by acetic acid solution of chitosan deposited by spray drying on cotton and polyester. The orange oil showed encapsulation efficacy of >90% when applied at 1:2 (w/w) ratio with chitosan. The storage stability of chitosan-coated oil was found to be much durable (lasted for months) compared with nonencapsulated oil (lasted only hours). Also, cotton fabric could retain the orange oil more than polyester fabric and more so with chitosan encapsulation [84]. Another functionality that PSs can offer to equip textiles is antiwrinkle finishing [87]. Table 11.4 accumulates some more studies addressing PSs applied as printing paste thickener and finishing agents to introduce functionalities in textiles.

TABLE 11.4

PS for Printing Paste Thickeners and Multifunctional Textiles Applications

Polysaccharide	Application Area	PS Applied on	PS Application Technique	PS Form	Special Feature	Reference
Tamarind gum	Printing paste thickener	Georgette fabric	–	Liquid	Excellent thickening, penetrability, stability, and thixotropy	[55]
Aloe vera	Printing paste thickener	Cotton, wool, polyester fabric	–	–	Good rubbing fastness and handle	[5]
Alginate	Printing paste thickener	Various textile fabric	–	Liquid	–	[55]
Chitosan	Printing paste thickener	Cellulose and viscose fabric	–	–	Improved shade depth and good wet, dry color fastness	[88]
				Application against micro-organism		
Chitosan	Antimicrobial	Cotton woven fabric (Shibeka, honey comb, crepe construction)	Solution dipping	*Pseudomonas aerugenosa*, *Staphylococcus aureus*, *Aspergillus niger*, and *Candida albicans*	Crepe construction of fabric showed maximum antibacterial performance as it absorbed maximum chitosan	[6]
Chitosan	Antimicrobial	Cotton fabric	Pad-dry-cure method	Staphylococcus aureus	The treated textiles showed tremendous (>99%) bacterial reduction	[61]
Chitosan	Antimicrobial	Polyester nonwoven mat	Nanofiber deposition by Electrospinning	*Staphylococcus aureus* and *Klebsiella pneumoniae*	The wettability of the nanofiber mat surface enhanced with introduction of PS; higher growth inhibition and tissue compatibility	[68]
Chitosan	Antimicrobial	Cotton fabric	Nanoparticle dispersion spray	–	Superhydrophobic textile surface	[80]
Chitosan	Antimicrobial	Polyester fabric	–	Staphylococcus aureus	Silver-loaded chitosan nanoparticle shows enhanced antibacterial character	[67]
Chitosan	Antimicrobial	Cotton woven fabric	LBL assembly	*Escherichia coli* and *Staphylococcus aureus*	The layer number demonstrated a positive correlation with the anti-microorganism performance	[73]

(Continued)

TABLE 11.4 (*Continued*)
PS for Printing Paste Thickeners and Multifunctional Textiles Applications

Polysaccharide	Application Area	PS Applied on	PS Application Technique	PS Form	Special Feature	Reference
Chitosan + alginate	Antimicrobial	Cellulose acetate (CA) fibrous mat	LBL assembly	*Escherichia coli* and *Staphylococcus aureus*	Can prevent food deterioration and extend shelf life	[72]
CH-HDH (chitosan derivative)	Antimicrobial	Cotton fabric	LBL assembly	*Staphylococcus aureus* and *Escherichia coli*	Showed deactivation of 100% bacteria with a minimum contact time of only 1 minute	[22]
Chitosan (N-halamine based)	Antimicrobial	Cotton fabric	Chemical crosslinking	*Staphylococcus aureus* and *Escherichia coli*	Showed improved crease recovery deactivation of 100% bacteria with a minimum contact time of only 5 minutes	[58]
Chitosan + (PAMAM)	Antimicrobial	Cotton fabric	Padding	Staphylococcus aureus	The modification of the chitosan improved the water solubility and chain mobility. Could perform antimicrobial activities at a more neutral pH	[62]
Chitosan	Antimicrobial	Silk and Cotton fabric	UV curing	*Escherichia coli*	Good antimicrobial activities were recorded even at a lower (2%) chitosan weight; was durable up to 5 wash cycles	[89]
Alginate	Antimicrobial	Cotton fabric	Pad-dry cure	*Staphylococcus aureus* and *Escherichia coli*	Demonstrated >99.99% of bacterial cell reduction in case of both gram-positive and negative bacteria. Durable even after 30 laundry wash cycles	[8]
Alginate	Antimicrobial	Cotton fabric	Immersion dry impregnation	Escherichia coli	Alginate impregnation does not impact the mechanical properties of the fabric; release of Cu (II) over a long period of 50H was observed	[71]

(Continued)

TABLE 11.4 (*Continued*)

PS for Printing Paste Thickeners and Multifunctional Textiles Applications

Polysaccharide	Application Area	PS Applied on	PS Application Technique	PS Form	Special Feature	Reference
Alginate	Antimicrobial	Cotton fabric	Pad-dry-cure	*Escherichia coli, Staphylococcus aureus*, and *Pseudomonas aeruginosa*	Ag NPs on the cotton fabric surface was deposited physically without binders; Ag particles show effective activities against 16 species of bacteria	[69]
Chitosan	Antimicrobial	Polyester-cotton fabric	Pad-dry-cure	*Staphylococcus aureus* and *Escherichia coli*	–	[63]
Alginate	Antimicrobial	Polyester fabric	Dipping and curing	*Escherichia coli*	Showed excellent laundering durability even after 5 cycles, a remarkable 99.9% reduction in bacterial colony was reported [5.7X105–165]	[59]
				UPF value		
Alginate	UV resistance	Cotton fabric	Two-stage surface impregnations	Maximum 377 after dyed with Procion Turquoise H-EXL dyestuff	–	[79]
Aloe vera+chitosan	UV resistance	Cotton woven fabric	Pad-dry-cure	Value was found 10 times higher than the untreated sample	The surface coating did not affect the air, water, and thermal properties of the sample, thus comfortability remained unaffected	[78]
Carboxymethyl chitosan (CMCS)	UV resistance	Cotton fabric	Plasma pretreatment and finishing	Maximum value of 50+ was recorded for ZnO/CMCS composite	The finished fabric displayed an excellent wash durability (up to 30 cycles) of UV protection application	[90]
				Contact angle and sliding angle		
Alginate	Water repellency	Cotton fabric	One-step dip-coating process	CA 158.2° and SA<10°	Water repellency behavior was well in effect up to 7 days, and a swelling absorbency of 124.1% was observed	[91]

CH-HDH, 1-Hydroxymethyl-5,5-dimethylhydantoin chitosan; PAMAM, Polyamidoamine; UPF, Ultraviolet protection factor.

11.8 POLYSACCHARIDE-BASED TEXTILES FOR MEDICAL APPLICATIONS

The conquest for novel biomaterials in the medical arena is driven by emerging clinical complexities and ever-growing economic challenges. For instance, the emergence of novel drug delivery alone is predicted to push the medical industry from $1,430 billion to $2,015 billion by 2025 [92]. Biopolymers as textiles and flexible films have attracted much attention in the field of medical science. Textile-based transdermal therapy has gained popularity for the treatment of skin diseases [93]. Wound dressing is another topic that renders component to facilitate proper healing of the wounds. Natural biopolymers like PS have attracted attention in this area for many technical and economic advantages. A high degree of amino or carboxylic groups present in different PS allows better solubility, bioadhesion, and biorecognition with biological tissue. Further, positive and negative charges can attract mucosal layers and antigen agents on the cell surface [94]. Adding to that, availability, biocompatibility, and biodegradability make PS an ultimate choice for such operations.

11.8.1 THERAPEUTIC APPLICATIONS

Therapeutics mainly covers the biomedical areas concerned with wound healing, regulated drug delivery, and tissue engineering [95]. Textiles can be utilized as a stable, wash-protected tool for carrying the medicinal particles. These substances can be subjected to encapsulation inside PS-based enclosures and immobilized in donor layers on base textiles [96]. Thus, textiles in combination with drug careers can introduce a new evasive mechanism for effective drug delivery system (DDS). Natural PSs and their derivatives have been recognized in the medical field for their controlled drug release behavior [11]. The stimuli (heat, perspiration, moisture, vibration, etc.) responsive characteristics of PS can be exploited to facilitate effective drug release at various conditions. Stimuli-driven sol–gel transition is crucial for such functionalities [93]. Furthermore, highly porous network and well-dispersed structure can facilitate improved release characteristics of containing drugs [97]. The PSs capable of attaining the lower critical solution temperature (LCST) are better suited for controlled drug release as the upper critical solution temperature (UCST) attainable PSs much higher temperature than normal physiological temperature to change phase [93].

A textile-based PS nanoconjugate with dual-responsive (pH/temperature) drug release performance was reported by Chatterjee et al. Gallic acid as pharmaceutical matter was loaded with hyaluronic acid-chitosan oligosaccharide lactate nanoconjugate hydrogel. Experiment conducted in pH condition mimicking human skin (pH 6.4) demonstrated good mechanical behavior and drug release. The synthesized hydrogel can be applied for the drug release application through textile-based therapy [97]. In another recent study, Eskens et al. synthesized an alginate-methylcellulose (MC) gel as a career for epidermal growth components for skin regeneration application. The acquired thermoresponsiveness and viscosity were found to be suitable for both protective immobilization and direct release behavior. A tunable and effective delivery system is thus formed [98]. In a similar type of study, Tang et al. synthesized a chitosan/MC/salt blended hydrogel intended for tissue engineering application. The effect of different types of salts on the gelation behavior was investigated, and thus the potential of the gel as an injectable matter for tissue engineering application was evaluated. While applied as scaffold for tissue engineering, the gel accounted for good results in terms of cell viability and proliferation [99].

11.8.2 WOUND DRESSING

The primary functionality of a wound dressing is protection and prevention of further injuries. Alginate is a popular fibrous PS for targeted wound dressing application due to its enhanced wound healing performance along with high moisture adsorption and ion-exchange abilities [71]. The ion exchange between alginate and wound exudates forms a gel on the wound surface [69]. Wound dressings with antimicrobial functionality is also of special interest as they can bar microbial infection on the wound and prevent bacterial impact of prolonging wound healing [86]. Some of the other important features the PS contains are good cytocompatibility, appropriate water uptake capacity, suitable water transmission rate, good mechanical stability, etc. In a recent study, Alzarea et al. fabricated gentamicin sulfate (GS) loaded SA flexible film with targeted application for wound dressing. Solvent casting process was incorporated to fabricate the film which included sonication to remove air, pouring on polystyrene plate and peeling off the film in dried condition. The film demonstrated >80% release of the GS within the first 24 hours suggesting very good performance against infection. The antimicrobial activity against *S. aureus*, *E. coli*, and *P. aeruginosa* further added to its performance [100]. Chitosan-based wound dressings have also become quite popular for cell viability and antioxidant capacity. Colobatiu et al. recently reported bioactive compound (*Plantago lanceolata*, *Arnica montana*, *Tagetes patula*, *Symphytum officinale*, *Calendula officinalis*, and *Geum urbanum* extracts) loaded chitosan film as a functional wound dressing specifically for diabetic patients. The film was fabricated via solvent casting technique. Experiment was conducted both in vitro and in vivo (streptozotocin-induced diabetic rat model) conditions. The in vivo healing rate of both blank and bioactive compound loaded chitosan films demonstrated positive correlation with time (At 3-, 7-, and 14-days points, 29%, 61%, and 88% for blank chitosan and 41%, 81%, and 97% for bioactive compound loaded chitosan film). On top of that, the reported dressing provided a moist environment to allow proper hydration and closure for the wound [101]. However, in many cases, high hydrophilicity and low water stability may pose challenges for PS like chitosan. In such cases, combining with another PS like alginate may solve the situation. One such research work conducted by Sobczyk et al. presented a chitosan/alginate-based film that had properties suitable for wound dressing applications in terms of mechanical stability, absorption capacity, and water vapor flux. The film prepared by casting technique was loaded with oregano leaves (OR) or oregano essential oil (OEO) with superior pharmaceutical properties. Additionally, the film demonstrated antimicrobial activities against both *Staphylococcus aureus* and *Escherichia coli* reaffirming its claim as wound dressing material [10]. Finishing of textile-based dressings and gauzes with embedded encapsulated medicinal components is also a topic of great interest nowadays. Antunes et al. fabricated a chitosan-based nano/microcapsule textile coating for applying on cotton gauzes. The 2 µm sized encapsulation contained antimicrobial peptide-Dermicidin-1-L and was capable of controlled release. Alternate LBL deposition was done with alginate and chitosan before functionalizing on cotton wound dressing by immersion followed by sonication. The trial with *Staphylococcus aureus* and *Klebsiella pneumoniae* revealed a remarkable bacterial reduction rate of 75.33% and 99.86%, respectively [102].

Metallic NPs contained and capped by PSs can facilitate wound healing up to an extent. For example, the noble metal Cu is known for promoting collagen crosslinking for the formation of bone matrix and immune-mediated inflammatory response for the burn injuries [103,104]. Ag salts in nanocrystal form have been reported to show excellent antimicrobial activities [105]. Table 11.5 accumulates some more studies addressing PSs for medical textile applications.

11.9 POLYSACCHARIDES FOR TEXTILE INDUSTRIAL WASTEWATER TREATMENT

Textile is one of the most highly polluting industries of the world and is responsible for discharging a big volume of contaminated wastewater. The discharge may contain a wide variety of pollutants, encompassing inorganic (heavy metals, nutrients, etc.) to organic ones (dyes, oil, synthetic fragments, etc.) [12]. A plethora of chemicals for instance, acids, bases, detergents, salts, dyes, wetting agents, finishing agents, oxidants, mercerizing agents, etc. are commonly found. Among them, mostly dyes, aromatic particles (such as phenolic derivatives, polycyclic compounds, etc.), and heavy metal ions are detrimental due to their toxic nature [109]. Dyes comprise the majority of the organic pollutants accounting for an approximate 700,000 tons outputs annually [110]. Heavy metal (metalloids with densities >5 g/cm^3) ions represent the most hazardous class in the inorganic category causing diseases and disorders as they are easily accumulated in the living cells [111].

TABLE 11.5

PSs for Textile-Based Therapeutic and Wound Dressing Medical Applications

Polysaccharide	Medical Applications Area	PS Applied on	Special Feature	Reference
Alginate	Wound dressing	Cotton gauze	Compresses with variation in absorbing capacity is possible to produce which shows good healing property of against different wounds	[79]
		Carrying drug		
Methylcellulose	Drug delivery system	Bovine serum albumin (BSA)	Demonstrated temperature-dependent gel-shear thinning behavior suitable for rectal application	[106]
Methylcellulose	Drug delivery system	Ketorolac tromethamine	With the addition of various salts, the gelling temperature decreased below 37°C, and thus drug release was done within the duration 1.5–5 hours based on salt concentration	[107]
Alginate	Drug delivery system	Indomethacin	Alginate membrane which is hydrophobically modified and biomineralized is used for smart and sustainable drug release	[108]

Salt of heavy metals is usually found in the mordents applied during textile dyeing [112]. Under the backdrop of anticipated energy and water crisis, PS may become both a viable alternative to the existing energy-consuming purification systems and a mean to establish recycled wastewater as a source for usable water.

11.9.1 POLYSACCHARIDE ADSORBENTS

Adsorption technique for pollutant removal from wastewater is well practiced due to its cost effectiveness and technical simplicity. It offers additional benefit of removing contaminants present in very low concentration [113]. PSs have been explored well in recent years since they offer rich adsorptive ability besides being biocompatible [12]. Adding to that, an ideal adsorbent must be of natural origin, easily accessible, and modifiable, all of which are found in PSs. Porosity, surface area, surface charge, concentration, etc. are some of the drivers for good adsorption performance. Generally, wastewater concentration and contact time duration demonstrate a positive relation in adsorption activities. However, the adsorption is found to be faster at the beginning and slows down as time passes [109].

Adsorption (as shown in Figure 11.7a) is basically a surface phase transfer technique that operates via physical and chemical interactions of the participating elements. In other words, adsorbate matter is caught from the persisting phase on to a solid phase. Crosslinking may offer a good strategy to produce such stable adsorbents. Functional groups like $-OH$, $-NH_2$, $-COOH$, and $-SO_3H$ contained in various PS (chitin, chitosan, glycosaminoglycans, etc.) act as the adsorptive agents. PSs like chitin, chitosan being enriched in hydroxyl or amino functional group renders them suitable for such activities [114]. Chitosan shows greater regeneration and structural diversity (NPs, fibers, membranes, beads, gels, films, etc.), which makes its application area broader in versatile situations [115]. Another category of PSs (cellulose, starch, pectin, alginate, etc.) containing large amount of highly polar $-OH$ and $-COOH$ groups can effectively react with and modify groups like phosphate, amino, etc. as well as other water pollutants to facilitate a more efficient purification. Like chitosan, they also offer ample structural diversity [116–118].

Dragan and Loghin reported a composite bioadsorbent made of chitosan and starch-g-PAN (polyacrylonitrile) to remove Cu^{2+}, Co^{2+}, and Ni^{2+} HMIs from wastewater. Highly porous PS-based composite beads were prepared via cryogelation technique. The starch-g-PAN was dispersed in the chitosan solution followed by the formation of beads. The maximum adsorption capacity demonstrated by the

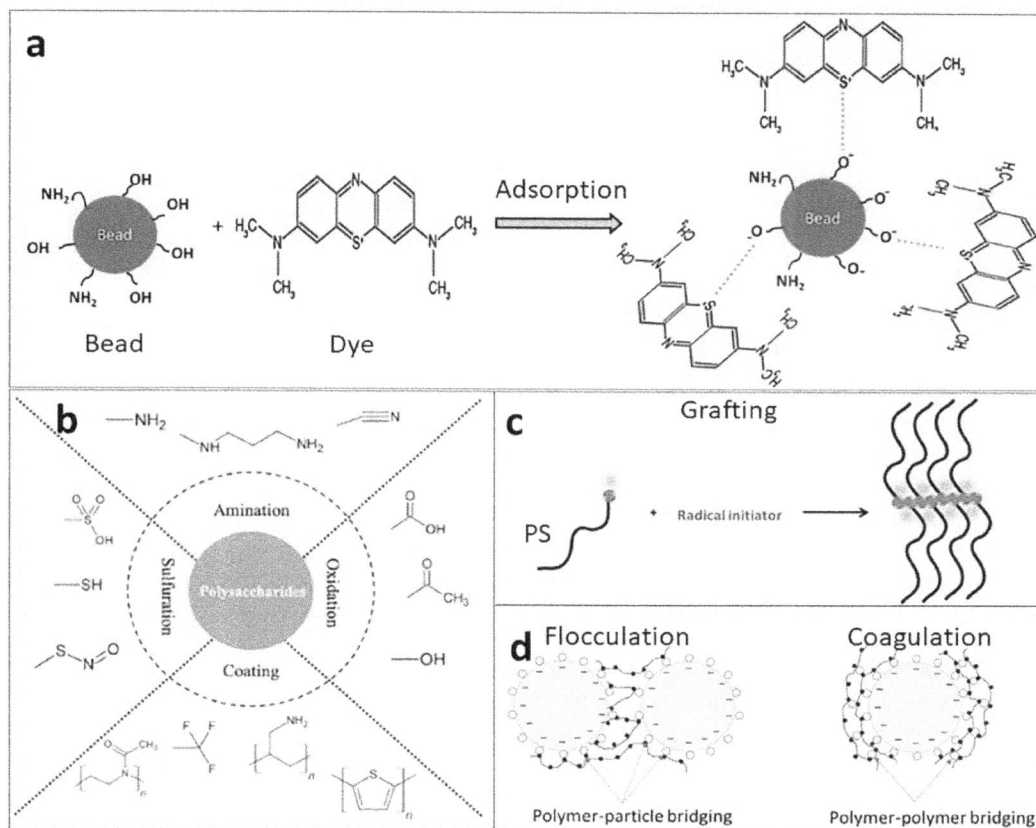

FIGURE 11.7 Pollutant removal techniques from textile wastewater: (a) Adsorption technique to remove dye from textile effluent, reused with permission from [135]. (b) Plasma modification techniques of PSs, reused with permission from [110], (c) 2021, Elsevier. (c) PS functionalization with polymers and radical initiator via grafting, reused with permission from [136], (c) 2018, Elsevier. (d) Flocculation and coagulation mechanism for wastewater pollutant removal, reused with permission from [137] (c) 2021, Elsevier.

bioadsorbents for Cu^{2+}, Co^{2+}, and Ni^{2+} HMIs were recorded 100.6, 83.25, and 74.01 mg/g, respectively [113]. Bo et al. conducted a research work on synthesizing SA-based adsorbents to remove Pb^{2+} ions from actual wastewater. The SA adsorbent was cofunctionalized with melamine and polyethyleneimine via sol–gel process to enhance mechanical behavior along with polarity and chemical resistance. The adsorbent was added with the Pb^{2+} containing aqueous solution in a conical flask. A high Pb^{2+} capture of 596.68 mg/g was reported [118]. Wu et al. synthesized a carboxymethyle chitosan hemicellulose resin (CMCH) to evaluate the potential to adsorb Ni (II), Cd (II), Cu (II), Mn (VII), Hg (II), and Cr (VI) from aqueous solution. Carboxymethyl chitosan (CMC) was thermally crosslinked with hemicellulose to prepare the final adsorbent before adding in the wastewater solution contained in conical flasks. The maximum adsorption capacities were found to be 362.3, 909.1, 333.3, 42.0, 28.2, and 49.0 mg g^{-1} for Ni (II), Cd (II), Cu (II), Mn (VII), Hg (II), and Cr (VI), respectively [111].

11.9.2 MODIFIED OR ACTIVATED POLYSACCHARIDES

Modification or activation is generally done to introduce functionalities to any matter in order to enhance performance for a targeted application. Physical or chemical crosslinking, plasma treatment, grafting other polymers [11], immobilization on supports, etc. are some techniques worth mentioning.

In case of PSs, untreated adsorbents sometimes require additional alterations to enhance performance. Some PSs (e.g., chitosan) have been reported to have flexible chemical structures that allow particular modifications more easily than others (e.g., starch) [119]. Plasma treatment is a popular ecofriendly technique as it does not require any solvent. It basically generates moieties on PS surface (up to a few millimeters depth) to alter the physical-chemical behavior without affecting the bulk properties. The introduced moieties act as the particular anchoring points for the pollutants. Since oxygen, nitrogen, sulfur, etc. can offer a lone pair of electrons, plasma treatment on PS can be carried out as oxidation, amination, sulfuration, or in the form of coating (as shown in Figure 11.7b). Apart from increasing adsorption capacity, plasma treatment allows selectivity toward any specific pollutant [110]. Copolymerizing by grafting (as shown in Figure 11.7c) is another technique that allows the incorporation of desired characteristics without affecting inherent properties. PSs can be modified by grafting other components with radical initiators using various elements and techniques like initiators, vinyl monomers, oxidizing agents, radiation technique, etc. [11,120]. Different surface adsorbents like activated carbons, carbon nanotubes, graphene oxides, etc. can easily be grafted on the PS surface for water treatment [121]. A big advantage of graft technique is the ability to regulate desired properties via percentage grafting [122]. Reactive groups in the PS structure can conjugate with other compounds like bentonite, kaolinite, oil palm ash, montmorillonite, polyurethane, zeolites, magnetite, etc. They are known for showing better adsorption capacity and durability in the acidic environment [119]. Nanotechnology has opened a whole new area for exploring the wastewater treatment techniques. PSs conjugated with nanomaterials are being studied to improve the wastewater remediation performance.

In a recent study, Nasiri et al. mentioned a PS-based adsorbent modified with iron-based particles to synthesize a novel magnetic adsorbent. The magnetic nanoadsorbent was made by grafting $CoFe_2O_4$ and activated carbon (AC) on methyl cellulose (MC) PS for the removal of reactive red 198 textile dye from wastewater. The synthesis technique involved mixing metallic particles ($FeCl_3.6H_2O$ and $CoCl_2.6H_2O$) in a 2:1 molar ratio with distilled water. After adding MC and AC, the solution was stirred, microwave irradiated, and dried to synthesize the adsorbent in the form of black precipitate. Experiment conducted on artificial and real wastewater revealed maximum dye removal efficiency of 92.2% and 78%, respectively. The specialty of such water treatment technique is the easy separation of the iron particles with an external magnetic field afterwards [121]. Kudal et al. synthesized a PS-based adsorbent by grafting poly(N-hydroxyethylacrylamide) (PHEAA) in pectin for the removal of a cationic dye (Rhodamine 6G) and Cu (II) and Hg (II) ions from aqueous solution. Further, magnetic NPs were diffused within the Pec-g-PHEAA to fabricate another PS-based nanocomposite (Pec-g-PHEAA/Fe_3O_4) adsorbent. Microwave irradiation and in situ diffusion techniques were applied to synthesize Pec-g-PHEAA and Pec-g-PHEAA/Fe_3O_4, respectively. The adsorbents were applied in the dye (10–500 mg L^{-1}) and metallic salt (20–1,000 mg L^{-1}) solution by immersion. Pec-g-PHEAA demonstrated maximum adsorption of 43.5, 237.1, and 228.6 mg/g for R6G, Cu (II), and Hg (II), respectively. On the other hand, Pec-g-PHEAA/Fe_3O_4 demonstrated a superior 57.2, 248.6, and 240.2 mg g^{-1} adsorption performance, respectively [109]. In a similar study, Nga et al. (2020) developed multifunctional chitosan film adsorbent grafted with MgO nanoparticles to remove reactive blue 19 dye from wastewater. The CS/MgO film was prepared by solvent casting followed by mild drying. The adsorbent film demonstrated temperature depended adsorption performance. For applied temperature 18, 28, and 38°C, the maximum adsorption was recorded 408.16, 485.43, and 512.82 mg g^{-1}, respectively [123].

11.9.3 BIOFLOCCULATION OR COAGULATION

Flocculation is a relatively simple technique for removing components from a solution by aggregating finely dispersed particles in to a large floc. This floc can be settled and separated easily [124]. Thus, flocculation has been applied in textile wastewater treatment for reducing turbidity, dissolved and suspended solids, dyes and auxiliaries, chemical oxygen demand (COD), etc. [125]. PSs are efficient flocculants among natural substances primarily due to their lower molecular weight.

Additional advantages of being stable, cheap, biodegradable, and reproducible make them a good choice [126]. Further, chemical modification can be done to enhance flocculation performance. PS-derived flocculants may act as the coagulating (neutralizing) or flocculating (bridging) components (as shown in Figure 11.7d) depending on the nature of the pollutant in the wastewater [13]. Some of the major bioflocculants are chitosan, starch, xanthan, cellulose, alginate, etc. that are all PSs [125]. Charges in the cationic PSs may flocculate the anionic suspended particles in the wastewater and vice versa [119].

Szygula et al. reported a chitosan coagulant-flocculant to treat textile wastewater containing a dye (Acid Blue 92). The dye solution was combined with chitosan solution and stirred following a predetermined parameter before filtering through a 1.2 μm filtration system. Interaction between the anionic sulfonic and cationic amino group in chitosan is responsible for sedimentation. A high intensity of dye removal (~99%) was reported at optimum condition [13]. Kono prepared water soluble celluloses with different DS. The cellulosic flocculants were derived by dissolving cellulose in urea/NaOH solution in the presence of 2,3-epoxypropyltrimethylammonium chloride (EPTMAC) reagent. When applied for the removal of anionic dyes, a great performance was achieved. For DS 0.56, 0.84, and 1.33, the yields were within the range of 65%–82% [127]. In another study, Pal et al. incorporated a tamarind kernel PS (TKP) based cationic flocculant to see the intensity of pollutant removal from a collected textile wastewater sample. TKP was mixed with N-3-chloro-2-hydroxypropyl trimethyl ammonium chloride (CHPTAC) and NaOH before stirring at 40°C–50°C. The flocculent was acquired by cooling down and precipitating the solution. When compared with a commercial flocculant, the TKP flocculant showed remarkably better result for various parameters [124]. Table 11.6 contains some more studies were PSs for removal of pollutants from textile wastewater were reported.

TABLE 11.6
PS in Textile Wastewater Purification Applications

Base Polysaccharide	Modifying/ Additional Components	Pollutant Name	Purifier Synthesis Technique	Pollutant Removal Capacity	Special Feature	Reference
Chitosan	Fe and AC	Cu^{2+}	Sol-gel	216.6 mg/g	Easy magnetic separation can retain 95% adsorption after 5th adsorption-desorption cycle	[128]
Chitosan	$CoFe_2O_4$ and graphene	Hg (II)	Ultrasonic stirring and crosslinking	361.0 mg/g	Easy magnetic separation and recovery using a little magnetic field gradient	[129]
Alginate	P (SA- co styrene) /organo-illite / smectite clay	Methylene blue	Sol-gel	1,843.46 mg/g	Organification improves the absorbency substantially	[130]
Gum ghatti	Poly(acrylic acid-co-acrylamide) and FeO_2	Rhodamine B	Crosslinking, sonication, stirring, and cooling	654.87 mg/g	Successful adsorption was achieved up to 3 adsorption- desorption cycles	[131]
Xanthan gum	$PAAM/SiO_2$	Methylene Blue (MB) and Methyl Violet (MV)	Sol-gel	497.5 mg/g (99.4%) for MB and 378.8 mg/g (99.1%) for MV	The high cationic dye uptake ability is associated with high dynamic volume caused by uniform SiO_2 distribution; good recyclability	[132]

(Continued)

TABLE 11.6 (*Continued*)
PS in Textile Wastewater Purification Applications

Base Polysaccharide	Modifying/ Additional Components	Pollutant Name	Purifier Synthesis Technique	Pollutant Removal Capacity	Special Feature	Reference
Guar gum	Poly(acrylamide)/ silica and	Reactive blue 4 (RB) and Congo red (CR)	Sol-gel	579.01 mg/g for RB and 233.24 mg/g (CR)	Spontaneous adsorption in the form of monolayer on homogeneous surface is observed	[133]
Starch	2-Chloro-4,6-isopropylamino-[1,3,5]-triazine	K-2BP and KN-B5	One-step "graft-to" method	2452.6 ± 23.9 (for K-2BP) and 792.7 ± 14.1 (for KN-B5) mg/g	The starch-based flocculant demonstrated pH sensitivity and stimulus responsiveness property	[134]

Abbreviation: AC, Activated Carbon; PAAM, Polyacrylamide.

11.10 CONCLUSION AND OUTLOOK

Apart from many inherent ecofriendly and robust properties, PS has many advantages to offer. Generally, lower dose of PS solution can facilitate dye removal by flocculation from textile wastewater. On top of that, it is possible to recover flocculated PSs with a desorption efficacy close to 100% [13]. PSs can also be employed as fillers in the composites to enhance the mechanical characteristics of the adsorbents. High performance further allows chitosan derivatives to be used as adsorption additives [119]. Good reducibility characterized by various functional groups in the chemical structure enables PSs to be a great protecting and stabilizing agent for various NPs in different conditions [105].

Nevertheless, PSs have some challenges. For textile printing paste thickener application, PS with poor rheological properties may lead to comptonization with print quality [55]. In case of coating, surface charge variation of nanoparticles can have a negative impact on the durability and wettability, particularly in the aqueous medium [80]. Further, uneven and porous surface may render difficulties during coating [73]. According to many other studies, due to the protection of lipopolysaccharides, PSs show reduced antibacterial intensity to gram-negative bacteria than gram-positive ones [72]. PSs have been reported to show better performances when integrated with NPs like TiO_2, ZnO_2, etc. [59,79]. Therefore, future scientists can look into this matter to exploit better performance out of PS. PSs may offer difficulties regarding regeneration and reutilization. Also, their mechanical characteristics are not the best and may demand to be used in combination with other materials [138]. Chitin demonstrates poor solubility and thus offers low adsorption capacity during wastewater purification [12]. Chitosan shows similar poor solubility behavior along with instability in acids and sophisticated extraction procedure [113]. On top of that, weak binding strength with textile fibers has been mentioned in some studies [57]. Most of the crosslinking agents used for binding PSs (chitosan) like glutaric aldehyde (GA) and formaldehyde derivatives are harmful for human beings [66]. Despite being a good adsorbent, alginate's application is limited by showcase of poor mechanical behavior, swelling capacity, and chemical resistance. Therefore, alteration of chemical structure and properties is carried out to provide proper flexibility and performance [118]. This mostly requires additional energy as well as chemical treatment. Achieving targeted performance in most environment is also a challenge. For example, chitosan shows antibacterial activities in acidic pH [6]. In the last few decades, an urgency has been noticed to find out sustainable alternative to the day-to-day synthetic products. While the existing traditional materials can barely support, they cannot match the performance the synthetics generate. As the demand is on the rise, so is the pollutants generated by these products. This can no longer be overlooked. Recent developments relating to PS and textile applications have proven to be a good solution to the problem, at least for the aforementioned textile areas. Different options within the PS category offer different advantages or at least similar performance to the existing ones. The summary of the functional textiles also upholds the promise of PS to be incorporated more rigorously for equipping textile with antimicrobial, UV resistant, insect repellent, hydrophobic, etc. functionalities. In the field of biomedical, PS can offer both facilities of being biocompatible along with properties like antimicrobial and infection free. PSs can also be applied separately for the removal of pollutants via absorbency, modifications with adsorbent materials, and bioflocculation techniques, among others. Despite some limitations, recent findings show they can be commercially applied to some extent. Nevertheless, this area requires more explorations to seal the gaps which still persist with the mentioned applications.

REFERENCES

[1] P. K. Deb, S. F. Kokaz, S. N. Abed, A. Paradkar, and R. K. Tekade, "Pharmaceutical and Biomedical Applications of Polymers," In *Basic Fundamentals of Drug Delivery*, Academic Press, 2019, pp. 203–267.

[2] P. R. Yaashikaa, P. Senthil Kumar, and S. Karishma, "Review on biopolymers and composites—Evolving material as adsorbents in removal of environmental pollutants," *Environ. Res.*, vol. 212, p. 113114, Sep. 2022, doi: 10.1016/J.ENVRES.2022.113114.

[3] N. Hernández, R. C. Williams, and E. W. Cochran, "The battle for the 'green' polymer. Different approaches for biopolymer synthesis: bioadvantaged vs. bioreplacement," *Org. Biomol. Chem.*, vol. 12, no. 18, pp. 2834–2849, May 2014, doi: 10.1039/C3OB42339E.

[4] R. S. Dassanayake, S. Acharya, and N. Abidi, "Biopolymer-based materials from polysaccharides: Properties, processing, characterization and sorption applications," *Adv. Sorption Process Appl.*, Nov. 2018, doi: 10.5772/INTECHOPEN.80898.

[5] F. Saad, A. L. Mohamed, M. Mosaad, H. A. Othman, and A. G. Hassabo, "Enhancing the rheological properties of aloe vera polysaccharide gel for use as an eco-friendly thickening agent in textile printing paste," *Carbohydr. Polym. Technol. Appl.*, vol. 2, p. 100132, Dec. 2021, doi: 10.1016/J.CARPTA.2021.100132.

[6] M. A. Ramadan, G. M. Taha, and W. Z. E. A. El- Mohr, "Antimicrobial and UV protection finishing of polysaccharide-based textiles using biopolymer and AgNPs," *Egypt. J. Chem.*, vol. 63, no. 7, pp. 2707–2716, Jul. 2020, doi: 10.21608/EJCHEM.2020.27968.2605.

[7] M. Navlani-García *et al.*, "Polysaccharides and metal nanoparticles for functional textiles: A review," *Nanomater.,*vol. 12, no. 6, p. 1006, Mar. 2022, doi: 10.3390/NANO12061006.

[8] H. W. Kim, B. R. Kim, and Y. H. Rhee, "Imparting durable antimicrobial properties to cotton fabrics using alginate-quaternary ammonium complex nanoparticles," *Carbohydr. Polym.*, vol. 79, no. 4, pp. 1057–1062, Mar. 2010, doi: 10.1016/J.CARBPOL.2009.10.047.

[9] S. M. H. Hosseini, F. Ghiasi, and M. Jahromi, "Nanocapsule formation by complexation of biopolymers," *Nanoencapsulation Technol. Food Nutraceutical Ind.*, pp. 447–492, Jan. 2017, doi: 10.1016/B978-0-12-809436-5.00012-4.

[10] A. de E. Sobczyk, C. L. Luchese, D. J. L. Faccin, and I. C. Tessaro, "Influence of replacing oregano essential oil by ground oregano leaves on chitosan/alginate-based dressings properties," *Int. J. Biol. Macromol.*, vol. 181, pp. 51–59, Jun. 2021, doi: 10.1016/J.IJBIOMAC.2021.03.084.

[11] D. Kumar, J. Pandey, V. Raj, and P. Kumar, "A review on the modification of polysaccharide through graft copolymerization for various potential applications," *Open Med. Chem. J.*, vol. 11, no. 1, p. 109, Oct. 2017, doi: 10.2174/1874104501711010109.

[12] X. Qi, X. Tong, W. Pan, Q. Zeng, S. You, and J. Shen, "Recent advances in polysaccharide-based adsorbents for wastewater treatment," *J. Clean. Prod.*, vol. 315, p. 128221, Sep. 2021, doi: 10.1016/J.JCLEPRO.2021.128221.

[13] A. Szyguła, E. Guibal, M. A. Palacín, M. Ruiz, and A. M. Sastre, "Removal of an anionic dye (Acid Blue 92) by coagulation-flocculation using chitosan," *J. Environ. Manage.*, vol. 90, no. 10, pp. 2979–2986, Jul. 2009, doi: 10.1016/J.JENVMAN.2009.04.002.

[14] O. Olatunji, *Series on Polymer and Composite Materials*. Springer.

[15] D. Gopalakrishnan, "Alginate Fibers—An Overview," *Fibre2Fashion*, 2006. [Online]. Available: https://www.fibre2fashion.com/industry-article/591/alginate-fibres-an-overview. [Accessed: 11-Apr-2022].

[16] M. Rinaudo, "Biomaterials based on a natural polysaccharide: alginate," *TIP*, vol. 17, no. 1, pp. 92–96, Jun. 2014, doi: 10.1016/S1405-888X(14)70322-5.

[17] Y. Qin, "The characterization of alginate wound dressings with different fiber and textile structures," *J. Appl. Polym. Sci.*, vol. 100, no. 3, pp. 2516–2520, May 2006, doi: 10.1002/APP.23668.

[18] J. Li, J. He, and Y. Huang, "Role of alginate in antibacterial finishing of textiles," *Int. J. Biol. Macromol.*, vol. 94, no. Pt A, pp. 466–473, Jan. 2017, doi: 10.1016/J.IJBIOMAC.2016.10.054.

[19] Y. J. Xu, L. Y. Qu, Y. Liu, and P. Zhu, "An overview of alginates as flame-retardant materials: Pyrolysis behaviors, flame retardancy, and applications," *Carbohydr. Polym.*, vol. 260, p. 117827, May 2021, doi: 10.1016/J.CARBPOL.2021.117827.

[20] X. Zheng *et al.*, "Robust ZIF-8/alginate fibers for the durable and highly effective antibacterial textiles," *Colloids Surfaces B Biointerfaces*, vol. 193, p. 111127, Sep. 2020, doi: 10.1016/J.COLSURFB.2020.111127.

[21] A. R. Shirvan, M. Shakeri, and A. Bashari, "Recent advances in application of chitosan and its derivatives in functional finishing of textiles," *Impact Prospect. Green Chem. Text. Technol.*, pp. 107–133, Jan. 2019, doi: 10.1016/B978-0-08-102491-1.00005-8.

[22] R. C. F. Cheung, T. B. Ng, J. H. Wong, and W. Y. Chan, "Chitosan: An update on potential biomedical and pharmaceutical applications," *Mar. Drugs*, vol. 13, no. 8, pp. 5156–5186, Aug. 2015, doi: 10.3390/MD13085156.

[23] I. Younes and M. Rinaudo, "Chitin and chitosan preparation from marine sources. Structure, properties and applications," *Mar. Drugs*, vol. 13, no. 3, pp. 1133–1174, Mar. 2015, doi: 10.3390/MD13031133.

[24] S. O. Majekodunmi, "Current development of extraction, characterization and evaluation of properties of chitosan and its use in medicine and pharmaceutical industry," *Am. J. Polym. Sci.*, vol. 6, no. 3, pp. 86–91, 2016, doi: 10.5923/j.ajps.20160603.04.

[25] L. X. Lixia Huang and G. Yang, "Chitosan application in textile processing," *Curr. Trends Fash. Technol. Text. Eng.*, vol. 4, no. 2, pp. 1–3, Aug. 2018, doi: 10.19080/CTFTTE.2018.04.555635.

[26] S. A. Qamar, M. Ashiq, M. Jahangeer, A. Riasat, and M. Bilal, "Chitosan-based hybrid materials as adsorbents for textile dyes-A review," *Case Stud. Chem. Environ. Eng.*, vol. 2, p. 100021, Sep. 2020, doi: 10.1016/J.CSCEE.2020.100021.

[27] Britannica, "Starch—Definition, Formula, Uses, & Facts," *Encyclopedia Britannica*. [Online]. Available: https://www.britannica.com/science/starch. [Accessed: 11-Apr-2022].

[28] S. W. Horstmann, K. M. Lynch, and E. K. Arendt, "Starch characteristics linked to gluten-free products," *Foods*, vol. 6, no. 4, p. 29, Apr. 2017, doi: 10.3390/FOODS6040029.

[29] J. BeMiller and R. Whistler, *Starch: Chemistry and Technology*, 3rd ed. Elsevier Science, 2009.

[30] J. A. Radley, "The textile industry," *Industrial Uses of Starch and Its Derivatives*, pp. 149–197, 1976, doi: 10.1007/978-94-010-1329-1_4.

[31] L. Kong and G. R. Ziegler, "Fabrication of pure starch fibers by electrospinning," *Food Hydrocoll.*, vol. 36, pp. 20–25, May 2014, doi: 10.1016/J.FOODHYD.2013.08.021.

[32] G. Liu, Z. Gu, Y. Hong, L. Cheng, and C. Li, "Electrospun starch nanofibers: Recent advances, challenges, and strategies for potential pharmaceutical applications," *J. Control. Release*, vol. 252, pp. 95–107, Apr. 2017, doi: 10.1016/J.JCONREL.2017.03.016.

[33] Z. Zhu and S. Shen, "Effect of amphoteric grafting branch on the adhesion of starch to textile fibers," *J. Adhes. Sci. Technol.*, vol. 28, no. 17, pp. 1695–1710, Sep. 2014, doi: 10.1080/01694243.2014.913514.

[34] X. D. Shi *et al.*, "Fabrication, interaction mechanism, functional properties, and applications of fish gelatin-polysaccharide composites: a review," *Food Hydrocoll.*, vol. 122, p. 107106, Jan. 2022, doi: 10.1016/J.FOODHYD.2021.107106.

[35] B. M. Eid and N. A. Ibrahim, "Recent developments in sustainable finishing of cellulosic textiles employing biotechnology," *J. Clean. Prod.*, vol. 284, p. 124701, Feb. 2021, doi: 10.1016/J.JCLEPRO.2020.124701.

[36] Q. Liang, W. Pan, and Q. Gao, "Preparation of carboxymethyl starch/polyvinyl-alcohol electrospun composite nanofibers from a green approach," *Int. J. Biol. Macromol.*, vol. 190, pp. 601–606, Nov. 2021, doi: 10.1016/J.IJBIOMAC.2021.09.015.

[37] L. Zhou, H. Zhou, and X. Yang, "Preparation and performance of a novel starch-based inorganic/organic composite coagulant for textile wastewater treatment," *Sep. Purif. Technol.*, vol. 210, pp. 93–99, Feb. 2019, doi: 10.1016/J.SEPPUR.2018.07.089.

[38] J. Li, Y. Fan, G. Huang, and H. Huang, "Extraction, structural characteristics and activities of *Zizylphus vulgaris* polysaccharides," *Ind. Crops Prod.*, vol. 178, p. 114675, Apr. 2022, doi: 10.1016/J.INDCROP.2022.114675.

[39] B. Singh Kaith, Rohit, and R. Kumar, "Psyllium polysaccharide-based hydrogels as smart biomaterials: Review," *Mater. Today Proc.*, vol. 53, pp. 244–246, Jan. 2022, doi: 10.1016/J.MATPR.2022.01.051.

[40] A. R. Madgulkar, M. R. P. Rao, and D. Warrier, "Characterization of Psyllium (Plantago ovata) polysaccharide and its uses," *Polysaccharides Bioactivity Biotechnol.*, pp. 871–890, Jan. 2015, doi: 10.1007/978-3-319-16298-0_49.

[41] E. Bozaci, E. Akar, E. Ozdogan, A. Demir, A. Altinisik, and Y. Seki, "Application of carboxymethylcellulose hydrogel based silver nanocomposites on cotton fabrics for antibacterial property," *Carbohydr. Polym.*, vol. 134, pp. 128–135, Dec. 2015, doi: 10.1016/J.CARBPOL.2015.07.036.

[42] P. B. S. Albuquerque, W. F. de Oliveira, P. M. dos Santos Silva, M. T. dos Santos Correia, J. F. Kennedy, and L. C. B. B. Coelho, "Skincare application of medicinal plant polysaccharides—A review," *Carbohydr. Polym.*, vol. 277, 118824, Feb. 2022, doi: 10.1016/J.CARBPOL.2021.118824.

[43] X. Li *et al.*, "Preparation, structural analysis, antioxidant and digestive enzymes inhibitory activities of polysaccharides from *Thymus quinquecostatus* Celak. leaves," *Ind. Crops Prod.*, vol. 175, p. 114288, Jan. 2022, doi: 10.1016/J.INDCROP.2021.114288.

[44] J. Fang, Z. Wang, P. Wang, and M. Wang, "Extraction, structure and bioactivities of the polysaccharides from *Ginkgo biloba*: A review," *Int. J. Biol. Macromol.*, vol. 162, pp. 1897–1905, Nov. 2020, doi: 10.1016/J.IJBIOMAC.2020.08.141.

[45] Z. Yin, D. Sun-Waterhouse, J. Wang, C. Ma, G. I. N. Waterhouse, and W. Kang, "Polysaccharides from edible fungi Pleurotus spp.: advances and perspectives," *J. Futur. Foods*, vol. 1, no. 2, pp. 128–140, Dec. 2021, doi: 10.1016/J.JFUTFO.2022.01.002.

[46] M. Xie *et al.*, "Anti-hypertensive and cardioprotective activities of traditional Chinese medicine-derived polysaccharides: A review," *Int. J. Biol. Macromol.*, vol. 185, pp. 917–934, Aug. 2021, doi: 10.1016/J.IJBIOMAC.2021.07.008.

[47] Z. Xiao, Q. Deng, W. Zhou, and Y. Zhang, "Immune activities of polysaccharides isolated from *Lycium barbarum* L. What do we know so far?," *Pharmacol. Ther.*, vol. 229, 107921, Jan. 2022, doi: 10.1016/J.PHARMTHERA.2021.107921.

[48] M. S. Abdelrahman, S. H. Nassar, H. Mashaly, S. Mahmoud, D. Maamoun, and T. A. Khattab, "Review in textile printing technology," *Egypt. J. Chem.*, vol. 63, no. 9, pp. 3465–3479, Sep. 2020, doi: 10.21608/EJCHEM.2020.23726.2418.

[49] B. Zhang *et al.*, "Synthesis and characterization of carboxymethyl potato starch and its application in reactive dye printing," *Int. J. Biol. Macromol.*, vol. 51, no. 4, pp. 668–674, Nov. 2012, doi: 10.1016/J.IJBIOMAC.2012.07.003.

[50] C. M. Obele, M. E. Ibenta, J. L. Chukwuneke, and S. C. Nwanonenyi, "Carboxymethyl cellulose and cellulose nanocrystals from cassava stem as thickeners in reactive printing of cotton," *Cellulose*, vol. 28, no. 4, pp. 2615–2633, Jan. 2021, doi: 10.1007/S10570-021-03694-0.

[51] H. W. Lin, C. P. Chang, W. H. Hwu, and M. Der Ger, "The rheological behaviors of screen-printing pastes," *J. Mater. Process. Technol.*, vol. 197, no. 1–3, pp. 284–291, Feb. 2008, doi: 10.1016/J.JMATPROTEC.2007.06.067.

[52] Q. Li, G. Chen, T. Xing, and S. Miao, "Dry transfer printing of silk and cotton with reactive dyes and mixed polysaccharide thickeners," *Color. Technol.*, vol. 134, no. 3, pp. 222–229, Jun. 2018, doi: 10.1111/COTE.12337.

[53] F. An, K. Fang, X. Liu, C. Li, Y. Liang, and H. Liu, "Rheological properties of carboxymethyl hydroxypropyl cellulose and its application in high quality reactive dye inkjet printing on wool fabrics," *Int. J. Biol. Macromol.*, vol. 164, pp. 4173–4182, Dec. 2020, doi: 10.1016/J.IJBIOMAC.2020.08.216.

[54] R. P. Singh and P. M. Davidson, "Food Additive," *Encyclopedia Britannica*. [Online]. Available: https://www.britannica.com/topic/food-additive#ref502208. [Accessed: 12-Feb-2022].

[55] L. Wang, R. Li, J. Shao, and Z. Wang, "Rheological behaviors of carboxymethyl tamarind gum as thickener on georgette printing with disperse dyes," *J. Appl. Polym. Sci.*, vol. 134, no. 26, Jul. 2017, doi: 10.1002/APP.45000.

[56] R. Fijan, M. Basile, S. Šostar-Turk, E. Žagar, M. Žigon, and R. Lapasin, "A study of rheological and molecular weight properties of recycled polysaccharides used as thickeners in textile printing," *Carbohydr. Polym.*, vol. 76, no. 1, pp. 8–16, Mar. 2009, doi: 10.1016/J.CARBPOL.2008.09.027.

[57] Shahid-ul-Islam and B. S. Butola, "Recent advances in chitosan polysaccharide and its derivatives in antimicrobial modification of textile materials," *Int. J. Biol. Macromol.*, vol. 121, pp. 905–912, Jan. 2019, doi: 10.1016/J.IJBIOMAC.2018.10.102.

[58] X. Cheng, K. Ma, R. Li, X. Ren, and T. S. Huang, "Antimicrobial coating of modified chitosan onto cotton fabrics," *Appl. Surf. Sci.*, vol. 309, pp. 138–143, Aug. 2014, doi: 10.1016/J.APSUSC.2014.04.206.

[59] D. Mihailović *et al.*, "Functionalization of polyester fabrics with alginates and TiO2 nanoparticles," *Carbohydr. Polym.*, vol. 79, no. 3, pp. 526–532, Feb. 2010, doi: 10.1016/J.CARBPOL.2009.08.036.

[60] I. M. El-Nahhal, J. Salem, R. Anbar, F. S. Kodeh, and A. Elmanama, "Preparation and antimicrobial activity of ZnO-NPs coated cotton/starch and their functionalized ZnO-Ag/cotton and Zn(II) curcumin/ cotton materials," *Sci. Reports*, vol. 10, no. 1, pp. 1–10, Mar. 2020, doi: 10.1038/s41598-020-61306-6.

[61] W. Ye, M. F. Leung, J. Xin, T. L. Kwong, D. K. L. Lee, and P. Li, "Novel core-shell particles with poly (n-butyl acrylate) cores and chitosan shells as an antibacterial coating for textiles," *Polymer (Guildf).*, vol. 46, no. 23, pp. 10538–10543, Nov. 2005, doi: 10.1016/J.POLYMER.2005.08.019.

[62] B. Klaykruayat, K. Siralertmukul, and K. Srikulkit, "Chemical modification of chitosan with cationic hyperbranched dendritic polyamidoamine and its antimicrobial activity on cotton fabric," *Carbohydr. Polym.*, vol. 80, no. 1, pp. 197–207, Mar. 2010, doi: 10.1016/J.CARBPOL.2009.11.013.

[63] S. Muzaffar *et al.*, "Enhanced mechanical, UV protection and antimicrobial properties of cotton fabric employing nanochitosan and polyurethane based finishing," *J. Mater. Res. Technol.*, vol. 11, pp. 946–956, Mar. 2021, doi: 10.1016/J.JMRT.2021.01.018.

[64] M. Sumithra and N. Vasugi Raaja, "Micro-encapsulation and nano-encapsulation of denim fabrics with herbal extracts," *Indian J. Fibre Text. Res.*, vol. 37, no. 4, pp. 321–325, 2012.

[65] H. F. S. Gafri, F. Mohamed Zuki, M. K. Aroua, and N. A. Hashim, "Mechanism of bacterial adhesion on ultrafiltration membrane modified by natural antimicrobial polymers (chitosan) and combination with activated carbon (PAC)," *Rev. Chem. Eng.*, vol. 35, no. 3, pp. 421–443, 2019, doi: 10.1515/ revce-2017-0006.

[66] S. F. Grgac, A. Tarbuk, T. Dekanic, W. Sujka, and Z. Draczynski, "The chitosan implementation into cotton and polyester/cotton blend fabrics," *Materials (Basel).*, vol. 13, no. 7, Apr. 2020, doi: 10.3390/ MA13071616.

[67] S. W. Ali, S. Rajendran, and M. Joshi, "Synthesis and characterization of chitosan and silver loaded chitosan nanoparticles for bioactive polyester," *Carbohydr. Polym.*, vol. 83, no. 2, pp. 438–446, Jan. 2011, doi: 10.1016/J.CARBPOL.2010.08.004.

[68] K. H. Jung *et al.*, "Preparation and antibacterial activity of PET/chitosan nanofibrous mats using an electrospinning technique," *J. Appl. Polym. Sci.*, vol. 105, no. 5, pp. 2816–2823, Sep. 2007, doi: 10.1002/APP.25594.

[69] M. K. Zahran, H. B. Ahmed, and M. H. El-Rafie, "Surface modification of cotton fabrics for antibacterial application by coating with AgNPs-alginate composite," *Carbohydr. Polym.*, vol. 108, no. 1, pp. 145–152, Aug. 2014, doi: 10.1016/J.CARBPOL.2014.03.005.

[70] M. H. Kudzin and M. Giełdowska, "Poly (lactic acid)/zinc/alginate complex material : Preparation and antimicrobial properties," *Antibiotics*, vol. 10, p. 1327, 2021, doi: https://doi.org/10.3390/ antibiotics10111327.

[71] S. K. Bajpai, M. Bajpai, and L. Sharma, "Copper nanoparticles loaded alginate-impregnated cotton fabric with antibacterial properties," *J. Appl. Polym. Sci.*, vol. 126, no. S1, pp. E319–E326, Oct. 2012, doi: 10.1002/APP.36981.

[72] W. Huang, H. Xu, Y. Xue, R. Huang, H. Deng, and S. Pan, "Layer-by-layer immobilization of lysozyme-chitosan-organic rectorite composites on electrospun nanofibrous mats for pork preservation," *Food Res. Int.*, vol. 48, no. 2, pp. 784–791, Oct. 2012, doi: 10.1016/J.FOODRES.2012.06.026.

[73] A. R. Shirvan, N. H. Nejad, and A. Bashari, "Antibacterial finishing of cotton fabric via the chitosan/ TPP self-assembled nano layers," *Fibers Polym,* vol. 15, no. 9, pp. 1908–1914, Oct. 2014, doi: 10.1007/ S12221-014-1908-Y.

[74] X. Cheng, R. Li, X. Li, M. M. Umair, X. Ren, and T. S. Huang, "Preparation and characterization of antimicrobial cotton fabrics via N-halamine chitosan derivative/poly(2-acrylamide-2-methylpropane sulfonic acid sodium salt) self-assembled composite films," *J. Ind. Text.*, vol. 46, no. 4, pp. 1039–1052, Oct. 2015, doi: 10.1177/1528083715612232.

[75] H. M. C. Azeredo and K. W. Waldron, "Crosslinking in polysaccharide and protein films and coatings for food contact—A review," *Trends Food Sci. Technol.*, vol. 52, pp. 109–122, Jun. 2016, doi: 10.1016/J. TIFS.2016.04.008.

[76] R. Devasia, A. Painuly, D. Devapal, and K. J. Sreejith, "Continuous fiber reinforced ceramic matrix composites," In *Fiber Reinforced Composites*, Woodhead Publishing, 2021, pp. 669–751.

[77] T. Tsuzuki and X. Wang, "Nanoparticle coatings for UV protective textiles," *Res. J. Text. Appar.*, vol. 14, no. 2, pp. 9–20, May 2010, doi: 10.1108/RJTA-14-02-2010-B002/FULL/XML.

[78] M. I. H. Mondal and J. Saha, "Antimicrobial, UV resistant and thermal comfort properties of chitosan- and aloe vera-modified cotton woven fabric," *J. Polym. Environ.*, vol. 27, no. 2, pp. 405–420, Jan. 2019, doi: 10.1007/S10924-018-1354-9.

[79] M. Gorenšek and V. Bukošek, "Zinc and alginate for multipurpose textiles," *Acta Chim. Slov.*, vol. 53, no. 2, pp. 223–228, 2006.

[80] N. A. Ivanova and A. B. Philipchenko, "Superhydrophobic chitosan-based coatings for textile processing," *Appl. Surf. Sci.*, vol. 263, pp. 783–787, Dec. 2012, doi: 10.1016/J.APSUSC.2012.09.173.

[81] J. A. B. Valle, R. de C. S. C. Valle, A. C. K. Bierhalz, F. M. Bezerra, A. L. Hernandez, and M. J. Lis Arias, "Chitosan microcapsules: Methods of the production and use in the textile finishing," *J. Appl. Polym. Sci.*, vol. 138, no. 21, p. 50482, Jun. 2021, doi: 10.1002/APP.50482.

[82] N. Misni, Z. M. Nor, and R. Ahmad, "Microencapsulation of citrus grandis peel oil using interfacial precipitation chemistry technique for repellent application," *Iran. J. Pharm. Res. IJPR*, vol. 18, no. 1, p. 198, 2019.

[83] A. Cerempei, "Aromatherapeutic textiles," *Act. Ingredients from Aromat. Med. Plants*, Mar. 2017, doi: 10.5772/66510.

[84] Y. Li *et al.*, "Properties of chitosan-microencapsulated orange oil prepared by spray-drying and its stability to detergents," *J. Agric. Food Chem.*, vol. 61, no. 13, pp. 3311–3319, Apr. 2013, doi: 10.1021/JF305074Q/ASSET/IMAGES/JF305074Q.SOCIAL.JPEG_V03.

[85] N. Singh, M. Yadav, S. Khanna, and O. Sahu, "Sustainable fragrance cum antimicrobial finishing on cotton: Indigenous essential oil," *Sustain. Chem. Pharm.*, vol. 5, pp. 22–29, Jun. 2017, doi: 10.1016/J.SCP.2017.01.003.

[86] A. Cerempei, E. I. Muresan, and N. Cimpoesu, "Biomaterials with controlled release of geranium essential oil," *J. Essent. Oil Res.*, vol. 26, no. 4, pp. 267–273, Jul. 2014, doi: 10.1080/10412905.2014.910711.

[87] L. Wang *et al.*, "Amphiphilic alginate stabilized UV-curable polyurethane acrylate as a surface coating to improve the anti-wrinkle performance of cotton fabrics," *Prog. Org. Coatings*, vol. 162, p. 106595, Jan. 2022, doi: 10.1016/J.PORGCOAT.2021.106595.

[88] Z. A. Raza, F. Anwar, and S. Abid, "Sustainable antibacterial printing of cellulosic fabrics using an indigenous chitosan-based thickener with distinct natural dyes," *Int. J. Cloth. Sci. Technol.*, vol. 33, no. 6, pp. 914–928, 2021, doi: 10.1108/IJCST-01-2020-0005/FULL/XML.

[89] M. Periolatto, F. Ferrero, and C. Vineis, "Antimicrobial chitosan finish of cotton and silk fabrics by UV-curing with 2-hydroxy-2-methylphenylpropane-1-one," *Carbohydr. Polym.*, vol. 88, no. 1, pp. 201–205, Mar. 2012, doi: 10.1016/J.CARBPOL.2011.11.093.

[90] C. Wang *et al.*, "Cotton fabric with plasma pretreatment and ZnO/Carboxymethyl chitosan composite finishing for durable UV resistance and antibacterial property," *Carbohydr. Polym.*, vol. 138, pp. 106–113, Mar. 2016, doi: 10.1016/J.CARBPOL.2015.11.046.

[91] C. Zheng *et al.*, "Superhydrophobic and flame-retardant alginate fabrics prepared through a one-step dip-coating surface-treatment," *Cellul. 2021 289*, vol. 28, no. 9, pp. 5973–5984, May 2021, doi: 10.1007/S10570-021-03890-Y.

[92] V. Gopinath, S. M. Kamath, S. Priyadarshini, Z. Chik, A. A. Alarfaj, and A. H. Hirad, "Multifunctional applications of natural polysaccharide starch and cellulose: An update on recent advances," *Biomed. Pharmacother.*, vol. 146, p. 112492, Feb. 2022, doi: 10.1016/J.BIOPHA.2021.112492.

[93] S. Chatterjee and P. C. L. Hui, "Review of applications and future prospects of stimuli-responsive hydrogel based on thermo-responsive biopolymers in drug delivery systems," *Polym.*, vol. 13, no. 13, p. 2086, Jun. 2021, doi: 10.3390/POLYM13132086.

[94] T. Miao, J. Wang, Y. Zeng, G. Liu, and X. Chen, "Polysaccharide-based controlled release systems for therapeutics delivery and tissue engineering: From bench to bedside," *Adv. Sci.*, vol. 5, no. 4, p. 1700513, Apr. 2018, doi: 10.1002/ADVS.201700513.

[95] N. P. Patil *et al.*, "Algal polysaccharides as therapeutic agents for atherosclerosis," *Front. Cardiovasc. Med.*, vol. 5, p. 153, Oct. 2018, doi: 10.3389/FCVM.2018.00153/BIBTEX.

[96] Medical Design Briefs, "Therapeutic Textiles Seamlessly Promote Wellness," 01-Feb-2012. [Online]. Available: https://www.medicaldesignbriefs.com/component/content/article/mdb/features/global-innovations/12821. [Accessed: 28-Apr-2022].

[97] S. Chatterjee, P. C. leung Hui, E. Wat, C. wai Kan, P. C. Leung, and W. Wang, "Drug delivery system of dual-responsive PF127 hydrogel with polysaccharide-based nano-conjugate for textile-based transdermal therapy," *Carbohydr. Polym.*, vol. 236, p. 116074, May 2020, doi: 10.1016/J.CARBPOL.2020.116074.

[98] O. Eskens, G. Villani, and S. Amin, "Rheological investigation of thermoresponsive alginate-methylcellulose gels for epidermal growth factor formulation," *Cosmetics*, vol. 8, no. 1, p. 3, Dec. 2020, doi: 10.3390/COSMETICS8010003.

[99] Y. Tang *et al.*, "Production and characterisation of novel injectable chitosan/methylcellulose/salt blend hydrogels with potential application as tissue engineering scaffolds," *Carbohydr. Polym.*, vol. 82, no. 3, pp. 833–841, Oct. 2010, doi: 10.1016/J.CARBPOL.2010.06.003.

[100] A. I. Alzarea *et al.*, "Development and characterization of gentamicin-loaded arabinoxylan-sodium alginate films as antibacterial wound dressing," *Int. J. Mol. Sci.*, vol. 23, no. 5, p. 2899, Mar. 2022, doi: 10.3390/IJMS23052899.

[101] L. Colobatiu *et al.*, "Evaluation of bioactive compounds-loaded chitosan films as a novel and potential diabetic wound dressing material," *React. Funct. Polym.*, vol. 145, p. 104369, Dec. 2019, doi: 10.1016/J.REACTFUNCTPOLYM.2019.104369.

[102] L. Antunes, G. Faustino, C. Mouro, J. Vaz, and I. C. Gouveia, "Bioactive microsphere-based coating for biomedical-textiles with encapsulated antimicrobial peptides (AMPs)," *Ciência Tecnol. dos Mater.*, vol. 26, no. 2, pp. 118–125, Jul. 2014, doi: 10.1016/J.CTMAT.2015.03.006.

[103] M. H. Malakyan *et al.*, "Pharmacological and haematological results of rat skin burn injury treatment with Cu(II)2(3,5-diisopropylsalicylate)4," *Inflammo. Pharmacol.*, vol. 12, no. 4, pp. 321–351, Dec. 2004, doi: 10.1163/1568560043696209.

[104] H. Tapiero, D. M. Townsend, and K. D. Tew, "Trace elements in human physiology and pathology. Copper," *Biomed. Pharmacother.*, vol. 57, no. 9, pp. 386–398, Nov. 2003, doi: 10.1016/S0753-3322(03)00012-X.

[105] H. E. Emam and H. B. Ahmed, "Polysaccharides templates for assembly of nanosilver," *Carbohydr. Polym.*, vol. 135, pp. 300–307, Jan. 2016, doi: 10.1016/J.CARBPOL.2015.08.095.

[106] A. Aprodu, J. Mantaj, B. Raimi-Abraham, and D. Vllasaliu, "Evaluation of a methylcellulose and hyaluronic acid hydrogel as a vehicle for rectal delivery of biologics," *Pharmaceutics*, vol. 11, no. 3, Mar. 2019, doi: 10.3390/PHARMACEUTICS11030127.

[107] M. Bhowmik, M. K. Bain, L. K. Ghosh, and D. Chattopadhyay, "Effect of salts on gelation and drug release profiles of methylcellulose-based ophthalmic thermo-reversible in situ gels," *Pharm. Dev. Technol.*, vol. 16, no. 4, pp. 385–391, Aug. 2011, doi: 10.3109/10837451003774369.

[108] J. Shi, Z. Zhang, W. Qi, and S. Cao, "Hydrophobically modified biomineralized polysaccharide alginate membrane for sustained smart drug delivery," *Int. J. Biol. Macromol.*, vol. 50, no. 3, pp. 747–753, Apr. 2012, doi: 10.1016/J.IJBIOMAC.2011.12.003.

[109] P. Kulal and V. Badalamoole, "Magnetite nanoparticle embedded Pectin-graft-poly(N-hydroxyethylacrylamide) hydrogel: Evaluation as adsorbent for dyes and heavy metal ions from waste water," *Int. J. Biol. Macromol.*, vol. 156, pp. 1408–1417, Aug. 2020, doi: 10.1016/J.IJBIOMAC.2019.11.181.

[110] C. M. Laureano-Anzaldo, M. E. González-López, A. A. Pérez-Fonseca, L. E. Cruz-Barba, and J. R. Robledo-Ortíz, "Plasma-enhanced modification of polysaccharides for wastewater treatment: A review," *Carbohydr. Polym.*, vol. 252, p. 117195, Jan. 2021, doi: 10.1016/J.CARBPOL.2020.117195.

[111] S. P. Wu, X. Z. Dai, J. R. Kan, F. Di Shilong, and M. Y. Zhu, "Fabrication of carboxymethyl chitosan-hemicellulose resin for adsorptive removal of heavy metals from wastewater," *Chinese Chem. Lett.*, vol. 28, no. 3, pp. 625–632, Mar. 2017, doi: 10.1016/J.CCLET.2016.11.015.

[112] M. B. Kasiri, "Application of chitosan derivatives as promising adsorbents for treatment of textile wastewater," *Impact Prospect. Green Chem. Text. Technol.*, pp. 417–469, Jan. 2019, doi: 10.1016/B978-0-08-102491-1.00014-9.

[113] E. S. Dragan and D. F. A. Loghin, "Fabrication and characterization of composite cryobeads based on chitosan and starches-g-PAN as efficient and reusable biosorbents for removal of Cu²⁺, Ni²⁺, and Co²⁺ ions," *Int. J. Biol. Macromol.*, vol. 120, pp. 1872–1883, Dec. 2018, doi: 10.1016/J.IJBIOMAC.2018.10.007.

[114] D. Zhou, L. Zhang, J. Zhou, and S. Guo, "Cellulose/chitin beads for adsorption of heavy metals in aqueous solution," *Water Res.*, vol. 38, no. 11, pp. 2643–2650, Jun. 2004, doi: 10.1016/J.WATRES.2004.03.026.

[115] X. Tong, W. Pan, T. Su, M. Zhang, W. Dong, and X. Qi, "Recent advances in natural polymer-based drug delivery systems," *React. Funct. Polym.*, vol. 148, p. 104501, Mar. 2020, doi: 10.1016/J.REACTFUNCTPOLYM.2020.104501.

[116] D. Kundu, S. K. Mondal, and T. Banerjee, "Development of β-cyclodextrin-cellulose/hemicellulose-based hydrogels for the removal of Cd(II) and Ni(II): Synthesis, kinetics, and adsorption aspects," *J. Chem. Eng. Data*, vol. 64, no. 6, pp. 2601–2617, Jun. 2019, doi: 10.1021/ACS.JCED.9B00088/SUPPL_FILE/JE9B00088_SI_001.PDF.

[117] H. Mondal, M. Karmakar, P. K. Chattopadhyay, A. Halder, and N. R. Singha, "Scale-up one-pot synthesis of waste collagen and apple pomace pectin incorporated pentapolymer biocomposites: Roles of waste collagen for elevations of properties and unary/ ternary removals of Ti(IV), As(V), and V(V)," *J. Hazard. Mater.*, vol. 409, p. 124873, May 2021, doi: 10.1016/J.JHAZMAT.2020.124873.

[118] S. Bo *et al.*, "Efficiently selective adsorption of Pb(II) with functionalized alginate-based adsorbent in batch/column systems: Mechanism and application simulation," *J. Clean. Prod.*, vol. 250, p. 119585, Mar. 2020, doi: 10.1016/J.JCLEPRO.2019.119585.

[119] P. Nechita, "Applications of chitosan in wastewater treatment," *Biol. Act. Appl. Mar. Polysaccharides*, Jan. 2017, doi: 10.5772/65289.

[120] P. Kumar, A. L. Ganure, B. B. Subudhi, and S. Shukla, "Preparation and characterization of pH-sensitive methyl methacrylate-g-starch/hydroxypropylated starch hydrogels: in vitro and in vivo study on release of esomeprazole magnesium," *Drug Deliv. Transl. Res. 2015 53*, vol. 5, no. 3, pp. 243–256, Mar. 2015, doi: 10.1007/S13346-015-0221-7.

[121] A. Nasiri, S. Rajabi, and M. Hashemi, "CoFe2O4@Methylcellulose/AC as a new, green, and eco-friendly nano-magnetic adsorbent for removal of reactive red 198 from aqueous solution," *Arab. J. Chem.*, vol. 15, no. 5, p. 103745, May 2022, doi: 10.1016/J.ARABJC.2022.103745.

[122] S. Mishra, S. Sinha, K. P. Dey, and G. Sen, "Synthesis, characterization and applications of polymethyl-methacrylate grafted psyllium as flocculant," *Carbohydr. Polym.*, vol. 99, pp. 462–468, Jan. 2014, doi: 10.1016/J.CARBPOL.2013.08.047.

[123] N. K. Nga, N. T. Thuy Chau, and P. H. Viet, "Preparation and characterization of a chitosan/MgO composite for the effective removal of reactive blue 19 dye from aqueous solution," *J. Sci. Adv. Mater. Devices*, vol. 5, no. 1, pp. 65–72, Mar. 2020, doi: 10.1016/J.JSAMD.2020.01.009.

[124] S. Pal, S. Ghosh, G. Sen, U. Jha, and R. P. Singh, "Cationic tamarind kernel polysaccharide (Cat TKP): A novel polymeric flocculant for the treatment of textile industry wastewater," *Int. J. Biol. Macromol.*, vol. 45, no. 5, pp. 518–523, Dec. 2009, doi: 10.1016/J.IJBIOMAC.2009.08.004.

[125] H. Salehizadeh, N. Yan, and R. Farnood, "Recent advances in polysaccharide bio-based flocculants," *Biotechnol. Adv.*, vol. 36, no. 1, pp. 92–119, Jan. 2018, doi: 10.1016/J.BIOTECHADV.2017.10.002.

[126] S. Pal, D. Mal, and R. P. Singh, "Cationic starch: An effective flocculating agent," *Carbohydr. Polym.*, vol. 59, no. 4, pp. 417–423, Mar. 2005, doi: 10.1016/J.CARBPOL.2004.06.047.

[127] H. Kono, "Cationic flocculants derived from native cellulose: Preparation, biodegradability, and removal of dyes in aqueous solution," *Resour. Technol.*, vol. 3, no. 1, pp. 55–63, Mar. 2017, doi: 10.1016/J.REFFIT.2016.11.015.

[128] J. Li et al., "Preparation and adsorption properties of magnetic chitosan composite adsorbent for Cu^{2+} removal," *J. Clean. Prod.*, vol. 158, pp. 51–58, Aug. 2017, doi: 10.1016/J.JCLEPRO.2017.04.156.

[129] Y. Zhang et al., "Preparation of novel cobalt ferrite/chitosan grafted with graphene composite as effective adsorbents for mercury ions," *J. Mol. Liq.*, vol. 198, pp. 381–387, Oct. 2014, doi: 10.1016/J.MOLLIQ.2014.07.043.

[130] Y. Wang, W. Wang, and A. Wang, "Efficient adsorption of methylene blue on an alginate-based nanocomposite hydrogel enhanced by organo-illite/smectite clay," *Chem. Eng. J.*, vol. 228, pp. 132–139, Jul. 2013, doi: 10.1016/J.CEJ.2013.04.090.

[131] H. Mittal and S. B. Mishra, "Gum ghatti and Fe$_3$O$_4$ magnetic nanoparticles based nanocomposites for the effective adsorption of rhodamine B," *Carbohydr. Polym.*, vol. 101, no. 1, pp. 1255–1264, Jan. 2014, doi: 10.1016/J.CARBPOL.2013.09.045.

[132] S. Ghorai, A. Sarkar, M. Raoufi, A. B. Panda, H. Schönherr, and S. Pal, "Enhanced removal of methylene blue and methyl violet dyes from aqueous solution using a nanocomposite of hydrolyzed polyacrylamide grafted xanthan gum and incorporated nanosilica," *ACS Appl. Mater. Interfaces*, vol. 6, no. 7, pp. 4766–4777, Apr. 2014, doi: 10.1021/AM4055657/SUPPL_FILE/AM4055657_SI_001.PDF.

[133] S. Pal et al., "Efficient and rapid adsorption characteristics of templating modified guar gum and silica nanocomposite toward removal of toxic reactive blue and Congo red dyes," *Bioresour. Technol.*, vol. 191, pp. 291–299, Sep. 2015, doi: 10.1016/J.BIORTECH.2015.04.099.

[134] K. Wang et al., "Evaluation of renewable pH-responsive starch-based flocculant on treating and recycling of highly saline textile effluents," *Environ. Res.*, vol. 201, p. 111489, Oct. 2021, doi: 10.1016/J.ENVRES.2021.111489.

[135] S. Biswas, T. U. Rashid, T. Debnath, P. Haque, and M. M. Rahman, "Application of Chitosan-Clay Biocomposite Beads for Removal of Heavy Metal and Dye from Industrial Effluent," *J. Compos. Sci.*, vol. 4, no. 1, p. 16, Feb. 2020, doi: 10.3390/JCS4010016.

[136] O. Garcia-Valdez, P. Champagne, and M. F. Cunningham, "Graft modification of natural polysaccharides via reversible deactivation radical polymerization," *Prog. Polym. Sci.*, vol. 76, pp. 151–173, Jan. 2018, doi: 10.1016/J.PROGPOLYMSCI.2017.08.001.

[137] V. Ajao, R. Fokkink, F. Leermakers, H. Bruning, H. Rijnaarts, and H. Temmink, "Bioflocculants from wastewater: Insights into adsorption affinity, flocculation mechanisms and mixed particle flocculation based on biopolymer size-fractionation," *J. Colloid Interface Sci.*, vol. 581, pp. 533–544, Jan. 2021, doi: 10.1016/J.JCIS.2020.07.146.

[138] T. Su et al., "Pullulan-derived nanocomposite hydrogels for wastewater remediation: Synthesis and characterization," *J. Colloid Interface Sci.*, vol. 542, pp. 253–262, Apr. 2019, doi: 10.1016/J.JCIS.2019.02.025.

12 Alginate-Based Nanocomposites for Smart Technology of Food Packaging

Md. Sohel Rana, Most. Afroza Khatun, Biplob Kumar Biswas, Shahanaz Parvin, Md. Wasikur Rahman, and M. Azizur R. Khan
Jashore University of Science and Technology

Sk Md Ali Zaker Shawon
Vanderbilt University

12.1 INTRODUCTION

The increased value of food packaged with polymeric nanocomposite benefits both consumers and producers. Using nanocomposite-based active and modified packaging, we can effectively increase the shelf life of food while preserving its early qualities and delivering them to consumers. Biopolymer modified with metal nanoparticles when blended with polymer frameworks improves quality characteristics. Thus advanced modification of sodium alginate (SA) can be achieved through monomer addition and irradiation (Rahman et al., 2016). Metal/metal oxide nanoparticles (AgNPs) are highly sensitive to acute temperature and exhibit low volatility and stability at normal conditions. They can be hosted in varied polymeric materials and stabilizing agents during polymer synthesis/blending. Ag-TiO$_2$ nanocomposites have strong ability to act as antimicrobial coating that halts, especially *E. coli* colony formation.

Alginates are unbranched 1,4-linked d-mannuronic acid copolymers that have a stereochemical gap at C-5 and are found in around 1,500 species and 265 taxa across the world. An alkaline extraction of sodium alginate results in the formation of sodium alginate as a solvent and the buildup of seaweed. Ion-exchange equilibrium with seawater-based treatment ions defines the internal structure of alginates (such as sodium, calcium, magnesium, and strontium). However, the use of nanoparticles on polymer composite dramatically improves qualities such as mechanical strength, aesthetic appearance, stiffness, oil and gas permeability, thermal stability, and electrical or thermal conductivity. Due to nanoparticles' great effectiveness, just 0.5%–5% of the polymer material's weight is typically utilized. So, the combination of nanoparticles with continuous monitoring of freshness is a major issue nowadays. Bionanocomposites based on sodium alginate are a promising material to develop novel packaging materials. In addition to reducing flammability, they also preserve the polymer matrix's translucency. Packaging materials including nanoparticles as nanosensors could inform the consumer when a food has gone bad.

Intelligent packaging materials can monitor the freshness, quality, and safety of food products over the lifespan of an individual. Many researchers are working to develop smart packaging materials that are able to detect changes in the quality of food while it's being stored.

12.2 ALGINATE: SOURCE, STRUCTURE AND CROSS LINKING

Alginates are 1,4-linked d-mannuronic acid copolymers having a stereo-chemical gap at C-5 that is not branched (Draget & Taylor, 2011). Chemical compounds function and reactivity depend heavily

DOI: 10.1201/9781003265054-12

on the presence of free hydroxyl and carboxyl groups. A variety of chemical reactions may be carried out on the functional groups (–OH and –COOH) of oxidized alginates, which may result in the formation of gels. Polymers of various molecular weights can be combined to increase the gel's flexibility. It is possible to assess the viscosity of an alginated solution by measuring its polymer content, MW distribution, pH, and the presence of G-blocks and M-block remnants in the solution. Alginate as an environmentally friendly food packaging material has the potential to be employed if meet the biocompatibility and biodegradability standards. Alginate may be found in approximately 1,500 species and 265 genera all over the world, depending on the coastline variety and composition of the sea water (Lee & Mooney, 2012). Some of the species and genera that can be found are Ascophyllum, Ecklonia, laminary, sargassum, turbinary, and macrocystis (Davis, Volesky, & Mucci, 2003), and a fibriller structure is both present in the algal cell walls (Kim et al., 2015) of alginic acid salt. The content and sequence of polysaccharide monomer units of alginate are influenced by a number of factors, including the time of harvest (Draget et al., 2002), the growth site, and the tissue age used for alginate preparation (Latifi, Nejad, & Babavalian, 2015).

The different mannuronic and guluronic acid (M/G) ratio varies significantly from 0.19 to 2.26, indicating that there are major differences in the makeup of species (Ramos et al., 2018). When selecting raw alginate sources, the G and M content of the alginate solution, gel, and biomaterials generated are important considerations since the characteristics of the alginate solution, gel, and biomaterials produced are dependent on the alginate material to be formed.

The science of sodium alginate is becoming increasingly obvious; in brown algae, the internal structure of alginates is defined by ion-exchange equilibrium with seawater-based treatment ions (such as sodium, calcium, magnesium, and strontium), which is determined by ion-exchange equilibrium with sodium, calcium, magnesium, and strontium (Pawar & Edgar, 2012). However, the ratios of mannuronic (M) and guluronic (G) acids cannot be used to calculate the quantity and size of these blocks for a given application of alginate composition and consistency. Because of the various saccharides conformations, the alginate polymer of the M and G chains is particularly scrupulous. Homopolymeric portions (MMMMMM or GGGGGG) or heteropolymeric areas (MGMGMG) with M-blocks and G-blocks in alternating sequence (Daemi & Barikani, 2012), respectively, make up the chain (Morris, Rees, & Thom, 1980). The uronic blocks sequence such as MG >MM>MG lowers stiffness. The polymer's chain lengthens and its inherent viscosity rises as a result of electrostatic repulsion between charged groups (Augst et al., 2006). When alginate was first discovered, it was considered that only the major G-blocks were involved in the formation of hydrogels, but subsequent research has shown that MG blocks are also involved in this process (Mørch et al., 2006). There has been a suggestion that the connection of extended rotational sequences at various MG/GM junction points reflects alginate gels' contraction (Donati et al., 2005). It is important to note that the choice of alginate material and crosslinking ion effects both the gel and ion binding properties of alginate (Saxena et al., 2022).

Although alginate are appealing for a variety of biological applications, they have limited mechanical characteristics when ionically crosslinked with divalent cations (Eiselt et al., 1999). Covalent crosslinking is used to compare the characteristics of alginate films to those of ionically crosslinked gels (Gasperini et al., 2014). This type of crosslinking is often shaped by the reaction between carboxylic groups (COOH) in alginate chains and a crosslinking particle containing necessary diamines (Gasperini et al., 2014). The use of crosslinking reagents should be thoroughly investigated since a considerable percentage of them can be dangerous, and unreacted crosslinkers should be totally excluded. How the polymers are cross-connected immensely affects how stress-relaxation acts in hydrogels. Stress is calmed basically by the breaking and ensuing renewal of ionic crosslinks in gels with ionic crosslinks, though stress is eased principally by the movement of water in gels with covalent crosslinks (Fan et al., 2011). Ionic crosslinking stresses cause the alginate gel to misshape plastically, yet covalent crosslinking causes a pressure unwinding that takes into consideration extensive versatile twisting (Kong et al., 2002).

It is possible to construct polymer network with a wide range of moduli by adjusting the chain length of the crosslinking particle or the crosslinking thickness (Gasperini et al., 2014). Polypeptides and polynucleotides are the most common synthetic polymers used in alginates, rather than biopolymers. The molecular weight of alginate is an average of the molecular weight of Na-alginate because of its dispersion. Due to short G-blocks, alginates include fragments of low molecular weight that do not contribute to the polymer matrix and hence do not contribute to its gel strength. When using alginate gel for high-tech applications, limited molecular mass distributions are recommended because of the risk of leakage (Draget et al., 2002). Sodium alginates are commercially available from 32,000 to 400,000 g mol^{-1} (Lee & Mooney, 2012). In general, a higher MW leads to greater interconnections and a greater strength and viscosity of the mechanical gel. The use of high MW alginate can have an effect on the real characteristics of its gels. In any instance, an alginate arrangement formed from a high MW polymer is extremely thick, which is typically unfavorable in handling (Kong et al., 2002; Draget & Taylor, 2011). High viscosity hampers the pre-gel/cell phase to preserve cell viability as a high solution viscosity will cause cells to be exposed to high shear forces during mixing. Cell membranes are highly labile and can cause damage or necrobiosis through their mixture (Boontheekul et al., 2005).

Thus by regulating the distribution and the molecular weight, the pregel solution, viscosity, and postgelling rigidity can be handled independently. Gel elasticity can be greatly enhanced by mixing a combination of different molecular weight polymers (Kong et al., 2002). So the addition of nanoparticle with carefully property modification and safety management alginate nanocomposite can be a possible alternative of traditional food packaging.

12.3 NANOCOMPOSITES

At the nanoscale, the combination of polymers with inorganic solids (ranging from clays to oxides) can result in the formation of nanocomposites, which are a type of heterogeneous or hybrid material. It has been discovered that their structures are more complicated than those of microcomposites. A particular property's structure, composition, interfacial interactions, and component parts all have substantial effect on how that feature manifests itself. The preparation of nanocomposites typically involves in-situ growth and polymerization of biopolymer and an inorganic matrix. This is the most prevalent process. Because of the remarkable potential of these cutting-edge materials, they are essential to a wide array of industries, ranging from the most rudimentary to the most complex manufacturing facilities. In addition to their positive effects on the environment, these have the potential to user in a wave of cutting-edge technological advancements and open up lucrative new markets for a wide variety of businesses, such as those in the transportation and building materials, electrical and electronic, food packaging, and logistics sectors. The use of nanoparticles significantly improves properties including mechanical strength, visual appearance, stiffness, oil and gas permeability, thermal stability, and electrical or thermal conductivity. Due to nanoparticle's high efficiency, just 0.5%–5% of the weight of the polymer material is normally used.

The surface area of a nanoparticle may be compared to its volume to understand how different it is from its bulk counterpart. Because of this, the polymer material's ability to hold nanoparticles in place shifts. This means that a composite can be enhanced by a multiple of nanoparticles. According to research, certain nanocomposites have a hardness of 1,000 times greater than their bulk polymer composite. Despite the fact that nanocomposites are becoming more and more prevalent, new uses are continually being discovered. Food packaging made from solid polymers, for example, may incorporate a broad variety of nanomaterials. As nanocomposite films that are both biodegradable and bio-based are being developed, there is a technological evolution that might compensate for some of the limitations of bio-based and biodegradable materials, for example, oxygen barrier characteristics, with active biopolymer nanocomposite materials. As an antibacterial or antioxidative nanomaterial is added to the nanocomposites, it may potentially eliminate the deficiencies between traditional plastic packaging materials and packed products in terms of shelf life, which is a critical factor in determining the quality of the product.

12.3.1 ALGINATE-BASED SILVER NANOCOMPOSITE

Silver nanoparticles are often used to functionalize polymeric materials for the purpose of bundling because of their excellent antibacterial properties against diseases, germs, and parasites (Quintero-Quiroz et al., 2020). The antibacterial impact of silver particles and silver nanoparticles is attributed to the activity of these particles. Because of the interaction between negatively and positively charged elements of the cell (sulfhydryl or disulfide clusters of catalysts and nucleic acid corrosive), the deformation of bacterial cell layers and cell dividers develops (Butkus et al., 2003; Feng et al., 2000). The increased surface area of silver nanoparticles, along with a greater potential for Ag particle discharge, is the critical variable in the antibacterial activity of silver nanoparticles, according to the researchers. Supposedly, the free revolutionary produced by silver nanoparticles will cause damage to the cell layer of microorganisms (Kim et al., 2007). The development of silver nanoparticles inside the cytoplasmic layer of microorganisms results in an increase in the porousness of the cell, resulting in the astonishing demise of microorganisms (Sondi & SalopekSondi, 2004). Silver particles combined with other inactive materials like as zeolite, nanoclay, and silicate (Incoronato et al., 2010; Egger et al., 2009) have been used as antibacterial specialists. The substitution of Na particles with Ag particles in zeolites results in the formation of silver zeolites. It has a broad spectrum of antibacterial action against a variety of microorganisms, including yeast, growth, mycelia, and germs. In this way, it is used as a significant antibacterial treatment; yet, it is shown to be lacking in antimicrobial action when tested against heat-resistant bacterial spores (Fernández et al., 2010). According to the results of an investigation, chitosan/Ag-zeolite exhibits significant antibacterial movement against gram-negative and gram-positive bacterial strains (Rhim et al., 2006). Silver silicate is produced with the use of a fire splash pyrolysis method. The silver nanoparticles produced as a result of the use of these biopolymers have been shown to have antibacterial properties. Nanochitosan combined with metallic nanoparticles such as Cu_2O, Mn_2O, Zn_2O, and Ag_2O has been shown to greatly increase antibacterial activity against *S. aureus*, *E. coli*, and *Salmonella choleraesuis*. The antibacterial properties of silver nanoparticles stacked with starch and chitosan have been demonstrated (Yoksan & Chirachanchai, 2010; Gupta & Tripathi, 2011). They are suitable for food bundling because of their material well-being and the extension in the timeframe of realistic utilization that these materials have seen (Tankhiwale & Bajpai, 2010; Zhu et al., 2009). There are a variety of factors that influence the effectiveness of antimicrobial workouts. Among them are the dimensions of the polymer network, the molecule size, the cooperation between the polymer network and the silver surface, the amount of molecule aggregation, the silver substance, and so on (Kim et al., 2007). A greater amount of absolute scattering of silver nanoparticles on polymeric surfaces is required for improved antibacterial movement. Copper particles have the ability to destroy infections and bacteria, and it is also a component of metallic catalysts, among other things.

12.3.2 ALGINATE BSILVER-MONTMORILLONITE (MMT) NANOCOMPOSITE

The antibacterial efficacy of polystyrene/silver silicate nanocomposites against *S. aureus* and *E. coli* is considerable (Egger et al., 2009). It is the exchanging of Na particles from ordinary MMT for Ag particles that causes the development of silver-montmorillonite to occur (Ag-MMT). Ag-MMT demonstrates strong antibacterial efficacy when tested against *Pseudomonas* spp (Valodkar et al., 2010). When it comes to delivering and settling silver nanoparticles, carbohydrates including waxy and dissolvable maize starch as well as sucrose are used. Many aromatic essential oils have been combined with montmorillonite (MMT) in biopolymers, such as montmorillonite-ginger essential oil (Yang et al., 2021), montmorillonite-rosemary essential oil (Krasniewska et al., 2020), and soy protein-montmorillonite-clove essential oil (Priyadarshi et al., 2021). Pires and colleagues investigated chitosan films that had been reinforced with sodium montmorillonite and infused with rosemary or ginger essential oils, respectively, as part of their investigation. The films were put through their paces with fresh poultry meat in packaging trials, and the results were promising. A reduction in lipid oxidation of meat was attributed to the inclusion of MMT alone, which the authors attributed to an enhanced barrier against UV radiation as well as a consequent rise in the meat's oxygen barrier (Pires et al., 2018).

Incorporating essential oils into the meat, in addition to montmorillonite, just decreased lipid oxidation and did not limit microbial development on the meat, according to the researchers (Dakal et al., 2016). The authors have also been investigated for their antibacterial characteristics in the perspective of the improvement of biodegradable materials, such as CMC/mucilage/ZnO (Siddiqi et al., 2018) and chitosan/cellulose acetate-phthalate/ZnO biocomposites (Shanmuganathan et al., 2019). However, by mixing metal oxides with essential oils, researchers were able to achieve an extra antibacterial effect in numerous trials. The following are examples of such research conducted in the recent several years: alginate-ZnO-essential oil (Abebe et al., 2020), WPI-cellulose nanofiber-TiO_2/rosemary essential oil (Stoimenov et al., 2002), and gelatin-ZnO nanorods-clove essential oil biopolymer nanocomposites (Sawai & Yoshikawa, 2004).

12.3.3 ALGINATE-BASED TITANIUM NANOCOMPOSITES

A polymer film of TiO_2 nanocomposite coating was developed and tested on cellulosic paper in 2016. The biopolymer nanocomposite coating improved the mechanical qualities and inhibited microbiological development on the surface of the paper packaging (Hajipour et al., 2012). Biodegradable coatings with chitosan integrating vermiculite nanoclay on a PET substrate have also been described (Youssef et al., 2015). Cost is a barrier to industrial adoption of biodegradable materials compared to conventional plastics. By reducing the quantity of biodegradable polymers necessary to achieve the desired qualities, nanotechnology may minimize production costs (Huang et al., 2019). Concerns about the biodegradability of biopolymer nanocomposite materials and how to adapt them for biodegradability need to be developed. Changing the crystallinity of various nanomaterials has been shown to modify the biodegradability and rates of microbial breakdown of biopolymers. The net chemical interaction between the biopolymer matrix and the nanoparticles might alter biodegradation (Khan et al., 2019) (Socas-Rodríguez et al., 2020). According to (Mishra et al., 2020) achieving compatibility between nanofillers and biopolymer matrix, and finding suitable processing techniques for biopolymer nanocomposite materials are the key challenges to commercialization (Socas-Rodríguez et al., 2020). As nanoparticles are liberated from biomaterials during degradation, Souza and Fernando warned of environmental pollution and ecotoxicity. It is unclear if nanoparticles discharged into the environment would bioaccumulate in the food chain or behave as pollutants (Khan et al., 2019). The release of gelatin ions, interactions with enzyme disulfide or sulfhydryl groups, DNA damage, and other cellular antioxidant activities of *Salmonella*, *Listeria*, *E. coli*, *S. aureus*, and *Bacillus cereus* are disrupted (Peighambardoust et al., 2019, Socas-Rodríguez et al., 2020). Silver-chitosan-starch inhibit the growth of *E. coli* and *S. aureus* (Yu et al., 2022). Silver-chitosan-starch nanocomposite inhibit the growth of *E. coli* and *S. aureus* (Yu et al., 2022), Silver-PVA-typhimurium based composite (Kumar et al., 2017) release ion and by ROS production and membrane failure disrupt *Salmonella typhimurium*, *E. coli*, and *S. aureus* (Bar-Ilan et al., 2009), TiO_2 enriched chitosan nanocomposite inhibit the growth of *E. coli* and *S. aureus*, *Bacillus* sp., *Lactobacillus platarum*, *Erwinia caratovora*, *Pichia jadini*, and *Pseudomonas fluorescens* are inactivated by polypropylene-TiO_2 nanocomposites (Peighambardoust et al., 2019; Osman et al. 2010), TiO_2-LDPE nanocmpositeprevent the growth of *E. coli* and *Pseudomonas* spp. (Kiss et al., 2008), CuO-LDPE disrupt key bacterial enzymes. Proton motive force, electron movement, energy control, or structural component production are all disrupted when essential oils (EO) and their active components interact with proteins and enzymes in the cell membrane of *E. coli* and *S. aureus* (Hua et al., 2021; Pavlicek et al., 2021; Van der Meulen et al., 2014).

The reaction mechanism of the silver-titanium dioxide (Ag/TiO$_2$) nanocomposite films with food matrix has been given in (Figure 12.2). The composite films provide antibacterial protection for carbohydrate, fat, or protein content of dietary products via photocatalytic interaction with biodegradable and active polymeric matrices for food packaging TiO$_2$/AgNPs. These nanoparticles release particular active species over sunlight (Rahman et al., 2022). A variety of metal oxides have been shown to exhibit antibacterial characteristics, including MgO, ZnO, and TiO$_2$. Antibacterial packaging films may be manufactured using these oxides because of their excellent stability and strong antimicrobial properties (Zhang et al., 2010). The antimicrobial mechanism of these oxides is related to the formation of very reactive oxygen species when exposure to UV radiation, wherein these oxides exhibit photocatalytic activity (Dong et al., 2013; Applerot et al., 2009). Zinc oxide nanoparticle-coated glass has been shown to have antibacterial properties against both gram-positive and gram-negative microorganisms (Applerot et al., 2009). Antibacterial activity against several germs or microbes is one of the distinctive qualities of TiO$_2$ that has made this substance particularly valuable in a wide range of industries, including paint, cosmetics, food, and food packaging products (Fujishima et al., 2000). Still, owing to its low photon-utilizing capacity and the necessity for an excitation source, this technology has limited practical utility (UV light). The antibacterial property of TiO$_2$ can be improved by doping it with metal oxides (SnO$_2$, Ag, or Fe$_3$)

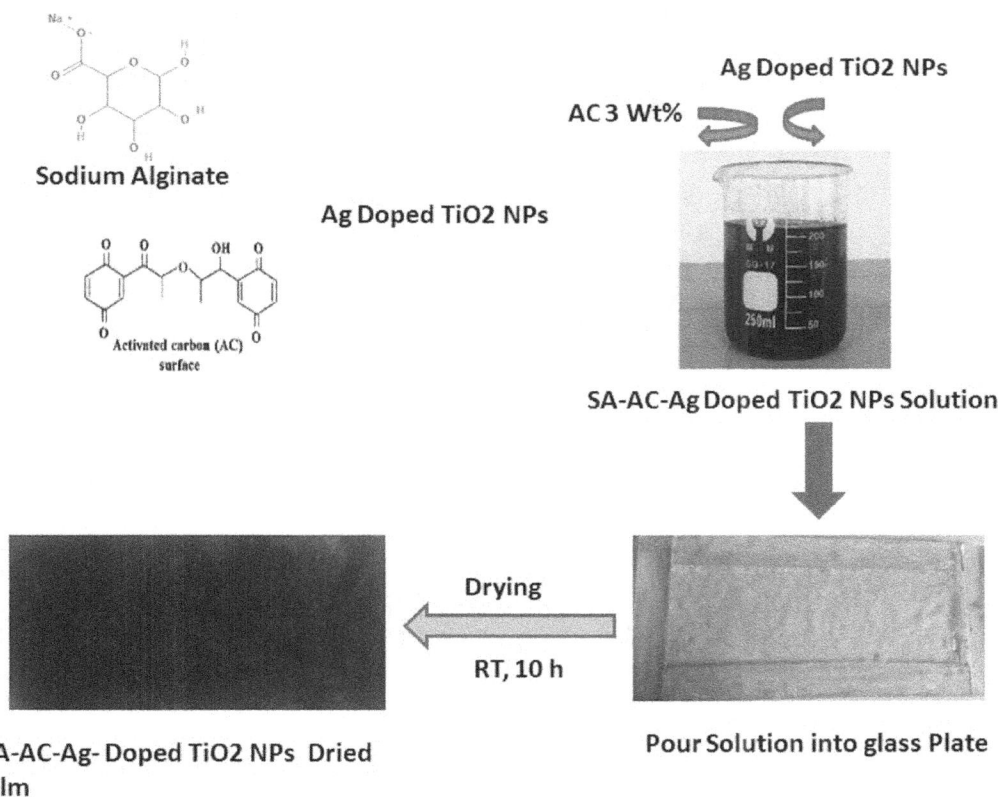

Sodium Alginate

Ag Doped TiO2 NPs

Activated carbon (AC) surface

AC 3 Wt%

Ag Doped TiO2 NPs

SA-AC-Ag Doped TiO2 NPs Solution

Drying

RT, 10 h

Pour Solution into glass Plate

SA-AC-Ag- Doped TiO2 NPs Dried Film

FIGURE 12.1 Sodium alginate-based silver-doped titanium nanocomposite film.

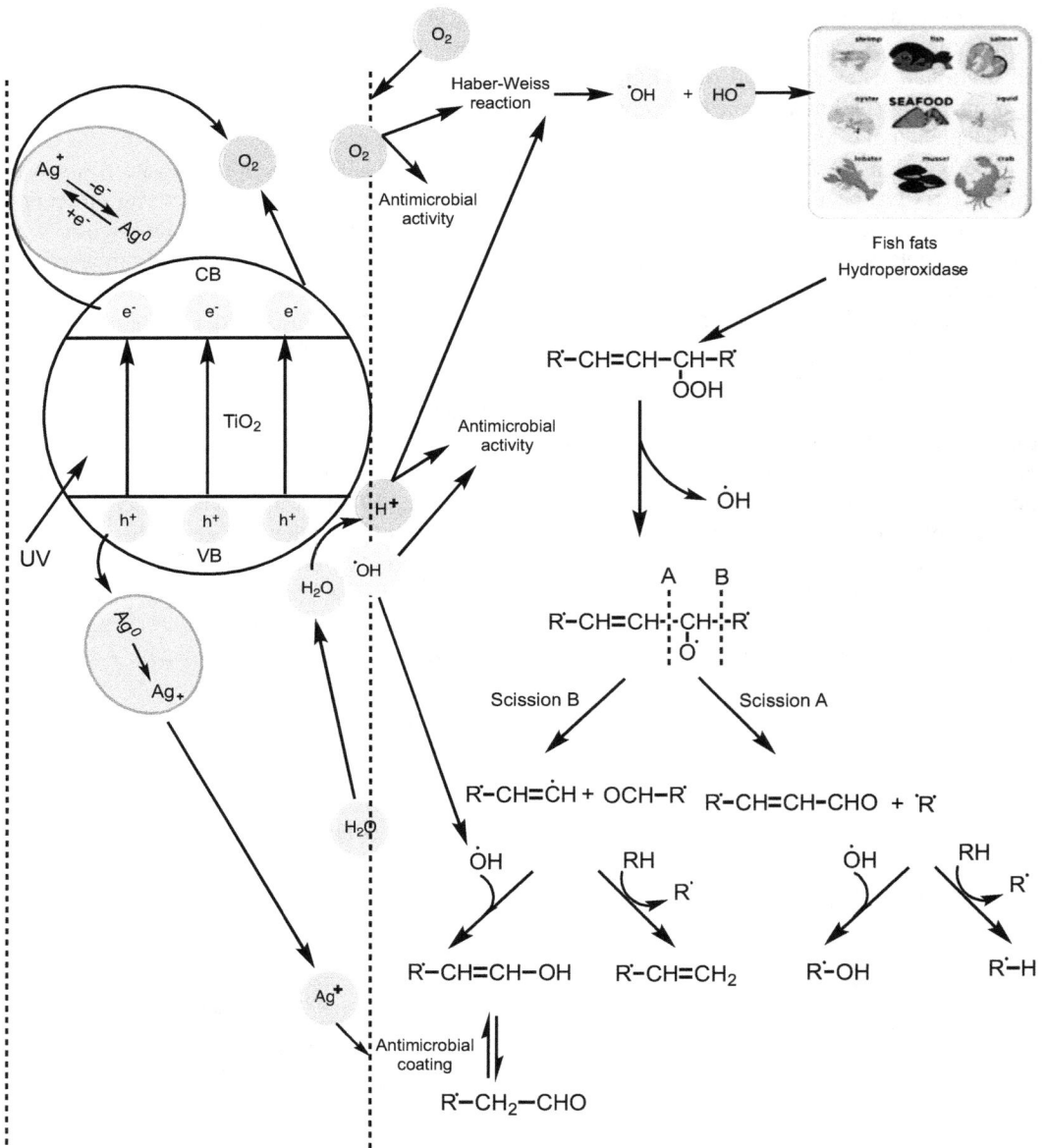

FIGURE 12.2 Nanocomposites (Ag-dopped-TiO$_2$) reaction with polymer food matrices (Rahman et al., 2022).

or metal ions (such as SnO$_2$, Ag, or Fe^{3+}) (Zhang et al., 2008). Microorganisms that cause food poisoning are better protected by polymeric packaging made of TiO$_2$-based oxides. Even when subjected to light with shorter wavelengths near the UV zone, it retains its stain, allergy, odor, and degradation resistance qualities. TiO$_2$ thin film may be prepared using a variety of methods, including magnetron sputtering, chemical vapor deposition, chemical spray pyrolysis, sol-gel process, evaporation, and electrodeposition. Using the sol-gel approach, a thin coating of very pure composite material may be produced in a single operation (Celik et al., 2006). Adding titanium dioxide (TiO$_2$) to the packing material increased mechanical and chemical stability in both the IR and visible ranges. These materials are also more biodegradable.

12.4 ALGINATE-BASED NANOCOMPOSITES APPLICATIONS IN SMART TECHNOLOGY

A growing trend in the food industry and the use of intelligent packaging (IP) and active packaging (AP) is increasing day by day. Through absorbing and diffusion systems, AP helps to extend the shelf life of food products. There are several important factors to consider when producing a food product, including cost, marketability, consumer acceptability, food safety, and organoleptic quality. Alginates may provide answers to many of the problems that are currently afflicting humanity and will likely occur in the near future. The biocompatibility of alginates, along with their availability and simplicity of preparation, play significant roles in their widespread use. Research into novel alginate-based materials with expanded capabilities and clearly defined features is necessary to meet the needs of specific application domains. The use of food packaging is one of the most important strategies for preserving and maintaining the quality of food. Food companies are rapidly incorporating new technology into their operations, such as intelligent packaging (IP) and AP. The use of IP systems in the food packaging environment improves the level of safety and the ability to issue warnings about potential risks. In 2024, the packaging industry in the United States is predicted to produce $6 billion in AP income and $3.45 billion in IP revenue, according to forecasts from the International Packaging Institute (Nano-enabled Packaging Market report). If existing packaging technologies (AP and IP) are to be used properly in the food industry, there are knowledge gaps that must be fulfilled.

12.4.1 Intelligent Packaging Technology

Intelligent packaging technology is a type of packaging that is designed to be environmentally friendly. IP is a new technology in the food packaging field. It has the potential to significantly enhance the traceability, safety, and nutritional value of food. Specifically, IP is defined as the science and technology that introduces communication tools for a food packaging system to monitor changes in the internal and external environmental conditions of the system, as well as changes in the packaged food, in order to communicate the status of a system to stakeholders in supply chains, such as producers and retailers, as well as consumers (Yam, 2012). IP supports decision-making, improves safety, and contributes to the improvement of quality by giving information, and alerting users to possible issues. An important feature of IP is the use of sensing and identification systems to provide information to the public about the composition of foodstuffs (Dekker et al., 2014).

12.4.1.1 Time Temperature Indicator (TTI)

Indicator films suitable for use in intelligent packaging have been fabricated using exclusively natural biopolymers. Simple and unobtrusive technologies, such time temperature indicators, can be put into intelligent packaging to alert consumers to any issues with the cleanliness and freshness of the products they are purchasing. Monitoring the long-term impact of temperature on food quality is done with time temperature indicators (TTIs). Food applications use both diffusion-based and microbiological TTI. In order to maximize the use of TTIs, it appears that an accurate model for correlating temperature changes with the rate at which a product rots is needed. Commercially available TTIs (9860 A-9860H, 9861A, and 9864C) have a wide range of time-temperature thresholds to meet the needs TTI indicators require specified activation energy (E_a) value in the diffusible material in order to work effectively.

The deterioration of food quality necessitates the rapid use of TTI measures. The TTIs are built on the principle of lactic acid vapor propagation. Lactulose is a monoprotic compound whose hydrogen atom is separated from the parent molecule by the acid's hydrogen bond. When hydronium reacts with the dye (In) mixture in a gel matrix, the acid form of the dye (HIn) is formed. The enzymatic response, that is, color change, was shown to be associated with the cumulative effects of the time-temperature variations on the food product under investigation during storage. Hsiao and Chang (2017) also created a microbial TTI for vacuum-packed grouper fish fillets, which was published in *Science Advances*. A flexible microbial TTI system was developed by Mataragas and colleagues (2019) using the microbe *Janthinobacterium* sp., which generates violet pigment through the synthesis of violacein during early development and may be used in a variety of applications. They used luminosity (L* factor) and Baranyi model to estimate the TTI endpoint.

Chen, Wang, and Jan (2014) suggested a low-cost colorimetric sensor array for detecting the freshness of chicken meat in a laboratory setting. During the next several years, it is projected that smart barcodes will become increasingly popular. The Japanese business TO-GENKYO has created a smart barcode that can be used to monitor the quality of meat. The sensors array had its own colorific fingerprint, which corresponded to the freshness of the chicken samples tested. TTIs based on diffusion are designed to operate across a wide range of temperatures and have a reasonably straightforward manufacturing process. However, the exudation of colored material has an influence on the accuracy and safety of such TTIs. Tekin and his college develop film by adding an extract of red beetroot (*Beta vulgaris* L.) that is rich in betalain to a mix film formed of alginate (A) and polyvinyl alcohol. This will result in the production of film (PVA). After spending a week being stored at temperatures of 4, 25, 40, and 60°C, TTI film was put through a series of tests to see how accurately it indicated color (Tekin et al., 2020).

12.4.1.2 Gas Indicator

Gas indicators can monitor alterations of metabolites such as CO_2, NH_3, H_2S, and trimethylamine (C_3H_9N) as quality indices. They are able to accurately monitor quality changes of food which are not visible to the consumer. The main problem of using pH-sensitive dyes in the packaging industry is that the dyes are strongly hydrophilic. Lyu et al. (2019) developed a (BTB)/tetrabutylammonium (TBA+) ion-paired dye and produced a CO_2-sensitive packaging film. During storage, metabolites like amines, aldehydes, ketones, and esters are produced as well as other low molecular weight substances responsible for off-flavors and sensory product rejection. Lee et al. (2019a) developed a freshness indicator for detecting chicken breast spoilage. The system was able to classify the samples into three categories: fresh, medium fresh, and spoiled. As the total volatile basic nitrogen (TVBN) content and CO_2 increased due to the proteins breakdown by the proteolytic enzymes produced by microorganisms, the color change from yellow to green suddenly occurred at day 7 and 4 at storage temperatures of 4°C and 10°C. The reduction reaction of platinum or palladium compounds can be applied for monitoring of ethylene gas. In such a system, molybdenum oxides are also used. This type of indicator is sensitive to UV light, other reducing materials, humidity, and temperature. Sodium alginate with calcium chloride can form ionic crosslinks. The pH indicator held up well against repeated washings even in moist conditions. Within a minute of introducing benzoquinone, the film's hue shifted from one extreme to the other at a concentration of 10–20 ppm of gases. Due to proton mobility, the color of colorimetric sensors changed from blue to yellow or yellow to blue when they came into contact with harmful gases revealed by MFC. Benzoquione facilitates proton transport. After 10 seconds, when the film came into touch with the hazardous vapors, the transformation began. In order to function properly, humidity is crucial. When humidity levels were high, its reaction time sped up.

Kim developed a dip-coated pH indicator-based film sensors. The main substrate was non-woven fabric wiper (WW-2109), and the polymeric binder was sodium alginate. Electron transfer agent benzoquinone improved reactivity. First, they mixed sodium alginate (1%), benzoquinone, and pH indicator (0.1 wt%). The solution was used to impregnate prepared films. Then they were impregnated with 10% calcium chloride. After making sensor, mass flow controller exposed harmful gases (20, 10 ppm) (MFC). Spectrophotometer was used to chart E*a*b* every 5 seconds to monitor color change (Kim et al., 2016).

12.4.1.3 Biological Sensor

Devices that can detect specific biological analyses and transform their presence or concentration into electrical, thermal, optical, or other signals are known as biosensors. The primary distinction between chemical sensors and biological sensors is found in the first component of the sensor, also known as the identification layer or the identification layer. A sensor that detects a change in the environment should mediate an irreversible reaction in order to be able to constantly monitor changes, which is especially important when the sensor is placed in a remote location. While the cost of the created biosensor should be reasonable, pricing is always a consideration. Some biosensors, in order to detect a spoiling point, must come into touch with the product's surface. The increasing need for packaged foods with a high level of safety has motivated academics and corporations to create biosensors in the intellectual property industry. The development of biosensors for the detection of food contamination, such as bacteria and poisons, has just recently begun. Research is needed to reduce the size of the sensor and simplify the detection methods. Yuan uses a biosensor based on sodium alginate, in which the synergistic increase in strength and flexibility is due to the strong hydrogen bond interact between graphene and SA. The proposed bio-based aerogels could be a strong contender in flexible strain sensors due to their combination of electronic conductivity and mechanical flexibility (Yuan et al., 2016). Kikuchi develop a sodium alginate-based biosensor to detect *E. coli* in human breast milk (Kikuchi et al., 2020).

12.4.1.4 Humidity Sensor

Electronic hygrometers are commonly used to make humidity sensors. Capacitance and/or resistance properties that may be measured changes in response to variations in relative humidity. In recent years, wheat gluten has been employed for monitoring of the packed food's relative humidity (Bibi et al., 2016). For their research in 2016, Gontard and Sorli used a thin wheat gluten protein as film as a model system. On the interdigital capacitor, sodium alginate a polymer that monitors relative humidity information and data collection (IDC) systems. In the case of packaged food goods, wheat gluten might be utilized to measure relative humidity levels. Using wheat gluten as a model, researchers investigated its capacity to interact with water molecules as well as its electrical and dielectric characteristics. The created system might be used in conjunction with radio frequency identification (RFID) systems to track and monitor the items' movement and condition. In order to securely employ humidity indicators in food packaging systems, researchers are concentrating on cobalt chloride-free alternatives (Konstantaki et al., 2006). It has also been developed and produced several metal chloride-based humidity indicators. The downside is that they are costly than cobalt (II) chloride and are also potentially hazardous in some situations (Qin et al., 2014).

Sodium alginate and carbonized lignin (CL) were used to make a humidity-sensitive composite film (SA). Repeatability, stability, and responserecovery of the CL/SA composite sensor at room temperature were studied. Maximum CL/SA composite film responsivity is 502,895.40% at 97% relative humidity. CL/SA composite film humidity sensors have ultrahigh sensitivity, minimal hysteresis, and consistent repeatability from 11% to 97% relative humidity (Yun et al., 2020).

12.4.1.5 Barcoding Technique

In the seventh decade of the 20th century, the barcoding technology or automated identification (Auto-ID) technique was developed to enable major retail and supermarket businesses. Characters and digits are used to encode barcodes on items. Scanners use optical barcodes to extract data from barcodes, which are subsequently sent to a centralized system. Barcodes are widely used these days due to their ease of use and low cost. One-dimensional (or conventional) barcodes typically consist of a series of narrow or wide black bars printed on a white backdrop. If the space between the bars is too large, then the bars are too close together. Using milelong lines, barcodes are measured in millimeters. Authenticity is ensured by measuring the blank distances to the right and left of the barcode character. There are several one-dimensional barcode formats that are widely used in the supermarket, including UPC, EAN, code 128, and code 39 (American barcode and RFID report, 2016). The 13 digits of one-dimensional food barcodes are typical. The first three digits of the barcode identify the country of manufacture (Manthou Vlachopoulou, 2001).

This barcode contains just around 2,953 bytes or 10–13 characters of information. A two-dimensional barcode format was born out of the necessity to store more information on the barcode

(Hadjila, Merzougui, & HadjIrid, 2018). Intermec (Everett, Washington, USA) created the first two-dimensional barcode format in 1988. This code's structure is made up of a series of white and black squares. A total of 4,296 alphanumeric and 7089 numeric characters can be stored in two-dimensional code. The most widely used two-dimensional codes are QR code, Data Matrix, and PDF 417. For the first time, a three-dimensional code called PM code was introduced in 2006 by content concept of Asia (Kuwana-shi in Mie-ken, Japan). It is, in fact, a three-dimensional QR code, with its third dimension being the color of the code. To acquire both the UPC code and the quality condition of food, Chen et al. (2017) proposed incorporating colorimetric dye sensor arrays into QR codes. A QR code that incorporates colorimetric dyes might be beneficial in tracking and evaluating meals. To help consumers to identify the quality, freshness level, or spoilage of the food that they purchase, smartphones, and mobile applications should be developed. Accurate food evaluation might be achieved through the use of machine learning algorithms in conjunction with picture processing techniques. However, the planned method was not implemented on a mobile app. To demonstrate the system's capacity to create complicated geometric patterns using a polymer mold, they simply built a mold featuring circles, squares, and triangles. A micro-pipette was used to drop three different kinds of dye-contained micro-beads onto the stamp designs. At long last, the tape with the micro-beads was fastened to the filter paper (as protective covering and a barrier). Some users, however, find it difficult to make use of this technology, raising concerns about its widespread adoption. Weber suggested an effective way of making barcodes using sodium alginate beds. Patient sera antibodies were used to examine the performance of barcoded beads (Weber et al., 2022).

12.4.1.6   Radio Frequency Identification (RFID)

To ensure the safety and quality of your food, RFID technology may be used to monitor the state of your products. RFID tags have a greater operating range than barcodes, and scanners may connect with several tags at once. RFID technology can make it easier to automate goods handling from warehouses to retail marketplaces. Products in the food industry and retail sector are fitted with an RFID tag (also known as an RFID label) for tracking purposes. Wireless signals are used to connect with the RFID reader on this gadget. Passive tags broadcast the signal in the same way as active ones, but they do so without relying on a battery. The sensor collects analogue data, which the CPU digitizes. The RFID tag may be found on the surface of a smart box called Mondi plus. NXP semiconductors (Eindhoven, Netherlands) has developed a wireless monitoring system for packaged food goods. For food manufacturers, intellectual property (IP) can help them better safeguard their products.

Organic and inorganic materials, as well as biosensors, may be harmful in TTIs, thus caution should be exercised while using them. Increased protection against data theft and counterfeiting is also necessary. System components used in AP include antioxidants, antimicrobials, and moisture controllers, as well as systems that absorb or emit CO_2, release or absorb taste, and absorb or release CO_2. Moisture may alter the texture, appearance, and microbiological activity of food products. We tested several methods for packaging and moisture absorber levels to see how they affected the shelf life of button mush. AP systems included various packaging films and four different moisture levels. Djatna suggested a nanocomposite of (Silver NPs—Reduced Graphene Oxide—Polyaniline) for use in RFID and its fabrication method (Djatna and Fahma, 2020).

In summary, the desire in developing new packaging systems owing to the safety and quality concerns along the food supply chain has pushed investigations on the IP systems and the commercialization of such innovations. Compared to previous approaches, IP has the capacity to give safer and higher quality food items to customers. Applying the IP systems such as TTIs, RFID tags, gas indicators, etc. can give solutions to food manufacturers to safeguard food goods in more convenient and secure methods. Nevertheless, the pricing of the marketed TTIs is rather pricey. The accuracy of TTIs is another challenge that needs additional study to fix. The TTIs including organic or inorganic elements as well as biosensors may cause toxicity issues owing to the migration into food. For this reason, toxicological studies should be established for such systems to alleviate safety worries of customers. The biggest challenge impacting the performance of the freshness indicators is responding to unrelated metabolites. The biosensors are unique concepts that were just presented in IP and more studies should be considered to address the hurdles in commercialization. The RFID systems are also pricey making this technology uneconomical for many food goods. Increasing the security of these systems against data theft and counterfeiting is also vital.

Further research on the IP advances will lead to further enhancement of the existing food packaging technologies, and developing new and safe mass production techniques of the IP systems will alleviate worries about the widespread usage of IP in the food business. AP technology is defined as "packaging in which ancillary components have been purposely added in or on either the packaging material or the package headspace to enhance the performance of the package system" (Janjarasskul & Suppakul, 2018; Robertson, 2012). The key AP systems include oxygen scavengers, ethylene absorbers, carbon dioxide absorbers/emitters, flavor releasing/absorbing systems, antioxidants, antimicrobials, and moisture controllers. The texture, appearance, and microbiological activity of food can all be affected by moisture. Preservatives, flavors, antimicrobials, and other additives can be found in gas absorbers and diffusion systems (emitters) now used in the AP industry. Recently, EO have been used to create active edible films with antibacterial and antioxidant characteristics (Al-Hashimi et al., 2020; Cai, Wang, & Cao, 2020; Li, Pei, Xiong & Xue, 2020).

The effectiveness of a multilayer film-based PET film with chitosan and alginate layers on shelf life extension and chicken breast meat quality was tested. It was determined that the AP film's ability to increase the shelf life of beef was due in part to its properties, especially its release rate of antimicrobial compounds. Citral (-CDCI) and transcinnamaldehyde (-CDCI) include bioactive chemicals (-CDTC). Li et al. (2019) AP films based on polypropylene were created and their antibacterial and antifungal properties examined for food applications. The development of edible chitosan-based antioxidative and antibacterial films for use in food packaging has resulted in better specifications. *Pistacia terebinthus* plant extracts were used to create CO_2 and moisture-absorbent pads that might preserve the quality of shiitake mushrooms Kaya et al. (2018). Sodium carbonate, calcium hydroxide, polyacrylate, and super absorbent polymer based Synergistic antibacterial polymer nanocomposites were created, and their ability to inhibit *Escherichia coli* growth in humus was established (Rhim et al., 2017). Thymol and carvacrol fish steaks in plastic pouches were tested for the effect of O_2 scavengers and antibacterial film on increasing their shelf life. A few drops of pure essential oil of ginger with a variety of packaging methods was used to keep pig patties from oxidizing during high-pressure processing, such as vacuum packing, oxygen scavenger packaging, and rosemary-based antioxidant (AP). Carnosic acid and the extract of rosemary based on polyamide an active antimicrobial packaging films have been developed by Bolumar, LaPena, Skibsted, and Orlien (2016) for a variety of fresh fruits and vegetables (Shemesh et al., 2016). Table 12.1 represent the change of indication factor over time for different food sample.

TABLE 12.1

The Indication of Smart Food Packaging in Different Food

Food Sample	Indication Type	References
Shrimp	pH	Liu et al. (2017)
Fish	Color	Zeng et al. (2019)
Milk	Color	Liu et al. (2017)
Dessert	pH	Nopwinyuwong et al. (2010)
Bacon	Oxygen indicators	Mills (2005)
Meat and apple	Enzyme	Liu et al. (2017)
Guava	pH	Kuswandi et al. (2013)
Chicken breast and silver carp	Gas	Zhai et al. (2020)
Green bell	pH	Chen et al. (2020)
Cereal	Circuit sensor	(Wang et al., 2015)

Research on the niger oxygen scavengers embedded in a PET matrix was tested for their effect on methanol oxidation in wine solutions at varying starting oxygen concentrations (1% and 3%). Oxygen scavengers designated 1osPET and 3osPET that are integrated into a PET matrix. In addition to Dombre et al. (2015), the quality attributes of button mushroom silica gel and spongy foam were evaluated using several AP systems, including diverse packaging films (clear PVC box and stretch PVC), and four degrees of moisture absorber at varying storage time levels. Karimi and Mosharraf are two of the most prominent figures in Iranian politics (2015). In order to find the best shelf-life conditions for button mushrooms, multiple AP systems, comprising various packaging films (stretch PVC and biaxially oriented polypropylene), as well as four levels of moisture absorber, were tested (Karimi et al., 2014). For the research of bread rotting induced by *Aspergillus niger*, active starch-clay nanocomposites were synthesized and studied for their physical and mechanical characteristics. Potassium sorbate in combination with cinnamon oil. Three Barzegars and one Azizi are included in this. Polypropylene ethylene vinyl alcohol copolymer bags with antibacterial characteristics were created and tested for salad packing. A blend of citral and oregano oil ($C_{10}H_{16}O$).

Protein isolate and gum acacia conjugates (via the Maillard reaction) were used to make emulsion-based antibacterial edible films (Muriel-Galet et al., 2012). Grapefruit essential oil was also encapsulated. Films' mechanical capabilities, water barrier qualities, surface hydrophobicity, thermal stability, and *E. coli* inhibitory action were examined. Various regulatory bodies, such as the European Food Safety Authority (European Food and Drug Administration in the United States), have regulations and standards that must be met while creating and manufacturing active systems (Restuccia et al., 2010). To conclude, sodium alginate-based nanocomposite would be a possible solution for smart packaging of food to retain its functional and nutritional quality.

12.5 ALGINATE-BASED NANOCOPOSITES APPLICATIONS IN INTELLIGENT FOOD PACKAGING MATERIALS

Studies have shown that sodium alginate nanocomposite is a potentially useful material for food packaging. Improved barriers; mechanical, thermal, and biodegradable qualities; and uses in active and intelligent food packaging are just a few of the ways its usefulness in food packaging has been demonstrated. However, the decreased particle size of nanomaterials and the fact that the chemical and physical characteristics of such nano-sized materials may be substantially different from those of their macroscale equivalents may provide difficulties for the use of nanocomposites in food packaging. Nanoparticles' potential risks to consumers need to be discussed, and a detailed examination into their properties is required, in addition to the quantification of data. To demonstrate the possibilities of intelligent packaging materials in the foodservice industry, we have included a number of examples of their use in the following section:

Food packaging is crucial important. The materials needed to package the food have tremendous effect on the safety of the food. In this section, we have discussed how and which materials could be chosen to package the different kinds of food.

12.5.1 SEAFOOD

Fish, shrimp, crabs, and lobsters, which include long-chain polyunsaturated fatty acid (PUFA), vitamins, and minerals, are a good source of nutrients for humans (Mohammadalinejhad, Almasi, & Moradi, 2020). Due to their short shelf lives, these foods are especially susceptible to chemical and microbiological degradation while being stored. The development of intelligent packaging materials to monitor the freshness, quality, and safety of seafood products over the lifespan of an individual is therefore of considerable interest. Among the many reasons why seafood deteriorates is the production of spoilage products by bacteria that are mostly volatile components such as ammonia, methylamine, dimethylamine, trimethylamine, and other related chemicals, together referred to as TVB-N products. As a result of the production of these substances, the pH of stored seafood products is altered. It's for this reason that many researchers are working to develop smart packaging materials that are able to detect changes in the quality of seafood while it's being stored. Anthocyanin extracts from mulberries (pH sensors) have been trapped in gelatin/PVA films designed to monitor the quality of fish (Zeng et al., 2019). A change in color from purple to grayish purple to dark green in the film was a sign of fish spoilage, according to the authors.

To monitor changes in shrimp quality during storage, curcumin and sulfur nanoparticles were incorporated into the pectin matrix, and a pH-responsive film was then formed (Ezati & Rhim, 2020). A pH-responsive color change in the film was observed as the shrimp's quality improved. Curcumin and nanoparticles in the film were also found to have antimicrobial activity against *E. coli* and *L. monocytogenes*. Volatility sensors embedded in food packaging have been used to monitor fish spoilage, which were reported to give a linear response to ammonium concentrations of 1.5 ppm (Bhadra, Narvaez, Thomson, & Bridges, 2015).

12.5.2 MEAT

In the same way as fish produces large quantities of TVB-N and pH changes under the influence of microbes, meat does the same. Thus, smart packaging materials that monitor changes in meat quality during storage may be utilized in the same way for fish products. For example, pH-sensitive dyes in packing materials can be used to gauge meat decomposition during storage (Zhang et al., 2019). Other microorganism-produced volatile chemicals, such as hydrogen sulfide and carbon dioxide, can also be detected by intelligent food packaging materials. Silver nanoparticles coated with gellan gum have been used to monitor the spoilage of chicken breast meat in real time using a colorimetric hydrogen sulfide (H_2S) sensor (Zhai et al., 2019).

Using the strong binding affinity of Ag for H_2S, this colorimetric sensor produces a color change from yellow to colorless. Microarrays of an *E. coli*-specific RNA-cleaving fluorogenic DNAzyme probe have been covalently attached to food packaging materials for real-time monitoring of *E. coli* contamination during the storage of raw beef (Youssef & El-Sayed, 2018). As pathogenic microorganisms are used to produce this type of intelligent food packaging, it may help to reduce the prevalence of foodborne diseases. Calcium alginate nanocomposite lowered shrinkage loss, drip and degree of off odor in beef, and also prolonged the muscle color (Williams et al., 2007). Gheorjhita utilized sodium alginate polymers, but they have since found widespread use in the food business, particularly the meat processing sector. Sodium alginate films are used to wrap meat items so that they don't lose weight or become ruined by light and oxygen (Gheorjhita et al., 2020). Marcos and others prepared alginate nanocomposite that delay the growth of *L. monocytogenes* in cooked ham (Marcos et al., 2007).

12.5.3 DAIRY PRODUCTS

Milk, cheese, and other dairy products are among the most widely consumed foods in the world. Milk is an excellent source of protein, fats, vitamins, minerals, and other trace elements (Roy, Priyadarshi, Ezati, & Rhim, 2022; Roy & Rhim, 2021). It is widely available and enjoyed over the world. When it comes to storing pasteurized milk, it is susceptible to spoiling (Aloui & Khwaldia, 2016). It is possible to monitor milk quality in the supply chain using intelligent food packaging in addition to the usual packaging function. While pH, microbial concentration, and gas concentration will all fluctuate as milk spoils, the indicator effect is currently achieved primarily through the monitoring of pH changes during storage. The freshness of milk was checked by the color change in the packaging, which was made using pH-sensitive anthocyanins and pigments from plants (Li, Wu, Wang, & Li, 2021; Moazami Goodarzi et al., 2020). The freshness of milk may be monitored by adding shikonin to gelatin/carrageenan-based films (Roy & Rhim, 2021). Shikonin also provides antibacterial and antioxidant characteristics to the packaging. Milk quality can also be monitored by the use of various index changes in food packaging.

In order to monitor volatile organic compounds produced by decaying bacteria in pasteurized milk and predict the shelf life of milk, Ziyaina et al. developed a colorimetric sensor based on SiO_2 nanoparticles and Schiff's reagent (Ziyaina, Rasco, Coffey, Ünlü, & Sablani, 2019). Another colorimetric sensor for intelligent milk packaging uses polydiacetylene/zinc oxide, which changes color depending on the change in lactic acid concentration during milk degradation. In addition to milk, we consume a lot of cheese, which contains a variety of nutrients including fat, vitamins, and inorganic salts. For the purpose of monitoring cheese quality, Bandyopadhyay and colleagues developed pH stickers based on essential oil and anthocyanins, which not only ensured that cheese was safe to eat but also protected the cheese's flavor from being damaged, making it easier for consumers to accept it.

12.5.4 FRUITS, VEGETABLES, AND OTHER FOODS

Food bundling materials for natural goods, veggies, and a variety of other food items have also been developed with innovative thinking. A colorimetric film containing bromophenol blue (sensor) and embedded in a bacterial cellulose matrix, for example, was used to determine the freshness of the fruit of the guava tree (Kuswandi, Maryska, Jayus, Abdullah, & Heng, 2013). As the guava became overripe, the color of the film changed from blue to green, which was attributed to the production of unstable natural atoms (such as acidic corrosive) that decreased the pH of the film. To determine the freshness of green pepper, a pH-sensitive pointer named combining methyl red and bromothymol blue (sensors) has been utilized to measure the pH of the pepper (Chen et al., 2020). When the pH is modified, this name undergoes a shading shift, which may be attributed to a variation in the carbon dioxide level in the bundle, which is linked to the deterioration of the green peppers (thus the name "green peppers"). Furthermore, pH-delicate color marker names have also been used for the continual quality monitoring of Thai pastry "brilliant droplets" to ensure that they are of the highest possible quality (Nopwinyuwong, Trevanich, & Suppakul, 2010). Gheorjhita utilized sodium alginate polymers film as coverings for fresh fruits and vegetables (Gheorjhita et al., 2020). Sodium alginate with silver nanocomposite increase shelf life of carrot and pear (Fayaz et al., 2009).

12.6 BIONANOCOMPOSITES AND SAFETY

Due to a lack of scientific study, it is difficult to assess the hazards associated with the use of bionanoparticles such as nanomaterials, and numerous nanoparticles that are physiologically active in the human body when dispersed elsewhere in the environment. There is some debate as to whether or not nanomaterials with very small particle sizes are more likely to facilitate the transfer of nanomaterials throughout the human body than those with larger particle sizes because of the smaller particle size, but nanomaterials with large surface areas have a higher activity, which in turn allows for greater interaction through cellular membranes, in addition to better ability (Ray et al., 2009). It has been documented in the literature that nanoparticles migrate from the packaging to the meal (Boyce et al., 2011). Food packed with polymer nanocomposite packaging materials is a major concern for customers, who are concerned about the impact of nanoparticles on human health from the mouth all the way to the intestines if they are ingested orally (Silvestre et al., 2011). We need to know how nanomaterials affect performance in the human body if we are to use them effectively. A nanocomposite based on polymer and nanoparticles was shown to slow the flow of some potentially dangerous substances, such as triclosan and caprolactam, into the food six times over (de Abreu et al., 2010).

12.7 IMPACT OF USING BIONANOCOMPOSITES ON HUMAN HEALTH

In most cases, polymer nanocomposite's beneficial qualities are well known, but the potential (eco-) toxicological features of nanomaterials, aside from their effects on human health, have received little attention up to this point. The main obstacle to employing nanocomposites is the rapid spread of nanomaterials-based consumer items, which highlights the need for better consideration of the probable biological impacts of nanoparticles (Bouwmeester et al., 2009). The effects of nanoparticles on the human body and the environment are becoming more well known in recent years. Using nanomaterials can lead to new allergies and dangerous strains, as well as an increase in the adsorption of nanoparticles by the environment.

12.8 FUTURE TRENDS AND CONCLUSION

Bionanocomposites (like alginate-nanocomposites) represent an inspiring route for creating new and innovative packaging materials. By adding appropriate nanoparticles such as zinc oxide (ZnO-NPs), titanium dioxide (TiO_2-NPs montmorillonite (MMT), and silver-titanium nanoparticles (Ag-NPs and Ti-NPs), it will be probable for fabricating films for packaging have good mechanical, barrier, and thermal performance. They can reduce flammability significantly and maintain the transparency of the polymer matrix. Inserted nanomaterials into packaging materials as nanosensors will alert the customer if food has gone deteriorated. Alginate-nanocomposites are a viable technology for "new" materials for near-future packaging applications, especially for flexible, fire resistance, antimicrobial, and transparent barrier packaging films. In conclusion, the area of bionanocomposites as packaging materials still needs scientific research and improvement in order to develop the shelf life, quality, and marketability of diverse packaging materials. What is possible in the near future, design and incorporation of various required functionalities? Antimicrobial, antibiotic, biodegradable, and combination reactions to environmental or chemical changes, for example.

Research on intelligent packaging (IP) still formative; however, it is intended that further research in this field would lead to more inexpensive, convenient, and environmentally friendly packaging options that will improve the food supply in terms of its nutritional value, safety, and long-term resilience. Moreover, bio-based nanostructure polymer nanocomposites might be developed in the near future. Look at nature's examples of packaging, skins, and structures with specific processes, and imagine if we could make synthetic equivalents.

ACKNOWLEDGEMENTS

This Work was funded by the Ministry of Science and Technology MoST), National Science and Technology (NST) Fellowship. Support from the Laboratory of Materials, Energy and Nanotechnology MENTECH) Labs Under the Grant number BAS-USDA-029.

REFERENCES

Abebe, B., Zereffa, E. A., Tadesse, A., & Murthy, H. C. A. (2020). A review on enhancing the antibacterial activity of ZnO: Mechanisms and microscopic investigation. *Nanoscale Research Letters*, *15*, 190.

Al-Hashimi, A. G., Ammar, A. B., Lakshmanan, G., Cacciola, F., & Lakhssassi, N. (2020). Development of a millet starch edible film containing clove essential oil. *Foods*, *9*, 184. https://doi.org/10.3390/foods9020184

Aloui, H., & Khwaldia, K. (2016). Natural antimicrobial edible coatings for microbial safety and food quality enhancement. *Comprehensive Reviews in Food Science and Food Safety*, *15*(6), 1080–1103. https://doi.org/10.1111/1541-4337.12226

Alves, D., Cerqueira, M. A., Pastrana, L. M. & Sillankorva, S. (2020). Entrapment of a phage cocktail and cinnamaldehyde on sodium alginate emulsion-based films to fight food contamination by *Escherichia coli* and *Salmonella Enteritidis*. *Food Research International*, *128*, 108791.

Applerot, G., Lipovsky, A., Dror, R., Perkas, N., Nitzan, Y., Lubart, R., & Gedanken, A. (2009). Enhanced antibacterial activity of nanocrystalline ZnO due to increased ROS-mediated cell injury. *Advanced Functional Materials*, *19*(6), 842–852.

Augst, A. D., Kong, H. J., & Mooney, D. J. (2006). Alginate hydrogels as biomaterials. *Macromolecular Bioscience, 6*(8), 623–633. https://doi.org/10.1002/MABI.200600069

Bar-Ilan, O., Albrecht, R. M., Fako, V. E., & Furgeson, D. Y. (2009). Toxicity assessments of multisized gold and silver nanoparticles in zebrafish embryos. *Small, 5,* 1897–1910.

Bhadra, S., Narvaez, C., Thomson, D. J., & Bridges, G. E. (2015). Non-destructive detection of fish spoilage using a wireless basic volatile sensor. *Talanta, 134,* 718–723. https://doi.org/10.1016/J.TALANTA.2014.12.017

Bibi, F., Guillaume, C., Vena, A., Gontard, N., & Sorli, B. (2016). Wheat gluten, a biopolymer layer to monitor relative humidity in food packaging: Electric and dielectric characterization. *Sensors and Actuators A: Physical, 247,* 355–367. https://doi.org/10.1016/j.sna.2016.06.017

Bolumar, T., LaPena, D., Skibsted, L. H., & Orlien, V. (2016). Rosemary and oxygen scavenger in active packaging for prevention of high-pressure induced lipid oxidation in pork patties. *Food Packaging and Shelf Life, 7,* 26–33. https://doi.org/10.1016/j.fpsl.2016.01.002

Boontheekul, T., Kong, H. J., & Mooney, D. J. (2005). Controlling alginate gel degradation utilizing partial oxidation and bimodal molecular weight distribution. *Biomaterials, 26*(15), 2455–2465. https://doi.org/10.1016/J.BIOMATERIALS.2004.06.044

Bouwmeester, H., Dekkers, S., Noordam, M. Y., Hagens, W. I., Bulder, A. S., De Heer, C., … & Sips, A. J. (2009). Review of health safety aspects of nanotechnologies in food production. *Regulatory Toxicology and Pharmacology, 53*(1), 52–62.

Boyce, J. A., Assa'ad, A., Burks, A. W., Jones, S. M., Sampson, H. A., Wood, R. A., … & Schwaninger, J. M. (2011). Guidelines for the diagnosis and management of food allergy in the United States: summary of the NIAID-sponsored expert panel report. *Journal of the American Academy of Dermatology, 64*(1), 175–192.

Butkus, M. A., Edling, L., & Labare, M. P. (2003). The efficacy of silver as a bactericidal agent: Advantages, limitations and considerations for future use. *Journal of Water Supply: Research and Technology AQUA, 52*(6), 407–416.

Cai, L., Wang, Y., & Cao, A. (2020). The physiochemical and preservation properties of fish sarcoplasmic protein/chitosan composite films containing ginger essential oilemulsions. *Journal of Food Process Engineering, 43*(10), e13495. https://doi.org/10.1111/jfpe.13495

Chen, C., Du, Y., Zuo, G., Chen, F., Liu, K., & Zhang, L. (2020). Effect of storage condition on the physicochemical properties of corn-wheat starch/zein edible bilayer films. *Royal Society Open Science, 7*(2), 1917777. https://doi.org/10.1098/rsos.191777

Chen, Y., Fu, G., Zilberman, Y., Ruan, W., Ameri, S. K., Zhang, Y. S., Miller, E., & Sonkusale, S. R. (2017). Low cost smart phone diagnostics for food using paperbased colorimetric sensor arrays. *Food Control, 82,* 227–232. https://doi.org/10.1016/j.foodcont.2017.07.003

Chen, Y.-Y., Wang, Y.-J., & Jan, J.-K. (2014). A novel deployment of smart cold chain system using 2G-RFID-Sys. *Journal of Food Engineering, 141,* 113–121. https://doi.org/10.1016/j.jfoodeng.2014.05.014

Daemi, H. & Barikani, M. (2012). Synthesis and characterization of calcium alginate nanoparticles, sodium homopolymannuronate salt and its calcium nanoparticles. *Scientia Iranica, 19*(6), 2023–2028. http://dx.doi.org/10.1016/j.scient.2012.10.005.

Dakal, T. C., Kumar, A., Majumdar, R. S., & Yadav, V. (2016). Mechanistic basis of antimicrobial actions of silver nanoparticles. *Frontiers in Microbiology, 7,* 1831.

Davis, T. A., Llanes, F., Volesky, B. and Mucci, A. (2003). Metal selectivity of Sargassum spp. and their alginates in relation to their α-L-guluronic acid content and conformation. *Environmental Science & Technology, 37*(2), 261–267.

da Silva, T. L., Da Silva Junior, A. C., Vieira, M. G. A., Gimenes, M. L., & Da Silva, M. G. C. (2014). Production and physicochemical characterization of microspheres made from sericin and alginate blend. *Chemical Engineering Transactions, 39*(Special Issue), 643–648. https://doi.org/10.3303/CET1439108

da Silva, T. L., Vidart, J. M. M., da Silva, M. G. C., Gimenes, M. L., & Vieira, M. G. A. (2017). Alginate and sericin: Environmental and pharmaceutical applications. *Biological Activities and Application of Marine Polysaccharides.* https://doi.org/10.5772/65257

de Abreu, D. A. P., Cruz, J. M., Angulo, I., & Losada, P. P. (2010). Mass transport studies of different additives in polyamide and exfoliated nanocomposite polyamide films for food industry. *Packaging Technology and Science: An International Journal, 23*(2), 59–68.

Djatna, T., & Fahma, F. (2020). A synthesis of AgNP-rGO-PANI nanocomposite and its use in fabrication of chipless RFID sensor: Current research progress. In *IOP Conference Series: Earth and Environmental Science* (Vol. 472, No. 1, p. 012027). IOP Publishing.

Dombre, C., Guillard, V., & Chalier, P. (2015). Protection of methionol against oxidation by oxygen scavenger: An experimental and modelling approach in wine model solution. *Food Packaging and Shelf Life, 3,* 76–87. https://doi.org/10.1016/j.fpsl.2015.01.002

Donati, I., Holtan, S., Mørch, Y. A., Borgogna, M., Dentini, M., & Skjåk-Bræk, G. (2005). New hypothesis on the role of alternating sequences in calcium-alginate gels. *Biomacromolecules*, 6(2), 1031–1040. https:// doi.org/10.1021/BM049306E

Dong, H., Snyder, J. F., Tran, D. T., & Leadore, J. L. (2013). Hydrogel, aerogel and film of cellulose nanofibrils functionalized with silver nanoparticles. *Carbohydrate Polymers*, 95(2), 760–767.

Draget, K. I., & Taylor, C. (2011). Chemical, physical and biological properties of alginates and their biomedical implications. *Food Hydrocolloids*, 25(2), 251–256. https://doi.org/10.1016/J. FOODHYD.2009.10.007

Draget, K. I., Smidsrød, O., & Skjåk-Bræk, G. (2005). Alginates from algae. Polysaccharides and polyamides in the food industry: properties, production, and patents, 1–30.

Eiselt, P., Lee, K. Y., & Mooney, D. J. (1999). Rigidity of two-component hydrogels prepared from alginate and poly(ethylene glycol)-diamines. *Macromolecules*, 32(17), 5561–5566. https://doi.org/10.1021/ MA990514M

Fan, L., Jiang, L., Xu, Y., Zhou, Y., Shen, Y., Xie, W., ... & Zhou, J. (2011). Synthesis and anticoagulant activity of sodium alginate sulfates. *Carbohydrate Polymers*, 83(4), 1797–1803.

Fayaz, A. M., Balaji, K., Girilal, M., Kalaichelvan, P., & Tvenkatesan, R. (2009). Mycobased synthesis of silver nanoparticles their incorporation into sodiumalginate films for vegetable fruit preservation. *Journal of Agricultural and Food Chemistry*, 57, 6246–6625.

Feng, Q. L., Wu, J., Chen, G. Q., Cui, F. Z., Kim, T. N., & Kim, J. O. (2000). A mechanistic study of the antibacterial effect of silver ions on *Escherichia coli* and *Staphylococcus aureus*. *Journal of Biomedical Materials Research*, 52(4), 662–668.

Fujishima, A., Rao, T. N., & Tryk, D. A. (2000). Titanium dioxide photocatalysis. *Journal of Photochemistry and Photobiology C: Photochemistry Reviews*, 1(1), 1–21. https://doi.org/10.1016/J.CARBPOL.2010.10.038

Gasperini, L., Mano, J. F., & Reis, R. L. (2014). Natural polymers for the microencapsulation of cells. *Journal of the Royal Society Interface*, 11(100), 20140817. https://doi.org/10.1098/RSIF.2014.0817

Ghaleh, A. S., Saghati, S., Rahbarghazi, R., Hassani, A., Kaleybar, L. S., Geranmayeh, M. H., Hassanpour, M., Rezaie, J. & Soltanzadeh, H. (2021). Static and dynamic culture of human endothelial cells encapsulated inside alginate-gelatin microspheres. *Microvascular Research*, 137, 104174.

Gheorghita, R., Gutt, G., & Amariei, S. (2020). The use of edible films based on sodium alginate in meat product packaging: An eco-friendly alternative to conventional plastic materials. *Coatings*, 10(2), 166.

Gupta, S. M., & Tripathi, M. (2011). A review of TiO2 nanoparticles. *Chinese Science Bulletin*, 56(16), 1639–1657.

Hadjila, M., Merzougui, R., & Hadj Irid, S. M. (2018). Detection of drug interactions via android smartphone: Design and implementation. *International Journal of Electrical and Computer Engineering*, 8(6), 5371–5380. https://doi.org/10.11591/ijece.v8i6

Hajipour, M. J., Fromm, K. M., Ashkarran, A. A., Jimenez de Aberasturi, D., de Larramendi, I. R., Rojo, T., Serpooshan, V., Parak, W. J., & Mahmoudi, M. (2012). Antibacterial properties of nanoparticles. *Trends in Biotechnology*, 30, 499–511.

Heising, J. K., Dekker, M., Bartels, P. V. & Van Boekel, M. A. J. S. (2014). Monitoring the quality of perishable foods: Opportunities for intelligent packaging. *Critical Reviews in Food Science and Nutrition*, 54(5), 645–654.

Hsiao, H.-I., & Chang, J.-N. (2017). Developing a microbial time-temperature indicator tomonitor total volatile basic nitrogen change in chilled vacuum-packed grouperfllets. *Journal of Food Processing and Preservation, 41*(5), e13158. https://doi.org/10.1111/jfpp.13158

Hua, Z., Yu, T., Liu, D., & Xianyu, Y. (2021). Recent advances in gold nanoparticles-based biosensors for food safety detection. *Biosensors & Bioelectronics, 179*, 113076.

Huang, D., Zhuang, Z., Wang, Z., Li, S., Zhong, H., Liu, Z., Guo, Z., & Zhang, W. (2019). Black phosphorus-Au filter paper-based three-dimensional SERS substrate for rapid detection of foodborne bacteria. *Applied Surface Science*, 497, 143825.

Incoronato, A. L., Buonocore, G. G., Conte, A., Lavorgna, M., & Del Nobile, M. A. (2010). Active systems based on silver-montmorillonite nanoparticles embedded into bio-based polymer matrices for packaging applications. *Journal of Food Protection*, 73(12), 2256–2262.

Janjarasskul, T., & Suppakul, P. (2018). Active and intelligent packaging: The indication of quality and safety. *Critical Reviews in Food Science and Nutrition*, 58(5), 808–831

Karimi, B., Mosharraf, L., & Ghazvini, H. (2014). The effect of active packaging on quantity, quality properties and shelf life of button mushroom. *Journal of Research and Innovation in Food Science and Technology*, 3(4), 347–360. https://www.sid.ir/en/journal/ViewPaper.aspx?id=506495

Karimi, N., & Mosharraf, L. (2015). The effect of active packaging by polyvinyl chloride film on the marketability of button mushroom. *Iranian Journal of Food Science and Technology, 12*(48), 68–77. https://www.sid.ir/En/Journal/ViewPaper.aspx?ID=403004

Kaya, M., Khadem, S., Cakmak, Y. S., Mujtaba, M., Ilk, S., Akyuz, L., ... & Deligoz, E. (2018). Antioxidative and antimicrobial edible chitosan films blended with stem, leaf and seed extracts of Pistacia terebinthus for active food packaging. *RSC Advances, 8*(8), 3941–3950. https://doi.org/10.1039/C7RA12070B

Khan, R., Rehman, A., Hayat, A., & Andreescu, S. (2019). Magnetic particles-based analytical platforms for food safety monitoring. *Magnetochemistry, 5*, 63.

Kikuchi, N., May, M., Zweber, M., Madamba, J., Stephens, C., Kim, U.,, and Mobed-Miremadi, M. (2020). Sustainable, alginate-based sensor for detection of Escherichia coli in human breast milk. *Sensors, 20*(4), 1145.

Kim, S. K. (2011). *Handbook of Marine Macroalgae: Biotechnology of Applied Phycology.* 2nd ed. John Wiley & Sons, 736 p.

Kim, J. S., Kuk, E., Yu, K. N., Kim, J. H., Park, S. J., Lee, H. J., ... & Cho, M. H. (2007). Antimicrobial effects of silver nanoparticles. *Nanomedicine: Nanotechnology, Biology and Medicine, 3*(1), 95–101.

Kim, S., Chon, J., Ko, J., & Chung, D. (2016). A fabrication of colorimetric sensor using pH indicator for determination of toxic gases. In Frontiers in Bioengineering and Biotechnology Conference Abstract: 10th World Biomaterials Congress. doi: https://doi.org/10.3389/conf.FBIOE.2016.01.02549

Kim, H. L., Jung, G. Y., Yoon, J. H., Han, J. S., Park, Y. J., Kim, D. G., Zhang, M. & Kim, D. J. (2015). Preparation and characterization of nano-sized hydroxyapatite/alginate/chitosan composite scaffolds for bone tissue engineering. *Materials Science and Engineering: C, 54*, 20–25.

Kiss, B., Br, T., Czifra, G., Tóth, B. I., Kertész, Z., Szikszai, Z., Kiss, Z., Juhász, I., Zouboulis, C. C., & Hunyadi, J. (2008). Investigation of micronized titanium dioxide penetration in human skin xenografts and its effect on cellular functions of human skin-derived cells. *Experimental Dermatology, 17*, 659–667.

Kong, H. J., Lee, K. Y., & Mooney, D. J. (2002). Decoupling the dependence of rheological/mechanical properties of hydrogels from solids concentration. *Polymer, 43*(23), 6239–6246. https://doi.org/10.1016/S0032-3861(02)00559-1

Konstantaki, M., Pissadakis, S., Pispas, S., Madamopoulos, N., & Vainos, N. A. (2006). Optical fiber long-period grating humidity sensor with poly (ethylene oxide)/cobalt chloride coating. *Applied Optics, 45*(19), 4567–4571.

Konuk Takma, D., & Korel, F. (2019). Active packaging films as a carrier of black cumin essential oil: Development and effect on quality and shelf-life of chicken breast meat. *Food Packaging and Shelf Life, 19*, 210–217. https://doi.org/10.1016/j.fpsl.2018.11.002

Krasniewska, K., Galus, S., & Gniewosz, M. (2020). Biopolymers-based materials containing silver nanoparticles as active packaging forfood applications—A review. *International Journal of Molecular Science, 21*, 698.

Kumar, V., Sharma, N., & Maitra, S. S. (2017). In vitro and in vivo toxicity assessment of nanoparticles. *International Nano Letters, 7*, 243–256.

Kuswandi, B., Maryska, C., Jayus, Abdullah, A., & Heng, L. Y. (2013). Real time on-package freshness indicator for guavas packaging. *Journal of Food Measurement and Characterization, 7*(1), 29–39. https://doi.org/10.1007/S11694-013-9136-5

Latifi, A. M., Nejad, E. S., & Babavalian, H. (2015). Comparison of extraction different methods of sodium alginate from brown alga *Sargassum* sp. localized in the southern of Iran. *Journal of Applied Biotechnology Reports, 2*(2), 251–255.

Lee, J. H., Jeong, D., & Kanmani, P. (2019). Study on physical and mechanical properties of the biopolymer/silver based active nanocomposite films with antimicrobial activity. *Carbohydrate Polymers, 224*, 115159.

Lee, K., Baek, S., Kim, D., & Seo, J. (2019). A freshness indicator for monitoring chickenbreast spoilage using a Tyvek(r) sheet and RGB color analysis. *Food Packaging and Shelf Life, 19*, 40–46. https://doi.org/10.1016/j.fpsl.2018.11.016

Lee, K. Y., & Mooney, D. J. (2012). Alginate: Properties and biomedical applications. *Progress in Polymer Science (Oxford), 37*(1), 106–126. https://doi.org/10.1016/J.PROGPOLYMSCI.2011.06.003

Li, C., Pei, J., Xiong, X., & Xue, F. (2020). Encapsulation of grapefruit essential oil inemulsion-based edible film prepared by plum (*Pruni domesticae semen*) seed proteinisolate and gum acacia conjugates. *Coatings, 10*(8), 784. https://doi.org/10.3390/COATINGS10080784

Lyu, J. S., Choi, I., Hwang, K.-S., Lee, J.-Y., Seo, J., Kim, S. Y., & Han, J. (2019). Development of a BTB-/TBA+ ion-paired dye-based CO_2 indicator and its application in a multilayered intelligent packaging system. *Sensors and Actuators B: Chemical, 282*, 359–365. https://doi.org/10.1016/j.snb.2018.11.073

Madima, N., Mishra, S.B., Inamuddin, I. & Mishra, A. K. (2020). Carbon-based nanomaterials for remediation of organic and inorganic pollutants from wastewater. A review. *Environmental Chemistry Letters*, *18*, 1169–1191.

Manthou, V., & Vlachopoulou, M. (2001). Bar-code technology for inventory and marketing management systems: A model for its development and implementation. *International Journal of Production Economics*, *71*(1), 157–164. https://doi.org/10.1016/S0925-5273(00)00115-8

Marcos, B., Aymerich, T., Monfort, J. M., & Garriga, M. (2007). Use of antimicrobialbiodegradable packaging to control *Listeria monocytogenes* during storage ofcooked ham. *The International Journal of Food Microbiology*, *120*, 152–158.

Mataragas, M., Bikouli, V. C., Korre, M., Sterioti, A., & Skandamis, P. N. (2019). Development of a microbial time temperature indicator for monitoring the shelf life of meat. *Innovative Food Science &Emerging Technologies*, *52*, 89–99. https://doi.org/10.1016/j.ifset.2018.11.003

Meghani, N., Dave, S., & Kumar, A. Introduction to nanofood. In *Nano-Food Engineering. Food Engineering Series*; Hebbar, U., Ranjan, S., Dasgupta, N., Mishra, R. K., Eds.; Springer: Cham, Switzerland, 2020.

Moazami Goodarzi, M., Moradi, M., Tajik, H., Forough, M., Ezati, P., & Kuswandi, B. (2020). Development of an easy-to-use colorimetric pH label with starch and carrot anthocyanins for milk shelf life assessment. *International Journal of Biological Macromolecules*, *153*, 240–247. https://doi.org/10.1016/J.IJBIOMAC.2020.03.014

Mohammadalinejhad, S., Almasi, H., & Moradi, M. (2020). Immobilization of *Echium amoenum* anthocyanins into bacterial cellulose film: A novel colorimetric pH indicator for freshness/spoilage monitoring of shrimp. *Food Control*, *113*, 107169. https://doi.org/10.1016/J.FOODCONT.2020.107169

Mørch, Ý. A., Donati, I., Strand, B. L., & Skjåk-Bræk, G. (2006). Effect of Ca^{2+}, Ba^{2+}, and Sr^{2+} on alginate microbeads. *Biomacromolecules*, *7*(5), 1471–1480. https://doi.org/10.1021/BM060010D

Morris, E. R., Rees, D. A. & Thom, D. (1980). Characterisation of alginate composition and block-structure by circular dichroism. *Carbohydrate Research*, *81*(2), 305–314. https://doi.org/10.1016/S0008?6215(00)85661?X.

Muriel-Galet, V., Cerisuelo, J. P., Lopez-Carballo, G., Lara, M., Gavara, R., & Hern´ andezMunoz, P. (2012). Development of antimicrobial films for microbiological control of packaged salad. *International Journal of Food Microbiology*, *157*(2), 195–201. https://doi.org/10.1016/j.ijfoodmicro.2012.05.002

Nopwinyuwong, A., Trevanich, S., & Suppakul, P. (2010). Development of a novel colorimetric indicator label for monitoring freshness of intermediate-moisture dessert spoilage. *Talanta*, *81*(3), 1126–1132. https://doi.org/10.1016/j.talanta.2010.02.008

Osman, I. F., Baumgartner, A., Cemeli, E., Fletcher, J. N., & Anderson, D. (2010). Genotoxicity and cytotoxicity of zinc oxide and titaniumdioxide in HEp-2 cells. *Nanomedicine*, *5*, 1193–1203.

Pavlicek, A., Part, F., Rose, G., Praetorius, A., Miernicki, M., Gazsó, A., & Huber-Humer, M. (2021). A European nano-registry as a reliable database for quantitative risk assessment of nanomaterials? A comparison of national approaches. *NanoImpact*, *21*, 100276.

Pawar, S. N., & Edgar, K. J. (2012). Alginate derivatization: A review of chemistry, properties and applications. *Biomaterials*, *33*(11), 3279–3305. https://doi.org/10.1016/J.BIOMATERIALS.2012.01.007

Peighambardoust, S. J., Peighambardoust, S. H., Pournasir, N., & Pakdel, P. M. (2019). Properties of active starch-based films incorporating acombination of Ag, ZnO and CuO nanoparticles for potential use in food packaging applications. *Food Packaging and Shelf Life*, *22*, 100420.

Pires, J. R. A., de Souza, V. G. L., & Fernando, A. L. (2018). Chitosan/montmorillonite bionanocomposites incorporated with rosemary and ginger essential oil as packaging for fresh poultry meat. *Food Packaging and Shelf Life*, *17*, 142–149.

Priyadarshi, R., Roy, S., Ghosh, T., Biswas, D., & Rhim, J.-W. (2021). Antimicrobial nanofillers reinforced biopolymer composite films foractive food packaging applications—A review. *Sustainable Materials and Technologies*, *32*, e00353.

Qin, Y., Howlader, M. M., Deen, M. J., Haddara, Y. M., & Selvaganapathy, P. R. (2014). Polymer integration for packaging of implantable sensors. *Sensors and Actuators B: Chemical*, *202*, 758–778.

Quintero-Quiroz, C., Botero, L. E., Zárate-Triviño, D., Acevedo-Yepes, N., Escobar, J. S., Pérez, V. Z., & Cruz Riano, L. J. (2020). Synthesis and characterization of a silver nanoparticle-containing polymer composite with antimicrobial abilities for application in prosthetic and orthotic devices. *Biomaterials Research*, *24*(1), 1–17.

Rahman, M. W., Rana, M. S., & Alam, M. J. (2022). Biodegradable and active polymeric matrices reinforced with silver-titania nanoparticles for state-of-the-art technology of food packaging. In *Radiation-Processed Polysaccharides*. Academic Press; pp. 75–89.

Rahman, M. W., Hossain, M. M., Alam, M. J., Dafader, N. C., Haque, M. E. (2013). Addition of transition metals to improve physico–mechanical properties of radiation vulcanised natural rubber latex films. *International Journal of Polymer Analysis and Characterization*, *18*, 479–487.

Ramos, P. E., Silva, P., Alario, M. M., Pastrana, L. M., Teixeira, J. A., Cerqueira, M. A., & Vicente, A. A. (2018). Effect of alginate molecular weight and M/G ratio in beads properties foreseeing the protection of probiotics. *Food Hydrocolloids*, *77*, 8–16.

Ray, P. C., Yu, H., & Fu, P. P. (2009). Toxicity and environmental risks of nanomaterials: challenges and future needs. *Journal of Environmental Science and Health Part C*, *27*(1), 1–35.

Restuccia, D., Spizzirri, U. G., Parisi, O. I., Cirillo, G., Curcio, M., Iemma, F., … & Picci, N. (2010). New EU regulation aspects and global market of active and intelligent packaging for food industry applications. *Food Control*, *21*(11), 1425–1435. https://doi.org/10.1016/j.foodcont.2010.04.028

Rhim, J. W., Hong, S. I., Park, H. M., & Ng, P. K. (2006). Preparation and characterization of chitosan-based nanocomposite films with antimicrobial activity. *Journal of Agricultural and Food Chemistry*, *54*(16), 5814–5822.

Roy, S., Priyadarshi, R., Ezati, P., & Rhim, J. W. (2022). Curcumin and its uses in active and smart food packaging applications—A comprehensive review. *Food Chemistry*, *375*, 131885. https://doi.org/10.1016/j.foodchem.2021.131885

Roy, S., & Rhim, J. W. (2021). Preparation of gelatin/carrageenan-based color-indicator film integrated with shikonin and propolis for smart food packaging applications. *ACS Applied Bio Materials*, *4*(1), 770–779. https://doi.org/10.1021/ACSABM.0C01353

Sawai, J., & Yoshikawa, T. (2004). Quantitative evaluation of antifungal activity of metallic oxide powders (MgO, CaO and ZnO) by an indirect conductimetric assay. *The Journal of Applied Microbiology*, *96*, 803–809.

Saxena, A., Sharda, S., Kumar, S., Kumar, B., Shirodkar, S., Dahiya, P., & Sahney, R. (2022). Synthesis of alginate nanogels with polyvalent 3D transition metal cations: Applications in urease immobilization. *Polymers*, *14*(7), 1277.

Shanmuganathan, R., Karuppusamy, I., Saravanan, M., Muthukumar, H., Ponnuchamy, K., Ramkumar, V. S., & Pugazhendhi, A. (2019). Synthesis of silver nanoparticles and their biomedical applications—A comprehensive review. *Current Pharmaceutical Design*, *25*, 2650–2660.

Shemesh, R., Krepker, M., Nitzan, N., Vaxman, A., & Segal, E. (2016). Active packaging containing encapsulated carvacrol for control of postharvest decay. *Postharvest Biology and Technology*, *118*, 175–182. https://doi.org/10.1016/j.postharvbio.2016.04.009

Siddiqi, K. S., Husen, A., & Rao, R. A. K. (2018). A review on biosynthesis of silver nanoparticles and their biocidal properties. *Journal of Nanobiotechnology*, *16*, 14.

Silvestre, C., Duraccio, D., & Cimmino, S. (2011). Food packaging based on polymer nanomaterials. *Progress in Polymer Science*, *36*(12), 1766–1782.

Socas-Rodríguez, B., Herrera-Herrera, A. V., Asensio-Ramos, M., & Rodríguez-Delgado, M. Á. (2020). Recent applications of magnetic nanoparticles in food analysis. *Processes*, *8*, 1140.

Sondi, I., & Salopek-Sondi, B. (2004). Silver nanoparticles as antimicrobial agent: A case study on E. coli as a model for Gram-negative bacteria. *Journal of Colloid and Interface Science*, *275*(1), 177–182.

Stoimenov, P. K., Klinger, R. L., Marchin, G. L., & Klabunde, K. J. (2002). Metal oxide nanoparticles as bactericidal agents. *Langmuir*, *18*, 6679–6686.

Tankhiwale, R., & Bajpai, S. K. (2010). Silver-nanoparticle-loaded chitosan lactate films with fair antibacterial properties. *Journal of Applied Polymer Science*, *115*(3), 1894–1900.

Tekin, E., Mazı, I. B., & Türe, H. (2020). Time temperature indicator film based on alginate and red beetroot (*Beta vulgaris* L.) extract: In vitro characterization. *Ukrainian Food Journal*, *9*(2), 344–480.

Valodkar, M., Bhadoria, A., Pohnerkar, J., Mohan, M., & Thakore, S. (2010). Morphology and antibacterial activity of carbohydrate-stabilized silver nanoparticles. *Carbohydrate Research*, *345*(12), 1767–1773.

van der Meulen, B., Bremmers, H., Purnhagen, K., Gupta, N., Bouwmeester, H., & Geyer, L. L. Nano-specific regulation. In *Governing Nano Foods: Principles-Based Responsive Regulation*; van der Meulen, B., Bremmers, H., Purnhagen, K., Gupta, N., Bouwmeester, H., Geyer, L., Eds.; Academic Press: San Diego, CA, USA, 2014; pp. 43–45.

Wang, S., Liu, X., Yang, M., Zhang, Y., Xiang, K., & Tang, R. (2015). Review of time temperature indicators as quality monitors in food packaging. *Packaging Technology and Science*, *28*(10), 839–867. https://doi.org/10.1002/PTS.2148

Weber, T. A., Metzler, L., Fosso Tene, P. L., Brandstetter, T., & Rühe, J. (2022). Single-color barcoding for multiplexed hydrogel bead-based immunoassays. *ACS Applied Materials & Interfaces, 14*(22), 25147–25154.

Yan, H., Chen, X., Li, J., Feng, Y., Shi, Z., Wang, X., & Lin, Q. (2016). Synthesis of alginate derivative via the Ugi reaction and its characterization. *Carbohydrate Polymers, 136*, 757–763. https://doi.org/10.1016/J.CARBPOL.2015.09.104

Yoksan, R., & Chirachanchai, S. (2010). Silver nanoparticle-loaded chitosan-starch based films: Fabrication and evaluation of tensile, barrier and antimicrobial properties. *Materials Science and Engineering: C, 30*(6), 891–897.

Youssef, A. M., & El-Sayed, S. M. (2018). Bionanocomposites materials for food packaging applications: Concepts and future outlook. *Carbohydrate Polymers, 193*, 19–27. https://doi.org/10.1016/j.carbpol.2018.03.088

Youssef, A. M., El-Sayed, S. M., Salama, H. H., El-Sayed, H. S., & Dufresne, A. (2015). Evaluation of bionanocomposites as packaging materialon properties of soft white cheese during storage period. *Carbohydrate Polymers, 132*, 274–285.

Yu, X., Zhong, T., Zhang, Y., Zhao, X., Xiao, Y., Wang, L., Liu, X., & Zhang, X. (2022). Design, preparation, and application of magnetic nanoparticles for food safety analysis: A review of recent advances. *Journal of Agricultural and Food Chemistry, 70*, 46–62.

Yuan, X., Wei, Y., Chen, S., Wang, P., & Liu, L. (2016). Bio-based graphene/sodium alginate aerogels for strain sensors. *RSC Advances, 6*(68), 64056–64064.

Yun, X., Zhang, Q., Luo, B., Jiang, H., Chen, C., Wang, S., & Min, D. (2020). Fabricating flexibly resistive humidity sensors with ultra-high sensitivity using carbonized lignin and sodium alginate. *Electroanalysis, 32*(10), 2282–2289.

Zeng, P., Chen, X., Qin, Y. R., Zhang, Y. H., Wang, X. P., Wang, J. Y., … & Zhang, Y. S. (2019). Preparation and characterization of a novel colorimetric indicator film based on gelatin/polyvinyl alcohol incorporating mulberry anthocyanin extracts for monitoring fish freshness. *Food Research International, 126*, 108604. https://doi.org/10.1016/J.FOODRES.2019.108604

Zhai, X., Li, Z., Shi, J., Huang, X., Sun, Z., Zhang, D., … & Wang, S. (2019). A colorimetric hydrogen sulfide sensor based on gellan gum-silver nanoparticles bionanocomposite for monitoring of meat spoilage in intelligent packaging. *Food Chemistry, 290*, 135–143. https://doi.org/10.1016/J.FOODCHEM.2019.03.138

Zhang, Y. S.., Chen, X., Qin, Y. R., Zhang, Y. H., Wang, X. P., Wang, J. Y., … & Zhang, Y. S. (2019). Preparation and characterization of a novel colorimetric indicator film based on gelatin/polyvinyl alcohol incorporating mulberry anthocyanin extracts for monitoring fish freshness. *Food Research International, 126*, 108604.

Ziyaina, M., Rasco, B., Coffey, T., Ünlü, G., & Sablani, S. S. (2019). Colorimetric detection of volatile organic compounds for shelf-life monitoring of milk. *Food Control, 100*, 220–226. https://doi.org/10.1016/J.FOODCONT.2019.01.018

13 Polysaccharide-Based Adsorbents for Water Treatment

Subrata Mondal
National Institute of Technical Teachers' Training
and Research (NITTTR) Kolkata

13.1 INTRODUCTION

Water is indispensable for life on the planet earth, and it is one of the global challenges for 21st century to have pure and drinkable water [1]. Water treatment is a process used to remove contaminants from the polluted water. Contaminated water contains physical, chemical, biological, and biochemical constituents. Examples of physical constitutes of the contaminated water are solids, turbidity, color, salinity, order, etc. Chemical constituents in the contaminated water can be subdivided into three category viz. organic, inorganic, and gases. Carbohydrate, fats, and proteins are few common examples of organic contaminants in the wastewater. Various heavy metals, silica, and salt are few examples of inorganic contaminants, and hydrogen sulfide, methane, and ammonia are some of the examples of gases contaminants in the wastewater. Whereas microorganism (such as bacteria, virus, etc.) and plants (such as algae, grass, etc.) are common biological constituents found in the contaminated water. The presence of heavy metals, dyes, other organic compounds, microorganisms, etc. are potential threat to the human life, aquatic life, and the environment [1]. Therefore, the elimination of various organic and inorganic toxic components from the wastewater is essential before it will be discharged to the environment.

Water treatment includes drinking water and wastewater treatments. Wastewater can be categorized into two major groups such as municipal wastewater and industrial wastewater. Municipal wastewater include household wastewater from the sink to the sewer. Whereas industrial wastewater include waste liquid from various industries such as textiles, chemical, food, pharmaceutical, leather, etc. Depending on the type of industries, contaminants of the wastewater will be significantly varied. Heavy metals and dyes are widely found contaminants in the wastewater. These are extremely harmful to environment and aquatic life due to their high toxicity and non-degradability in nature [2]. Depending on the source, water pollution can be categorized into two groups viz. point source and non-point source. Wastewater from industries, sewages, oil refineries, and power plant are a few examples for the point of source wastewater. Whereas non-point of source wastewater cannot be traced in a single discharge point and their contaminations are diffused across broad areas [1].

There are various water treatment techniques, such as coagulation, precipitation, membrane separation, oxidation, ion exchange, adsorption, activated sludge, etc. Water treatment method can be broadly classified into three groups, such as physical method, chemical method, and biological method. Adsorption is physical method of water treatment. Depending on the type of adsorbent, adsorption mechanism can be classified as specific adsorption, nonspecific adsorption, and ion exchange. Many of these wastewater treatment techniques have disadvantages like high operating and maintenance cost, etc. However, adsorption is one of the simplest, economical, and effective

DOI: 10.1201/9781003265054-13

water treatment techniques [1]. Adsorption is a separation process by which substances from liquid or gas are attached to the exterior and internal surface of a material which is known as adsorbent [3]. For the wastewater treatment, the ideal adsorbent should be economical, easily available, renewable in nature, and opportunity to modify the adsorbent surface in order to remove wide range of contaminants from the wastewater. Polysaccharide or polycarbohydrates are one of the most abundant carbohydrates found in the nature. They are polymeric carbohydrate macromolecules consists of monosaccharide linked together by the glycosidic linkages. This chapter presents a review on various polysaccharide-based adsorbents for the water treatment. The chapter focused on four major polysaccharides such as chitin, chitosan, starch, and cyclodextrin-based adsorbents for the wastewater treatment.

13.2 CHARACTERISTICS OF WASTEWATER

Rapid industrialization contributed to the environmental pollution [4]. A wide range of wastewater are produced depending on the various activities. Discharge of these untreated wastewater causes significant environmental concern. Depending on the source, wastewater majorly can be classified into domestic and industrial wastewater. Major pollutants of domestic wastewater are nitrogen, various heavy metals, phosphorus, detergents, pesticides, oil and grease, etc. [4]. With rapid economic development, increasingly domestic swages, industrial wastewater, and solid waste leachate are produced [5]. Characteristics of wastewater are influenced by the source of discharge [6–8]. For an example, wastewater from textile industries contained high concentration of organic and inorganic contaminants, such as salt, enzymes, microfibers, surfactants, residual dyes, etc. [9–11]. Not only the industrial discharged effluent, municipal wastewater also contains various emerging contaminants in the wastewater. Murphy et al. reported microplastic contamination in the municipal wastewater effluent. The source of microplastic is mainly from different personal care products which contain plastic beads. The study revealed different forms of microplastics such as flake, fiber, film, bead, foam, etc. However, two-thirds of them are flake microplastics. Their study also revealed different colors of the microplastics in the effluent and majority of them are red, blue, and green [12]. Contaminant in the wastewater can be categorized into biodegradable and non-biodegradable constitutes. Degradable contaminants include simple organic components, such as dead microorganism which undergo gradual microbial degradation. However, non-degradable contaminants are inert to the biological action and do not degrade with time, such as heavy metals. Further, various constituent exists in the wastewater can be organic or inorganic in nature. Industrial wastewater contains large concentration of non-biodegradable organic matter. Organic matter in the wastewater can be calculated as equivalent quantity, such as biological oxygen demand or chemical oxygen demand [13]. Other characterization of wastewater are total dissolve solid, total phosphorous content, pH, chloride, total nitrogen, order, color, temperature, total solids, total dissolve solid, dissolved oxygen, etc. [4,14].

13.3 ADSORBENTS FOR WASTEWATER TREATMENT

The term adsorption is simply defined as accumulation of substances on the interface of two phase's viz. solid–liquid or solid–gas. The adsorption can occur through physical and/or chemical reaction; therefore, the process is physiochemical in nature. Substances that accumulate on the interface are termed as adsorbate, whereas the solid on which adsorbates are accumulated has been termed as adsorbent [15]. The process formed a film of adsorbate on the interface of adsorbent and fluid. In water and wastewater treatment, adsorption technique is used for the separation of dissolved impurities [16]. Because of its simplicity and cost-effectiveness, adsorption techniques are considered as one of the most widely used wastewater treatment techniques. Different types of adsorbents are reported in the literature, and these could be broadly divided into six groups: natural adsorbents [17], biomass-based adsorbents [2], agricultural waste-based adsorbents [15], synthetic adsorbents [18], carbon-based adsorbents [19], and industrial waste-based adsorbents [20]. Adsorbents based on activated carbon are widely applied for the treatment of wastewater [21–23] due to their unique characteristics such as surface characteristics, surface functional groups, porosity, pore volume, pore morphology, etc. Adsorption process can be classified as physisorption and chemisorption. Pore morphology and specific surface are prime factors that influence physisorption, whereas chemisorption related to functional groups on the adsorbent surface [24]. A few key characteristics of a good adsorbents are as follows: (i) good compatibility with the adsorbate; (ii) high specific surface area; (iii) high adsorption capacity; (iv) excellent stability in the aqueous media; (v) able to regenerate; (vi) higher selectivity, (vii) low cost, etc. There are various process parameters which can influence the adsorption capacity of adsorbent viz. adsorbent dosage, pH, temperature, contact time, initial concentration of the toxic water, size and shape of the adsorbent, ionic strength, etc. [25].

13.4 VARIOUS POLYSACCHARIDE-BASED ADSORBENTS

Polysaccharides are a wide range of natural carbohydrate polymers that are joined by glycosidic linkages. Depending on the source of availability, polysaccharide can be categorized as microbial polysaccharide, animal polysaccharide, and plant polysaccharide. Animal polysaccharide contains a large number of surface functional groups viz. amino, hydroxyl, carboxylic, etc., and these functional groups can attach various toxic components from the wastewater. Example of animal polysaccharide is chitin/chitosan. Metabolites of plant cells formed the plant polysaccharides. Examples of plant polysaccharides are starch, cellulose, cyclodextrin, etc. Microbial polysaccharides are extracellular polymeric biomolecules synthesized by microorganism under the metabolic process. Examples of microbial polysaccharides are xanthan gum, dextran, etc. [25]. Polysaccharides are widely used as adsorbent due to their excellent adsorption capacity, renewability, low cost, etc. Outstanding adsorption behavior of polysaccharides are attributed to several factors: (i) the presence of hydroxyl groups of glucose units, which enhanced the hydrophilicity; (ii) the existence of significant number of other functional groups; and (iii) excellent chemical reactivity of these groups, allowing for the modification of the polysaccharide surface by using other molecules [26].

13.4.1 CHITIN AS ADSORBENTS

Chitin is the natural biopolymer of acetylated and non-acetylated glucosamine. It is the second most abundant biopolymer available in nature [27]. It can be extracted from crustacean shells (shrimp, lobster, crab shells), which are waste product of seafood industries. The constituent monosaccharide unit in chitin are linked together by β-1,4-glycosidic bonds. Chitin is non-toxic, biocompatible, biodegradable, ecofriendly, low-cost material, and can be easily filmable/spinnable forms that possess adequate mechanical properties. Chitin is insoluble in water and alkali but soluble in mineral acid and anhydrous formic acid [28]. Chitin has been widely applied as an adsorbent for the removal of different impurities from the wastewater due to their nice porous structure and the abundancy of functional groups on its surface [29–35]. Karthikeyan et al. described the removal of iron (III) by using chitin as adsorbent. They have studied various parameters on the iron (III) adsorption capacity of chitin viz. particle size, contact time, initial concentration of contaminant, dosage, influence of ionic strength, etc. Experimental results revealed that adsorption of iron (III) is maximum with least particle size of chitin, and adsorption increases with dosage of adsorbate and initial concentration of iron (III). Thermodynamic evaluation showed that adsorption is endothermic in nature, random, and beneficial [28]. Modification of chitin can enhance the adsorption of contaminants due to the higher number of functional groups on the adsorbent surface, increase specific surface area, morphological changes, etc. [36–39]. Yang et al. present arsenic [As (III)] removal by using thiol-modified chitin nanofiber adsorbent. Nanofiber has dimension of 6 nm in thickness and 24 nm in width with a few nm in length. In order to form more adsorption sites for arsenic (III), grafting of cysteine has been carried out. Maximum adsorption of arsenic (III) was 149 mg g^{-1} of thiol-modified chitin at pH 7.0, and this is significantly higher than the unmodified chitin. Significant increase of arsenic

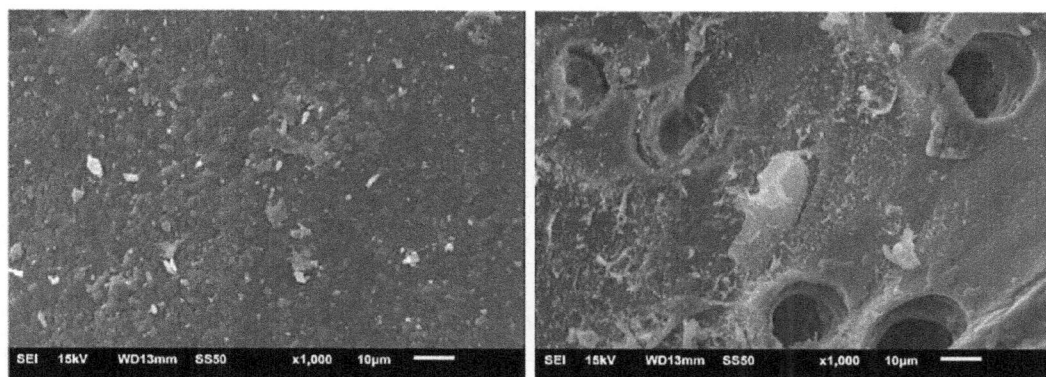

FIGURE 13.1 Scanning electron microscopic images of: (a) unmodified chitin and (b) ultrasound modified chitin. Reproduced with permission from Ref [38] © 2015 Elsevier Inc.

TABLE 13.1
Adsorption of Various Contaminants by Using Chitin as Adsorbent

Form of Chitin	Adsorbate	Max. Adsorption Capacity	Ref
Microparticle	Chlortetracycline	82.9%	[40]
Thiol modified nanofiber	As(III)	149 mg g^{-1}	[37]
Magnetite-modified chitin	Cu^{2+}	91.67%	[36]
USM chitin	Methylene blue	85%	[38]
Complex of PVA chitin nanofiber/Fe(III)	Methyl orange	810.4 mg g^{-1}	[41]
TEMPO-modified chitin	Cd(II)	207.9	[5]
Porous chitin sorbents	Methylene blue	79.8%	[42]
Lignin chitin film	Fe(III)	84%	[43]
Chitin nanofiber/nanowhisker based adsorbents	Reactive blue 19	1331 mg g^{-1}	[44]
Ultrasound-modified chitin	Cobalt	83.94 mg g^{-1}	[45]

USM: Ultrasonic surface modified; PVA: Poly(vinyl alcohol)

(III) adsorption by modified chitin is due to the higher surface area and abundancy of functional groups on nanofiber-based adsorbents [37]. Dotto et al. reported methylene blue dye adsorption by using ultrasound modified chitin. Ultrasound modification decreases the crystallinity content due to the rearrangement of chitins, as a result of more amorphous region for adsorption of contaminants. Microscopic images revealed rigid, smooth, and non-porous surface for the unmodified chitin, whereas porous rough surface for the ultrasound modified chitins (Figure 13.1). Solution diffusion through the pores of adsorbents increases the adsorption capacity [38]. Adsorption of various contaminants by using chitin or modified chitin as adsorbents has been tabulated in Table 13.1.

13.4.2 CHITOSAN AS ADSORBENTS

Chitosan is the deacetylated form of chitin. In the presence of concentrated alkali at high temperature, deacetylation of chitin is usually carried out [46]. Chitosan is a linear form of polysaccharide consists of randomly distributed β-(1,4)-linked d-glucosamine and N-acetyl-d-glucosamine. It has been made from shrimp and other crustacean shell after treatment with sodium hydroxide. Major characteristics of chitosan are linear polyamine, abundant hydroxyl and carboxylic functionalities on the surface, biocompatible, biodegradable, etc. The presence of abundant amino, hydroxyl, and carboxylic functional groups on the chitosan surface makes the biopolymer chitosan the most widely used animal polysaccharide for the preparation of adsorbents. Chitosan is a well-known biopolymeric adsorbent for the separation of contaminants from the wastewater, due to their outstanding characteristics such as renewability, availability of functional groups, cost-effective production, biodegradability, biocompatibility, ecofriendliness, etc. [47]. Szlachta et al. reported chitosan for the removal of uranium from real mine effluent. Chitosan exhibits adsorption capacity of uranium because of the electrostatic interaction between uranyl ions and functional groups of chitosan. Maximum separation capacity of uranium by the chitosan was $17.44 \, mg \, g^{-1}$ [48]. Chitosan can be modified with other functional molecules to effectively remove contaminants from the wastewater [49,50]. The process of chemically modifying the chitosan includes the attachment of various surface functional groups on chitosan, such as alkyl, carboxymethyl, phenolic molecules, etc. [46]. The processes modify chitosan without affecting inherent properties of chitosan. Few of the well-known methods for the chemical modification of chitosan are alkylation [51], acylation [52], sulfation [53], carboxyalkylation [54], etc. Al-Harby et al. presented novel uracil functionalized chitosan for the removal of Congo red dye. Modified chitosan contains free amino and hydroxyl active center in its main chains. $434.78 \, mg \, g^{-1}$ was the maximum monolayer coverage capacity of adsorbate [55]. Chitosan can also modify by incorporating other compounds. Chitosan modification can improve stability, durability, and adsorption capability of various adsorbates. Xu et al. present novel granular adsorbents based on titanium and chitosan, which was prepared by using sol-gel method. Amorphous TiO_x combined with chitosan by primary (C–O–Ti) and secondary (N–Ti) bonds (Figure 13.2). Adsorbents were studied for the removal of As(V). As(V) was adsorbed on the titanium chitosan surface by forming bonds with TiOH sites on the surface of the Ti-chitosan. Rapid small-scale column test exhibits adsorption of 165.6 microgram L^{-1}. Arsenic in groundwater has been effectively separated by approximately 126-bed volumes [56]. Yang et al. reported glutaral-dehyde crosslinked chitosan-coated Fe_3O_4 nanocomposite for the separation of methyl orange from the wastewater. Nanocomposite-based adsorbents showed approximately 98% removal of contaminants by adding 0.2 mg of cetyltrimethylammonium bromide to form mixed hemimicelles, which increase the adsorption capacity of nanocomposite [57]. Hassan et al. reported chitosan/silica/ZnO nanocomposite for the separation of methylene blue dye. The whole process for the preparation

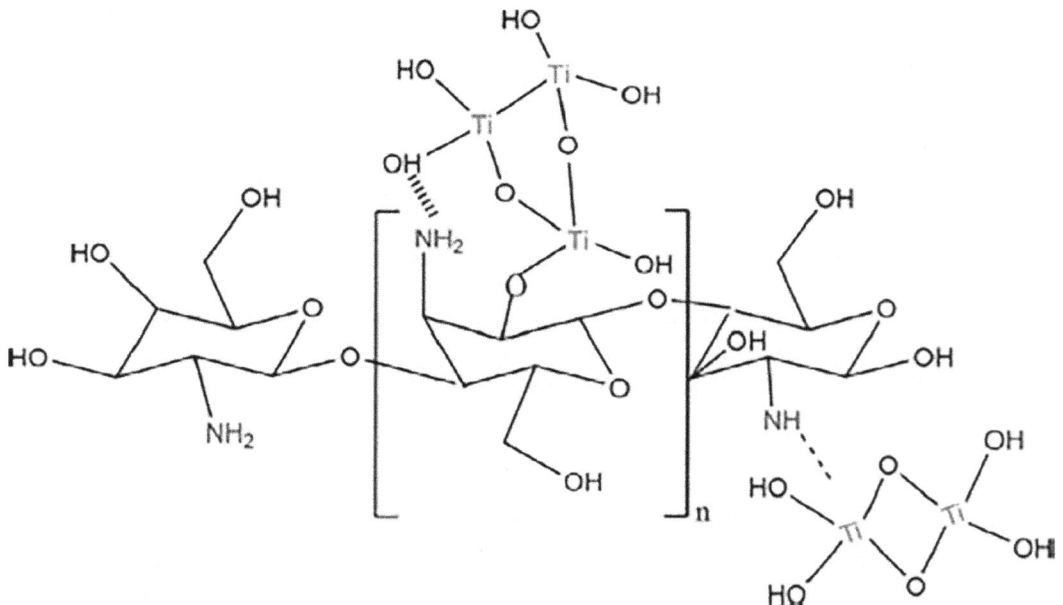

FIGURE 13.2 Interaction between titanium network and chitosan chains. Reproduced with permission from Ref [56] © 2021 The Research Center for Eco-Environmental Sciences, Chinese Academy of Sciences. Published by Elsevier B.V.

of nanocomposite has been depicted schematically in Figure 13.3. Hydroxyl and amino groups of chitosan and residual hydroxyl groups of silica play an important role in immobilization of ZnO in the network structure. Highest adsorption capacity of adsorbent was 293.2 mg g^{-1} at the slightly basic medium [58]. Table 13.2 summarizes adsorption of various adsorbate in the modified chitosan.

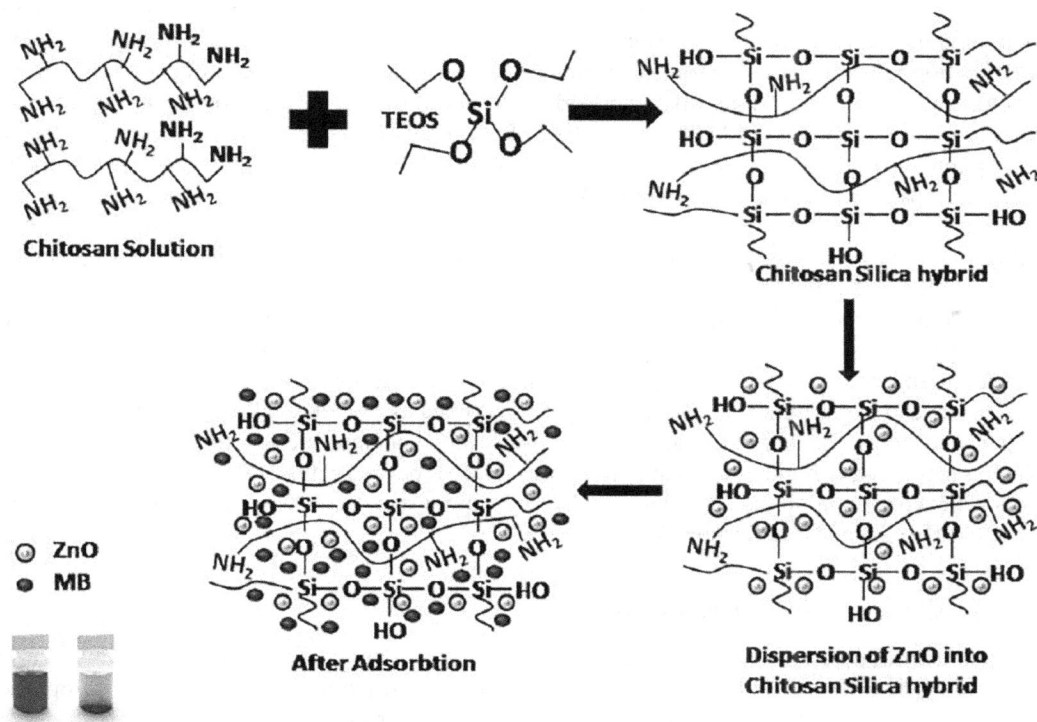

FIGURE 13.3 Schematic showing preparation of chitosan/silica/ZnO nanocomposites for the separation of methylene blue dye. Reproduced with permission from Ref [58] © 2019 Elsevier B.V.

TABLE 13.2
Modified Chitosan as Effective Adsorbents

Modified Chitosan	Adsorbate	Max. Adsorption Capacity	Ref
Polypropylene glycol modified chitosan	Cu(II)	661.8 mg g^{-1}	[59]
Hydroxyapatite/graphene oxide/chitosan bead	Copper ion	256.41 mg g^{-1}	[60]
Hydroxyapatite/graphene oxide/chitosan bead	Methylene blue dye	99 mg g^{-1}	[60]
Magnetic chitosan microsphere	Iodine	0.8087 mmol g^{-1}	[61]
Phosphate ligand contained chitosan/*Chlorella pyrenoidosa* composite	Uranium	1393.338 mg g^{-1}	[62]
Magnetic chitosan/Al$_2$O$_3$/Fe$_3$O$_4$ nanocomposite	Cd(II)	99.98%	[63]
Ionic chitosan/silica nanocomposite	Methylene blue dye	847.5 mg g^{-1}	[64]
Chitosan/diatomite composite	Pefloxacin	310.7 mg g^{-1}	[65]
Chitosan cellulose microsphere	Reactive black 5	214.36 mg g^{-1}	[66]
Sodium lignosulfonate/chitosan	Pb^{2+}	345 mg g^{-1}	[67]
Pyridine-2,6-dicarboxylic acid crosslinked chitosan	Cu(II)	2186 mmol g^{-1}	[68]
Salicylaldehyde and β-cyclodextrin-modified chitosan	Phenol	179.73 mg g^{-1}	[69]

13.4.3 STARCH AS ADSORBENTS

Starch is a natural polysaccharide extracted from various renewable vegetable sources and their byproducts. Recently, starch-based materials are widely applied as potential adsorbents for the separation of various toxic components from the wastewater [70]. Starch-based adsorbents containing amide, phosphate, amino, and carboxylic groups can be used to effectively separate various contaminants from the wastewater. Huang et al. reported novel starch-based adsorbents involving esterification with maleic anhydride and radical crosslinking with acrylic acid. The whole process for the preparation of starch-based adsorbent has been depicted schematically in Figure 13.4. The adsorbents were applied for the separation of toxic heavy metals, such as Hg(II) and Pb(II). Adsorption capacity of mercury and lead ions were 131.2 mg g^{-1} and 123.2 mg g^{-1}, respectively. Excellent separation capacity of heavy metals by the adsorbent is due to the interaction among metal ions and functional groups by physical and chemical means [71]. Haq et al. reported succinylated carboxymethyl starch-based adsorbents for the removal of phenol. The maximum adsorption efficiency of phenol was 0.324 g g^{-1} [72]. Ren et al. reported amino-functionalized starch-based adsorbents for the separation of uranyl ions. Separation capacity of functionalized starch was reported at 118.92 mg g^{-1}. Excellent adsorption capacity of adsorbents was due to the chelation and coordination between uranyl ions and –NH$_2$ groups on the surface of the adsorbents [73]. Starch-grafted polyacrylamide copolymer reinforced with Fe$_3$O$_4$ and graphene exhibited excellent adsorption capacity for the Ni(II) ions. The maximum adsorption capacity of nanocomposite adsorption was 295 mg g^{-1}, which is significantly higher than that of the starch-grafted polyacrylamide copolymer-based adsorbents. The adsorption on the adsorbents occurred in three steps such as surface adsorption, intraparticle diffusion, and equilibrium [74].

13.4.4 CYCLODEXTRIN AS ADSORBENTS

Cyclodextrins (CD) are cyclic oligosaccharides derived from the enzymatic degradation of starch [75]. The cyclic orientation of CD provides a truncated cone structure, which possesses hydrophilic rims, and on the interior, it is more lipophilic in nature [76]. Adsorption capacity of cyclodextrin-based adsorbents is influenced by various factors, such as surface charge, porosity, binding affinity, etc.

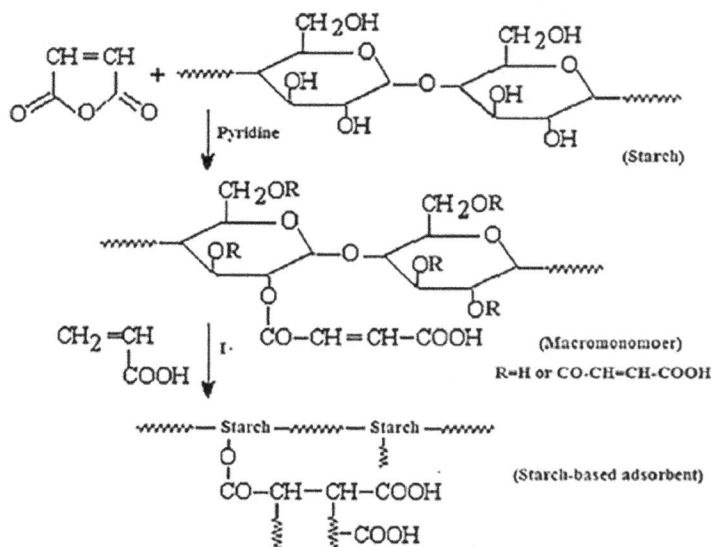

FIGURE 13.4 Schematic showing synthesis of starch-based adsorbents. Reproduced with permission from Ref [71] © 2011 Elsevier B.V.

Adsorption capacity of cyclodextrin can be enhanced by chemical modification of CD. Wei et al. reported improved adsorption efficiency of organic dye and metal ions for the imidazole modified β-cyclodextrin adsorbents [77]. Murcia-Salvador et al. reported the removal of direct blue 78 dyes by using epichlorohydrin modified β-cyclodextrin. Maximum adsorption capacity of dye molecules was observed by Freundlich isotherm, which was 23.47 mg g^{-1} [75]. Ye et al. reported adsorbent based on core/shell structure of β-cyclodextrin/diatomite, which was formed by emulsion polymerization of β-CD coated with diatomite. Maximum adsorption capacity of adsorbent for methylene blue reached at 271.7 mg g^{-1} [78]. Sikdar et al. reported novel adsorbent based on carboxymethyl-β-cyclodextrin attached chitosan impregnated with epichlorohydrin crosslinked β-cyclodextrin polymer. Composite adsorbents exhibited excellent separation capacity for the Cd(II). Separation capacity for the heavy metal was over 378 mg g^{-1}. Heavy metal ions are attached with composite adsorbent by chemisorption and electrostatic interaction (Figure 13.5). Free hydroxyl, amino, and carboxylic functionalities are responsible for the coordination with metal ions [79]. Cellulose nanofiber aerogel grafted with β-cyclodextrin exhibits excellent adsorption capacity of p-chlorophenol. Maximum adsorption capacity of phenol has been reached at 148 μmol g^{-1}. Increased adsorption capacity of phenol by the aerogel adsorbent is due to the excellent porous structure, higher specific surface area, and available functional groups [80]. Fan et al. reported a novel adsorbents based on β-cyclodextrin–chitosan functionalized with Fe$_3$O$_4$ nanoparticles for the separation of methyl blue. β-Cyclodextrin is grafted on the chitosan surface and grafted molecules formed a layer on the magnetic nanoparticles (Figure 13.6) [76].

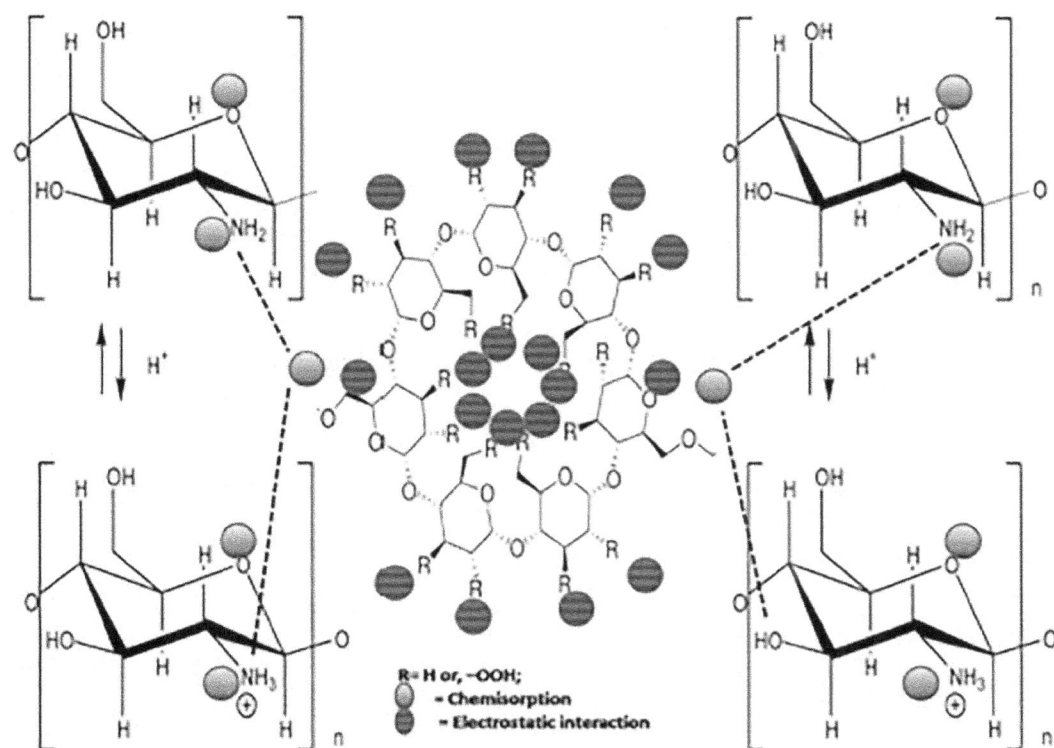

R = H or, –OOH;
○ = Chemisorption
● = Electrostatic interaction

FIGURE 13.5 Schematic showing possible interaction between cyclodextrin-based composite adsorbent and metal ions. Reproduced with permission from Ref [79] © 2017 Elsevier Ltd.

FIGURE 13.6 (a) Scanning electron microscopic image and (b) transmission electron microscopic images of β-cyclodextrin–chitosan modified Fe_3O_4 nanoparticles. Reproduced with permission from Ref [76] © 2011 Elsevier B.V.

13.5 SUMMARY

Water pollutions due to various toxic components remain a serious environmental concern. Adsorption is most widely used technique for the wastewater treatment due to their several advantages, such as cost effectiveness, ease of operation, recyclability, high efficiency, etc. The adsorption process is affected by various factors viz. adsorbent characteristics, various process parameters, and nature of contaminants. Major challenges to select the promising type of adsorbent are based on cost effectiveness, efficiency, selectivity, rapid kinetic, higher adsorption capacity, etc. Polysaccharides are promising adsorbents for wastewater treatment due to their high adsorption capacity, cost effectiveness, renewability, environmental friendliness, ease of surface modification, etc. Physical and chemical modification of polysaccharide possesses superior adsorption properties for the removal of a wide range of contaminants from the wastewater. This chapter presents a review and overview of various polysaccharide-based adsorbents for the wastewater treatments. An account of chitin and modified chitin are discussed for the separation of several contaminants from the wastewater. This is followed by chitosan as adsorbents has been elaborated. Chitosan can be modified by using physical, chemical, and enzymatic method to enhance the adsorption capacity of chitosan-based adsorbents. Starch-based adsorbents containing various functional groups for the separation of toxic components from the wastewater are discussed. Finally, the preparation of cyclodextrin-based adsorbents and their applications for the removal of various contaminants are discussed.

REFERENCES

1. N.B. Singh, G. Nagpal, S. Agrawal, Rachna, Water purification by using adsorbents: A review, *Environmental Technology & Innovation* 11 (2018) 187–240.
2. K. Ali, M.U. Javaid, Z. Ali, M.J. Zaghum, Biomass-derived adsorbents for dye and heavy metal removal from wastewater, *Adsorption Science & Technology* 2021 (2021) 9357509.
3. G. Crini, E. Lichtfouse, L.D. Wilson, N. Morin-Crini, Conventional and non-conventional adsorbents for wastewater treatment, *Environmental Chemistry Letters* 17(1) (2019) 195–213.
4. O.P. Sahu, P.K. Chaudhari, The characteristics, effects, and treatment of wastewater in sugarcane industry, *Water Quality Exposure and Health* 7(3) (2015) 435–444.

5. X.B. Sun, J.F. Zhu, Q.Y. Gu, Y.H. You, Surface-modified chitin by TEMPO-mediated oxidation and adsorption of Cd(II), *Colloids and Surfaces A-Physicochemical and Engineering Aspects* 555 (2018) 103–110.

6. B. Karagozoglu, A. Altin, M. Degirmenci, Flow-rate and pollution characteristics of domestic wastewater, *International Journal of Environment and Pollution* 19(3) (2003) 259–270.

7. W. Janczukowicz, A. Mielcarek, J. Rodziewicz, K. Ostrowska, T. Jozwiak, I. Klodowska, M. Kordas, Quality characteristics of wastewater from malt and beer production, *Rocznik Ochrona Srodowiska* 15 (2013) 729–748.

8. X. Hu, Y.F. Shi, J.L. Wang, Characteristics of municipal wastewater treatment by moving-bed biofilm reactor, *International Journal of Environment and Pollution* 37(2–3) (2009) 177–185.

9. Y. Mountassir, A. Benyaich, M. Rezrazi, P. Bercot, L. Gebrati, Wastewater effluent characteristics from Moroccan textile industry, *Water Science and Technology* 67(12) (2013) 2791–2799.

10. X. Xu, Q.T. Hou, Y.G. Xue, Y. Jian, L.P. Wang, Pollution characteristics and fate of microfibers in the wastewater from textile dyeing wastewater treatment plant, *Water Science and Technology* 78(10) (2018) 2046–2054.

11. D.A. Yaseen, M. Scholz, Textile dye wastewater characteristics and constituents of synthetic effluents: A critical review, *International Journal of Environmental Science and Technology* 16(2) (2019) 1193–1226.

12. F. Murphy, C. Ewins, F. Carbonnier, B. Quinn, Wastewater treatment works (WwTW) as a source of microplastics in the aquatic environment, *Environmental Science & Technology* 50(11) (2016) 5800–5808.

13. Y.Y. Choi, S.R. Baek, J.I. Kim, J.W. Choi, J. Hur, T.U. Lee, C.J. Park, B.J. Lee, Characteristics and biodegradability of wastewater organic matter in municipal wastewater treatment plants collecting domestic wastewater and industrial discharge, *Water* 9(6) (2017) 409.

14. C. Onet, A. Teusdea, A. Onet, E. Pantea, N.C. Sabau, V. Laslo, T. Romocea, E. Agud, Comparative study of dairy and meat processing wastewater characteristics, *Journal of Environmental Protection and Ecology* 19(2) (2018) 508–514.

15. M.M. Kwikima, S. Mateso, Y. Chebude, Potentials of agricultural wastes as the ultimate alternative adsorbent for cadmium removal from wastewater. *A review, Scientific African* 13 (2021) e00934.

16. S.B. Pillai, Adsorption in Water and Used Water Purification, in: J. Lahnsteiner (Ed.), *Handbook of Water and Used Water Purification*, Springer International Publishing, Cham, 2020, pp. 1–22.

17. A. Sihem, M. Bencheikh Lehocine, H.A. Miniai, Preparation and characterisation of an natural adsorbent used for elimination of pollutants in wastewater, *Energy Procedia* 18 (2012) 1145–1151.

18. M. Arvand, R. Shemshadi, A.A. Efendiev, N.A. Zeynalov, Synthetic polymers as adsorbents for the removal of Cd(II) from aqueous solutions, *Asian Journal of Chemistry* 23(6) (2011) 2445–2448.

19. C. Lu, H. Chiu, H. Bai, Comparisons of adsorbent cost for the removal of zinc (II) from aqueous solution by carbon nanotubes and activated carbon, *Journal of Nanoscience and Nanotechnology* 7(4–5) (2007) 1647–1652.

20. A. Bhatnagar, A.K. Minocha, B.H. Jeon, J.M. Park, Adsorptive removal of cobalt from aqueous solutions by utilizing industrial waste and its cement fixation, *Separation Science and Technology* 42(6) (2007) 1255–1266.

21. S. Abbaszadeh, S.R.W. Alwi, C. Webb, N. Ghasemi, Muhamad, II, Treatment of lead-contaminated water using activated carbon adsorbent from locally available papaya peel biowaste, *Journal of Cleaner Production* 118 (2016) 210–222.

22. A. Bagreev, H. Rahman, T.J. Bandosz, Study of regeneration of activated carbons used as H_2S adsorbents in water treatment plants, *Advances in Environmental Research* 6(3) (2002) 303–311.

23. E. Pagalan, M. Sebron, S. Gomez, S.J. Salva, R. Ampusta, A.J. Macarayo, C. Joyno, A. Ido, R. Arazo, Activated carbon from spent coffee grounds as an adsorbent for treatment of water contaminated by aniline yellow dye, *Industrial Crops and Products* 145 (2020) 111953.

24. C. Jiang, S. Cui, Q. Han, P. Li, Q. Zhang, J. Song, M. Li, Study on application of activated carbon in water treatment, *IOP Conference Series: Earth and Environmental Science* 237 (2019) 022049.

25. X.L. Qi, X.Q. Tong, W.H. Pan, Q.K. Zeng, S.Y. You, J.L. Shen, Recent advances in polysaccharide-based adsorbents for wastewater treatment, *Journal of Cleaner Production* 315 (2021) 128221.

26. G. Crini, Recent developments in polysaccharide-based materials used as adsorbents in wastewater treatment, *Progress in Polymer Science* 30(1) (2005) 38–70.

27. I. Anastopoulos, A. Bhatnagar, D.N. Bikiaris, G.Z. Kyzas, Chitin adsorbents for toxic metals: A review, *International Journal of Molecular Sciences* 18(1) (2017) 114.

28. G. Karthikeyan, N.M. Andal, K. Anbalagan, Adsorption studies of iron(III) on chitin, *Journal of Chemical Sciences* 117(6) (2005) 663–672.
29. U. Filipkowska, Adsorption and desorption of reactive dyes onto chitin and chitosan flakes and beads, *Adsorption Science & Technology* 24(9) (2006) 781–795.
30. C.H. Xiong, Adsorption of cadmium (II) by chitin, *Journal of the Chemical Society of Pakistan* 32(4) (2010) 429–435.
31. R. Andreazza, K.K.A. Heylmann, T.R.S. Cadaval, M.S. Quadro, S. Pieniz, F.A.D. Camargo, T.F. Afonso, C.F. Demarco, Copper adsorption by different extracts of shrimp chitin, *Desalination and Water Treatment* 141 (2019) 220–228.
32. G. Akkaya, I. Uzun, F. Guzel, Kinetics of the adsorption of reactive dyes by chitin, *Dyes and Pigments* 73(2) (2007) 168–177.
33. K.M. Pang, S. Ng, W.K. Chung, P.K. Wong, Removal of pentachlorophenol by adsorption on magnetite-immobilized chitin, *Water Air and Soil Pollution* 183(1–4) (2007) 355–365.
34. M. Yahyaei, F. Mehrnejad, H. Naderi-Manesh, A.H. Rezayan, Protein adsorption onto polysaccharides: Comparison of chitosan and chitin polymers, *Carbohydrate Polymers* 191 (2018) 191–197.
35. D. Schleuter, A. Gunther, S. Paasch, H. Ehrlich, Z. Kljajic, T. Hanke, G. Bernhard, E. Brunner, Chitin-based renewable materials from marine sponges for uranium adsorption, *Carbohydrate Polymers* 92(1) (2013) 712–718.
36. K.S. Wong, K.H. Wong, S. Ng, W.K. Chung, P.K. Wong, Adsorption of copper ion on magnetite-immobilised chitin, *Water Science and Technology* 56(7) (2007) 135–143.
37. R. Yang, Y. Su, K.B. Aubrecht, X. Wang, H.Y. Ma, R.B. Grubbs, B.S. Hsiao, B. Chu, Thiol-functionalized chitin nanofibers for As (III) adsorption, *Polymer* 60 (2015) 9–17.
38. G.L. Dotto, J.M.N. Santos, I.L. Rodrigues, R. Rosa, F.A. Pavan, E.C. Lima, Adsorption of methylene blue by ultrasonic surface modified chitin, *Journal of Colloid and Interface Science* 446 (2015) 133–140.
39. Z.C. Li, G.L. Dotto, A. Bajahzar, L. Sellaoui, H. Belmabrouk, A. Ben Lamine, A. Bonilla-Petriciolet, Adsorption of indium (III) from aqueous solution on raw, ultrasound- and supercritical-modified chitin: Experimental and theoretical analysis, *Chemical Engineering Journal* 373 (2019) 1247–1253.
40. M.S. Tunc, O. Hanay, B. Yildiz, Adsorption of chlortetracycline from aqueous solution by chitin, *Chemical Engineering Communications* 207(8) (2020) 1138–1147.
41. J. Ghourbanpour, M. Sabzi, N. Shafagh, Effective dye adsorption behavior of poly(vinyl alcohol)/chitin nanofiber/Fe(III) complex, *International Journal of Biological Macromolecules* 137 (2019) 296–306.
42. Y.L. Cao, Z.H. Pan, Q.X. Shi, J.Y. Yu, Modification of chitin with high adsorption capacity for methylene blue removal, *International Journal of Biological Macromolecules* 114 (2018) 392–399.
43. Y.Q. Duan, A. Freyburger, W. Kunz, C. Zollfran K, Lignin/chitin films and their adsorption characteristics for heavy metal ions, *ACS Sustainable Chemistry & Engineering* 6(5) (2018) 6965–6973.
44. L. Liu, R. Wang, J. Yu, L.J. Hu, Z.G. Wang, Y.M. Fan, Adsorption of reactive blue 19 from aqueous solution by chitin nanofiber-/nanowhisker-based hydrogels, *RSC Advances* 8(28) (2018) 15804–15812.
45. G.L. Dotto, J.M. Cunha, C.O. Calgaro, E.H. Tanabe, D.A. Bertuol, Surface modification of chitin using ultrasound-assisted and supercritical CO$_2$ technologies for cobalt adsorption, *Journal of Hazardous Materials* 295 (2015) 29–36.
46. S. Begum, N.Y. Yuhana, N.M. Saleh, N.H.N. Kamarudin, A. Sulong, Review of chitosan composite as a heavy metal adsorbent: Material preparation and properties, *Carbohydrate Polymers* 259 (2021) 117613.
47. A.M. Omer, R. Dey, A.S. Eltaweil, E.M. Abd El-Monaem, Z.M. Ziora, Insights into recent advances of chitosan-based adsorbents for sustainable removal of heavy metals and anions, *Arabian Journal of Chemistry* 15(2) (2022) 103543.
48. M. Szlachta, R. Neitola, S. Peraniemi, J. Vepsalainen, Effective separation of uranium from mine process effluents using chitosan as a recyclable natural adsorbent, *Separation and Purification Technology* 253 (2020) 117493.
49. N.F. Al-Harby, E.F. Albahly, N.A. Mohamed, Kinetics, isotherm and thermodynamic studies for efficient adsorption of Congo red dye from aqueous solution onto novel cyanoguanidine-modified chitosan adsorbent, *Polymers* 13(24) (2021) 4446.
50. X.Y. Cui, Y.J. Wang, Y.M. Yan, Z.L. Meng, R.H. Lu, H.X. Gao, C.P. Pan, X.L. Wei, W.F. Zhou, Phenylboronic acid-functionalized cross-linked chitosan magnetic adsorbents for the magnetic solid-phase extraction of benzoylurea pesticides, *Journal of Separation Science* 45(4) (2022) 908–918.
51. Y. Kurita, A. Isogai, Reductive N-alkylation of chitosan with acetone and levulinic acid in aqueous media, *International Journal of Biological Macromolecules* 47(2) (2010) 184–189.
52. A.A. Golyshev, Y.E. Moskalenko, Y.A. Skorik, Comparison of the acylation of chitosan with succinic anhydride in aqueous suspension and in solution, *Russian Chemical Bulletin* 64(5) (2015) 1168–1171.

53. F. Karadeniz, M.Z. Karagozlu, S.Y. Pyun, S.K. Kim, Sulfation of chitosan oligomers enhances their anti-adipogenic effect in 3T3-L1 adipocytes, *Carbohydrate Polymers* 86(2) (2011) 666–671.
54. Y.A. Skorik, A.V. Pestov, M.I. Kodess, Y.G. Yatluk, Carboxyalkylation of chitosan in the gel state, *Carbohydrate Polymers* 90(2) (2012) 1176–1181.
55. N.F. Al-Harby, E.F. Albahly, N.A. Mohamed, Synthesis and characterization of novel uracil-modified chitosan as a promising adsorbent for efficient removal of Congo red dye, Polymers 14(2) (2022) 271.
56. Z.B. Xu, Y.Q. Yu, L. Yan, W. Yan, C.Y. Jing, Asenic removal from groundwater using granular chitosan-titanium adsorbent, *Journal of Environmental Sciences* 112 (2022) 202–209.
57. D.Z. Yang, L.B. Qin, Y.L. Yang, Efficient adsorption of methyl orange using a modified chitosan magnetic composite adsorbent, *Journal of Chemical and Engineering Data* 61(11) (2016) 3933–3940.
58. H. Hassan, A. Salama, A.K. El-ziaty, M. El-Sakhawy, New chitosan/silica/zinc oxide nanocomposite as adsorbent for dye removal, *International Journal of Biological Macromolecules* 131 (2019) 520–526.
59. Z. Ji, Y.S. Zhang, H.C. Wang, C.R. Li, Polypropylene glycol modified chitosan composite as a novel adsorbent to remove Cu(II) from wastewater, *Tenside Surfactants Detergents* 58(6) (2021) 486–489.
60. N.V. Hoa, N.C. Minh, H.N. Cuong, P.A. Dat, P.V. Nam, P.H.T. Viet, P.T.D. Phuong, T.S. Trung, Highly porous hydroxyapatite/graphene oxide/chitosan beads as an efficient adsorbent for dyes and heavy metal ions removal, *Molecules* 26(20) (2021) 6127.
61. X. Li, D.L. Zeng, Z.Y. He, P. Ke, Y.S. Tian, G.H. Wang, Magnetic chitosan microspheres: An efficient and recyclable adsorbent for the removal of iodide from simulated nuclear wastewater, *Carbohydrate Polymers* 276 (2022) 118729.
62. W.J. Liu, Q.L. Wang, H.Q. Wang, Q. Xin, W. Hou, E.M. Hu, Z.W. Lei, Adsorption of uranium by chitosan/Chlorella pyrenoidosa composite adsorbent bearing phosphate ligand, *Chemosphere* 287 (2022) 132193.
63. F. Karimi, A. Ayati, B. Tanhaei, A.L. Sanati, S. Afshar, A. Kardan, Z. Dabirifar, C. Karaman, Removal of metal ions using a new magnetic chitosan nano-bio-adsorbent: A powerful approach in water treatment, *Environmental Research* 203 (2022) 111753.
64. A. Salama, R.E. Abou-Zeid, Ionic chitosan/silica nanocomposite as efficient adsorbent for organic dyes, *International Journal of Biological Macromolecules* 188 (2021) 404–410.
65. M.R. Abukhadra, S.M. Ibrahim, J.S. Khim, A.A. Allam, J.S. Ajarem, S.N. Maodaa, Enhanced decontamination of pefloxacin and chlorpyrifos as organic pollutants using chitosan/diatomite composite as a multifunctional adsorbent: equilibrium studies, *Journal of Sol-Gel Science and Technology* 99(3) (2021) 650–662.
66. L.Z. Qiao, S.S. Wang, T. Wang, S.S. Yu, S.H. Guo, K.F. Du, High-strength and low-swelling chitosan/cellulose microspheres as a high-efficiency adsorbent for dye removal, *Cellulose* 28(14) (2021) 9323–9333.
67. J. Pan, J.W. Zhu, F.L. Cheng, Preparation of sodium lignosulfonate/chitosan adsorbent and application of Pb2+ treatment in water, *Sustainability* 13(5) (2021) 2997.
68. I.O. Bisiriyu, R. Meijboom, Adsorption of Cu(II) ions from aqueous solution using pyridine-2,6-dicarboxylic acid crosslinked chitosan as a green biopolymer adsorbent, *International Journal of Biological Macromolecules* 165 (2020) 2484–2493.
69. A.O. Francis, M.A.A. Zaini, I.M. Muhammad, S. Abdulsalam, U.A. El-Nafaty, Physicochemical modification of chitosan adsorbent: A perspective, *Biomass Conversion and Biorefinery*, 13 (2023) 5557–5575.
70. I. Ihsanullah, M. Bilal, A. Jamal, Recent developments in the removal of dyes from water by starch-based adsorbents, *Chemical Record*, (2022), https://doi.org/10.1002/tcr.202100312.
71. L. Huang, C.M. Xiao, B.X. Chen, A novel starch-based adsorbent for removing toxic Hg(II) and Pb(II) ions from aqueous solution, *Journal of Hazardous Materials* 192(2) (2011) 832–836.
72. F. Haq, H.J. Yu, L. Wang, L.S. Teng, S. Mehmood, M. Haroon, A. Bilal Ul, M.A. Uddin, S. Fahad, D. Shen, Synthesis of succinylated carboxymethyl starches and their role as adsorbents for the removal of phenol, *Colloid and Polymer Science* 299(11) (2021) 1833–1841.
73. Z.Q. Ren, C. Liu, B.Y. Zhang, M.S. Wu, Y. Tan, X. Fang, P.F. Yang, L.X. Liu, Preparation of amino-functionalized starch-based adsorbent and its adsorption behavior for uranyl ions, *Journal of Radioanalytical and Nuclear Chemistry* 328(3) (2021) 1253–1263.
74. S.H. Hegazy, S.K. Mohamed, Starch-graft-polyacrylamide copolymer/Fe3O4/graphene oxide nanocomposite: Synthesis, characterization, and application as a low-cost adsorbent for Ni (II) from aqueous solutions, *Journal of Polymer Research* 28(2) (2021) 49.
75. A. Murcia-Salvador, J.A. Pellicer, M.I. Fortea, V.M. Gomez-Lopez, M.I. Rodriguez-Lopez, E. Nunez-Delicado, J.A. Gabaldon, Adsorption of direct blue 78 using chitosan and cyclodextrins as adsorbents, *Polymers* 11(6) (2019) 1003.

76. L.L. Fan, Y. Zhang, C.N. Luo, F.G. Lu, H.M. Qiu, M. Sun, Synthesis and characterization of magnetic beta-cyclodextrin-chitosan nanoparticles as nano-adsorbents for removal of methyl blue, *International Journal of Biological Macromolecules* 50(2) (2012) 444–450.

77. Q.K. Wei, L. Bai, X.M. Qin, C.Y. Hu, L. Li, W. Jiang, F. Song, Y.Z. Wang, Contrastive study on beta-cyclodextrin polymers resulted from different cavity-modifying molecules as efficient bi-functional adsorbents, *Reactive & Functional Polymers* 154 (2020) 104686.

78. Y.M. Ye, J.F. Shi, Y.H. Zhao, L.W. Liu, M. Yan, H.T. Zhu, H. Zhang, Y. Wang, J. Guo, Y.Z. Wang, J. Sun, Facile preparation of core-shell structure beta-cyclodextrin/diatomite as an efficient adsorbent for methylene blue, *European Polymer Journal* 136 (2020) 109925.

79. M.T. Sikder, M. Jakariya, M.M. Rahman, S. Fujita, T. Saito, M. Kurasaki, Facile synthesis, characterization, and adsorption properties of Cd (II) from aqueous solution using beta-cyclodextrin polymer impregnated in functionalized chitosan beads as a novel adsorbent, *Journal of Environmental Chemical Engineering* 5(4) (2017) 3395–3404.

80. F. Zhang, W.B. Wu, S. Sharma, G.L. Tong, Y.L. Deng, Synthesis of cyclodextrin-functionalized cellulose nanofibril aerogel as a highly effective adsorbent for phenol pollutant removal, Bioresources 10(4) (2015) 7555–7568.

14 Polysaccharides in Energy Storage

Nizam P. A. and Sabu Thomas
Mahatma Gandhi University

14.1 INTRODUCTION

With energy being a big problem in the 21st century and the depletion of fuels and the environment occurring at the same time, researchers have dug deep into sustainable energy sources as well as ecologically acceptable materials for energy applications. The expanding population and the resulting need to fulfill their requirements, along with industrial developments, emit a high level of pollutants into the environment, having a significant influence on climate conditions and ecological balances. The use of early kinds of energy, such as fossil fuels, and technologies such as combustion engines, must be curtailed, and environmentally friendly energies must be developed.

Sustainable energy sources, defined as those that can be replenished or utilized for today's needs without compromising or depleting future generations, are being researched. This encompasses solar, tidal, wind, mechanical, and other forms of energy. These types of energy are environmentally friendly and do not pollute the environment. Various ways are used to produce power from various sources. The key problem in this regard is the technology for storing these energies, which necessitates energy storage devices with high efficiency, cyclability, and charging-discharging rates. Supercapacitors, batteries, fuel cells, and nanogenerators are all popular energy storage systems, with the latter two also producing electricity. Fuel cells use hydrogen as a fuel to create electricity, with the only byproduct being a waste vapor, whereas nanogenerators generate power through mechanical movement or strain. This energy storage device composition, which comprises electrodes, cathodes, separators, binders, and electrolytes, is largely made of synthetic polymeric materials that may be handled as polluting waste after use. It is imperative that they be replaced with biofriendly and biodegradable polymers.

Polysaccharides are biopolymers composed of simple sugar monosaccharides joined together by glycosidic bonds. Polysaccharides have a linear or highly branched molecular structure and can be composed of the same (homopolysaccharide) or diverse (heteropolysaccharide) monosaccharide units. They are taken from plants as well as living beings and include cellulose, starch, chitosan, alginates, and so on. Their abundance, low cost, biodegradability, mechanical qualities, and interactions with water make them ideal for sustainable materials. Their chemical structure also allows for additional alteration, allowing them to be tailored for a wide range of uses. They are employed in various domains such as water purification, electronics (Pai et al., 2022), therapeutics, and other industries. These materials have been used in energy storage systems as electrodes, electrolytes, separators, binders, and so on.

This chapter discusses various polysaccharides such as cellulose, starch chitosan, and alginates in energy storage applications such as supercapacitors, batteries, fuel cells, and nanogenerators. The history of energy storage along with the development of bio-based sustainable materials for their applications are briefly described.

DOI: 10.1201/9781003265054-14

14.2   HISTORY AND EVOLUTION OF ENERGY STORAGE

Energy storage may be traced back to the beginning of civilizations when wood and charcoal were used as energy sources. Charcoal was the primary driving force of ancient civilizations because it works as a biomass energy store of solar power. The process was simple when the energy stored in charcoal or woods is fired to cook food, to bring warmth and brightness. Later, the discovery of coal, a fossil fuel that had been buried for millions of years, revolutionized the history of energy since these materials store solar energy at a high density. They were also employed as a fuel in steam engines, which marked the beginning of the first industrial revolution, and later to generate electricity.

Petroleum, extracted from the biodegradable organic materials, was another source of fuel which were extracted massively by the beginning of the 20th century. Many additional by-products, such as synthetic fibers, polymers, and plastics, were produced from this petroleum. Natural gas, coal, and oil are all forms of natural energy storage for solar energy. This context may be explained as follows: plants absorb solar energy through photosynthesis and develop into wood, and these solar energies are stored in these materials before being transformed into fossil fuels, which then release the solar energy in the form of heat (Liu et al., 2010).

Later in the 1970s, electric generators and motors were invented, and thus electrical energy became the important secondary energy source of consumed energy. Traditionally, electricity is created via solar power, fuel-burning power, hydropower, nuclear power, wind power, biopower systems, tidal power, and other means, and it is essential in practically every aspect of our lives, from lighting, cooling, and heating to entertainment, communication, and transportation. With the growing population and advancements in technologies, there arises a need to design storage devices to store these energies. Since the late 1800s, energy-storing applications were researched where a large double-layer electrode in polyelectrolyte designs was invented. Traditional energy storage devices like supercapacitors, batteries, etc. gained popularity due to their characteristics like longer life cycle, power density, energy density, etc. Because of these amazing benefits, they have become a prospective supplier for digital cameras, industrial equipment, electronic devices, and other applications. Furthermore, the fast-growing market for portable electronic devices such as mobile phones and laptop computers, as well as their development trend of being small, lightweight, and flexible, have resulted in an ever-increasing and urgent demand for ecofriendly electrochemical energy storage and conversion systems (Jadhav, Mane and Shinde, 2020).

14.3   AN OVERVIEW OF ENERGY STORAGES

Energy storage devices are being researched as there is a growing need to store renewable energies. Depletion of fossil fuels has lowered their consumption for generating electricity, and new efficient and environmentally friendly energy resources such as solar, wind, and tidal are being exploited. These arose the demands for energy storage devices like batteries and supercapacitors. Let's have a peek into the different devices used for energy storage applications.

Lithium-ion batteries (LIBs) are particularly attractive for energy storage because of characteristics such as high Columbic efficiency, high-energy densities, minimal self-discharge, and so on. Their common applications include aircraft, electrical equipment, hybrid electric cars, and so forth. Traditional battery systems based on the Li^+ intercalation method are inefficient in producing large charge capacities and hence require modification (Kim et al., 2019). Therefore, new technologies are being investigated based on lithium-rich, transition-rich oxides, Li-sulfur, organic electrodes, Li-air, etc. Supercapacitors are another type of energy storage device that resembles the working of a battery, employs a thin dielectric high surface area material as an electrode, and broadcasts capacitance those are several highs in magnitude when compared with traditional capacitors. They bridge the gaps between energy density and power density of electrolytic capacitors and conventional batteries. Batteries have a low power density but a high-energy density of 100–265 Wh kg^{-1}. Supercapacitors

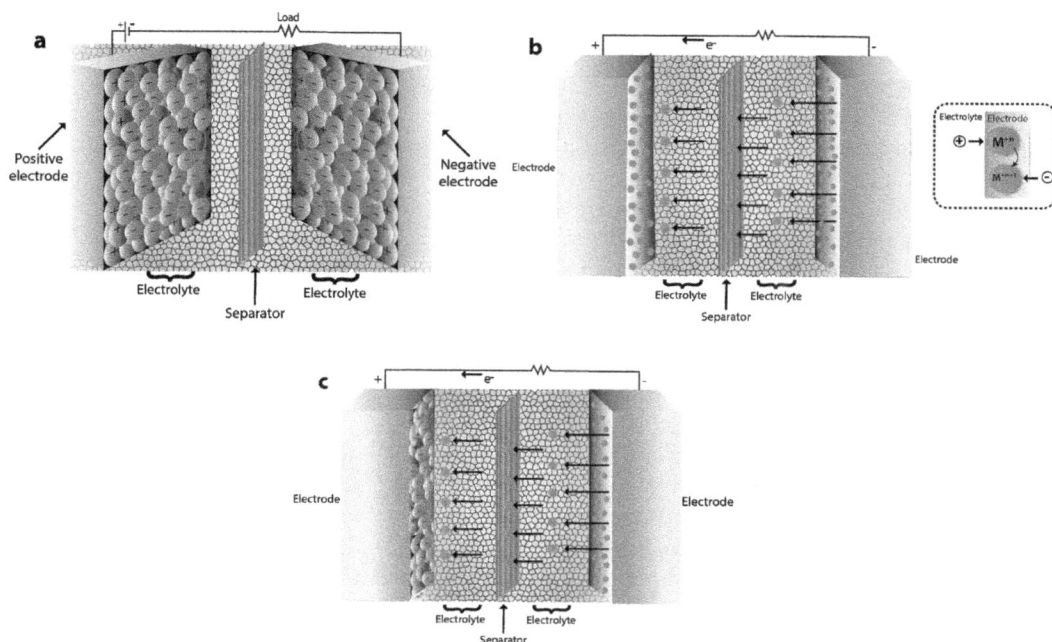

FIGURE 14.1   (a) EDLC type (electrodes are parallel, drawn outward to picturize double-layer formation). (b) Pseudocapacitor. (c) Hybrid.

have good energy density as well as high-power density which are 100 times greater than batteries (Xie et al., 2018).

Three types of supercapacitors are distinguished. Electrochemical double layer (EDLC), pseudocapacitors, and hybrid (Figure 14.1). The EDLC type has no charge transfer between electrolytes and electrodes and operates in a non-faradaic manner with no chemical reactions. When a voltage is supplied, charges collect at the electrode/electrolyte interface and are stored on the pores of the opposite charge electrodes through ion diffusion. Because of the great reversibility of the storing charge, EDLCs may run for numerous charge-discharge cycles without experiencing morphological or volume changes. For electroneutrality balance, a double layer of opposing charges is formed on the electrodes. Charging entails moving electrons from negative electrodes to positive electrodes in conjunction with an external supply in which anions in the electrolyte shift toward positive electrodes and cations shift toward negative electrodes. The discharge procedure is the inverse of this procedure. The electrolyte concentration remains constant as energy is stored in the double-layer interface. The most prevalent electrode materials are activated carbon (Zachariah, Nizam, Grohens and Kalarikkal, 2021) and graphene, which have outstanding features such as large surface area, high porosity, and suitable electrolyte solutions (Lokhande, Chavan and Pandey, 2020).

Pseudocapacitors (PC) function in the opposite way as ELDCs, with reversible redox processes between electrolytes and electrodes that create and transfer charges across the double layer. The pseudocapacitance of pseudocapacitors is governed by thermodynamic considerations. The process is faradaic and may be broken down into three steps: reversible adsorption, redox reactions at metal oxides, and doping/de-doping at the electrodes. Due to faradaic reactions that occur at the electrode as well as the near-surface, this kind has a higher energy density and specific capacitance than the EDLC. Electrodes are often made of a high-energy combination of conducting polymer and metal oxide/hydroxide. Polypyrrole (Ppy), polyaniline (PANI), MnO_2, RuO_2, $Co(OH)_2$, and other common electrode materials include Ppy, PANI, and MnO_2. The power density of PC is less than EDLC as faradaic processes are slower. Furthermore, the electrodes are prone to volume change when charging and discharging, reducing the life lifetime progressively (Lokhande, Chavan and Pandey, 2020). The low power density of PC and the low specific capacitance of EDLCs are overcome by employing a hybrid system of EDLC and PC in which faradaic and non-faradaic processes operate concurrently. Capacitance and cell voltages may be increased by using appropriate electrode materials, resulting in better power and energy densities, as well as longer lifespan and stability. Based on the arrangement, three types of hybrid supercapacitors are being addressed: composites (produced by mixing EDLC and PC electrodes), asymmetric (redox reaction at one end, and non-faradaic at the other end), and battery (a mixture of battery and supercapacitors electrodes).

A fuel cell is another form of energy storage device that generates electricity from fuels such as hydrogen. When hydrogen is used as a fuel, it produces electricity, as well as by-products, such are water and heat, making it more ecologically friendly. Fuel cells function similarly to batteries but do not require charging. An electrolyte separates two electrodes in a fuel cell system, the anode receives hydrogen fuel, whereas the cathode receives air or oxygen. The hydrogen atoms are disintegrated into electrons and protons by a catalyst at the anode. Positive protons can pass throw the electrolyte membrane but electron flow is restricted. They flow through the electrolyte to the cathode, whereas electrons pass through a circuit, producing electricity and accumulating at the cathode. At the cathode, hydrogen proton, electron, and oxygen molecules combine to make water. The process produces water and heat as by-products (Dong et al., 2013).

Nanogenerators, which turn minuscule mechanical motions into electricity, are another type of energy generation and storage technology. Nanogenerators of three sorts are being developed: triboelectric, piezoelectric, and pyroelectric nanogenerators. The first two ways convert mechanical movements or energies into electricity, whereas the third uses thermal energy to generate electricity. Two dielectric materials are attached to two electrodes, which when exposed to external strain generate a charge when they come into touch and a voltage when they move up generating a gap. The load is balanced by passing it through an external circuit, which balances the electrostatic level. Piezoelectric movement is the movement of two materials upward toward the position. Triboelectric effects are produced when these two materials glide or brush against one other. The application of polysaccharides on these energy storage devices is briefly described in the following sections (Kim et al., 2021).

14.4 POLYSACCHARIDES IN SUPERCAPACITORS

In the past decade, there has been a growing interest in biopolymers under polysaccharides due to their potential and characteristics such as structural diversity, abundance, physicochemical properties, and low cost. Polysaccharides that are often used include starch, chitosan, alginates, cellulose, lignin, and others. They are used directly for conductive composites or as a template for high surface area supercapacitor material. 'The primary focus of supercapacitor research is to enhance the surface areas and conductivity of the electrode. With properties such as pore size, high chemical stability, and environmental endurance, porous carbon is a suitable electrode material for supercapacitors. The electrochemical performance of these materials is determined by the raw material from which the carbon is generated as well as the method of activation.

14.4.1 CELLULOSE

Cellulose is the most prevalent biopolymer on the planet, having unique properties such as low cost, renewability, biodegradability, biocompatibility, and flexibility. Their physiochemical and structural properties enable them to be modified in a variety of ways, allowing them to be used in a wide range of applications such as water purification (Nizam et al., 2020), electronics (Nizam et al., 2021), composite parts, reinforcements (Rose Joseph et al., 2021), etc. The fibrous structure and porous nature of cellulose improve the encapsulation of different nanoparticles to their surface, increasing their applications in supercapacitors (Selvaraj et al., 2020). Cellulose is an insulator by nature; however, their aforementioned qualities allow for tunability with other conducting materials, enabling them to be used in supercapacitor applications.

Before we go into cellulose-based composites, let's have a look at cellulose-based carbon electrodes. Wang et al. demonstrated in their work that carbon fabricated using cellulose as a raw material has a high surface area and porosity. The preparation process and temperature have a significant impact on their surface characteristics. Spray drying a $Zn(EDA)_2^{2+}$/cellulose solution at a certain pressure and nozzle temperature was performed. The calcination at a higher temperature eliminates the volatile elements, which are then washed away with acid. The calcination temperature determines the zinc size and the final product, so temperatures were monitored and studies were conducted at various temperatures. The result revealed the influence of temperature, which enhances the surface area up to 1,000°C and then begins to demote above that temperature, implying the effect of temperature on the preparation of carbon materials (Wang et al., 2018). Moreover, carbon-based fibers with uniform diameter are also an excellent electrode material which is developed using the carbonization technique using cellulose lignin as raw materials exhibiting a specific capacitance of 346.6 F g^{-1} (Cao et al., 2020).

A simple scalable approach for producing graphene-induced cellulose paper was reported. The sonicated graphene suspension was vacuum filtered through cellulose filter paper on both sides until the suspension became transparent. The three-dimensional woven network of cellulose and graphene provides a flexible, mechanically resilient electrode with high power performance, specific capacitance, and cyclic stability (Weng et al., 2011). The presence of a conducting filler in a cellulose matrix enhances specific capacitance, but what if the filler is a nonconductive graphene derivate? The graphene oxide, which is normally not conductive, was mixed with cellulose, and the resulting composite had intriguing features. A capacitance of 17.3 F g^{-1} was measured, as well as an increase in current density in the cv curve and galvanostatic charging-discharging time. These findings are the result of increased porosity in the matrix, which improves surface area and consequently electrode properties (Kafy et al., 2017).

Previous research has demonstrated that a single system of either carbonaceous material or conducting polymer does not match the requirements of a supercapacitor electrode due to its low energy density. To illustrate, cellulose nanofibers (CNFs)/PANI aerogels were investigated for supercapacitors and demonstrated a specific capacitance of 59.26 mF cm^{-2} (Wang et al., 2019). These low values are addressed by employing a different matrix scheme of more than one filler. A cellulose and graphene oxide scaffold aerogel was produced, and PANI nanoclusters were synthesized inside the scaffold in situ. This combination improves the electrode properties, yielding a high areal-specific capacitance of 1,218 mF cm^{-2} (Li et al., 2020). Polypyrrole, another conducting polymer, produced comparable results when a composite of cellulose and reduced graphene oxide was synthesized through wet spinning and in-situ polymerization of polypyrrole. These composites achieved a specific capacitance of 391 F g^{-1}, demonstrating the benefit of using two fillers in supercapacitor applications (Sheng et al., 2019).

Inorganic nanomaterials are excellent materials for reinforcements (Nizam, Binumol and Thomas, 2021) for supercapacitor applications. Graphene-based inorganic nanoparticles were discussed earlier. Other inorganic metal oxide nanoparticles and carbon nanotubes (CNT) are also employed for electrodes in supercapacitors. Multiwalled carbon nanotubes wrapped around CNF along with TiO_2 nanotubes exhibited a discharge capacity of 62.5 mF cm^{-2}. TiO_2 has a distinct advantage in ultrafast charge-discharge processes via a pseudocapacitive charging mechanism. The inclusion of MWCNT in the composite inhibits TiO_2 agglomeration, functioning as a suitable dispersion agent for metal oxides (Wang and Li, 2015). Metal-organic frames a class of inorganic porous material linked with organic ligands and inorganic clusters via coordination bonds are attractive materials for fabricating electrodes for supercapacitors. These materials have the extra advantage of turnability with predetermined architecture and properties for various applications. Their high surface area and electrical conductivity make them an ideal candidate for electrochemical storage applications. CNF is employed as a template for the synthesis of the MOF due to its high porosity. These nano-paper composites have a high electrical conductivity of up to 100 S cm^{-1} with exceptional mechanical properties and porosity. Furthermore, the constructed supercapacitor exhibits extraordinarily high cycle stability, with capacitance retentions of more than 99% after 10,000 continuous charge/discharge cycles (Zhou et al., 2019).

Cellulose is also employed as a separator, electrolyte as well as a binder for supercapacitors. The primary goal of a separator is to prevent contact between the two electrodes to prevent a short circuit. An ideal separator meets the properties such as high thermal and chemical stability, porosity, hydrophilicity, mechanical properties, etc. The pores in the cellulose enhance a higher electrolyte uptake as well as ionic conductivity which are prerequisites for an ideal separator. These points were confirmed in a study where cellulose was dissolved in a solvent and films were cast using the phase inversion method. The fabricated films exhibited excellent mechanical properties, hydrophilicity, thermal stability, porosity (59%), electrolyte absorption (330%), etc. A charge-discharge efficiency of 99% at 1 A g^{-1} was attained. This implies the future of cellulose for separator applications (Teng et al., 2020). Their further applications in energy storage are discussed in the following sections.

14.4.2   CHITOSAN

Chitosan is a semicrystalline, cationic polysaccharide found abundantly in nature. They are produced by partial deacetylation of chitin (more than 70%), another bio-based polymer. The degree of deacetylation plays an important role in chitosan formed where higher DD manifests chitosan with high chemical stability, superior biological activity, and good solubility in an acidic medium (Roy, Tahmid and Rashid, 2021). These materials are extensively researched for their use as electrodes and electrolytes due to their ionic conductivity and excellent mechanical strength. The ionic conduction is a result of the lone-pair electron in -OH, NH$_2$, and C-O-C groups in the structure.

Chitosans are employed as a precursor for the production of porous carbon electrodes for supercapacitors. They demonstrate great competitive excellence in the energy storage industry due to their exceptional porosity, high surface area, and low production cost. Porous carbon alone doesn't work well as an electrode and requires further modifications such as doping with heteroatoms. The modifications improve the wettability, electronic conductivity, as well as specific capacitance. Two commonly employed doping routes include artificial doping where the activated carbons are treated with several types of reagents to form functional groups on the surface of the carbon. Self-co-doping is an efficient method where the carbon precursor act as a host for heteroatoms. The structural features of chitosan with amino groups permit self-doped heteroatoms, thus providing a better uniform dispersion of these atoms in the carbon matrix. Hierarchical porous carbons provide better performance as an electrode material. The micropores in the carbon are not that efficient for electrolytes thus limiting ion transportation. A combined, mesoporous, microporous, and microporous carbon is considered efficient for supercapacitors. Nitrogen-doped chitosan precursor-based porous carbon is produced by microwave carbonization of ZnCl$_2$. By modulating the ZnCl$_2$ concentration, varied porosity was obtained and exhibited a specific capacitance of 435 F g^{-1} (Abbasi, Antunes and Velasco, 2019; Li et al., 2017). Such modifications with heteroatoms are also carried out in cellulose for synthesizing porous carbons.

Chitosan is also used as a composite for supercapacitor electrodes. Chitosan-conducting polymer, chitosan-graphene/other inorganic metal oxides, and chitosan-conducting polymer-graphene/metal oxide are examples of typical binary and tertiary systems. Because of the nanostructures, these composites often have a high pseudocapacitive and porosity. Conductive polymers and nanoparticles generally give electric energy via a faradaic reaction redox process, and carbon scaffolds offer an electron channel for this created energy to flow. PANI, although a great conductor, lacks cyclic stability due to weak shrinkage and swelling. These drawbacks can be mitigated by attaching PANI to supporting materials like graphene or metal oxides. These materials must fulfill characteristics such as excellent electrical conductivity and an appropriate structural channel for electron transport. Graphene performs well in this function for PANI, albeit it tends to clump or agglomerate, resulting in poor PANI graphene dispersion. This system may be enhanced further by employing chitosan as a graphene stabilizing agent, resulting in a more uniform dispersion of the electrode. The specific capacitance of these composite electrodes was 340 F g^{-1} (Jang et al., 2021).

Chitosan is also used as an electrolyte in supercapacitors. 1-ethyl-3-methyl imidazolium tetra-fluoroborate and chitosan-based gel electrolytes with outstanding mechanical characteristics and ionic liquid retentions were produced for EDLC. Despite the use of a gel-state electrolyte, charge-discharge investigations revealed that a model EDLC cell with a chitosan gel had a high discharge capacitance, implying the strong affinity of chitosan for the activated carbon electrode decreases the electrode-electrolyte interfacial resistance. During 5,000 cycles, these composites retained a coulombic efficiency of greater than 99.9% (Yamagata et al., 2013). Another gel electrolyte based on carbonylated chitosan in HCl that was generated using the phase separation procedure had exceptional flexibility, an electrolyte absorption rate of 742.0 wt.%, and the highest ionic conductivity of 8.69 10^{-2} S cm^{-1}. The current progress in the development of chitosan-based supercapacitors is much more visible than it was previously. Research and studies in the right areas will be able to overcome the existing limitations of manufactured supercapacitors, ensuring environmental safety and allowing human society to make a significant leap in energy storage technology (Yang et al., 2019).

14.4.3 STARCH

Starch, like other polysaccharides, has gained a lot of interest due to its availability, renewability, low cost, and so on, and these materials are used in applications including bi-sensors, packaging, supercapacitors, and so on. As two sides of a coin starch also possess disadvantages like inherent properties such as low water solubility, shear resistance, high viscosity, poor mechanical properties, etc. These inherent qualities are frequently addressed by reinforcing with extra fillers or adding co-biopolymers. Starch, like the previously stated polysaccharides, is employed in supercapacitor applications as a precursor for porous carbon electrodes because of its carbon-rich composition, low cost, and degradable properties.

Porous carbon nanofibers produced from starch, electrospun in PVA, and coated with cobalt oxide have a high specific capacitance of 137 F g^{-1}. The superior porosity nature of the composite, as well as the synergetic impact of carbon and cobalt oxide, contributes to the electrode's performance (Kebabsa et al., 2020). Dialdehyde starch is employed as a reducing agent in the incorporation of GO with PANI. When graphene aggregates in a PANI/graphene composite, it precipitates. This is solved by utilizing starch as a reducing agent, and the resulting composite has a specific capacitance of 499 F g^{-1} (Wu et al., 2014). Starch and GO are combined to create a green hybrid binder. The electrodes for supercapacitors were then built utilizing a standard and industrial-ready manufacturing procedure that simply required water as an ecological solvent. The advantage of employing starch as green support for GO in supercapacitor applications is demonstrated by a specific capacitance of 174 F g^{-1} (Rapisarda, Marken and Meo, 2021).

14.4.4 ALGINATES

Abundant and non-toxic marine biomass resources, such as sodium alginate (SA), have been exploited to develop energy storage materials employing innovative synthesis methodologies. SA is a natural anionic polysaccharide of high modulus derived from abundant brown seaweed that is constituted of the monomers (1,4)-D-mannuronic acid (M) and (1,3)-L-guluronic acid (G). SA is an excellent precursor for supercapacitors due to their high amount of hydroxyl and carboxyl functional groups, which provide some pseudocapacitance. The –COO- groups of SA monomers may be chelated with metal ions such as Cu^{2+}, Fe^{3+}, Co^{2+}, and Cr^{3+} by an ion-exchange process. Alginates like other polysaccharides can be used to fabricate environmentally friendly aerogel of porous carbon. In one study, copper-induced carbon aerogel from alginates prepared using ion-exchange, freeze-dried, and one-step carbonization exhibits a surface area of $230\,m^2g^{-1}$. The synergetic effect of Cu along with the porous structure enhances the electrochemical properties in carbon aerogel with a specific capacitance of $415\,F\,g^{-1}$ (Zhai et al., 2019). A similar result was observed when nitrogen doped in SA porous carbon aerogels were prepared (Zhao et al., 2020).

Conducting polymers also adhere with SA for the study of electrode materials. PANI-based SA composite synthesized using template-induced method exhibited good reversible stability, electrochemical discharge capacitance, and faster oxidation and reduction responses. The composite showed an enhanced retention life, which can be attributed to the high surface area of the nanoelectrode which facilitates high electrode/electrolyte interfaces and shorter path lengths for ion transport (Patil et al., 2016). Furthermore, the sodium alginate hydrogels also proved to be a good recipient for electrode material, where pyrrole mixed in aqueous SA and polymerized to form polypyrrole/SA hydrogel broadcasted good electrode properties (Huang et al., 2017). Polysaccharides in the supercapacitor applications are vast but can be simplified into carbon-based, conductive polymer-based, and metal oxide-based.

14.5 POLYSACCHARIDES IN BATTERY

Polysaccharides were important components of early battery designs and are still widely employed in a wide range of commercial applications today. Except for energy and power densities, batteries and supercapacitors are nearly identical. When it comes to polysaccharides, they are used as separators, binders, composites, and electrolytes. Polysaccharides are used in both batteries and superconductors since both systems have the same electrode, separator, electrolyte, and so on. Li-ion batteries are an exceedingly promising power solution in most electronic products. Because lithium is the lightest and most electropositive metal, it can be used in battery applications with high-energy density and terminal voltages. A typical Li-ion battery has two electrodes, one positive and one negative, with the positive being a metal oxide of lithium ($LiCoO_2$) with a layer or tunnel structure and the negative being graphitic carbon or sulfur. They are encased in conductive materials such as aluminum, copper, and so on. Binders are usually polymer compounds, such as PVDF, that improve conductivity, and the electrolytes are lithium salts in organic solvents. A typical polyethylene microporous separator sheet separates the negative and positive electrodes. The positive electrode is oxidized while the negative electrode is reduced throughout the charging process. The procedure entails the migration of Li ions onto the negative point and gets intercalated at the cathode. The compensating electrons travel through the external circuit and are accepted at the host end to balance the reaction. During discharge, this process is reversed (Jabbour et al., 2013).

Polysaccharides are used in battery applications as a precursor for carbon-based electrodes, as a composite with graphene, and as separators and binders with conducting polymer and metal. Cellulose is used as a separator material in batteries to separate the anode and the cathode. An excellent separator must have strong mechanical properties, as well as tunable thickness and porosity, as well as good chemical stability and electrolyte wettability. Their abundance of added functionality and wettability owing to the presence of hydroxyl groups on the surface make them ideal for use in separator applications.

A composite of cellulose/polysulfonamide (PSA) was fabricated as a separator as PSA is known for its excellent mechanical, thermal, chemical, and dielectric properties. A simple paper-making process was employed to fabricate the composite, with cellulose matrix to reduce the cost and PSA to impart mechanical and other thermal and chemical properties. Because of the synergetic action between the electrolyte and the composite material, the electrolyte-soaked cellulose/PSA composite membrane demonstrated improved ion conductivity (Xu et al., 2014). Cellulose nanofibrils soaked in ethanol manifest an areal density of $7.1\,g\,m^{-2}$ and thickness of $12\,\mu m$ along with uniform pores structure and high porosity and mechanical properties. An electrolyte uptake of 281% along with excellent electrolyte affinity and wettability four times higher, along with electrochemical stability up to 4.8v with good cyclic performance were reported (Sheng et al., 2020). Carboxymethyl cellulose is used as a binder for cathodes and anodes in Li-ion batteries. This helps the insertion and extraction of Li ions between cathode and anode and enhances the charge-discharge capability of whole Li/Ion batteries. After 200 cycles, the CMC-Li cathode exhibited a reversible capacity of 175 mAh g^{-1} and increased electrochemical performance. This demonstrates that the CMC-Li binder may generate an effective electrically conductive network as well as a stable cathode interface structure (Lei et al., 2019).

Lithium-sulfur is another type of energy storage device with high theoretical capacity as well as theoretical energy-specific capacity. Sulfur is an abundant material with no toxicity and low cost. Their main disadvantage and restriction for commercialization is their dissolution of lithium polysulfide intermediates into the organic electrolyte during the electrochemical reaction. Polysulfide soluble in electrolyte diffuses out of the cathode and moves toward the metallic lithium to form short-chain polysulfide which reduces the active lithium contaminating the whole system (Figure 14.2). This effect commonly referred to as the shuttle effect lowers the efficiency, cycle life, self-discharge, etc., leading to a failure. Studies have revealed that the hydroxyl and amine groups in chitosan can reduce this shuttle effect of polysulfide by capturing them. When chitosan was used as a separator for the Li-S battery, an increased cyclic performance and reversible capability were observed. The improved separator, which contains chitosan as an addition, catches more polysulfide and reactivates the trapped active components. Given the safety, ease of preparation of chitosan, and environmental friendliness, this simple functional addition gives key hints in the development of lithium-sulfur batteries and may be widely used in high-performance lithium-sulfur batteries (Chen et al., 2015). Chitosan derivatives formed by the carboxymethylation of chitosan and carboxymethyl chitosan are used as a binder for Si and Li-ion batteries. The characterization techniques suggest a hydrogen-bonding interaction between the hydroxylated Si surface and hydroxyl groups of carboxylate chitosan. The electrode fabricated using carboxylate chitosan as a binder broadcast a very high discharge capacity, better rate performance, and better cycling efficiency compared to traditional binders such as PVDF (Sun et al., 2017).

Starch is an inexpensive, renewable, and biological material employed in battery applications as a precursor for carbon. Potato starch is a refined starch with spherical morphologies, which were employed to make anode. Hard carbon spherules were fabricated using potato starch using a one-step program process in an inert atmosphere. This process retains their spherules morphology, and as an anode, they exhibited good reversible capacity, stable cyclic performance, and rate capability (Li, Chen and Wang, 2011). Because of its remarkable capacity and tolerable working potential, silicon offers significant promise as a replacement for graphite, which is presently utilized commercially as an anode material in LIBs. Si incorporated in carbon from corn starch was studied and revealed excellent electrical conductivity accommodating a large volume change and aggregation prevention during each cycle. This development in desirable cheap, biodegradable polymer binders from starch has led to the improved performance of the Si anodes (Kwon et al., 2020). Proper binding materials improve the life of a battery by increasing electrochemical performance. Sustainable

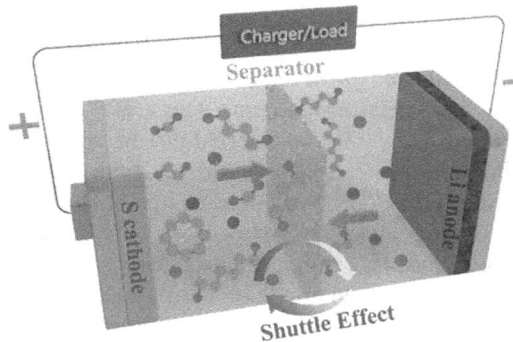

FIGURE 14.2 Charging and discharging in a Li-S battery. Reprinted from Zhang et al. (2021).

binders normally that are dissolved in an environmentally friendly medium like water are being investigated. Traditional binder PVDF forfeits such properties and is dissolved in complex, toxic, and aqueous solvents. Tapioca starch derived from cassava roots along with polyethylene glycol was tested for Si electrodes in Li-ion batteries. Tapioca starch is a water-soluble, biodegradable polymer composed of amylose units with glycosidic bonds. Other commonly used biodegradable binders, such as CMC, are brittle, and peeling from the substrate is tough which is the nature of linear polymer chains. Moderately branched amylopectin groups present in starch enhance stress relaxation during drying making them a better binder than CMC. Starch/Peg exhibited better mechanical hardness and elastic modulus compared to PVDF binders, which are attributed to better adhesion between Si nanoparticles and the binders. High coulombic efficiency and reversible capacity retention were observed in starch-based binders (Hapuarachchi et al., 2020).

Alginates are mostly employed as separators and binders in battery applications. SA binders with aluminum and barium cations doped were analyzed as a binder for Li-rich batteries. This type of doping into SA reduces the capacity decay and voltage fading. Binders based on SA can maintain a stable structure in electrodes during the cyclic process due to their vicious nature. Moreover, they form a coating on the electrode, which arrests the etching of active materials into the electrolytes. Such kind of doped SA exhibits remarkable capability along with high-capacity retention and suppressed voltage decay (Zhang et al., 2018). SA is investigated as a binder for sulfur-based cathode as well. They exhibited a discharge capacity of 508 mAh·g^{-1} and 66% capacity retention after 50 cycles. Electrochemical experiments show that the Na-alginate sulfur cathode has better kinetic properties, lower resistance, and greater cycle stability than the PVDF sulfur cathode. These findings indicate that Na-alginate is a suitable sulfur cathode binder in lithium-sulfur battery applications (Bao et al., 2013). Alginate dressings due to their electrolyte retention/adsorption capabilities and thermal shrinkage are introduced into the lithium-ion battery as a separator. The alginate dressing constitutes 85% calcium alginate and 15% of sodium carboxymethyl cellulose. Calcium alginate is insoluble in water and organic solvents which contribute to the skeleton and the Na-CMC functions as a sticky agent for the separator. Their surface chemistry of carbonyl and hydroxyl groups allows them to dope with other chemical agents, where polyacrylic acid (PAA) and Li$^+$ ions doped broadcasted a good electrochemical performance (Dai et al., 2020). All of the major polysaccharides are promising for battery applications mainly as separators and binders.

14.6 POLYSACCHARIDES IN FUEL CELLS

A fuel cell is a promising and growing field in future energy technologies in which chemical bonds in the fuel, primarily hydrogen, are directly converted into electricity via chemical reactions via oxidation. When compared to combustion technologies, the pollution emission is negligible because the by-products are water vapor and heat. In terms of efficiency, fuel cells surpass other techniques of energy conversion—depending on the kind of fuel, the electrical efficiency of a typical cell is in the range of 40%–60%, and total efficiency can reach 80%–90% (Daszkiewicz et al., 2021). An electrolyte material separates the porous anode and cathode of the fuel cells, where the input fuels are passed through the anode and air or oxygen along with the cathode, which then catalyzes the hydrogen and dissociates into electrons and ions. Ions travel down the electrolyte to opposite charged electrodes, while electrons go via the external circuit to generate electricity Figure 14.3. They are often classed according to the electrolyte utilized. Fuel cells are classified into six categories. Molten carbonate fuel cell (MCFC), alkaline fuel cell (AFC), proton exchange membrane fuel cell (PEMFC), solid oxide fuel cell (SOFC), phosphoric acid fuel cell (PAFC), and biofuel cells are all examples of fuel cells. Figure 14.4 compares the characteristics of each of these cells. All of these fuel cells are based on hydrogen and are also referred to as hydrogen cells. Direct-methanol fuel cells (DMFCs), direct-ethanol fuel cells (DEFCs), reformed methanol fuel cells (RMFCs), and direct-formic acid fuel cells (DFAFCs) are other types of non-hydrogen cells that use methanol,

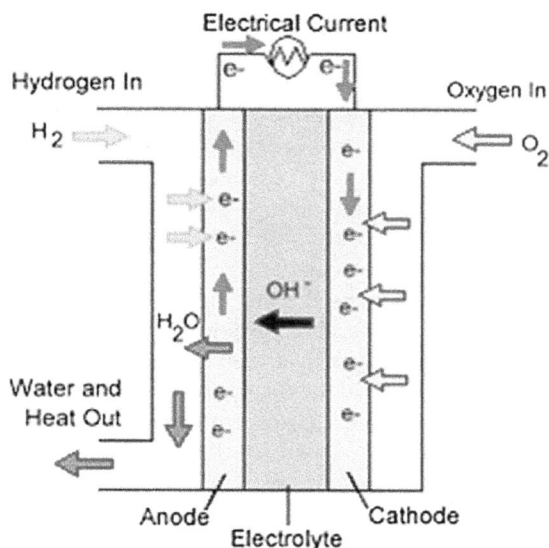

FIGURE 14.3 A working of a fuel cell by Vaghari et al. (2013) licensed under CC.BY 2.0, reprinted from Springer Sustainable Chemical Processes.

FIGURE 14.4 Comparison of various types of the hydrogen-based fuel cell by Vaghari et al. (2013) licensed under CC.BY 2.0, reprinted from Springer Sustainable Chemical Processes.

ethanol, and formic acids as electron donors in the anode. Fuel cells are well known for their efficiency, silent operation, high-energy density, and environmental friendliness.

To overcome the disadvantages of liquid electrolytes, polymer membranes are (Danyliv et al., 2021) utilized as an electrolyte in fuel cells. Some desirable characteristics for fuel cell membranes include high proton conductivity, impermeability to fuel gas or liquid, low electrical conductivity, good mechanical toughness in both the dry and hydrated states, hydrolytic stability, and so on. Nanocellulose is a potentially new membrane material for fuel cells that is less costly and has a lower environmental impact than Nafion or Aquivion. Polysaccharides have thermomechanical qualities but lack proton conductivity, which protic ionic liquids can provide. Cellulose nanocrystals and protic ionic liquids composite have a higher ionic conductivity of 10^{-4} to 10^{-3} S cm^{-1} at temperatures ranging from 120°C to 160°C.

Cellulose nanofibers are modified with sulfonation to improve the ionic conductivity and paper-based membranes were fabricated as a membrane for fuel cells. They exhibited good current density and power density which symbolize the use of cellulose in fuel cells as an electrolytic membrane(Bayer et al., 2021).

Chitosan, the *N*-deacetylated derivative of chitin is a promising membrane material. The presence of free amine and hydroxyl functional groups on the chitosan backbone allows for numerous chemical modifications to adapt chitosan for specific applications such as polymer electrolyte membranes. Chitosan-based membranes are easily produced, have high hydrophilicity, and are thermally and chemically stable. Anionic poly(acrylic acid) coupled with cationic chitosan membranes were studied by ionically crosslinking to a fabricated polyelectrolyte complex. This study showed that this type of membrane exhibits high proton conductivity, high ion-exchange capacity, low methanol permeability, and sufficient mechanical and thermal stability (Smitha, Sridhar and Khan, 2004). Chitosan is a low conductive material, with three hydrogen atoms bonded to their structure, which doesn't be mobilized to enhance proton conduct under an electric field. When it is dissolved in acetic acids CH_3COO^-, H^+ or H_3O^+ ions will be dispersed solvent which can be mobilized upon the application of an electric field. If the latter two positive ions become more mobile than the former proton condition occurs. Chitosan membranes crosslinked in acid such as sulfuric acid can be used in proton exchange membranes. These membranes outperform Nafion 117 in water uptake experiments where they absorb 60% water (Mukoma, Jooste and Vosloo, 2004).

Polysaccharides find application in microbial fuels, a type of biofuel cell, where electricity is produced from organic matter by microorganisms as biocatalysts. This environment benign process works by oxidizing the microbes in the anode compartment producing electrons and protons whereby electrons are passed through an external circuit producing electricity and collected at the cathode. Protons move through the internal membrane and accumulate at the cathode whereby oxygen gets reduced. Electricity can be produced from various waste sources or marine sediments, sludge, or other sources under mild conditions. Starch-processed wastewaters containing oxygen were used as a source for the anode in such a fuel cell to generate electricity. Maximum power density and voltage output were acquired along with coulombic efficiency of 8% (Lu et al., 2009). It is demonstrated that specially designed anode like platinum electrodes by a coating of poly(tetrafluoroaniline) and biocatalyst clostridium butyricum electricity can be produced from starch (Niessen, Schröder and Scholz, 2004). Along with starch, other polysaccharides such as cellulose and chitosan are also employed as fuel for microbial fuel cells and are exhibiting promising results for future commercialization.

14.7 POLYSACCHARIDES IN NANOGENERATORS

Nanogenerators are ambient and simple next-generation electronic devices with versatile characteristics for a wide range of applications. They employ mechanical movements or energy found in the environment into electric energy exploiting piezoelectric, electrostatic, electromagnetic, and triboelectric effects. This sort of energy can be used for wearable electronics, self-charging, and wireless devices. During a mechanical deformation, a change in polarization occurs in piezoelectric materials producing to generate electricity. Polysaccharides such as chitosan, bacterial cellulose, and cellulose are used as a material for piezoelectric generators. A drawback of this type of generator is its mechanical excitation range is 60–100 Hz, below which very low-voltage values are acquired. The piezoelectric in wood is known for decades which simultaneously leads cellulose for the same effect, generation of polarization upon mechanical stress. Cellulose nanocrystals, as well as nanofibers, are evaluated as piezoelectric material and showed good results. The piezoelectric response of the CNC films was attributed to the asymmetric crystalline structure of the cellulose crystals. The piezoelectric constant of CNC was comparable with that of a reference piezoelectric metal oxide confirming their future in nanogenerators (Csoka et al., 2012). Chitin has good piezoelectric properties which can be attributed to the non-centrosymmetric crystal structure of both α- and β-chitin polymorphs through which intrinsic molecular polarization arises. Chitosan is better as it can be solubilized when compared to chitin. Due to their frequency range, there are of less interest when compared to triboelectric generators.

Triboelectric nanogenerators (TENGs) can harvest energy in low frequencies 10 Hz and below from water motion, human motion, animal organs, and other small frequency sources. TENGs work on the principle of electrostatic and triboelectric phenomena, whereas the latter is also known as contact esterification which occurs when two materials are brought together and separated. When such a process is happening, electrons from a high Fermi-level surface try to transfer to a low-surface Fermi level. They are intended to alternately touch and separate two materials with opposing triboelectric polarities to force-induced electrons between the electrodes. TENG output voltage and current are proportional to the triboelectric charge density on the surface, and output power is related to charge density. Polysaccharides due to their diverse properties are being employed in TENGs applications. The functional group present in the polysaccharides enhances their performances due to their electron-withdrawing/gaining tendency. Triboelectric charge density can be lowered or increased by altering the functional groups, where a fivefold improvement in efficiency and surface charge were observed when CNF was modified as nitro-CNF (Yao et al., 2016).

Surface area is another factor, where a larger contact area, as well as the surface roughness, promotes tribo outputs. Nanocellulose fibers with high surface area and roughness due to fiber morphologies are ideal for such applications which can be further modified by the incorporation of fillers or nanowires. One of the main drawbacks of polysaccharides in nanogenerator application is their moisture absorption from the atmosphere, which hinders charge transfer and leads to a lower triboelectrification (Torres and De-la-Torre, 2021). Cellulose due to its surface area and abundant oxygen groups in its structure exhibits a high tendency to lose electrons and become positive. Therefore, cellulose films can be paired with negative triboelectric materials. The observed triboelectric voltage output was found to be a function of CNF film size, owing to time-dependent triboelectric discharging behavior. The output electric energy was determined to be in the range of 0.01–0.16 mJ as a function of the mechanical impact frequency using normal capacitor charging (Lei et al., 2019). Modification of CNF with nitro and methyl group by chemical reaction renders them with opposite and tribo-polarity. The surface charge density of nitro-CNF and methyl-CNF was -85.8 and 62.5 μC m^{-2}, respectively, which is much higher than the pure CNF. This type of modification further enhances their employment in TBNGs, thus creating sustainable energy applications (Yao et al., 2017).

Chitosan also has proved to be an excellent material for triboelectric generators. Their pristine form exhibited triboelectric efficiency but the electrical output is low. This can be improved by surface engineering the chitosan by modifying the functional groups present on the surface. Chitosan is dissolved in citric acid where two hydrolysis reactions occur: hydrolysis of N-acetyl linkages and glycosidic linkages. A three-dimensional hybrid network structure of chitosan/citric acid is formed by a crosslinking of the two reactants by hydrogen bonding. Such modified chitosan has demonstrated high efficiency in TENG for self-powered devices, paving the way for new technologies that make use of abundant bio-derived materials for the economically feasible and environmentally friendly production of functional devices in electronics, energy, and sensor applications (Yao et al., 2017).

14.8 CONCLUSION

Polysaccharides are intriguing materials in the field of energy storage applications, such as supercapacitors, batteries, fuel cells, and nanogenerators. Their abundance, low cost, renewability, and biodegradable properties aid their use in different applications. Their surface chemistry with functional groups further enhances the fabrication of composites as well as modification, thus making these materials an ideal candidate for energy storage devices.

REFERENCES

Abbasi, H., Antunes, M. and Velasco, J. I. (2019) 'Recent advances in carbon-based polymer nanocomposites for electromagnetic interference shielding', *Progress in Materials Science*, 103(October 2017), pp. 319–373. doi: 10.1016/j.pmatsci.2019.02.003.

Bao, W. et al. (2013) 'Enhanced cyclability of sulfur cathodes in lithium-sulfur batteries with Na-alginate as a binder', *Journal of Energy Chemistry*, 22(5), pp. 790–794. doi: 10.1016/S2095-4956(13)60105-9.

Bayer, T. et al. (2021) 'Spray deposition of sulfonated cellulose nanofibers as electrolyte membranes in fuel cells', *Cellulose*, 28(3), pp. 1355–1367. doi: 10.1007/s10570-020-03593-w.

Cao, Q. et al. (2020) 'Novel lignin-cellulose-based carbon nanofibers as high-performance supercapacitors', *ACS Applied Materials and Interfaces*, 12(1), pp. 1210–1221. doi: 10.1021/acsami.9b14727.

Chen, Y. et al. (2015) 'Chitosan as a functional additive for high-performance lithium-sulfur batteries', *Journal of Materials Chemistry A*, 3(29), pp. 15235–15240. doi: 10.1039/c5ta03032c.

Csoka, L. et al. (2012) 'Piezoelectric effect of cellulose nanocrystals thin films', *ACS Macro Letters*, 1(7), pp. 867–870. doi: 10.1021/mz300234a.

Dai, D. et al. (2020) 'Modified alginate dressing with high thermal stability as a new separator for Li-ion batteries', *Chemical Communications*, 56(45), pp. 6149–6152. doi: 10.1039/d0cc01729a.

Danyliv, O. et al. (2021) 'Self-standing, robust membranes made of cellulose nanocrystals (CNCs) and a protic ionic liquid: toward sustainable electrolytes for fuel cells', *ACS Applied Energy Materials*, 4(7), pp. 6474–6485. doi: 10.1021/acsaem.1c00452.

Daszkiewicz, P. et al. (2021) 'Fuel cells based on natural polysaccharides for rail vehicle application', *Energies*, 14(4). doi: 10.3390/en14041144.

Dong, S. et al. (2013) 'Nanostructured transition metal nitrides for energy storage and fuel cells', *Coordination Chemistry Reviews*, 257(13–14), pp. 1946–1956. doi: 10.1016/j.ccr.2012.12.012.

Hapuarachchi, S. N. S. et al. (2020) 'Mechanically robust tapioca starch composite binder with improved ionic conductivity for sustainable lithium-ion batteries', *ACS Sustainable Chemistry and Engineering*, 8(26), pp. 9857–9865. doi: 10.1021/acssuschemeng.0c02843.

Huang, H. et al. (2017) '3D nanostructured polypyrrole/sodium alginate conducting hydrogel from self-assembly with high supercapacitor performance', *Journal of Macromolecular Science, Part B: Physics*, 56(8), pp. 532–540. doi: 10.1080/00222348.2017.1342951.

Jabbour, L. et al. (2013) 'Cellulose-based Li-ion batteries: a review', *Cellulose*, 20(4), pp. 1523–1545. doi: 10.1007/s10570-013-9973-8.

Jadhav, V. V, Mane, R. S. and Shinde, P. V (2020) *Bismuth-Ferrite-Based Electrochemical Supercapacitors*. Springer.

Jang, S. et al. (2021) 'A hierarchically tailored wrinkled three-dimensional foam for enhanced elastic supercapacitor electrodes', *Nano Letters*, 21(16), pp. 7079–7085. doi: 10.1021/acs.nanolett.1c01384.

Kafy, A. et al. (2017) 'Porous cellulose/graphene oxide nanocomposite as flexible and renewable electrode material for supercapacitor', *Synthetic Metals*, 223, pp. 94–100. doi: 10.1016/j.synthmet.2016.12.010.

Kebabsa, L. et al. (2020) 'Highly porous cobalt oxide-decorated carbon nanofibers fabricated from starch as free-standing electrodes for supercapacitors', *Applied Surface Science*, 511(January), p. 145313. doi: 10.1016/j.apsusc.2020.145313.

Kim, T. et al. (2019) 'Lithium-ion batteries: outlook on present, future, and hybridized technologies', *Journal of Materials Chemistry A*, 7(7), pp. 2942–2964. doi: 10.1039/C8TA10513H.

Kim, W. G. et al. (2021) 'Triboelectric nanogenerator: structure, mechanism, and applications', *ACS Nano*, 15(1), pp. 258–287. doi: 10.1021/acsnano.0c09803.

Kwon, H. J. et al. (2020) 'Nano/microstructured silicon-carbon hybrid composite particles fabricated with corn starch biowaste as anode materials for Li-Ion batteries', *Nano Letters*, 20(1), pp. 625–635. doi: 10.1021/acs.nanolett.9b04395.

Lei, C. et al. (2019) 'Fabrication of metal-organic frameworks@cellulose aerogels composite materials for removal of heavy metal ions in water', *Carbohydrate Polymers*, 205, pp. 35–41. doi: 10.1016/j.carbpol.2018.10.029.

Li, W., Chen, M. and Wang, C. (2011) 'Spherical hard carbon prepared from potato starch using as anode material for Li-ion batteries', *Materials Letters*, 65(23–24), pp. 3368–3370. doi: 10.1016/j.matlet.2011.07.072.

Li, B. et al. (2017) 'Nitrogen doped and hierarchically porous carbons derived from chitosan hydrogel via rapid microwave carbonization for high-performance supercapacitors', *Carbon*, 122, pp. 592–603. doi: 10.1016/j.carbon.2017.07.009.

Li, Y. et al. (2020) 'Green synthesis of free standing cellulose/graphene oxide/polyaniline aerogel electrode for high-performance flexible all-solid-state supercapacitors', *Nanomaterials*, 10(8), pp. 1–18. doi: 10.3390/nano10081546.

Liu, C. et al. (2010) 'Advanced materials for energy storage', *Advanced Materials*, 22(8), pp. 28–62. doi: 10.1002/adma.200903328.

Lokhande, P. E., Chavan, U. S. and Pandey, A. (2020) *Materials and Fabrication Methods for Electrochemical Supercapacitors: Overview*, *Electrochemical Energy Reviews*. Springer Singapore. doi: 10.1007/s41918-019-00057-z.

Lu, N. et al. (2009) 'Electricity generation from starch processing wastewater using microbial fuel cell technology', *Biochemical Engineering Journal*, 43(3), pp. 246–251. doi: 10.1016/j.bej.2008.10.005.

Mukoma, P., Jooste, B. R. and Vosloo, H. C. M. (2004) 'Synthesis and characterization of cross-linked chitosan membranes for application as alternative proton exchange membrane materials in fuel cells', *Journal of Power Sources*, 136(1), pp. 16–23. doi: 10.1016/j.jpowsour.2004.05.027.

Niessen, J., Schröder, U. and Scholz, F. (2004) 'Exploiting complex carbohydrates for microbial electricity generation - A bacterial fuel cell operating on starch', *Electrochemistry Communications*, 6(9), pp. 955–958. doi: 10.1016/j.elecom.2004.07.010.

Nizam, P. A. et al. (2020) 'Mechanically robust antibacterial nanopapers through mixed dimensional assembly for anionic dye removal', *Journal of Polymers and the Environment*, 28(4), pp. 1279–1291. doi: 10.1007/s10924-020-01681-3.

Nizam, P. A. et al. (2021) *Nanocellulose-Based Composites*, *Nanocellulose Based Composites for Electronics*. Elsevier Inc. doi: 10.1016/b978-0-12-822350-5.00002-3.

Nizam, P. A., Binumol, T. and Thomas, S. (2021) 'Overview of Nanostructured Materials for Electromagnetic Interference Shielding', in Suji Mary Zachariah, S. T. (ed.) *Nanostructured Materials for Electromagnetic Interference Shielding*. 1st Edition. Taylor & Francis, p. 11. Available at: Nizam, P.A., Binumol, T. and Thomas, S., 2021. Overview of Nanostructured Materials for Electromagnetic Interference Shielding. *Nanostructured Materials for Electromagnetic Interference Shielding*, pp. 99–109.

Pai, A. R. et al. (2022) 'Recent progress in electromagnetic interference shielding performance of porous polymer nanocomposites - a review', *Energies*, 15(11), p. 3901.

Patil, D. S. et al. (2016) 'Facile preparation and enhanced capacitance of the Ag-PEDOT:PSS/polyaniline nanofiber network for supercapacitors', *Electrochimica Acta*, 213, pp. 680–690. doi: 10.1016/j.electacta.2016.07.156.

Rapisarda, M., Marken, F. and Meo, M. (2021) 'Graphene oxide and starch gel as a hybrid binder for environmentally friendly high-performance supercapacitors', *Communications Chemistry*, 4(1). doi: 10.1038/s42004-021-00604-0.

Rose Joseph, M. et al. (2021) 'Development and characterization of cellulose nanofibre reinforced Acacia nilotica gum nanocomposite', *Industrial Crops and Products*, 161(December 2020), p. 113180. doi: 10.1016/j.indcrop.2020.113180.

Roy, B. K., Tahmid, I. and Rashid, T. U. (2021) 'Chitosan-based materials for supercapacitor applications: a review', *Journal of Materials Chemistry A*, 9(33), pp. 17592–17642. doi: 10.1039/d1ta02997e.

Selvaraj, T. et al. (2020) 'The recent development of polysaccharides biomaterials and their performance for supercapacitor applications', *Materials Research Bulletin*, 126(February), p. 110839. doi: 10.1016/j.materresbull.2020.110839.

Sheng, N. et al. (2019) 'Polypyrrole@TEMPO-oxidized bacterial cellulose/reduced graphene oxide macrofibers for flexible all-solid-state supercapacitors', *Chemical Engineering Journal*, 368(February), pp. 1022–1032. doi: 10.1016/j.cej.2019.02.173.

Sheng, J. et al. (2020) 'Ultra-light cellulose nanofibril membrane for lithium-ion batteries', *Journal of Membrane Science*, 595(October), p. 117550. doi: 10.1016/j.memsci.2019.117550.

Smitha, B., Sridhar, S. and Khan, A. A. (2004) 'Polyelectrolyte complexes of chitosan and poly(acrylic acid) as proton exchange membranes for fuel cells', *Macromolecules*, 37(6), pp. 2233–2239. doi: 10.1021/ma0355913.

Sun, R. et al. (2017) 'Highly conductive transition metal carbide/carbonitride(MXene)@polystyrene nanocomposites fabricated by electrostatic assembly for highly efficient electromagnetic interference shielding', *Advanced Functional Materials*, 27(45), pp. 1–11. doi: 10.1002/adfm.201702807.

Teng, G. et al. (2020) 'Renewable cellulose separator with good thermal stability prepared via phase inversion for high-performance supercapacitors', *Journal of Materials Science: Materials in Electronics*, 31(10), pp. 7916–7926. doi: 10.1007/s10854-020-03330-w.

Torres, F. G. and De-la-Torre, G. E. (2021) 'Polysaccharide-based triboelectric nanogenerators: a review', *Carbohydrate Polymers*, 251(July 2020), p. 117055. doi: 10.1016/j.carbpol.2020.117055.

Vaghari, H. et al. (2013) 'Recent advances in application of chitosan in fuel cells', *Sustainable Chemical Processes*, 1(1), pp. 1–12. doi: 10.1186/2043-7129-1-16.

Wang, F. and Li, D. (2015) 'Foldable and free-standing 3D network electrodes based on cellulose nanofibers, carbon nanotubes and elongated TiO2 nanotubes', *Materials Letters*, 158, pp. 119–122. doi: 10.1016/j.matlet.2015.06.008.

Wang, C. et al. (2018) 'Cellulose-derived hierarchical porous carbon for high-performance flexible supercapacitors', *Carbon*, 140, pp. 139–147. doi: 10.1016/j.carbon.2018.08.032.

Wang, D. C. et al. (2019) 'Supramolecular self-assembly of 3D conductive cellulose nanofiber aerogels for flexible supercapacitors and ultrasensitive sensors', *ACS Applied Materials and Interfaces*, 11(27), pp. 24435–24446. doi: 10.1021/acsami.9b06527.

Weng, Z. et al. (2011) 'Graphene-cellulose paper flexible supercapacitors', *Advanced Energy Materials*, 1(5), pp. 917–922. doi: 10.1002/aenm.201100312.

Wu, W. et al. (2014) 'A facile one-pot preparation of dialdehyde starch reduced graphene oxide/polyaniline composite for supercapacitors', *Electrochimica Acta*, 139, pp. 117–126. doi: 10.1016/j.electacta.2014.06.166.

Xie, J. et al. (2018) 'Puzzles and confusions in supercapacitor and battery: theory and solutions', *Journal of Power Sources*, 401(December 2017), pp. 213–223. doi: 10.1016/j.jpowsour.2018.08.090.

Xu, Q. et al. (2014) 'Cellulose/polysulfonamide composite membrane as a high performance lithium-ion battery separator', *ACS Sustainable Chemistry and Engineering*, 2(2), pp. 194–199. doi: 10.1021/sc400370h.

Yamagata, M. et al. (2013) 'Chitosan-based gel electrolyte containing an ionic liquid for high-performance non-aqueous supercapacitors', *Electrochimica Acta*, 100, pp. 275–280. doi: 10.1016/j.electacta.2012.05.073.

Yang, H. et al. (2019) 'Biopolymer-based carboxylated chitosan hydrogel film crosslinked by HCl as gel polymer electrolyte for all-solid-sate supercapacitors', *Journal of Power Sources*, 426(April), pp. 47–54. doi: 10.1016/j.jpowsour.2019.04.023.

Yao, C. et al. (2016) 'Triboelectric nanogenerators and power-boards from cellulose nanofibrils and recycled materials', *Nano Energy*, 30(June), pp. 103–108. doi: 10.1016/j.nanoen.2016.09.036.

Yao, C. et al. (2017) 'Chemically functionalized natural cellulose materials for effective triboelectric nanogenerator development', *Advanced Functional Materials*, 27(30), pp. 1–7. doi: 10.1002/adfm.201700794.

Zachariah, S. M., Nizam, P. A., Grohens, Y., Kalarikkal, N. and Thomans, S. (2021) 'Carbon Materials Potential Agents in Electromagnetic Interference Shielding', in Zachariah, S. M. and Thomas, S. (ed.) *Nanostructured Materials for Electromagnetic Interference Shielding*. 1st Editio. Taylor & Francis, p. 14.

Zhai, Z. et al. (2019) 'Green and facile fabrication of Cu-doped carbon aerogels from sodium alginate for supercapacitors', *Organic Electronics*, 70(April), pp. 246–251. doi: 10.1016/j.orgel.2019.04.028.

Zhang, S. J. et al. (2018) 'Sodium-alginate-based binders for lithium-rich cathode materials in lithium-ion batteries to suppress voltage and capacity fading', *ChemElectroChem*, 5(9), pp. 1321–1329. doi: 10.1002/celc.201701358.

Zhang, Z. et al. (2021) 'Cellulose-based material in lithium-sulfur batteries: a review', *Carbohydrate Polymers*, 255(December 2020), p. 117469. doi: 10.1016/j.carbpol.2020.117469.

Zhou, S. et al. (2019) 'Cellulose nanofiber @ conductive metal-organic frameworks for high-performance flexible supercapacitors', *ACS Nano*, 13(8), pp. 9578–9586. doi: 10.1021/acsnano.9b04670.

Zhao, Y. et al. (2020) 'Facile preparation of N-O codoped hierarchically porous carbon from alginate particles for high performance supercapacitor', *Journal of Colloid and Interface Science*. doi: 10.1016/j.jcis.2019.12.027.

15 Polysaccharides for Agricultural Applications
A Growing Presence on the Farms

Enock Siankwilimba
University of Zambia

Bhasha Sharma
Shivaji College, University of Delhi

Md Enamul Hoque
Military Institute of Science and Technology (MIST)

15.1 INTRODUCTION

The building blocks of any civilization are composed of substances that have served a variety of purposes to improve both human and animal existence since the earth was first created. Materials have been important in many facets of human development since the advent of agriculture more than 10,000 years ago (Fabrice 2019). Over the past millennium, humankind has made significant advancements to support their constantly shifting livelihoods, as seen by the domestication of crops and animals from the wild into domestic management and control (Zeder 2015). Various technological developments have led to the creation of many foods and animal feed products for humans and animals. As a result, even in the current environment, it is still possible to create new technologies and push the boundaries of science in order to manipulate crops and animals to increase accessibility, affordability, availability, and utilization of different forms. Polysaccharides derived from plants and animals have been one of the essential components that have fuelled human development since the beginning of time in order to nourish and sustain an ever-increasing human population (Souza et al. 2022). The significance, origin, and processing of polysaccharides in order to maximize their use in various parts of life have all been extensively studied. The need for starch-based products has influenced the evolution of agriculture and the global economy. Some chains of starch can be consumed immediately by humans and animals, while others need to be processed in the industrial setting in order to be combined with other goods. According to other studies, many polysaccharides or starch chains are being wasted daily, even though both animals and humans need to consume them in large quantities. According to Souza et al. (2022), biological materials or biomass should be potentialized in the industry so that each resource obtained has a specific use and does not produce waste. According to Mohammed et al. (2021), the continued need to create novel materials as a result of the previous work has prompted current research into a wide variety of biological and synthetic chemicals and compounds that routinely utilize polysaccharides. This has led to the production and usage of these materials in several technological areas, such as agriculture, medical, engineering, and business development. According to research, polysaccharide materials have many uses throughout the world, but even when they are being used historically, most farmers don't seem to be aware of them. Others are utterly ignorant of polysaccharides and are not employing them

DOI: 10.1201/9781003265054-15

efficiently or sustainably. Some report that polysaccharides are being debated and found daily and that their applications transform their lives. This chapter attempted to outline and explain the use of polysaccharides for agricultural applications and ascertain their growing presence on the farm to increase production and productivity as they proceed through the resilience pathway to end poverty.

15.1.1 Polysaccharides from Industrial Biomass in a Wasted World

The most prevalent biological substance in the world, polysaccharides help provide new solutions for sustainability in terms of the economy, society, and the environment. Despite their diversity and great promise, their agro-industrial application is currently very limited, but it has recently risen dramatically due to research and development. According to Souza et al. (2022), the use of these biopolymers increased by approximately 17% between 2017 and 2021, with the global market topping $10 billion in the first half of 2021 (Souza et al. 2022). FMR15151A (2021) forecasts that the use of polysaccharides and oligosaccharides in "clean label" and sustainable improvements in the food and beverage sectors will result in a global market of more than $22 billion by 2030 and a compound annual growth rate (CAGR) of more than 5%. For instance, the extraction of palm kernels can include up to 66% carbohydrates, including highly soluble (>95%) mannose, galactose, glucose, arabinose, xylose, and rhamnose.

According to Asim et al. (2015), agriculture and forestry goods produce 30%–40% of waste that might be used in processing with added value. It implies that low-density natural fibers can be utilized for practical application sales. For objects with low load-bearing capacity, grass fiber, for example, can be a great replacement. In addition to low density, cheap cost, low salary, non-carcinogenicity, and biodegradability, it also benefits from a few other features. Asim et al. (2015) contend that scientists, engineers, and even farmers are very interested in finding new sources of raw materials that have the same physical and mechanical properties as synthetic fibers. Generally, natural fibers are also appealing because of how cheap they are, how good they are for the environment, how safe they are for human and animal health, how flexible they are, how old the plants are, how easy they are to collect, and where they grow (Abed et al. 2016). Natural fibers are the best choice for a sustainable supply because they are less expensive, less dense, require less processing, don't pose any health risks, and have better mechanical and physical properties. They are also renewable resources, which makes them the best choice. Because natural fibers tend to soak up water, chemicals are used to change their surface properties (Bhardwaj et al., 2021). Expensive fiber-reinforced synthetic polymers have an effect on the environment. Pineapple, kenaf, coir, abaca, sisal, cotton, jute, bamboo, banana, palmyra, talipot, hemp, and flex are among the plant fibers that could be exploited as raw materials in a variety of industries, as shown in Table 15.1. Commonly grown crops in developing nations, such as pineapple leaf fiber, sugar cane, and cassava grains, are among the agricultural waste products. According to Imam et al. (2021), pineapple (*Ananas comosus*) is one of the most important tropical fruits in the world, following citrus and banana, as indicated in Table 15.1. Pineapple leaves are processed into natural fibers and are categorized as fruit waste. Commercially, pineapple fruits are highly significant. PALF is chemically composed of ash, lignin, and holocellulose (70%–82% of PALF is holocellulose) (1.1%). Pineapple (PALF), which possesses exceptional mechanical qualities, can be used in low-density polyethylene (LDPE) composites, biodegradable plastic composites, and reinforced polymer composites (Amanor, 2019). When gathered and processed, the leaves and outer layers of pineapple trees have proven beneficial in this regard.

Asim et al. (2015) estimate that between 30% and 40% of the waste materials produced by agriculture and forestry goods could be utilized in value-added processing. These materials can be used for impact resistance, processing viscoelastic behavior, and tensile and flexural strength. It suggests that low-density natural fibers can also be used in real-world situations. For simple load-bearing items, grass fiber, for instance, is a great substitute. It also benefits from a variety of other characteristics, such as low density, low cost, low wage, non-carcinogenicity, and biodegradability (Shih et al. 2014). Discovering new sources of raw materials whose physical and mechanical

TABLE 15.1

Displays the Various Types of Sources and Requirements of Polysaccharides Worldwide

Provider of Fiber	International Output (10 tons)	Fraction Used
Bamboo	10,000	Stalk
Banana	2,000	Fruit
Abaca	70	Stalk
Cotton lint	18,500	Stalk
Elephant grass	Plentiful	Stalk
Broom	Plentiful	Stalk
Coir	100	Stalk
Kenaf	770	Stalk
Flax	810	Stalk
Jute	2,500	Stalk
Pineapple	Plentiful	Leaf
Linseed	Plentiful	Fruit
Cassava	Plentiful	Tuber
Nettles	Plentiful	Stalk
Caroa	–	Leaf
Sunhemp	70	Stalk
Sisal	380	Stalk
Oil palms fruit	Plentiful	Fruit
Sugar cane bagasse	75,000	Stalk
Maize	Plentiful	Fruit and stalk
Rice husk	Plentiful	Fruit/grain
Rice straw	Plentiful	Stalk
Wood	1,750,000	Stalk
Palm rah	Plentiful	Stalk
China jute	–	Stalk
Roselle	250	Stalk
Ramie	100	Stalk

Source: Data adopted from Asim et al. (2015) and many other sources.

qualities resemble synthetic fibers is of enormous interest to scientists, engineers, and even farmers (Kumar Sinha 1982). When selecting raw materials, it is vital to evaluate various other criteria that directly impact the acceptability of natural fibers, such as cost, environmental friendliness, absence of health hazards, high degrees of flexibility, plant age, ease of collecting, and regional availability (Asim et al. 2015). Natural fibers are a superior option for a sustainable supply chain due to their low cost, low density, minimal processing, absence of health hazards, and high mechanical and physical qualities (Nicolle et al. 2021). In addition, they are renewable resources. The fundamental disadvantage of natural fibers, hygroscopicity, necessitates the application of chemical surface modification. The environment is impacted by expensive synthetic fiber-reinforced plastics. Several plant fibers, including pineapple, kenaf, coir, abaca, sisal, cotton, jute, bamboo, banana, palmyra, talipot, hemp, and flex, could be utilized as raw materials in a wide range of industries (Kengkhetkit & Amornsakchai 2012). Waste products from the agriculture industry include pineapple leaf fiber, sugar cane, and cassava grains, which are frequently cultivated in underdeveloped nations (Amanor 2012; Suvedi et al. 2017). One of the most important tropical fruits in the world, the pineapple (*Ananas comosus*) ranks third after citrus and banana. Pineapple leaves are regarded as fruit waste and are utilized to manufacture natural fibers. The industry for pineapple fruit is enormous. The bulk of PALF consists of ash, lignin, and holocellulose (1.1%) (Nicolle et al. 2021) although Asim et al. (2015) revealed that it has 70%–82% of holocellulose. Pineapple (PALF), a material with superior mechanical properties, can be used to create composites of low-density polyethylene (LDPE), biodegradable plastic, and polymer with reinforcement (Nicolle et al. 2021). Composites' viscoelastic behavior, processing, tensile strength, flexural strength, and impact resistance are all affected by the fiber length, matrix ratio, and fiber arrangement (Asim et al. 2015).

Cassava, recently discovered as one of the crops with abundant starch and carbs, currently provides the world with much-needed polysaccharide products for use at home and in the workplace (Otun et al. 2022; Zezza et al. 2021). Cassava (*Manihot esculenta* Crantz) is the fourth-best source of calories or dietary carbs for around 800 million individuals worldwide (Otun et al. 2022). Cassava roots contain 30%–40% more dry matter than other types of roots, such as yams and potatoes. Even though cassava produces much-needed starch, it is challenging to cultivate for smallholder farmers due to biotic and abiotic factors such as climate change and temperature changes (Mwiinde et al. 2022). The development of cassava varieties with a higher starch content has been identified as a significant area for development. Despite the problems, recent studies indicate that there are a range of strategies available to increase the quality and output of cassava starch (Adéyèmi et al. 2020). According to Zidenga et al. (2017), scientists isolated and characterized cassava gene homologs involved in processes such as the conversion of assimilated carbon to sucrose in photosynthetic cells, the transport of sucrose to storage organs through the phloem, the conversion of sucrose to starch, and the breakdown of starch into simple sugars in order to improve the function of starch. Karlström et al. (2016) assert that by altering the activity of the genes, the molecular and functional characterization of the genes involved in these processes can greatly increase the amount of starch in cassava. Additionally, it is now possible to genetically modify cassava roots to produce more starch of a higher quality and with greater yields due to advances in the understanding of starch production and the identification of the genes responsible for it (Ahimbisibwe et al. 2020). It is believed that hairpin dsRNAs targeting the 1,4-alpha-glucan-branching enzyme (be1) genes were constitutively expressed in transgenic cassava, resulting in the manufacture of starches containing up to 50% amylose (Otun et al. 2022). In addition, Otun et al. (2022) report that by editing genes related to the starch synthesis pathway, researchers were able to create cassava mutants known as MESSIII-1 and MESSIII-2 derived from the MESSIII genes. This finding prompted the investigation of the role of genes in regulating the formation of amylopectin glucan in cassava. Using CRISPR/Cas9 to directly change the two genes responsible for producing amylose, PTST1 and GBSS, the amylose content of root starch can be lowered or completely removed, according to the research by Zsögön et al. (2018).

The production of animal feeds has expanded significantly because of the new commercialization of cassava, increasing the crop's yield by 11.5%–33%. However, the use of cassava in animal feed has forced researchers to focus on how the cassava plant's cell walls affect the availability of nutrients, particularly starch, as well as efforts to abolish the presence of cyanogenic glycosides in all of its parts. Using an exogenous non-starch polysaccharide degrading enzyme product (NSPase) and sun-drying cassava chips for a few days until the moisture content is reduced to 100–140 mg may improve the shelf life of the product by releasing the volatile hydrogen cyanide (Staack et al. 2019). According to Staack et al. (2019), the use of enzymes in animal feed reduces variation in raw material quality and permits the use of a greater diversity of raw materials, even those that are difficult to digest. This decreases the variation in nutritional value across batches of identical ingredients. Staack is aware that enzymes can help form prebiotic oligomers by acting on cell wall polysaccharides, which are advantageous to an animal's digestive health, and that they can help reduce the viscosity of feed raw materials containing high concentrations of NSP. Previous research has linked the breakdown of homogalacturonan and the release of 1,4-D-galactan and 1,5-L-arabinan to a reduction in cassava viscosity (Staack et al. 2019). Cassava has lately proved its utility in the production of animal feed in Zambia. In recent years, large quantities of cassava have been purchased from contract farmers by foreign companies like Zambia Brewery and Zhongkai International to produce ethanol, ethanol for industrial use, glue, and food for animals. For instance, the presence of ethanol in hand sanitizers has aided in the fight against COVID-19 (Silimina 2021). The production of large, healthy cassava that can be used in homes and sold as agreed upon for the industrial process requires a large number of farmers, as can be seen in Figure 15.1. In light of COVID-19 restrictions and the conflict between Ukraine and Russia, businesses have recently been employed in the production of biofertilizers and in assisting to lower the cost of agricultural production (Zeufack et al. 2022). As can be seen in Figure 15.2, where farmers are observing a demonstration for educational purposes, extension services are necessary for cassava cultivators to have healthy plants and good tubers.

Mr. Chen, managing director of Zhongkai International, was quoted in Silimina (2021)'s study as saying, "Our ultimate product is ethanol, which is why we require this crop (cassava)." "For instance, the cassava from Mansa District contains 75% starch, and we discovered that Zambia produces the most cassava in Africa," said Chen Guiping, CEO of Zhongkai International, in an interview with ChinAfrica. "Before creating the business, we visited all African nations and discovered that Zambian produce had the greatest starch content; hence, we built the facility in Zambia." Chen disclosed that his company uses over 200 tons of dried cassava and that a farmer may easily earn K5,000 ($231) from just 1 lima (0.49 acre) of land, which can go a long way toward supporting their way of life. Mr. Chen stated that his company required cassava in order to manufacture methylated spirits, hand sanitizers, cooking gel, and ethanol for blending with gasoline as fuel.

FIGURE 15.1 Pineapple farm and plants with a farmer holding a pineapple in his hand.

FIGURE 15.2 A farm of cassava cultivated by smallholder farmers.

15.1.2 Polysaccharides from Cassava Production and Marketing in Zambia

Over the years, Zambia just like many countries in Southern Africa has been battling with economic diversification moving away mineral-dependent nation to agriculture. Within the agricultural sector, maize, cotton, and wheat have been extensively grown by smallholder and commercial farmers as major suppliers of polysaccharide products. The government has outlined the function of the private sector as the engine of agricultural diversification in the face of climate change through Zambia Vision 2030 and national agricultural policy documents (MoF 2006; SNAP 2016). The country has strategically positioned itself to explore the production of cassava as the alternative crop to supply starch to the population and the industry in light of the rising cost of production for the maize crop due to global inflation rates, the Ukrainian-Russian conflict, and the lost market value for cotton globally (Chilufya & Mulendema 2019; McCord et al. 2015; Oswake 2021; Sekwati 2012; Zeufack et al. 2022).

With unpredictable climate change which interface with droughts and floods, the country could find itself in polysaccharides shortage to support the industry whose animals and human growing population (Hussain et al. 2020). Therefore, cassava being a drought-tolerant crop due to its physiological genetically designed has been promoted and has attracted many processors for animals and human feed/food, alcohol/ethanol, exports/imports, paper, and charcoal fines binder among many different uses (Reincke et al. 2018). Therefore, cassava being a drought-tolerant crop due to its physiological genetically designed has been promoted and has attracted many processors for animals and human feed/food, alcohol/ethanol, exports/imports, paper, and charcoal fines binder among many different uses (Reincke et al. 2018). Based on the cassava value chain assessment done by Mwansakilwa et al. in 2011 and a review of other relevant literature, the largest market for cassava flour or dry cassava (dc) chips in Zambia is thought to be 3,530,019 MT. This number was figured out using government policies on biofuels and the Cassava Expansion Strategy 2020–2024 for market development and diversification (Szyniszewska et al., 2021). The active/licensed alcohol/ethanol market, worth 102,250 MT dry cassava, and the exports/imports market, worth 49,506 MT dry cassava, are the two second-largest end markets. With 2,544 MT of dry cassava, the combined paper and charcoal binding markets are incredibly modest(Szyniszewska et al., 2021).

15.1.3 Processing of Cassava Ethanol to Create Ethanol-Based Products

According to Mwansakilwa et al. (2021) report, the beverage and bioethanol sectors can use cassava to create a diversity of products. The report states that due to restricted capacity, Zambian Breweries (ZB) uses HQCF to produce beer (Eagle Lager), which only needs a maximum of 4,000 dry metric tonnes. The establishment of ethanol production facilities in each of the five major cassava-producing provinces is suggested by the ethanol production business model. The three major companies currently using cassava for the manufacturing of bioethanol are Sunbird, Zhongkai International, and Thomro Investments, with Thomro Investments operating a prototype plant with a daily capacity of 2 MT. The goal of Sunbird's ethanol production is to provide 120 million litres of biofuel yearly, or around 10% of Zambia's gasoline imports.

At a minimum yield of 6 MT of dry cassava per hectare, Zambia's farming community would make US$ 346.1 million just from the provision of bioethanol feedstock. According to conversion from the Cassava value chain study, a tonne of dry cassava may yield roughly 330 L at a cost of the US $0.72 per L (or K12.68/lt at the exchange rate K17.61/dollar posted at the Bank of Zambia website on April 5, 2022) (Mwansakilwa et al. 2021). 146,000 MT of dried cassava per year is the start-up market with a ready offtake for the E10 ethanol blending announced by the Zambian government.

According to estimates, ethanol products will require 3,529,686 MT of dry cassava (hydrous ethanol), 1.924 million MT of hydrous ethanol to replace gasoline in transportation, 61,920 MT of hydrous ethanol to replace kerosene, 7,000 MT of hydrous ethanol to replace non-biodegradable plastics and similar materials, and 14,510 MT of hydrous ethanol for sanitizers to replace chlorine.

Government will create an enabling environment, but private sector investment will be used to produce ethanol. The agribusiness center and the ethanol processors will enter into agreements for the ongoing supply of raw cassava and/or cassava chips/starch for ethanol production. Now that the ethanol production methods are reasonably understood locally, a dedicated fund will be formed to support the bioenergy/biofuels industry. A production facility for ethanol is thought to cost $5 million to start up.

Regarding government assistance, there is a legal framework, a national energy policy, and a national strategy (the Zambia Cassava Sector Development Strategy (2020–2024)) that all support the use of ethanol for transportation, the replacement of charcoal, and the production of bread using some wheat–cassava composite flour. One of the end markets for cassava in the Zambia Cassava Sector Development Strategy 2020–2024 is ethanol for transportation and other purposes (Chikoti et al., 2019). The trends in automotive technology that are moving away from fossil fuels and toward greener alternatives provide additional encouragement. The e-biofuel cell is one of the best technologies for Zambia, which has a lackluster infrastructure and uses electric vehicles as one example.

The profitability of ethanol production in Zambia is dependent on the production setup, according to an economic analysis of the industry. Cassava Gross Margin Analysis Based on Traditional Production and Family Labour Only (TFL) and Gross Margin Analysis with Subsidized Mechanization both show good returns for ethanol producers (SM). According to estimates by Mwansakilwa et al., the Cassava Gross Margin Analysis on ethanol production using traditional Economic Labour Prices (TEL) indicated a loss of 4.1% at the current level of cassava productivity (2021).

15.1.4 CARBOHYDRATES AND FIBER ARE TWO MACRONUTRIENTS

Humans and other non-ruminant animals have difficulty metabolizing fiber for energy or other nutritional needs, according to research (Turck et al. 2021). Due to the fact that fiber adds bulk to meals, it can satisfy hunger without adding many calories to the diet. There is an abundance of scientific evidence demonstrating the multiple health benefits of long-term dietary fiber consumption, including its ability to reduce the incidence of colon and other cancers, despite the ongoing debates. Recent microbiome twin studies that examined the relationships between diet and gut microbial community structure and function discovered that differences in our gut microbiome ecology influence our propensity for obesity or malnutrition and that diet, not applied probiotics, was the single most significant indicator of gut health (Pino et al. 2022). These studies evaluate the intestinal microbiota/microbiome of malnourished or malnourished concordant or discordant twins residing in numerous developing nations and who provided samples soon before, during, and after therapy. Specific gut bacteria, primarily Bifidobacteria, digest dietary fiber or other unabsorbed carbohydrates to create short-chain saturated fatty acids. These short-chain fatty acids may trigger apoptosis, inhibit 3-hydroxy-3-methylglutaryl coenzyme-A reductase (HMG-CoAR), and boost mineral absorption, hence decreasing LDL production and avoiding colon cancer (Grothmann and Patt 2005). Individual sugars such as sucrose and fructose, as well as polymeric carbohydrates such as starches and fructans, can be produced by plants (Hyeon et al. 2020). The biosynthesis of these molecules is sufficiently well understood that crops can be genetically altered to create polysaccharides that aren't ordinarily produced, and their properties can be bioengineered. In sugar beet and potato, polymeric carbohydrates such as fructans, inulins, and amylase (resistant starch) have been created without affecting growth or phenotypic variety, according to study (Yuan et al. 2021). Using a similar method, soybean cultivars having oligo-fructan components that raise populations of helpful bacterial species in the intestines of humans and animals while suppressing the growth of bad bacteria are being developed (Asghar et al. 2021; la Peña-Armada et al. 2021). According to Li et al. (2022), polysaccharides are used in agriculture to improve the structure and texture of soil. A study was undertaken to develop a brand-new sort of multifunctional soil conditioner utilizing the 4-arm star-shaped polymer-modified mesoporous MCM-41 as a urea transporter and toxic material scavenger. A silane coupling agent was utilized to couple the star-shaped polymer with MCM-41. According to the results, modified MCM-41 performed around 50% better than raw MCM-41 in terms of its effect on urea release. The crop's growth was successfully boosted by 20%–25%, while loading efficiency (LE) and encapsulation efficiency (EE) increased by around 39% and 42%, respectively. In addition, the 4-arm polymer-modified mesoporous MCM-41 possesses excellent methylene blue adsorption characteristics. Due to its unique structure, the modified MCM-41 exhibited high biocompatibility and negligible biological toxicity. The results suggest further that the multifunctional soil conditioner is appropriate for use in agricultural management strategies to improve the soil environment and promote crop growth. Due to its increased and sustained kinetic energy, the soil conditioner was effective in terms of prolonged release, absorption, and crop growth stimulation.

Wang et al. (2021) looked into the effectiveness of pH-responsive gel spheres for controlled release of humic acid (CSGCHs) using an integrated instillation technique with a composite material made of sodium alginate (SA) and charcoal-activated carbon (CAC) as a carrier. The goal of the study was to find out how well pH-responsive gel spheres work for slow release, changing pH, and improving soil. During soil remediation trials, it was found that humic acid (CSGCH) was effective in repairing a variety of soil types. After 50 days of cleanup, the amount of nutrients and organic matter in the soil went up a lot, while the pH and salt content of the salty soils went down by 15.2% and 29.8%, respectively. In the plant experiment, CSGCH was shown to make plants grow faster. This means that the soil conditioner that was made could be used in agriculture to improve soil conditions and make plants grow faster.

It is obvious that polysaccharides play a critical role in maintaining and preparing soils for increased agricultural output and productivity, which benefits both the economy and the environment. Even though these results are encouraging, it is crucial that farmers utilize climate-smart agricultural techniques and inputs to maximize results through soil cultivation and use over generations (Jagustović et al. 2021; FAO 2013). According to Figure 15.3, polysaccharides are used to increase the soil conditioning as way of giving life to plants and living animals and ultimately reducing human poverty.

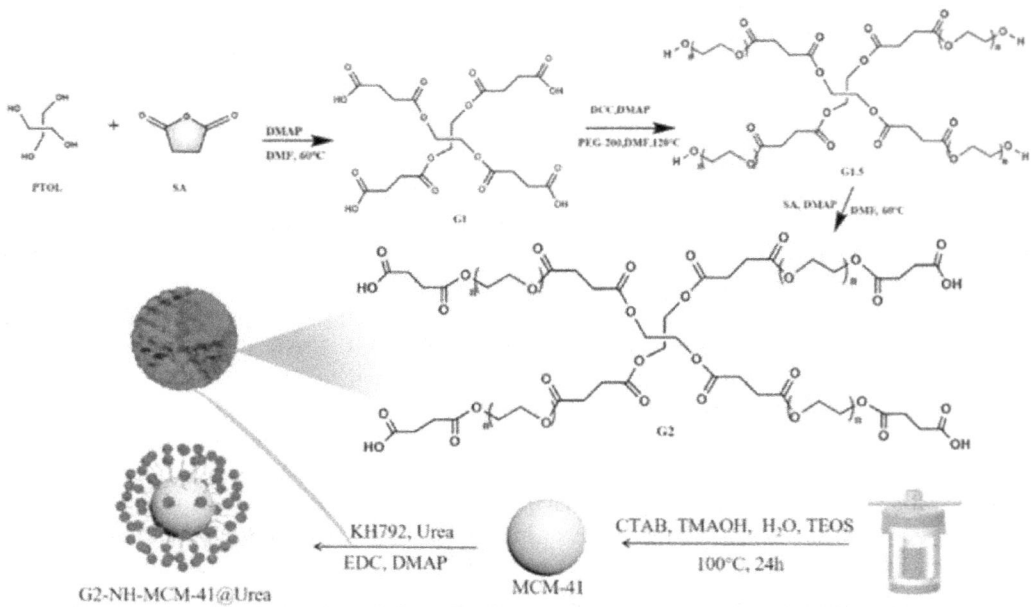

FIGURE 15.3 Graphical representation of the soil conditioner polysaccharide (Li et al. 2022).

As shown in Table 15.1, cellulose and lignin make up the majority of the chemical components of fibers like coir, banana, pineapple leaf, sisal, palmyra, sun hemp, and others. Both lignin and cellulose are constituents of natural fibers; these celluloses have numerous fibrils running along them and are linked by hydrogen bonds, which give them strength and flexibility.

15.1.5 THE FUNDAMENTALS OF SUPERABSORBENT HYDROGELS (SH)

The use of "superabsorbent hydrogels" (SH), made of soil polysaccharides, has been beneficial for soil conditioning. Hydrophilic polymers that are crosslinked chemically or physically form three-dimensional (3-D) matrices called hydrogels. These matrices can be linear or have multiple branches. According to Baloch et al. (2020), they have the ability to absorb enormous amounts of water or biological fluids over a predetermined period of time.

Numerous studies have demonstrated that SH can maintain network stability even when it is bloated in a crosslinked structure, ensuring SH stability in a range of media and situations and being the cause of these characteristics (Baloch et al. 2021; Khan et al. 2021). The crosslinking of hydrogels can be done via chemical or physical means. According to Souza et al. (2022), the key characteristic of chemically crosslinked SH is the establishment of irreversible covalent connections between the polymeric chains. As a result, chemical hydrogels have been made using a range of crosslinking processes, including radical polymerization, the reactivity of complimentary groups, grafting reactions, and enzymes. Physically crosslinked hydrogels have interactions between their crosslinks that are caused by H-bonds, Van der Waals forces, and electrostatic forces. These interactions hold together the polymeric chains. In the last type of hydrogel, crosslinking can be undone, and the matrix can be destroyed under certain conditions. According to Rivas et al. (2021), the type of crosslinking used to make SH has a substantial effect on many key variables, the final properties of SH, and, consequently, the possible applications for SH. The crosslinking approach affects a variety of properties, such as the ability to absorb water, swelling kinetics, mechanical and rheological properties, rate of deterioration, porosity, and toxicity (Rivas et al. 2021). SH synthesis must be tuned to the ultimate application to produce materials that exhibit (or are capable of exhibiting) desired responses, such as fast swelling, degradability, porosity, and so on. Obviously, more criteria may need to be met to guarantee specific features. For instance, Rivas et al. (2021) observed that the polymer type can improve the hydrophilicity and (bio)degradability of the SH matrix. Frequently, acrylic monomers and polymers are used as the primary building blocks in the production of high-performance SH that meets the required specifications. Polyacrylamide, poly(acrylic acid), and other polyacrylates have been used for a very long time to produce high-performance superabsorbent materials (Rivas et al. 2021). As environmental concerns have grown, the use of polymers derived from free oil has expanded. According to Etaemadi Baloch et al. (2021), the use of certain monomers (or polymers) in SH synthesis permits the fabrication of stimuli-responsive materials that can react differently depending on the medium's conditions. According to Guilherme et al. (2015), SH constructed from thermos-responsive polymers such as poly(N-isopropylacrylamide) and poly(N,N-diethylacrylamide) can be hydrophobic or hydrophilic, making it stimuli-responsive and known as "smart soft material that responds to external (environmental) stimuli." The development of innovative SH exploiting polysaccharides such as pectin, cashew gum, Arabic gum, starch, chitosan, and chitin appears to have been motivated by, among other factors, low cost, abundance, renewability, and biodegradability. Molasses, a byproduct of the sugar cane industry, can be used in the construction of farms and agricultural soils (Kibiti & Ndegwa 2016). Figure 15.4 demonstrates how and where molasses is combined with maize grain and fed to animals as an energy source (Siankwilimba 2019). Molasses has been utilized in numerous countries as a supplement to animal feed, as depicted in Figure 15.4.

Molasses is frequently used as the primary ingredient in road construction in nations with sugarcane plantations instead of bituminous materials (Kibiti & Ndegwa 2016; Prudhvi & Rao 2017; Solomon et al. 2020). A byproduct of the refining of sugar cane is molasses, a syrupy liquid that is

FIGURE 15.4 A cattle herd at a feedlot consuming molasses, a byproduct of sugar cane.

viscous, dark brown, and generated in large quantities. Another name for it is treacle. It contains resinous and inorganic components, making it unfit for human consumption. According to Prudhvi and Rao (2017), adding 10% molasses to a soil sample causes the cohesiveness to rise from 0.25 to 0.6 and the friction angle to rise from 9° to 19°. After being treated with 10% molasses and lime, the maximum dry density of the soil increased from 1.89 to 1.933 gm cm^{-3}. The optimal moisture content of the soil went from 10.0% to 12.0% after the addition of molasses and lime. The soil's coefficient of permeability was decreased by adding 10% molasses and lime, going from 4.566E-05 to 2.06E-05. According to these results, stabilizing soil with molasses and lime increased the soil's strength properties by 7%–10%. Hareru and Ghebrab (2020) also said that using bio-asphalt in asphalt mixtures is now seen as one of the best ways to use less bitumen, which is good for the environment, the economy, and people's health in farm blocks. Even though studies have shown that bio-asphalts are sensitive to moisture, which can have a big effect on the durability of hot-mix asphalt (HMA) pavements, they are still used as a bitumen substitute to build pavements and roads, especially when sugarcane molasses is used instead of bitumen because it is better for the environment and costs less to build with. Hareru and Ghebrab (2020) say that it makes asphalt pavements more resistant to heat and cuts greenhouse gas emissions by up to 30%. Hareru and Ghebrab (2020)'s study asserts that 4.7% and 13%–15% for bitumen with a 30-viscosity grade, 0%–3% and 9% for bitumen with a 60%–70% penetration grade, and 0%–10% and 5%–20% for bitumen with a 50%–70% penetration grade were some of the numbers used in research studies to find the best % of molasses to replace bitumen when making HMA. Molasses can be used in place of bitumen to improve the performance of the pavement. As shown in Figures 15.3 and 15.4, the production of sugar could have helped build highways and many agricultural areas. Compared to regular gravel roads without molasses, these roads are much, much better, as shown in Figure 15.5.

Using Zambia as a case study, Figure 15.6 demonstrates how sugar companies compete to generate one of the essential polysaccharides utilized by humans and animals in agricultural development.

15.1.6 CELLULOSE A FORM OF POLYSACCHARIDE

Cellulose is a low-cost biopolymer made from renewable resources such as the walls of wood and plant cells, bacteria, algae, and, bizarrely, tunicates, the only known organisms that contain cellulose. According to a study, the annual industrial production in the globe is expected to exceed 1.5 million tons, with the cotton and wood industries, which consume 98% and 90% of pulp,

FIGURE 15.5 A vehicle traveling along a road made of molasses next to a sugar cane plantation.

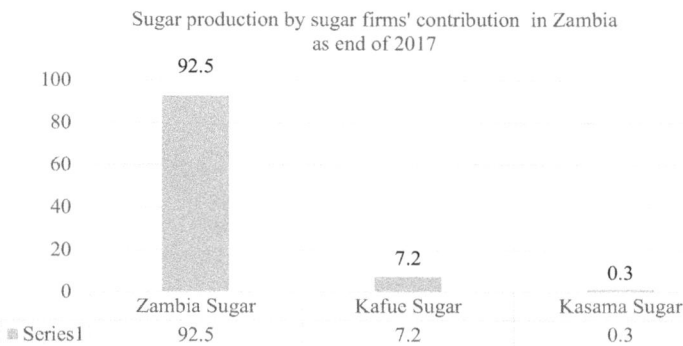

FIGURE 15.6 Contribution of sugar cane production by sugar firms in Zambia that are also used at farm level.

respectively, accounting for the majority of this output (Souza et al. 2022). Souza et al. (2022) say that cellulose is the most common polymer in nature. It has a high molecular weight and a strong tendency to form fibers with crystals. It is made of glucose monomers linked at positions 1–4. Based on how the carbon skeleton and hydrogen bonds are set up, there are four polymorphic forms: I, II, III, and IV. The five basic sources of cellulose are wood, plants and their byproducts, algae, animals, and microbes (Zuo et al. 2012). Tertiary sources include processed waste products from the usage, transformation, and conversion of ethanol. Secondary sources are things like bark and leaves that come directly from the food industry, agriculture (Siankwilimba et al. 2021), or forestry (Siankwilimba et al. 2021). Traditional or primary resources are used to make textile fibers, paper, wood for building, and ethanol. The initial three levels of cellulose origin are plant-based. Due to their chemical makeup, many plant remains can separate cellulose molecules. However, not all agricultural waste has a high concentration of cellulose, which makes extraction difficult or impossible in some cases, like with bagasse. On the other hand, pseudo-stems, stalks, maize cobs, and barks are useful agricultural supplies because they have more than 40% cellulose. Several factors, such as storage, moisture, pollution, volume, distance from industrial processing facilities, other possible uses, like compost or fertilizer, local resources, and logistics facilities, could help or hurt the process of extracting cellulose for biotechnology applications. The Brazilian Biomass Industries Association (Fărcaş et al. 2021) gave a way to figure out the residual factor in the study. Therefore,

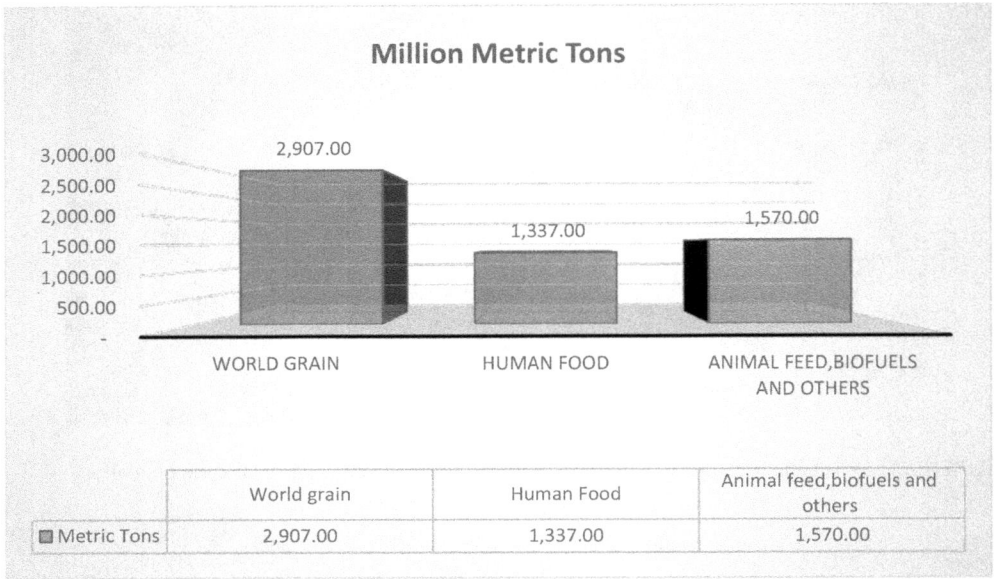

FIGURE 15.7 Distribution of the global grains between humans and animals, biofuels, and other uses.

grain waste may be classified as cellulosic biomass. According to Fărcaş et al. (2021), rice had the lowest residual factor, but maize had the lowest cellulose content, with just 18%, despite having the highest amount of cellulose, up to 44%. Cereals have different qualities based on their species and how they are made, such as whether they are milled dry or wet. Rice and barley are pearled, but maize and wheat are ground wet and dry, respectively. Maize waste consists of biomasses such as cob, husk, germ, bran, and gluten (Fărcaş et al. 2021). Compared to the other cereals, this crop had the highest annual output and residual amounts, at about 1,200 and 700 million metric tons, respectively. This demonstrates its excellent viability in the extraction of cellulose and hemicellulose. Considering this, extensive study has linked cellulose to the textile mills where clothing and other materials are produced (Alexander 2016). On this foundation, animal wool is not exceptional. According to recent report from The Economist website (2022) of 2,907, million metric tons, nearly half of all grain is either burned as biofuel or consumed by animals, meaning that humans do not consume most of the world's grain as shown in Figure 15.7. As such cereals remain, the global sources of polysaccharides used in fuels and animal feeds and the rest human consumption.

15.1.7   CHITOSAN/CHITIN USE IN AGRICULTURE

Over 20 years ago, it was discovered that cell wall-produced oligosaccharides might be used to control plant growth, defense, and development. Natural compounds called chitin and chitosan are crucial in agriculture and a few other industries. Because they strengthen plants' defences against invasive microorganisms, they are also known as "defence boosters." Chitin and chitosan have been demonstrated in numerous studies to activate defence mechanisms in plants, promoting seed germination, plant growth, and yields (Salama 2020; Khan et al. 2021; Luo et al. 2022). For the past 100 years, chemical fertilizers and insecticides have been the mainstays of crop security. The chemical pesticide, on the other hand, has two sides to it because it has the ability to both heal and destroy. Overuse of pesticides and fertilizers helps farmers produce more, but these practices have negative effects on biological diversity, agricultural systems, and public health. They can also breed resistant strains of organisms (Marczak et al. 2020; Jastrzębska & Vasilchenko 2021). The bioactivities of chitosan and its derivatives, which include triggering a defence mechanism in plants and animals as well as encouraging plant and animal growth, have long been acknowledged as being significant in agricultural applications. For instance, according to Kumar et al. (2020), various inputs, including seeds, were coated with carboxymethyl, hydroxyethyl-chitin, and depolymerized chitin before being sown in the ground. According to Kumar et al. (2020), some studies revealed that chitin and chitosan addition increased the plant's dry weight by 8% and increased crop production by 12%. In order to get over some of chitosan's technological drawbacks, like brittleness, limited elongation, and flexibility, coatings appear to be created by mixing it with other biotechnologically active substances (Siankwilimba et al., 2023). According to research (Duncan et al. 2021; Martey et al. 2021), composite coatings consisting of chitosan, poly(vinyl) alcohol, lignin nanoparticles, and other polysaccharides exhibit better plastic resilience than their constituent components. According to Berliana et al. (2020), chitosan biopolymer can be utilized to protect food goods, promote plant growth, and provide abiotic and biotic stress tolerance in various horticulture commodities. Chitin and chitosan research has been prioritized since 1990 because of its outstanding properties, including biodegradation in the human body, control of plant diseases, immunological, antibacterial, antifungal, and wound-healing action. Chitin and chitosan are thus gaining popularity as cutting-edge functional materials in agricultural development. The chitosan biopolymer is a viable option for the induction of plant defensive systems and the stimulation of plant growth, both of which are crucial in agricultural applications, despite their intriguing biological characteristics. The evocative effects of chitin and chitosan fragments promote the accumulation of phytoalexins, the production of lignin and cellulose, pathogen-related proteins, and proteinase inhibitors in host plants in response to microbial infections (Berliana et al. 2020). Chitosan's use in agriculture for plant defence and yield improvement is mostly dependent on how this glucosamine polymer impacts the molecular biology and biochemistry of the plant cell. In the 20th century, chitosan was employed in agriculture, especially in agricultural arenas all over the world, to feed the population that was constantly expanding by increasing productivity per area of land used for food production (Sangiorgio et al. 2021). The discussion is furthered by Kumar et al. (2020), who state that a special modified form of chitosan called chitosan nanoparticles has been used as a practical delivery system for pesticides, fertilizers, herbicides, and micronutrients for crop growth promotion through balanced and sustained nutrition. Chitosan was discovered to be capable of triggering defense against more than 60 diseases in a variety of plants, according to research by Asghar et al. (2021). Chitosan promotes the accumulation of phytoalexins, which trigger antifungal responses and strengthen defense against upcoming infections. Orchid leaf spot disease has been shown to be significantly less severe when chitosan is sprayed on the plants. Chitosan has the potential to stimulate the synthesis of plant hormones like gibberellins. In addition, a signaling pathway associated with auxin synthesis that is unrelated to tryptophan may be used by chitosan to promote growth and development. On the other hand, chitosan has conflicting results in terms of growth and development. According to Berliana et al. (2020)'s findings, chitosan significantly enhanced A's height. The results showed that the number of leaves increased by a lot at 10 ppm, and hybridus increased by up to 19.59%. However, chitosan treatment had no effect on chlorophyll content, fresh or dry weight, even though the dosage response peaked at 10 ppm. Chitosan therapy might be a good way to help many crops deal with drought stress, especially in developing countries where climate change is wreaking havoc in unpredictable ways (Mwiinde et al. 2022). Chitosan also decreased transpiration rates by 36.66%–66.26% (Maulu et al. 2021; Siankwilimba et al. 2022). Recent research by Stasińska-Jakubas and Hawrylak (2022) shows that chitin and chitosan can be made from insects and some types of fungus as well as seafood. According to the study, traditional chemical processes usually need close to 30 kg of wet shrimp shells to make 1 kg of chitin. On the other hand, traditional chemical processes usually need up to 1.5 kg of chitin and many acidic and alkaline substances to make 1 kg of chitosan. Chitin is a polysaccharide that doesn't dissolve in water. This is because the oxygen of the acetamido group interacts with nearby NH or OH function groups (Teixeira-Costa & Andrade 2021). Teixeira-Costa and Andrade (2021) also said that the amount of N-acetylation in chitin chains has a big effect on its insolubility in water, which limits its ability to swell and, as a result, its use in industry. Allomorphs, pogonophoran, and vestimentiferan worms produce the chitin that is present in squid pens. Even though there are more chitin forms, it is uncommon to find them produced by worms like pogonophoran and vestimentiferan worms and combined with proteins (Khan et al. 2021). The different orientations of the microfibrils in the three crystalline phases make it possible to make a wide range of functional materials that can be used for medical, anti-inflammatory, antibacterial, and immunological purposes. When chitin is extracted and deacetylated to make chitosan, dangerous or toxic solvents are often used, processing yields are limited, toxic wastewater effluent is made, and processing costs are high. As long-term solutions to these problems, green strategies like using deep eutectic solvents and ionic liquids have been put in place (Teixeira-Costa & Andrade 2021). Like other polysaccharides made from residues, chitin's biological functions depend on its molecular weight and level of deacetylation. For example, higher levels of deacetylation make reactions stronger. The physicochemical properties of chitin and chitosan can be affected by where the raw materials come from, when they are harvested, and how they are processed (Nwe et al. 2014). Pure chitosan only has a few industrial uses, so it can be made more useful by making its structure chemically functional and crosslinking it to many molecules. Both chitin and chitosan are utilized in the biomedical industry to administer medications. Saberi Riseh et al. (2021) say that both biopolymers are used to make hypocholesterolemic dietary supplements, to control the transfer of drugs, and to take allergenic proteins out of certain meals. In tissue engineering, they help heal and regrow epithelial tissue in both humans and cattle. They are also used in antibacterial, anti-inflammatory, sunscreen, and anti-aging cosmetics (Guarner et al. 2004). Chitosan has a wide range of applications in food technology, including extending the shelf life of foods like bread and apples. Its ability to inhibit microbial growth and reduce starch retrogradation in both food preparation and the creation of coating packaging with bioactive properties makes these applications possible.

Polysaccharides contain inherent storage qualities, such as starch, or structural properties, such as cellulose, which offer physical structure and stability as shown in Figure 15.8. Positively charged polysaccharides (chitosan) and negatively charged polysaccharides, such as heparin alginate, pectin, and hyaluronic acid, can be categorized based on polyelectrolyte. Glycosaminoglycans (GAGs) as stated in Figure 15.8 are extensively investigated major components of the cell surface and the cell-extracellular matrix (ECM). Heparin, heparan sulfate, hyaluronan, chondroitin sulfate, dermatan

FIGURE 15.8 Grouping of polysaccharides based on the kind of monosaccharide constituents and their physiological effects. Source: Mohammed et al. (2021) and author computation.

sulfate, and keratan sulfate are the most prominent GAGs polysaccharides in mammalian tissues. As shown in Figure 15.8, there are numerous sources of polysaccharides, including plants like soyabeans and maize, microbes, algae, and mammals, such as cattle, goats, and sheep. Expected to their physicochemical features, they are susceptible to physical and chemical alterations that result in increased properties; this is the fundamental notion underlying their wide range of biological and pharmacological uses. According to Mohammed et al. (2021), chemical modifications of polysaccharides, such as sulfation, phosphorylation, and carboxymethylation, are highly effective procedures for modifying and changing the biological properties of polysaccharides, making them suitable for drug delivery systems in various drug preparations because they can be more stable, non-toxic, and biodegradable. Furthermore, Mohammed et al. (2021) discussed that chemical changes such as grafting, crosslinking, complexation, and covalent coupling increase the potential for drug transport and, thus, enhance the therapeutic efficacy. They may also serve as a suitable substitute for certain excipients or synthetic polymers. It is argued that traditional polysaccharides have been used in vaccines by virtue of them containing pollutants and impurities in their heterogeneous combinations; thus, the chemically synthesized polysaccharides-based vaccination can solve these deficiencies (Khan et al. 2021; Khalina et al. 2015; Teixeira-Costa & Andrade 2021).

Suffice to note that stability, hydrophilicity, and biodegradability, as well as other features such as the diversity of physicochemical properties of natural polysaccharides, serve as the basis for its vast array of biological features (Teixeira-Costa & Andrade 2021). It has been demonstrated that the biochemical and physical characteristics of polysaccharides have contributed to their biological uses. Additionally, polysaccharides are indispensable macromolecules that almost occur in all living organisms and have important functions in the body. Mohammed et al. (2021) stated that they are getting increasing attention seeing as they exhibit a wide range of biological and functional properties, including antitumor, immunomodulatory, antimicrobial, antioxidant, anticoagulant, anti-diabetic, antiviral, and hypoglycemia activities, making them one of the most promising candidates in the biomedical and pharmaceutical fields.

15.1.8 Development of Agriculture Using Inulin

The naturally occurring polysaccharide inulin is present in over 30,000 different plant species, and the most widely used commercial sources are cereals, tubers, and roots (Souza et al. 2022; Kilic et al., 2021). It might be due to its flavor neutrality and health benefits. A remarkable biofortification ingredient is inulin, which is categorized as a dietary fiber. Cardullo et al. (2022) list chicory, asparagus, garlic, banana, onion, rye, barley, and wheat as the foods that typically contain this polysaccharide's residual sources. They also mention that it can be discovered in unexpected places, such as aça seeds. Let's just say that enzymatically produced inulin can also be obtained from potatoes that have undergone genetic modification. As a result, it's possible to find food residues from foods like onions (18%) and garlic (28% of dry mass), which have a little bit more inulin (Souza et al. 2022). Numerous studies claim that inulin is a unique oligo- or polysaccharide because each sugar ring is a part of its backbone, which consists of linear fructose chains connected by glycosidic connections created by 2–100 monomeric units (Branca et al. 2022). According to some studies, it has more flexibility than pyranose rings because the majority of its chains are furanose groups (Melilli et al. 2020). Due to its lengthy polymeric chain and glycosidic connections in the -(2-1) position, which render it insoluble at body temperatures and resistant to salivary amylase and intestinal enzymes, inulin has the potential to be used in the food and pharmaceutical industries (Souza et al. 2022).

Because it has up to ten fructose units that are useful in agriculture, inulin is called a fructo-oligosaccharide (FOS) and is considered a prebiotic (Su et al. 2007). Rakhesh et al. (2015) say that many plants use inulin to store carbohydrates. Because of this, it can be found in the parts of fruits and vegetables that are left over after processing, such as husks, seeds, stems, and bagasse. The majority of the time, these by-products are a result of processing techniques like pulping, washing, sieving, and food bleaching, all of which are necessary for the processing and preservation of food (Padalino et al. 2017). Rakhesh et al. (2015) say that the biomedical and agricultural fields are interested in biotechnological applications based on inulin, especially for nutritional or medical purposes. According to many studies, inulin is used to replace fats and sugars in foods because it has health benefits (Raccuia & Melilli 2004; Nejati et al. 2017). This makes it possible to make ice cream and other high-fiber foods with fewer calories (Nejati et al. 2017). This is probably because inulin doesn't get broken down in the digestive system, which makes it possible to make functional foods. There have been many studies on fructan extraction, but few papers (Valluru & Van Den Ende 2008; Branca et al. 2022) have been written about how to improve the extraction process. It looks like inulin has been used as an ingredient in bioactive functional meals to help bifidobacteria and other good microorganisms grow in the gut microbiota, which is good for humans. It is used in the pharmaceutical industry as a bioactive carrier. In this case, inulin can help get more of the drug to the last part of the digestive system and stop it from breaking down in the upper part of the digestive system, which delays the drug's early breakdown. In this method, when the inulin gets to the colon, bacteria from the local microbiome break it down. This makes it easier for the body to absorb minerals and ions from food. In relation to the aforementioned advantages, inulin concentration enhances product texture. But because of its physical and chemical properties, high concentrations of inulin may change the way products feel. Inulin may have an impact on the sensory qualities of numerous products. The physicochemical significance of inulin depends on the extent of its polymerization. Inulin's short-chain oligosaccharides are sweeter and more soluble than its long-chain oligosaccharides. It has qualities that are similar to those of other sugars and can make the tongue feel better. So, a prebiotic is a part of food that can't be digested but is good for the host because it promotes the growth and/or activity of one or a small number of bacteria in the colon, which is good for human health (Schuster 2005; Dahiya & Nigam 2012). The International Scientific Association for Probiotics and Prebiotics (ISAPP) defined prebiotics in 2008 as "a selectively fermented ingredient that causes specific changes in the composition and/or activity of the gastrointestinal microbiota, thereby benefiting host health" (Abed et al. 2016). Abed et al.'s (2016) research states that in order for something to be classified as having a prebiotic effect, it must first meet the following requirements: (i) Fermentation by bacteria in the gut; (ii) Resistance to the acid in the stomach; (iii) Breakdown by enzymes in mammals; and (iv) the ability to boost the growth and/or activity of gut bacteria that are good for health. A product is said to be symbiotic if it has both probiotics and prebiotics. This is because probiotics and prebiotics work together to keep the bacteria in the intestines in balance. According to this theory, the term "symbiotic" can only be used if the prebiotic has a positive effect on the growth of the probiotic. Liu et al. (2020) say that probiotics are living microorganisms that help the host, especially animals and people, when given in the right amounts. Studies show that the survival of probiotic bacteria is always in question when they are in a hostile environment. So, probiotic bacteria are often put in tiny capsules to keep them alive during processing, storage, and delivery to specific parts of the digestive tract. In the field of probiotic microencapsulation, polysaccharide is one of the walls that are used most often. On the other hand, traditional polysaccharides can't handle these problems because the demand for probiotic microcapsules is growing and there are new ways to use them. By keeping a healthy gut microbiota, you can avoid problems like gastrointestinal infections, inflammatory bowel disorders, and even cancer (Dahiya & Nigam 2022). Lactobacilli like Lactobacillus acidophilus and bifidobacteria have been the most popular probiotics for a long time, according to research (Gule & Geremew 2022). Before it can help people's health, a probiotic must meet a few conditions: (i) it must have good technological properties so that it can be made and added to food without losing its viability and functionality or making the food taste or feel bad; (ii) it must survive the upper gastrointestinal tract and reach its site of action alive; and (iii) it must be able to do its job in the environment of the gut (Pino et al. 2022). But Abed et al. (2016) found that probiotic bacteria are often very sensitive to the harsh environment they live in, such as acid, high temperatures, and high oxygen levels. Campos et al. (2021) say that microencapsulation technology is often used to improve survival in harsh environments and functionality in the human gastrointestinal tract. This is done to make up for these problems.

Campos et al. (2021) say that microencapsulation is a process that coats or encases bioactive compounds in a substance called wall material to make microcapsules in the millimeter to micron range. Microencapsulation is used to protect the active ingredient from the outside world until it is ready to be used. The wall material is one of the most important components in probiotic bacteria microencapsulation. Polysaccharides and proteins are two of the most common wall components used in probiotic microencapsulation. Polysaccharides and probiotics are related in a true sense. A polysaccharide is a big molecular polymer composed of several smaller monosaccharides, such as glucose (Stasińska-Jakubas & Hawrylak-Nowak 2022). There are two types of polysaccharides: homopolysaccharides and heteropolysaccharides (Stasińska-Jakubas & Hawrylak-Nowak 2022). Polysaccharides can take many different shapes, depending on the monosaccharides that are added to the chains. The following polysaccharide characteristics are relevant to the basis of probiotic microencapsulation: (i) an ion-induced gelation property, which allows the polysaccharide to create a cross-linked hydrogel structure by interacting with a variety of ions, such as Ca^{2+} with alginate and pectin and K+ with carrageenan (Jeong et al. 2012). Yao et al. (2020) say that traditional polysaccharides can't keep up with the demand for probiotic microencapsulation because there are so many new uses for them, and the demand is going up. Scientists are actively searching for novel polysaccharide kinds that are ideal for microencapsulating probiotics as a result of this discovery. The good news is that research is already being done on the idea of putting probiotic bacteria inside of polysaccharides. According to Nguyen and Wang (2019), the plants Abelmoschus esculentus in okra and Linum usitatissimum in linseed create exopolysaccharides like mucilages, while Paenibacillus jamilae produces jamilan. Probiotics have been used to keep and improve the health of animals that have been stressed by diseases, droughts, or other things that can be seen as stresses.

15.1.9 ALGAE AND THE ENRICHMENT OF CARBOHYDRATES

Macroalgae are an attractive source of bioactive polysaccharides with industrial and innovative food applications (Cottrell et al. 2020). They include a variety of fascinating and frequently unique polysaccharides that are being researched for potential biomedical and dietary applications. However, a much more extensive application would be to employ seaweed sugars as an alternative fuel source. This application can be accelerated by utilizing photosynthesis, nature's energy cycle, and the resulting plant biomass (Giglio et al. 2018). In order to gain the benefits of carbon-neutral biofuel, plastics, and healthcare, society must move from a hydrocarbon to a carbohydrate economy (Husmann et al. 2018). Marine biomass is an often-overlooked source of carbohydrates that might be a significant source of renewable energy, despite the fact that macroalgae are excellent solar energy converters and can rapidly create huge quantities of biomass. According to Kraan (2012), for example, macroalgae's proximate composition is influenced by high ammonia and nitrate concentrations, resulting in a shift toward increased protein and typically lower levels of carbohydrates such as starch or dietary fiber. It may be amusing to vary the amount and kind of carbs in algae, although it is unknown how this affects the carbohydrate composition. The red carrageenophyte *Kappaphycus alvarezii* exhibited increased gel strength of the carrageenan with an increase in ammonia, revealing the considerable impact of ammonium supplementation on the carrageenan content in an otherwise nitrogen-depleted environment (Balasubramanian et al. 2021).

15.1.10 OPPORTUNITY FOR CIRCULAR ECONOMY

The great opportunity exists for all market players to harness the polysaccharide business globally in today's world. There is business for all to recycle the plastic that have caused a lot of air, water, and land pollution due waste material. Produced from the farms and forest, then to industry, these materials call for urgent need to work on non-fossil and the degradable substances. Marczak et al. (2020) have reported that the most common non-fossil plastics as of right now are those that are mechanically recycled, followed by those that are bio-based or biodegradable whose economic and environmental importance cannot be over emphasized. In recent years, products such as bottles made of polyethylene terephthalate (PET) and polyethylene (PE) are being successfully recovering these materials through mechanical procedures, which can be used to create recycled beverage containers (bottle-to-bottle recycling) (Pettersen et al. 2020).

However, due to safety worries about contamination, employing recycled plastics in food-grade products is extremely difficult. This problem can be overcome by advanced recycling, which turns recycled materials back into hydrocarbons and chemical precursors that can be used as feedstocks by other processes(Hallack et al. 2022). Advanced recycling provides a supplementary option to broaden the recycling landscape and includes innovations like pyrolysis, gasification, solvolysis, and microwave (Islam et al. 2021). As a result, it is anticipated to become more crucial in meeting circular economy goals and commitments and contributing to the expansion of the types, quantities, and qualities of plastic trash that can be recycled (Sheina & Babenko 2014). Even though rigid polyethylene (PE) and rigid PET resin have good mechanical recycling rates, some sophisticated recycling machines may take a variety of polymers, including mixed plastics with possibly higher contamination (Pettersen et al. 2020).

Many experts, including Peng et al. (2022), believe that advanced recycling will keep expanding and play a critical role in meeting the demand for recycled polymers if the current sustainability momentum continues at the same rate and if national restrictions are lifted. To meet 4%–8% of the total polymer demand by 2030, for instance, such a scenario would call for advanced recycling to require more than $40 billion in capital investments over the ensuing 10 years. Even though this may only represent a negligible portion of the entire plastics industry, it would still represent a significant increase from its current share of less than 1% and would result in an annual increase in technological potential of more than 20% until 2030 (Peng et al. 2022).

15.2 CONCLUSION

Polysaccharides are essential for both on- and off-farm use, and they have been identified as materials with the potential to transform biorefineries all over the world. Depending on how they are combined with other goods along the value chain, polysaccharides can significantly contribute to a country's and the farmer's economy being more environmentally friendly. Despite the differences in their methodologies, all the examined polysaccharides serve similar roles in the manufacturing of materials, functional foods, nutraceuticals, cosmetics, and pharmaceuticals as drug and medication carriers. Additionally, they are properly utilized in the agricultural industry, particularly when it comes to the control and prevention of various diseases and pests that have a financial impact on the productivity and output of the farmers.

Chitosan accelerated green plant development in greenhouse settings, resulting in plants that were heavier, taller, and had more leaves. The rate at which the leaves transpired was slowed down by the application of chitosan. Chitosan may increase plant productivity, especially in areas suffering from drought. Through regional trade, international trade, and import and export, agro-industrial waste is dispersed widely around the world. It is possible to find creative ways to optimize their use in other industries. Agro-industrial methods can be used to convert biomass into a wide range of chemical components.

The abundance of animal and plant biomass makes it possible to estimate the quantity of residues rich in polysaccharides, as this study has shown. Due to the sizeable amounts of waste that are imported and exported from important crops like rice, wheat, and maize, there are a wide variety of potential uses for cellulose extraction (Nugraha et al., 2023). The primary producers of leftovers are also the primary growers of these crops. For use in the livestock and seafood industries, respectively, chitin, chitosan, and hyaluronic acid are extracted from crustacean shells, fish eyeballs, and chicken crests. Pectin and inulin have both been extracted from chicory roots and citrus waste. Because they help to strengthen their immune systems, probiotic products are essential in stressful situations for both humans and animals. To expand this area of the economy, a variety of interventions are required to encourage innovation among all market participants. It has been established, however, that this approach is incompatible with the idea of sustainability due to the abundance of other readily accessible, reasonably priced energy sources, such as solar and wind energy. Increasing the isolation of polysaccharides with properties capable of producing bioactive natural products and using agro-industrial waste as a source of valuable materials for industrial use are both necessary.

REFERENCES

Abed, Sherif M., Abdelmoneim H. Ali, Anwar Noman, and Amr M. Bakry. 2016. "Inulin as Prebiotics and Its Applications in Food Industry and Human Health; A Review." *International Journal of Agriculture Innovations and Research* 5 (1): 88–97.

Adéyèmi, A. D., A. P. P. Kayodé, I. B. Chabi, O. B. O. Odouaro, M. J. R. Nout, and A. R. Linnemann. 2020. "Screening Local Feed Ingredients of Benin, West Africa, for Fish Feed Formulation." *Aquaculture Reports* 17. https://doi.org/10.1016/j.aqrep.2020.100386

Ahimbisibwe, B. P., J. F. Morton, S. Feleke, A. Alene, T. Abdoulaye, K. Wellard, E. Mungatana, A. Bua, S. Asfaw, and V. Manyong. 2020. "Household Welfare Impacts of an Agricultural Innovation Platform in Uganda." *Food and Energy Security* 9 (3). https://doi.org/10.1002/fes3.225

Alexander, Rachel. 2016. "Sustainability in Global Production Networks: Rethinking Buyer-Driven Governance." PQDT - UK & Ireland. https://search.proquest.com/dissertations-theses/sustainability-global-production-networks/docview/1837034003/se-2?accountid=41849.

Amanor, K. S. 2012. "Global Resource Grabs, Agribusiness Concentration and the Smallholder: Two West African Case Studies." *Journal of Peasant Studies* 39 (3–4): 731–749. https://doi.org/10.1080/03066150.2012.676543

Amanor, K. S. 2019. "Global Value Chains and Agribusiness in Africa: Upgrading or Capturing Smallholder Production?" *Agrarian South* 8 (1–2): 30–63. https://doi.org/10.1177/2277976019838144

Asghar, Sajid, Ikram Ullah Khan, Saad Salman, Syed Haroon Khalid, Rabia Ashfaq, and Thierry F. Vandamme. 2021. "Plant-Derived Nanotherapeutic Systems to Counter the Overgrowing Threat of Resistant Microbes and Biofilms." *Advanced Drug Delivery Reviews*. https://doi.org/10.1016/j.addr.2021.114019.

Asim, M., K. Abdan, M. Jawaid, M. Nasir, Z. Dashtizadeh, M. R. Ishak, M. E. Hoque, and Y. Deng. 2015. "A Review on Pineapple Leaves Fibre and Its Composites." *International Journal of Polymer Science* 2015. https://doi.org/10.1155/2015/950567

Balasubramanian, S., N. G. G. Domingo, N. D. Hunt, M. Gittlin, K. K. Colgan, J. D. Marshall, A. L. Robinson, I. M. L. Azevedo, S. K. Thakrar, M. A. Clark, C. W. Tessum, P. J. Adams, S. N. Pandis, and J. D. Hill. 2021. "The Food We Eat, the Air We Breathe: A Review of the Fine Particulate Matter-Induced Air Quality Health Impacts of the Global Food System." *Environmental Research Letters* 16 (10). https://doi.org/10.1088/1748-9326/AC065F

Baloch, F. E., Afzali, D., and Fathirad, F. 2021. "Design of acrylic acid/nanoclay grafted polysaccharide hydrogels as superabsorbent for controlled release of chlorpyrifos." *Applied Clay Science* 211, 106194. https://doi.org/10.1016/j.clay.2021.106194

Berliana, A. I., C. D. Kuswandari, B. P. Retmana, A. Putrika, and S. Purbaningsih. 2020. "Analysis of the Potential Application of Chitosan to Improve Vegetative Growth and Reduce Transpiration Rate in Amaranthus Hybridus." *IOP Conference Series: Earth and Environmental Science* 481 (1). https://doi.org/10.1088/1755-1315/481/1/012021.

Bhardwaj, Abhishek Kumar, Shanthy Sundaram, Krishna Kumar Yadav, and Arun Lal Srivastav. 2021. "An Overview of Silver Nano-Particles as Promising Materials for Water Disinfection." *Environmental Technology & Innovation* 23 (August): 101721. https://doi.org/10.1016/j.eti.2021.101721.

Branca, Ferdinando, Sergio Argento, Anna Maria Paoletti, and Maria Grazia Melilli. 2022. "The Physiological Role of Inulin in Wild Cardoon (Cynara Cardunculus L. Var. Sylvestris Lam.)." *Agronomy* 12 (2). https://doi.org/10.3390/AGRONOMY12020290.

Campos, Amanda C., Andre L. Silva, Aderbal M. A. Silva, Jaime M. Araujo Filho, Tatiane Costa, José M. Pereira Filho, Juliana P. F. Oliveira, and Leilson R. Bezerra. 2021. "Dietary Replacement of Soybean Meal with Lipid Matrix-Encapsulated Urea Does Not Modify Milk Production and Composition in Dairy Goats." Animal Fee*d Science and Technology* 274 (April): 114763. https://doi.org/10.1016/J.ANIFEEDSCI.2020.114763.

Cardullo, Nunzio, Vera Muccilli, Vita Di Stefano, Sonia Bonacci, Lucia Sollima, and Maria Grazia Melilli. 2022. "Spaghetti Enriched with Inulin: Effect of Polymerization Degree on Quality Traits and α-Amylase Inhibition." *Molecules (Basel, Switzerland)* 27 (8). https://doi.org/10.3390/MOLECULES27082482.

Chikoti, P. C., Mulenga, R. M., Tembo, M., and Sseruwagi, P. 2019. "Cassava mosaic disease: a review of a threat to cassava production in Zambia." *Journal of Plant Pathology* 101(3): 467–477. https://doi.org/10.1007/s42161-019-00255-0

Chilufya, W., and N. Mulendema. 2019. "Beyond Maize: Exploring Agricultural Diversification in Zambia." *International Institute for Environment and Development*. HIVOS. https://www.iied.org/beyond-maize-exploring-agricultural-diversification-zambia.

Cottrell, R. S., J. L. Blanchard, B. S. Halpern, M. Metian, and H. E. Froehlich. 2020. "Global Adoption of Novel Aquaculture Feeds Could Substantially Reduce Forage Fish Demand by 2030." *Nature Food* 1 (5): 301–308. https://doi.org/10.1038/S43016-020-0078-X

Dahiya, Divakar, and Poonam Singh Nigam. 2022. "The Gut Microbiota Influenced by the Intake of Probiotics and Functional Foods with Prebiotics Can Sustain Wellness and Alleviate Certain Ailments like Gut-Inflammation and Colon-Cancer." *Microorganisms* 10 (3): 665. https://doi.org/10.3390/MICROORGANISMS10030665

Duncan, Sylvia H., Ajay Iyer, and Wendy R. Russell. 2021. "Impact of Protein on the Composition and Metabolism of the Human Gut Microbiota and Health." *Proceedings of the Nutrition Society* 80 (2): 173–185. https://doi.org/10.1017/S0029665120008022

Etemadi Baloch, Fatemeh, Daryoush Afzali, and Fariba Fathirad. 2021. "Design of Acrylic Acid/Nanoclay Grafted Polysaccharide Hydrogels as Superabsorbent for Controlled Release of Chlorpyrifos." *Applied Clay Science* 211. https://doi.org/10.1016/j.clay.2021.106194

Fabrice, Teletchea. 2019. "Fish Domestication: An Overview." *Animal Domestication*, July. https://doi.org/10.5772/INTECHOPEN.79628.

FAO. 2013. Climate Smart Agriculture Sourcebook. https://www.fao.org/3/i3325e/i3325e.pdf.

Fărcaş, Anca, Georgiana Dreţcanu, Teodora Daria Pop, Bianca Enaru, Sonia Socaci, and Zoriţa Diaconeasa. 2021. "Cereal Processing By-Products as Rich Sources of Phenolic Compounds and Their Potential Bioactivities." *Nutrients*. Multidisciplinary Digital Publishing Institute. https://doi.org/10.3390/nu13113934.

FMR1515 A. 2021. "Humectants Market Growth by Product (Alpha-Hydroxy Acids and Polysacchar." *Fact Market Research*. https://www.bccresearch.com/partners/fact-market-research/humectants-market.html

Giglio, L., L. Boschetti, D. P. Roy, M. L. Humber, and C. O. Justice. 2018. "The Collection 6 MODIS Burned Area Mapping Algorithm and Product." *Remote Sensing of Environment* 217: 72–85. https://doi.org/10.1016/J.RSE.2018.08.005

Grothmann, Torsten, and Anthony Patt. 2005. "Adaptive Capacity and Human Cognition: The Process of Individual Adaptation to Climate Change." *Global Environmental Change* 15 (3): 199–213. https://doi.org/10.1016/j.gloenvcha.2005.01.002

Guarner, Jeannette, Bill J. Johnson, Christopher D. Paddock, Wun Ju Shieh, Cynthia S. Goldsmith, Mary G. Reynolds, Inger K. Damon, et al. 2004. "Monkeypox Transmission and Pathogenesis in Prairie Dogs." *Emerging Infectious Diseases* 10 (3): 426–431. https://doi.org/10.3201/EID1003.030878.

Guilherme, Marcos R, Fauze A. Aouada, André R. Fajardo, Alessandro F. Martins, Alexandre T. Paulino, Magali F. T. Davi, Adley F. Rubira, and Edvani C. Muniz. 2015. "Superabsorbent Hydrogels Based on Polysaccharides for Application in Agriculture as Soil Conditioner and Nutrient Carrier: A Review." *European Polymer Journal* 72: 365–385. https://doi.org/10.1016/j.eurpolymj.2015.04.017

Gule, Thandile T., and Akewake Geremew. 2022. "Dietary Strategies for Better Utilization of Aquafeeds in Tilapia Farming." *Aquaculture Nutrition* 2022 (January): 1–11. https://doi.org/10.1155/2022/9463307

Hallack, E., N. M. Peris, M. Lindahl, and E. Sundin. 2022. "Systematic Design for Recycling Approach - Automotive Exterior Plastics." *Procedia CIRP* 105: 204–209. https://doi.org/10.1016/J.PROCIR.2022.02.034

Hareru, Werku, and Tewodros Ghebrab. 2020. "Rheological Properties and Application of Molasses Modified Bitumen in Hot Mix Asphalt (HMA)." *Applied Sciences* 10 (6): 1931. https://doi.org/10.3390/APP10061931

Husmann, K., S. Rumpf, and J. Nagel. 2018. "Biomass Functions and Nutrient Contents of European Beech, Oak, Sycamore Maple and Ash and Their Meaning for the Biomass Supply Chain." *Journal of Cleaner Production* 172: 4044–4056. https://doi.org/10.1016/j.jclepro.2017.03.019

Hussain, S., A. Hussain, J. Ho, O. A. E. Sparagano, and U.-R. Zia. 2020. "Economic and Social Impacts of COVID-19 on Animal Welfare and Dairy Husbandry in Central Punjab, Pakistan." *Frontiers in Veterinary Science* 7. https://doi.org/10.3389/FVETS.2020.589971/EPUB

Hyeon, Hyejin, Jiu Liang Xu, Jae Kwang Kim, and Yongsoo Choi. 2020. "Comparative Metabolic Profiling of Cultivated and Wild Black Soybeans Reveals Distinct Metabolic Alterations Associated with Their Domestication." *Food Research International* 134 (August). https://doi.org/10.1016/j.foodres.2020.109290

Imam, M. F., W. Wan, N. A. Khan, M. H. Raza, M. A. A. Khan, and M. Yaseen. 2021. "Effectiveness of Agricultural Extension's Farmer Field Schools (FFS) in Pakistan: The Case of Citrus Growers of Punjab Province." *Ciência Rural* 51 (9). https://doi.org/10.1590/0103-8478cr20200807

Islam, A., S. H. Teo, Y. H. Taufiq-Yap, C. H. Ng, D. V. N. Vo, M. L. Ibrahim, M. M. Hasan, M. A. R. Khan, A. S. M. Nur, and M. R. Awual. 2021. "Step Towards the Sustainable Toxic Dyes and Heavy Metals Removal and Recycling from Aqueous Solution - A Comprehensive Review." *Resources, Conservation and Recycling* 175: 105849.

Jagustović, Renata, George Papachristos, Robert B. Zougmoré, Julius H. Kotir, Aad Kessler, Mathieu Ouédraogo, Coen J. Ritsema, and Kyle M. Dittmer. 2021. "Better before Worse Trajectories in Food Systems? An Investigation of Synergies and Trade-Offs through Climate-Smart Agriculture and System Dynamics." *Agricultural Systems* 190 (May): 1–15 103131. https://doi.org/10.1016/j.agsy.2021.103131

Jastrzębska, Agnieszka M., and Alexey S. Vasilchenko. 2021. "Smart and Sustainable Nanotechnological Solutions in a Battle against COVID-19 and Beyond: A Critical Review." *ACS Sustainable Chemistry and Engineering. American Chemical Society.* https://doi.org/10.1021/acssuschemeng.0c06565.

Jeong, Wooyoung, Jinyoung Kim, Suzie E. Ahn, Sang In Lee, Fuller W. Bazer, Jae Yong Han, and Gwonhwa Song. 2012. "AHCYL1 Is Mediated by Estrogen-Induced ERK1/2 MAPK Cell Signaling and MicroRNA Regulation to Effect Functional Aspects of the Avian Oviduct." *PLoS ONE* 7 (11). https://doi.org/10.1371/journal.pone.0049204.

Karlström, Amanda, Fernando Calle, Sandra Salazar, Nelson Morante, Dominique Dufour, and Hernán Ceballos. 2016. "Biological Implications in Cassava for the Production of Amylose-Free Starch: Impact on Root Yield and Related Traits." *Frontiers in Plant Science* 7 (MAY2016). https://doi.org/10.3389/FPLS.2016.00604/FULL

Kengkhetkit, N., and T. Amornsakchai. 2012. "Utilisation of Pineapple Leaf Waste for Plastic Reinforcement: 1. A Novel Extraction Method for Short Pineapple Leaf Fibre." *Industrial Crops and Products* 40 (1): 55–61. https://doi.org/10.1016/J.INDCROP.2012.02.037

Khalina Abdan, M. Jawaid, M. Nasir, M. R. Ishak, M. Enamul Hoque, and Yulin Deng are among those named Asim. 2015. "A Review on Pineapple Leaves Fibre and Its Composites." *International Journal of Polymer Science.* https://doi.org/10.1155/2015/950567

Khan, Elena, Kadir Ozaltin, Andres Bernal-Ballen, and Antonio Di Martino. 2021. "Renewable Mixed Hydrogels Based on Polysaccharide and Protein for Release of Agrochemicals and Soil Conditioning." *Sustainability (Switzerland)* 13 (18). https://doi.org/10.3390/su131810439.

Kilic, Talip, Heather Moylan, John Ilukor, Clement Mtengula, and Innocent Pangapanga-Phiri. 2021. "Root for the Tubers: Extended-Harvest Crop Production and Productivity Measurement in Surveys." *Food Policy* 102 (July). https://doi.org/10.1016/j.foodpol.2021.102033

Kraan, Stefan. 2012. "Algal Polysaccharides, Novel Applications and Outlook." In: Chuan-Fa Chang (ed.) *Carbohydrates - Comprehensive Studies on Glycobiology and Glycotechnology.* InTech, Chapter 22, pp. 460–532. https://doi.org/10.5772/51572

Kumar, Vijayalakshmi, K. Sangeetha, P. Ajitha, S. Aisverya, S. Sashikala, and P. N. Sudha. 2020. "Chitin and Chitosan: The Defense Booster in Agricultural Field." In *Handbook of Biopolymers*, pp. 93–134. Jenny Stanford Publishing; Taylor and Francis Group https://doi.org/10.1201/9780429024757-5

Kumar Sinha, M. 1982. "A Review of Processing Technology for the Utilisation of Agro-Waste Fibres." *Agricultural Wastes* 4 (6): 461–475. https://doi.org/10.1016/0141-4607(82)90041-5

la Peña-Armada, Rocío De, María José Villanueva-Suárez, Antonio Diego Molina-García, Pilar Rupérez, and Inmaculada Mateos-Aparicio. 2021. "Novel Rich-in-Soluble Dietary Fibre Apple Ingredient Obtained from the Synergistic Effect of High Hydrostatic Pressure Aided by Celluclast(r)." *LWT* 146 (July). https://doi.org/10.1016/j.lwt.2021.111421

Li, Shuhong, Ye Zhang, Kailing Xiang, Jiacheng Chen, and Jincheng Wang. 2022. "Designing a Novel Type of Multifunctional Soil Conditioner Based on 4-Arm Star-Shaped Polymer Modified Mesoporous MCM-41." *Colloids and Surfaces A: Physicochemical and Engineering Aspects* 648. https://doi.org/10.1016/j.colsurfa.2022.129137

Liu, Huan, Mingyong Xie, and Shaoping Nie. 2020. "Recent Trends and Applications of Polysaccharides for Microencapsulation of Probiotics." *Food Frontiers* 1 (1): 45–59. https://doi.org/10.1002/FFT2.11.

Luo, Lei, Zifeng Yang, Jingyi Liang, Yu Ma, Hui Wang, Chitin Hon, Mei Jiang, et al. 2022. "Crucial Control Measures to Contain China's First Delta Variant Outbreak." *National Science Review* January. https://doi.org/10.1093/nsr/nwac004

Marczak, Daria, Krzysztof Lejcuś, Joanna Grzybowska-Pietras, Włodzimierz Biniaś, Iwona Lejcuś, and Jakub Misiewicz. 2020. "Biodegradation of Sustainable Nonwovens Used in Water Absorbing Geocomposites Supporting Plants Vegetation." *Sustainable Materials and Technologies* 26 (December). https://doi.org/10.1016/j.susmat.2020.e00235

Martey, Edward, Prince M. Etwire, Desmond Sunday Adogoba, and Theophilus Kwabla Tengey. 2021. "Farmers' Preferences for Climate-Smart Cowpea Varieties: Implications for Crop Breeding Programmes." *Climate and Development*. https://doi.org/10.1080/17565529.2021.1889949

Maulu, Sahya, Oliver Jolezya Hasimuna, Bornwell Mutale, Joseph Mphande, and Enock Siankwilimba. 2021. "Enhancing the Role of Rural Agricultural Extension Programs in Poverty Alleviation: A Review." *Cogent Food and Agriculture* 7 (1). https://doi.org/10.1080/23311932.2021.1886663

McCord, P. F., M. Cox, M. Schmitt-Harsh, and T. Evans. 2015. "Crop Diversification as a Smallholder Livelihood Strategy within Semi-Arid Agricultural Systems Near Mount Kenya." *Land Use Policy* 42: 738–750. https://doi.org/10.1016/j.landusepol.2014.10.012

Melilli, M. G., F. Branca, C. Sillitti, S. Scandurra, P. Calderaro, and V. Di Stefano. 2020. "Germplasm Evaluation to Obtain Inulin with High Degree of Polymerization in Mediterranean Environment." *Natural Product Research* 34 (1): 187–191. https://doi.org/10.1080/14786419.2019.1613402

M'Ndegwa, Julius Kibiti. 2016. "Diversifying the Use of Molasses Towards Improving the Infrastructure and Economy of Kenya." *Civil and Environmental Research* 8 (11): 37–42. www.iiste.org.

MoF. 2006. "A prosperous Middle-income Nation By 2030."

Mohammed A. S. A., M. Naveed, and N. Jost. 2021. "Polysaccharides; Classification, Chemical Properties, and Future Perspective Applications in Fields of Pharmacology and Biological Medicine (A Review of Current Applications and Upcoming Potentialities)." *Journal of Polymers and the Environment* 29 (8): 2359–2371. https://doi.org/10.1007/s10924-021-02052-2

Mwiinde, A. M., E. Siankwilimba, M. Sakala, F. Banda, and C. Michelo. 2022. "Climatic and Environmental Factors Influencing COVID-19 Transmission-An African Perspective." *Tropical Medicine and Infectious Disease* 7 (12): 433. https://doi.org/10.3390/tropicalmed7120433

Nejati, Roghayeh, Hamid Reza Gheisari, Saeid Hosseinzadeh, and Mehdi Behbod. 2017. "Viability of Encapsulated Lactobacillus Acidophilus (LA-5) in UF Cheese and Its Survival under in Vitro Simulated Gastrointestinal Conditions." *International Journal of Dairy Technology* 70 (1): 77–83. https://doi.org/10.1111/1471-0307.12312

Nguyen, Van Bon, and San Lang Wang. 2019. "Production of Potent Antidiabetic Compounds from Shrimp Head Powder via Paenibacillus Conversion." *Process Biochemistry* 76 (January): 18–24. https://doi.org/10.1016/J.PROCBIO.2018.11.004

Nicolle, L., C. M. A. Journot, and S. Gerber-Lemaire. 2021. "Chitosan Functionalization: Covalent and Non-Covalent Interactions and Their Characterization." *Polymers* 13 (23). https://doi.org/10.3390/polym13234118

Nugraha, Achmad T., Gunawan Prayitno, Faizah A. Azizi, Nindya Sari, Izatul Ihsansi Hidayana, Aidha Auliah, and Enock Siankwilimba. 2023. Structural Equation Model (SEM) of Social Capital with Landowner Intention. *Economies* 11: 127. https://doi.org/10.3390/economies11040127

Nwe, Nitar, Tetsuya Furuike, and Hiroshi Tamura. 2014. "Isolation and Characterization of Chitin and Chitosan from Marine Origin." *Advances in Food and Nutrition Research* 72: 1–15. https://doi.org/10.1016/B978-0-12-800269-8.00001-4

Oswake, P. 2021. COVID-19 and the Challenge of Developing Productive Capacities in Zambia. https://www.worldometers.info/coronavirus/country/zambia/

Otun, S., A. Escrich, I. Achilonu, M. Rauwane, J. Alexis Lerma-Escalera, J. R. Morones-Ramírez, and L. Rios-Solis. 2022. "The Future of Cassava in the Era of Biotechnology in Southern Africa." *Critical Reviews in Biotechnology*. https://doi.org/10.1080/07388551.2022.2048791

Padalino, Lucia, Cristina Costa, Amalia Conte, Maria Grazia Melilli, Carla Sillitti, Rosaria Bognanni, Salvatore Antonino Raccuia, and Matteo Alessandro Del Nobile. 2017. "The Quality of Functional Whole-Meal Durum Wheat Spaghetti as Affected by Inulin Polymerization Degree." *Carbohydrate Polymers* 173 (October): 84–90. https://doi.org/10.1016/j.carbpol.2017.05.081

Peng, Z., T. J. Simons, J. Wallach, and A. Youngman. 2022. *Advanced Recycling - Opportunities for Growth*. May.

Pettersen, M. K., M. S. Grøvlen, N. Evje, and T. Radusin. 2020. "Recyclable Mono Materials for Packaging of Fresh Chicken Fillets: New Design for Recycling in Circular Economy." *Packaging Technology and Science* 33 (11): 485–498. https://doi.org/10.1002/PTS.2527

Pino, Alessandra, Bachir Benkaddour, Rosanna Inturri, Pietro Amico, Susanna C. Vaccaro, Nunziatina Russo, Amanda Vaccalluzzo, et al. 2022. "Characterization of Bifidobacterium Asteroides Isolates." *Microorganisms* 10 (3): 655. https://doi.org/10.3390/MICROORGANISMS10030655

Prudhvi, M., and M. Kameswar Rao. 2017. "Stabilization of Gravel Soil by Using Molasses-Lime." *International Journal of Latest Engineering and Management Research (IJLEMR)* 02 (06): 1–6. www.ijlemr.com.

Raccuia, S. A., and M. G. Melilli. 2004. "Cynara Cardunculus L., a Potential Source of Inulin in the Mediterranean Environment: Screening of Genetic Variability." *Australian Journal of Agricultural Research* 55 (6): 693–698. https://doi.org/10.1071/AR03038

Rakhesh, Nisha, Christopher M. Fellows, and Mike Sissons. 2015. "Evaluation of the Technological and Sensory Properties of Durum Wheat Spaghetti Enriched with Different Dietary Fibres." *Journal of the Science of Food and Agriculture* 95 (1): 2–11. https://doi.org/10.1002/jsfa.6723

Reincke, K., E. Vilvert, A. Fasse, F. Graef, S. Sieber, and M. A. Lana. 2018. "Key Factors Influencing Food Security of Smallholder Farmers in Tanzania and the Role of Cassava as a Strategic Crop." *Food Security* 10 (4): 911–924. https://doi.org/10.1007/S12571-018-0814-3

Rivas, María Ángeles, Rocío Casquete, Alberto Martín, María De Guía Córdoba, Emilio Aranda, María José Benito, Frédéric J Tessier, Paul B. Tchounwou, Angela F. Cunha, and Adolfo Suárez. 2021. "Strategies to Increase the Biological and Biotechnological Value of Polysaccharides from Agricultural Waste for Application in Healthy Nutrition." *Mdpi.ComPaperpile* 18: 5937. https://doi.org/10.3390/ijerph18115937.

Saberi Riseh, Roohallah, Marzieh Ebrahimi-Zarandi, Mozhgan Gholizadeh Vazvani, and Yury A. Skorik. 2021. "Reducing Drought Stress in Plants by Encapsulating Plant Growth-Promoting Bacteria with Polysaccharides." *Mdpi.ComPaperpile.* https://doi.org/10.3390/ijms222312979

Salama, Suzy Munir. 2020. "Nutrient Composition and Bioactive Components of the Migratory Locust (Locusta Migratoria)." African Edible Insects as Alternative Source of Food, Oil, Protein and Bioactive Components, January, 231–239. https://doi.org/10.1007/978-3-030-32952-5_16

Sangiorgio, P., A. Verardi, S. Dimatteo, A. Spagnoletta, S. Moliterni, and S. Errico. 2021. "Valorisation of Agri-Food Waste and Mealworms Rearing Residues for Improving the Sustainability of Tenebrio Molitor Industrial Production." *Journal of Insects as Food and Feed*, October 1–16. https://doi.org/10.3920/JIFF2021.0101

Schuster, Heinz Georg. 2005. "Complex Adaptive Systems." In: Radons, G., Just, W., Häussler, P. (eds) *Collective Dynamics of Nonlinear and Disordered Systems*, Springer Berlin Heidelberg, pp. 359–69. https://doi.org/10.1007/3-540-26869-3_16.

Sekwati, L. 2012. "Economic Diversification : The Case of Botswana." *Revenue Watch* 14 (3): 501–525.

Sheina, S., and L. Babenko. 2014. "Municipal Solid Waste Management in Russia: Practices and Challenges." *Advanced Materials Research* 864–867 (11): 1989–1992. https://doi.org/10.4028/www.scientific.net/AMR.864-867.1989

Shih, Y. F., W. C. Chang, W. C. Liu, C. C. Lee, C. S. Kuan, and Y. H. Yu. 2014. "Pineapple Leaf/Recycled Disposable Chopstick Hybrid Fibre-Reinforced Biodegradable Composites." *Journal of the Taiwan Institute of Chemical Engineers* 45 (4): 2039–2046. https://doi.org/10.1016/J.JTICE.2014.02.015

Siankwilimba, E., Mumba, C., Hang'ombe, B.M., et al. 2023. "Bioecosystems towards sustainable agricultural extension delivery: effects of various factors." *Environment, Development and Sustainability*. https://doi.org/10.1007/s10668-023-03555-9

Siankwilimba, E. 2019. "Effects of Climate Change Induced Electricity Load Shedding on Small Holder Agricultural Enterprises in Zambia: The Case of Five Southern Province Districts." *Journal of Agriculture and Research*. https://ijrdo.org/index.php/ar/article/download/3184/2500/

Siankwilimba, Enock, Jacqueline Hiddlestone-mumford, Hang Mudenda, Chisoni Mumba, and Md Enamul Hoque. 2022. "COVID-19 and the Sustainability of Agricultural Extension Models." *Visnav.InPaperpile* 3 (January): 1–20. https://visnav.in/ijacbs/article/covid-19-and-the-sustainability-of-agricultural-extension-models/

Siankwilimba, E., E. S. Mwaanga, J. Munkombwe, C. Mumba, and B. M. Hang'ombe. 2021. "Effective Extension Sustainability in the Face of COVID-19 Pandemic in Smallholder Agricultural Markets." *International Journal for Research in Applied Science and Engineering Technology* 9: 865–878. https://doi.org/10.22214/jraset.2021.39403

Silimina, Derrick. 2021. "Zambia Raises Production of Its Second-Largest Crop to Diversify Food Basket, Boost Agricultural Income– ChinAfrica." April 26, 2021. https://www.chinafrica.cn/Homepage/202103/t20210326_800241579.html

SNAP. 2016. Second National Agricultural Policy Ministry. Ministry of Agriculture.

Solomon, S., G. P. Rao, and M. Swapna. 2020. "Impact of COVID-19 on Indian Sugar Industry." *Sugar Tech* 22 (4): 547–551. https://doi.org/10.1007/s12355-020-00846-7

Souza, Márcio Araújo de, Isis Tavares Vilas-Boas, Jôse Maria Leite-da-Silva, Pérsia do Nascimento Abrahão, Barbara E. Teixeira-Costa, and Valdir F. Veiga-Junior. 2022. "Polysaccharides in Agro-Industrial Biomass Residues." *Polysaccharides* 3 (1): 95–120. https://doi.org/10.3390/polysaccharides3010005

Staack, Larissa, Eduardo Antonio Della Pia, Bodil Jørgensen, Dan Pettersson, and Ninfa Rangel Pedersen. 2019. "Cassava Cell Wall Characterization and Degradation by a Multicomponent NSP-Targeting Enzyme (NSPase)." *Scientific Reports* 9 (1): 1–11. https://doi.org/10.1038/s41598-019-46341-2

Stasińska-Jakubas, Maria, and Barbara Hawrylak-Nowak. 2022. "Protective, Biostimulating, and Eliciting Effects of Chitosan and Its Derivatives on Crop Plants." *Molecules.* MDPI. https://doi.org/10.3390/molecules27092801

Su, P., A. Henriksson, and H. Mitchell. 2007. "Prebiotics Enhance Survival and Prolong the Retention Period of Specific Probiotic Inocula in an in Vivo Murine Model." *Journal of Applied Microbiology* 103 (6): 2392–2400. https://doi.org/10.1111/J.1365-2672.2007.03469.X

Suvedi, M., R. Ghimire, and M. Kaplowitz. 2017. "Revitalizing Agricultural Extension Services in Developing Countries: Lessons from Off-Season Vegetable Production in Rural Nepal." *Journal of the International Society for Southeast Asian Agricultural Sciences* 23 (1): 1–11.

Szyniszewska, A. M., Chikoti, P. C., Tembo, M., Mulenga, R., Gilligan, C. A., van den Bosch, F., and McQuaid, C. F. 2021. "Smallholder cassava planting material movement and grower behavior in Zambia: implications for the management of cassava virus diseases." *Phytopathology®* 111(11), 1952–1962. https://doi.org/10.1094/PHYTO-06-20-0215-R

Teixeira-Costa, Barbara E., and Cristina T. Andrade. 2021. "Chitosan as a Valuable Biomolecule from Seafood Industry Waste in the Design of Green Food Packaging." *Biomolecules* 11 (11). https://doi.org/10.3390/biom11111599

The Economist. 2022. Most of the world's grain is not eaten by humans | The Economist. Most of the World's Grain. https://www.economist.com/graphic-detail/2022/06/23/most-of-the-worlds-grain-is-not-eaten-by-humans

Turck, Dominique, Jacqueline Castenmiller, Stefaan De Henauw, Karen Ildico Hirsch-Ernst, John Kearney, Alexandre Maciuk, Inge Mangelsdorf, et al. 2021. "Safety of Frozen and Dried Formulations from Migratory Locust (Locusta Migratoria) as a Novel Food Pursuant to Regulation (EU) 2015/2283." *EFSA Journal* 19 (7). https://doi.org/10.2903/J.EFSA.2021.6667

Valluru, Ravi, and Wim Van Den Ende. 2008. "Plant Fructans in Stress Environments: Emerging Concepts and Future Prospects." *Journal of Experimental Botany* 59 (11): 2905–2916. https://doi.org/10.1093/JXB/ERN164

Wang, Weicong, Keqi Qu, Xinrui Zhang, Min Teng, and Zhanhua Huang. 2021. "Integrated Instillation Technology for the Synthesis of a PH-Responsive Sodium Alginate/Biomass Charcoal Soil Conditioner for Controlled Release of Humic Acid and Soil Remediation." *Journal of Agricultural and Food Chemistry* 69 (45): 13386–13397. https://doi.org/10.1021/acs.jafc.1c04121

Yao, Mingfei, Jiaojiao Xie, Hengjun Du, David Julian McClements, Hang Xiao, and Lanjuan Li. 2020. "Progress in Microencapsulation of Probiotics: A Review." *Comprehensive Reviews in Food Science and Food Safety* 19 (2): 857–874. https://doi.org/10.1111/1541-4337.12532

Yuan, Ping chuan, Tai li Shao, Jun Han, Chun yan Liu, Guo dong Wang, Shu guang He, Shi xia Xu, Si hui Nian, and Kao shan Chen. 2021. "Burdock Fructooligosaccharide as an α-Glucosidase Inhibitor and Its Antidiabetic Effect on High-Fat Diet and Streptozotocin-Induced Diabetic Mice." *Journal of Functional Foods* 86 (November). https://doi.org/10.1016/j.jff.2021.104703

Zeder, Melinda A. 2015. "Core Questions in Domestication Research." *Proceedings of the National Academy of Sciences of the United States of America* 112 (11): 3191–3198. https://doi.org/10.1073/PNAS.1501711112

Zeufack, A. G., C. Calderon, A. Kabundi, M. Kubota, V. Korman, D. Raju, K. Girma Abreha, W. Kassa, and S. Owusu. 2022. Africa's Pulse, No. 25, April 2022 (Vol. 25, Issue April). World Bank. https://doi.org/10.1596/978-1-4648-1871-4

Zezza, A., A. Martuscelli, P. Wollburg, S. Gourlay, and T. Kilic. 2021. "Viewpoint: High-Frequency Phone Surveys on COVID-19: Good Practices, Open Questions." *Food Policy* 105: 102153. https://doi.org/10.1016/j.foodpol.2021.102153

Zidenga, Tawanda, Dimuth Siritunga, and Richard T. Sayre. 2017. "Cyanogen Metabolism in Cassava Roots: Impact on Protein Synthesis and Root Development." *Frontiers in Plant Science* 8 (February): 220. https://doi.org/10.3389/FPLS.2017.00220/BIBTEX

Zsögön, A., T. Čermák, E. R. Naves, M. M. Notini, K. H. Edel, S. Weinl, L. Freschi, D. F. Voytas, J. Kudla, and L. E. P. Peres. 2018. "De Novo Domestication of Wild Tomato Using Genome Editing." *Nature Biotechnology* 36 (12): 1211–1216. https://doi.org/10.1038/NBT.4272

Zuo, Shengpeng, Guobin Liu, and Ming Li. 2012. "Genetic Basis of Allelopathic Potential of Winter Wheat Based on the Perspective of Quantitative Trait Locus." *Field Crops Research* 135 (August): 67–73. https://doi.org/10.1016/j.fcr.2012.07.005

16 Polysaccharide-Based Fluorescent Materials for Sensing and Security Applications

Akhil Padmakumar and Drishya Elizebath
CSIR-National Institute for Interdisciplinary
Science and Technology (CSIR-NIIST)
Academy of Scientific and Innovative Research (AcSIR)

Jith C. Janardhanan
CSIR-National Institute for Interdisciplinary
Science and Technology (CSIR-NIIST)

Rakesh K. Mishra
National Institute of Technology, Uttarakhand (NITUK)

Vakayil K. Praveen
CSIR-National Institute for Interdisciplinary
Science and Technology (CSIR-NIIST)
Academy of Scientific and Innovative Research (AcSIR)

16.1   INTRODUCTION: POLYSACCHARIDES

Polysaccharides are biopolymers of monosaccharides, which are linked covalently by glycosidic bonds with a general formula of $(C_6H_{10}O_5)_n$, where n varies from 40 to 3,000 and has various degrees of branching (Zong et al., 2012). Nature synthesizes polysaccharides for various functions such as energy storage, structural support, and gelling agents forming the intercellular matrix (Mozammil Hasnain et al., 2019). They are usually found in various species of bacteria (cellulose, dextran, levan, xanthan, gellan, and polygalactosamine), fungi (chitin, chitosan, elsinan, pollulan, and yeast glucans), algae (alginates, carrageenans, agar, galactans, and fucoidan), plants (cellulose, starch, hemicellulose, pectin, glucomannan, gums, and mucilage), and animals (cellulose, chitin, chitosan, glycosaminoglycans, heparin, and hyaluronic acid).

These polysaccharides, obtained from differing sources, have differences in molecular weight with low, intermediate, and high molecular weights having differing polydispersity, position, and stereochemistry of the glycosidic bond as well as on the structure (linear or branched), functional group present (monofunctional, or polyfunctional), and properties such as water solubility, chirality, crystallinity, and so on. The inherent properties of the polysaccharides, such as biocompatibility, non-toxicity, low cost, ease of processing, and high sorption capacity, along with ease of availability, have made polysaccharides a hot topic of research in developing next-generation materials

DOI: 10.1201/9781003265054-16

for various biological and industrial applications (Shariatinia, 2019). The intermolecular and intra-molecular hydrogen bond-forming moieties of the polysaccharides, and covalent modification of the different functional groups (hydroxyl, carboxyl, and aldehyde groups), are further utilized to develop nanomaterials and nanocomposites or nanohybrids, improving their thermal expansion coefficient, mechanical strength, flame-retardant capability, super-hydrophobicity, optical performance, stimuli responsiveness, adsorption, solubility, and so on, making them ideal for various industrial applications such as food packaging, dye removal, electrode materials in flexible sensors, batteries, organic thin-film transistors, supercapacitors, triboelectric nanogenerators, organic light-emitting diodes (OLEDs), tissue bioelectronics, and other flexible electronics (Li et al., 2014; Mobarak et al., 2015; Nešić et al., 2019; Pal et al., 2021; Wang et al., 2020; Wu et al., 2018; Zhao et al., 2021). The improved immunoregulatory, anti-inflammatory, anti-tumor, anti-virus, anti-oxidant activities, and mechanical properties of these functionalized polysaccharides make them useful for medical applications such as tissue engineering, bone repair, wearable sensors, drug delivery, healthcare monitoring, and theranostics (Fu et al., 2019; Peng et al., 2016; Shariatinia, 2019; Zong et al., 2012). The details regarding different additives and methods for preparing various polysaccharide-based nanocomposites and their applications have been described in Table 16.1 (Zheng et al., 2015).

TABLE 16.1

Different Strategies for the Preparation of Polysaccharide Nanocomposites and Their Applications

Polysaccharide	Additive	Composite Method	Applications
Cellulose	Ag	Chemical or UV reduction of Ag on cellulose	Antibacterial
	CNTs	Mixing	Energy storage devices
		Coaxial electrospinning	Electronic devices, energy storage
	Graphene	Solution Mixing	Electronics devices
	CaCO$_3$	*In situ* CaCO$_3$ preparation on cellulosic fibers	Reinforcing fillers in industrial polyethylene matrixes
	CdS	*In situ* CdS preparation on regenerated cellulose	Energy production: photocatalytic H$_2$ production
	TiO$_2$	In situ TiO$_2$ preparation in the presence of cellulosic fibers	Industrial papermaking
	ZnO	*In situ* ZnO preparation in cellulosic fibers	Multisource energy conversion
	Lysostaphin	Electrospinning	Wound healing
	Cibacron blue F3GA	Covalent coupling of dye to electrospun cellulose	Bovine serum albumin affinity purification
	Poly(N-vinylcaprolac tam)	Electrospinning	Protein affinity purification
	Epoxy resin	Film casting, dip-coating	Thermal conductivity
	Xylan-rich hemicelluloses	Solution mixing and film casting	Tensile strength and thermal stability
	Organic rectorite, chitosan, sodium alginate	Layer-by-layer techniques	Antibacterial
	Quaternized poly (1,2-butadiene)-block-poly(dimethylamino ethyl methacrylate)	Ionic assembly	Biomimetic nanocomposites
	Hydroxyapatite	Biomimetic technique	Bone tissue engineering.
	Polyvinyl Alcohol	Solution mixing, thermal processing (cycling)	Tissue engineering
	Pullulan	Film casting	Food packaging, electronic devices

(Continued)

TABLE 16.1 (*Continued*)

Different Strategies for the Preparation of Polysaccharide Nanocomposites and Their Applications

Polysaccharide	Additive	Composite Method	Applications
Heparin	Ag	Reduction of Ag onto diaminopyridinylated heparin	Antibacterial
	Au	Reduction of Au onto diaminopyridinylated heparin	Antibacterial
		Reduction of Au by NaBH$_4$ onto heparin-dihydroxyphenylalanine	Liver-specific CT imaging
		Au-thiol linkage	Imaging and induced cancer cell apoptosis
	C.N.T.s	Coupling of activated heparin to nanotubes	Blood compatibility nanodevices
	Fe$_3$O$_4$	Co-precipitation	Targeted drug delivery
	Fibrin	Dehydrative coupling by ethyl(dimethylaminopropyl) carbodiimide/*N*-Hydroxysuccinimide (EDC/NHS)	Drug delivery: bone morphogenetic protein 2 (BMP-2)
	Poly(glycolide-*co*-lactide) and pluronic	Solvent-diffusion method	Drug delivery: vascular endothelial growth factor (VEGF)
	Poly(L-lactide-*co*-ε-caprolactone)	Co-electrospinning	Vascular tissue engineering
	Gelatin, Ca–P/ poly(3-hydroxybutyrate-co-3-hydroxyvalerate)	Coating Ca–P/polyhydroxybutyrate with gelatin, dehydrative coupling of gelatin and heparin by EDC/NHS	Bone tissue regeneration
Chitosan	Ag	Ascorbic acid reduction on lactose-modified chitosan	Antibacterial
	Ag, Polyvinylpyrrolidone	Dip-coating and thermal reduction	Antibacterial
	[Fe(pz){M(CN)$_4$}] (M=Ni, Pd, Pt)	*In situ* preparation in the presence of chitosan	Spin-crossover properties
	Collagen	Electrospinning	Tissue engineering
	Heparin	Solution mixing	Tissue engineering
	Bacterial cellulose	Solution mixing, film casting	Food packaging and electronic displays
	Alginate	Electrospinning	Tissue engineering
	Monoclonal antibody	Ionotropic gelation	Monoclonal antibody (MAb) delivery
Pectin	Fe$_3$O$_4$	Co-precipitation and direct encapsulation	Cu^{2+} removal
Starch	ZnO	*In situ* preparation in soluble starch	Antibacterial and UV protection cotton fabrics
	Polyaniline	*In situ* chemical oxidative polymerization of aniline	Removal of reactive dyes from synthetic effluent
	Cellulose	Solution mixing and freeze-drying	Packaging materials and biomedical materials
Hyaluronan	Ag	Ag reduced with hyaluronan	Antibacterial
	Au	Ag reduced with hyaluronan	Antibacterial
	CNT	Solution mixing and coating	Nicotinamide adenine dinucleotide (NADH) biosensing
	Gelatin	Electrospinning	Tissue engineering

(Continued)

TABLE 16.1 (*Continued*)

Different Strategies for the Preparation of Polysaccharide Nanocomposites and Their Applications

Polysaccharide	Additive	Composite Method	Applications
Alginate	Fe(pz){M(CN)$_4$}] (M=Ni, Pd, Pt)	*In situ* preparation in alginate	Spin-crossover properties
	ZnO	*In situ* preparation in alginate	Antibacterial
	Cellulose nanocrystals, chitin whiskers, platelet-like starch nanocrystals, CaCl$_2$	Mixing and adsorption	Drug release
Guar	Au	Au reduction by guar gum	Aqueous ammonia sensor
	Montmorillonite	Solution intercalation	Drug delivery
Starch/chitosan	Ag	Reduction of crosslinked tapioca dialdehyde starch-chitosan	Functional hydrogels
Chitosan/ heparin	Activated carbon beads	EDC/NHS coupling of chitosan and heparin and carbon beads coating	Removal of chemotherapeutic, doxorubicin
	Fe$_3$O$_4$	*In situ* preparation of Fe$_3$O$_4$, EDC/NHS coupling of chitosan and heparin	Low-density lipoprotein removal
	Fe$_3$O$_4$, Au, Tween 80	Coating by ionic interaction	Magnetic resonance imaging with a tumor-targeting characteristic
	Bovine jugular veins	Self-assembly with EDC/NHC chemistry	Tissue engineering
Chitosan/ hyaluronan	Heparin	Ionotropic gelation	Drug delivery
Hyaluronan/ heparin	Steel	Covalently bonded	Drug-eluting stents
Cellulose and chitin whiskers, platelet-like starch	Cyclodextrin/polymer inclusion	Solution mixing	Drug release

Source: Adapted with permission from Zheng and Lindhart (2015) © 2014 Elsevier Ltd.

Polysaccharide-based fluorescent nanomaterials developed either through the covalent functionalization of the polysaccharide with the fluorophore or using the polysaccharide as the substrate have been utilized as photosensitizers and for applications such as bioimaging, UV shielding, and chemical sensing, and for anti-counterfeiting applications such security ink and so on (Jia et al., 2019a; Nawaz et al., 2021; Tian et al., 2016). Such fluorescence-based polysaccharides act through a synergy of various non-covalent interactions such as electrostatic interactions, hydrogen bonding, π-stacking, and van der Waals interactions, in addition to the anchoring effect and dilution effect arising from the polysaccharide, providing an advantage of developing materials from fluorophores possessing aggregation-induced emission (AIE) and aggregation caused quenching (ACQ) alike, helping in the development of highly fluorescent materials (Tian et al., 2016). This chapter discusses exciting aspects of the fluorescent materials based on polysaccharides used for sensing and anti-counterfeiting applications with the help of selected examples.

16.2 POLYSACCHARIDE-BASED FLUORESCENT MATERIALS FOR SENSING

Chemical and biological recognition of various analytes are of great interest as they are byproducts in the chemical industry, food industry, dyeing process, fine chemical engineering, and biological progress and are essential markers of environmental contamination and human diseases increased in the past years. Accumulating various metal ions in the body can lead to Parkinson's, Minamata, Pinks, Alzheimer's, and Wilson's diseases, and affect the muscle, kidney, central nervous system, metabolism rate, growth, and immune-system development. Various analytical techniques include inductively coupled plasma mass spectroscopy (ICP-MS), atomic absorption and emission spectrometry, X-ray fluorescence spectrophotometry, atomic absorption, and emission spectrometry, neutron activation analysis, etc., have been utilized to detect these analytes. However, the use of expensive instruments, hazardous sample preparation, and the requirement of highly trained operators hinder the real-time use of these sophisticated techniques (Basabe-Desmonts et al., 2007).

Hence, optical sensors that display signals in the form of changes in color or fluorescence upon interaction with analyte binding have been extensively utilized for the detection of analytes due to their sensitivity, selectivity, cost-effectiveness, operational simplicity, and rapid response of this technique compared to conventional spectroscopic and chromatographic analytical methods (Li et al., 2017). Over time, fluorescence-based sensors utilizing fluorescent nanomaterials, such as organic fluorophores, nanoclusters, carbon dots, quantum dots, and semiconducting polymer dots, have been extensively investigated due to their high sensitivity, selectivity, and ease of use. However, the applicability of these fluorescent materials is limited due to poor biocompatibility, toxicity, and water solubility. Therefore, the search for ecofriendly sensors with simple and smart detection is extremely important. Hence, the low toxicity, biocompatibility, and functionalizability of polysaccharide-based fluorescent sensors make them good candidates for replacing traditional fluorescent sensors (Li et al., 2017; Nawaz et al., 2021).

16.2.1 Nitroaromatic Sensing

Nitroaromatics (NA) threaten the human population and environment regarding pollution and security as many explosives belong to this category. In this scenario, selective detection and monitoring of NAs are of significant interest. Fluorescent dye-grafted polysaccharide derivatives can be utilized as a fluorescent sensing platform for the selective detection of NAs. The surface functionalities such as hydroxy, carboxy, etc., facilitate the covalent modification of cellulose with fluorescent dyes and lead to the development of diverse polysaccharide-based fluorescent materials. Generally, on exposure to NAs, the fluorescence emission of the polysaccharide-based fluorescent materials undergoes quenching due to the high affinity of electron-deficient NAs to electron-rich fluorescent sensor moieties and consequent photoinduced electron/energy transfer from the latter to the former. Zhou, Nishio, and coworkers have developed a fluorescent cellulose derivative (AC-AET-DMANM) comprising of aminoethanethiol modified allyl cellulose backbone and 4-dimethylamine-1,8-naphthalimide fluorescent signaling moiety, which exhibited selective detection of 2,4-dinitrophenylhydrazine (DNH) and trinitrophenol (TNP) over other NAs in both aqueous and non-aqueous solvent medium (Figure 16.1a and b). Due to photoinduced electron transfer (PET) and fluorescence resonance energy transfer (FRET) from signaling fluorescent dye to NA analyte, DNH and TNP selectively quenched the fluorescence of AC-AET-DMANM with good selectivity (Hu et al., 2018).

Zhang, Chen, and coworkers demonstrated that the hydroxyl groups of hydrophilic polymer chains shorten the equilibrium response time by facilitating the fast diffusion of NA molecules onto a solid film matrix. They synthesized pyrene functionalized hydrophilic fluorescent polymer (poly(2-hydroxyethyl methacrylate-co-pyrene-butyric acid poly(HEMA-*co*-PyMA)) with different weight% (0.67 & 13.5 wt%) of pyrene, which was found to be sensitive to trinitrotoluene (TNT). Poly(HEMA-*co*-PyMA)) forms a stable charge transfer complex with electron-deficient TNT, and hence, causing a quenching in the emission intensity of the pyrene polymer (Figure 16.1c and d). The higher wt% of pyrene (13.5 mol%; poly 2) can extend the fluorescence emission quenching response

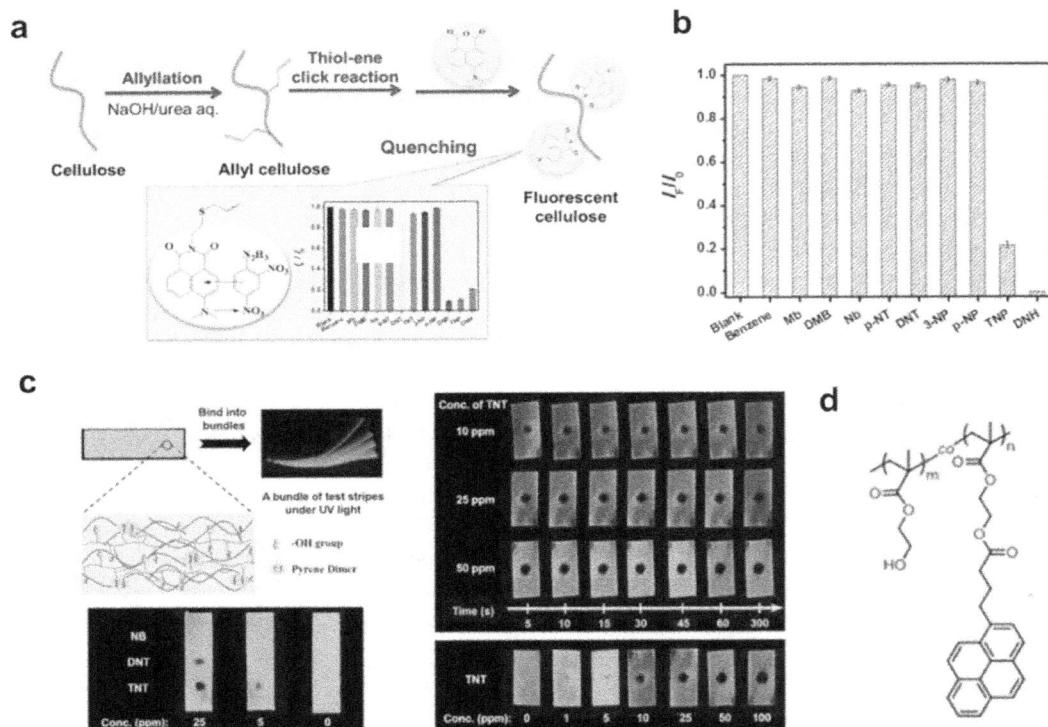

FIGURE 16.1 (a) Schematic representation for the synthesis and NA sensing properties of AC-AET-DMANM. (b) Plot showing relative fluorescence intensity of AC-AET-DMANM toward various NAs, including TNP and DNH. a and b are reproduced with permission from Hu et al. (2018) © 2017 American Chemical Society. (c) Photographs of cellulose paper test strips containing (poly(2-hydroxyethyl methacrylate-co-pyrene-butyric acid, poly(HEMA-co-PyMA) and their fluorescence response toward NAs at various duration. (Taken under 365 nm UV light illumination.) (d) Chemical structure of (poly(2-hydroxyethyl methacrylate-co-pyrene-butyric acid, poly(HEMA-co-PyMA) c and d are reproduced with permission from Lu et al. (2017) © 2017 American Chemical Society.

in the visible region by forming excimers through intramolecular and intermolecular aggregations. Cellulose paper absorbed with poly 2 can act as a fluorescence sensing platform for the selective ultra-fast onsite naked-eye detection of TNT (Lu et al., 2017). Since TNT is more electron deficient than dinitrotoluene (DNT) and nitrobenzene (NB), the charge transfer complex formed between TNT and pyrene is more stable than that with DNT and DB, making poly 2 TNT selective.

16.2.2 GAS SENSING

Fluorescent dye-infused cellulose paper can be utilized as a fluorescent gas sensing platform. Meldrum and coworkers developed bis(4-pyridyl)-dineopentoxyl-p-phenylenedivinylene (P4VB) infused cellulose paper strips as fluorescent CO_2 gas sensors (Wang et al., 2020). Here water vapor is necessary for detecting CO_2 in terms of red spectral shift. The carbonic acid formed from CO_2 and water will protonate the pyridyl moiety of P4VB, and weakly bound P4VB-carbonic acid perturbed the HOMO-LUMO energy gap of P4VB and caused a spectral shift.

Baker and coworkers demonstrated that bacterial cellulose ionogel could serve as a platform for sensing ammonia on doping with pH-responsive trihexyltetradecylphosphonium8-hydroxypyrene-1,3,6-trisulfonate ([$P_{14,6,6,6}$]$_3$[HPTS]in the presence of trihexyltetradecylphosphonium bis-[(trifluoromethyl)sulfonyl]imide [$P_{14,6,6,6}$][Tf$_2$N] ionic liquid (Figure 16.2) (Smith et al., 2017). Due to the photo-acidic

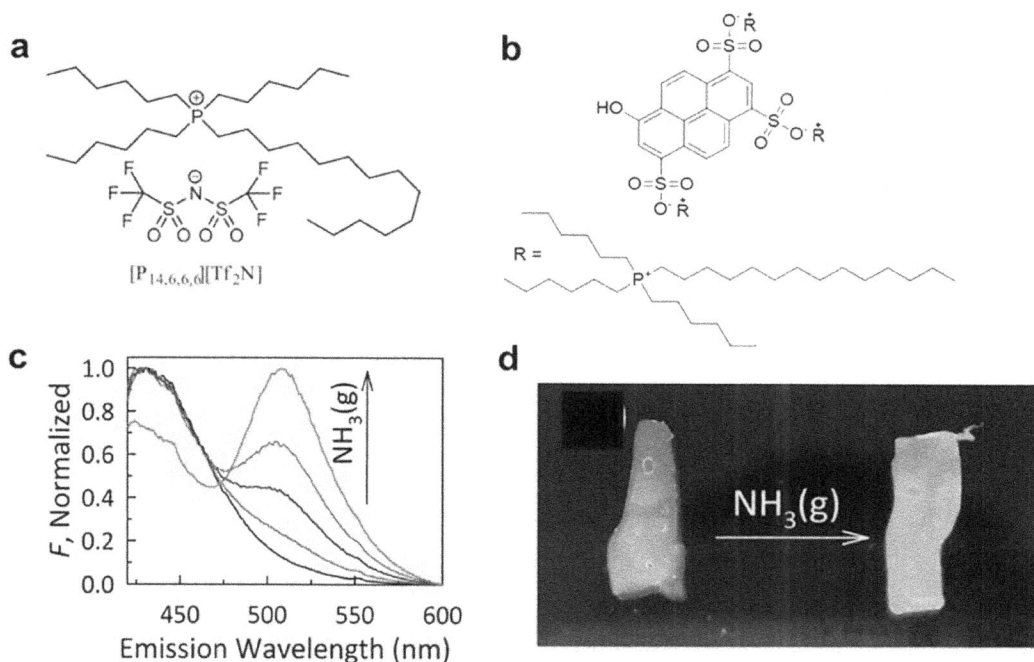

FIGURE 16.2 Chemical structure of (a) trihexyltetradecylphosphonium bis-[(trifluoromethyl)sulfonyl] imide [P$_{14,6,6,6}$][Tf$_2$N] ionic liquid and (b) trihexyltetradecylphosphonium 8-hydroxypyrene-1,3,6-trisulfonate [P$_{14,6,6,6}$]$_3$[HPTS]. (c) Normalized steady-state emission spectra of [P$_{14,6,6,6}$][Tf$_2$N]]-bacterial cellulose ionogel comprising the fluorogenic probe [P$_{14,6,6,6}$]$_3$[HPTS] on exposure to various amounts of NH$_3$ gas. (d) Photograph of fluorescence emission response of [P$_{14,6,6,6}$]$_3$[HPTS]-containing bacterial cellulose ionogel toward NH$_3$ taken under UV light. Reproduced with permission from Smith et al. (2017) © 2017 American Chemical Society.

nature, on exposure to ammonia, HPTS-doped bacterial cellulose ionogel showed a clear fluorescence profile difference before and after ammonia exposure. Ammonia exposure induces a two-state fluorescence (peaks at 511 and 445 nm) response of HPTS, and I_{511}/I_{445} seems to increase with ammonia concentration, which can be easily identified even using the naked eye upon excitation using a UV lamp. By exploiting the chemistry of reduction of azide functionality to amine functionality, bacterial cellulose ionogel was utilized as a fluorescence sensor for gaseous detection of H$_2$S by incorporating pro-fluorescent 8-azidopyrene-1,3,6-trisulfonic acid molecule into the entrapped fluid phase of a bacterial cellulose ionogel. On contact with H$_2$S, the fluorescence intensity was increased, which can be attributed to the formation of fluorescent aminopyrene-1,3,6-trisulfonic acid upon reduction of azide functionality 8-azidopyrene-1,3,6-trisulfonic acid with H$_2$S.

16.2.3 Metal Ion Sensing

Covalent grafting or simple incorporation of metal ion-sensitive fluorescent probes into cellulose matrix have been utilized to develop a fluorescent sensing platform for detecting metal ions. Ertekin and coworkers have developed ethyl cellulose nanofibrous thin films comprising Cu^{2+} responsive fluoroionophore 2-{[(2-aminophenyl)imino]methyl}-4,6-di-*tert*-butylphenol (DMK) for selective detection of Cu^{2+} (Figure 16.3c) (Kacmaz et al., 2015). On the interaction of DMK, which is encapsulated within an ethyl cellulose matrix with Cu^{2+} in the presence of anionic additive potassium *tetrakis*-(4-chlorophenyl)borate, fluorescence quenching was observed in DMK with picomolar sensitivity. It can be attributed to the ion-exchange mechanism where Cu^{2+} is selectively extracted from solution to fiber by displacing K$^+$ into solution. Zhang and coworkers developed

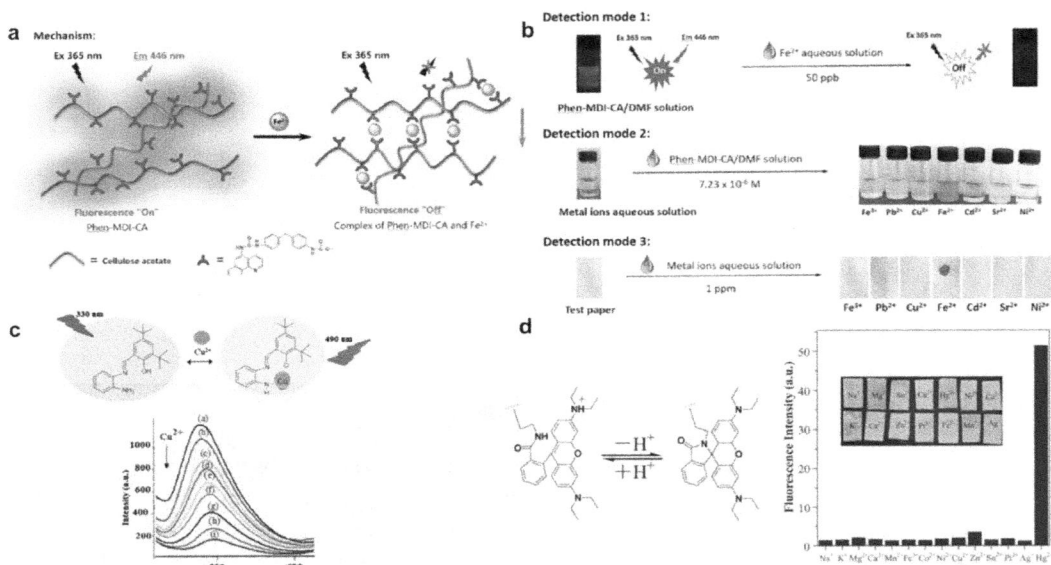

FIGURE 16.3 (a) General schematic representation for Fe^{2+} sensing of Phen-MDI-CA, (b) fluorescence change and color change in solution and solid-state upon the interaction of Phen-MDI-CA with Fe^{2+} metal ion solution. Fig. 3a and 3b are reproduced with permission from Nawaz et al. (2018) © 2017 American Chemical Society. (c) Sensing mechanism and fluorescence spectral response of ethyl cellulose nanofibrous thin films comprising off fluoroionophore 2-{[(2-aminophenyl)imino]methyl}-4,6-di-tert-butylphenol toward Cu^{2+} ions. Reproduced with permission from Kacmaz et al. (2015) ©2014 Elsevier BV and (d) ring-opening mechanism of rhodamine spirocycle and photograph showing selectivity of cellulose-g-RD toward Hg^{2+} metal ion. Reproduced with permission from Xu et al. (2013) © 2013 The Royal Society of Chemistry.

a cellulose-based sensing platform (Phen-MDI-CA) for the rapid and selective detection of Fe^{2+} by covalent grafting of Fe^{2+} responsive fluorescent probe 1,10-phenanthroline-5-amine onto cellulose framework by employing 4,4′-methylene diphenyl diisocyanate crosslinker (Figure 16.3a and b) (Nawaz et al., 2018). Phenanthroline can easily form complex with Fe^{2+} ions, and when Phen-MDI-CA interacts with Fe^{2+} ions, quenching of blue fluorescence was observed with a sensitivity of 50 ppb. Interestingly, the red color of the non-fluorescent Fe-(Phen-MDI-CA) complex facilitates easy naked-eye detection of Fe^{2+} due to its insoluble nature with red color. Wang and coworkers utilized cellulose nanofiber (CNF) hydrogels functionalized with fluorescent carbon dots (CDs) for detecting Fe^{3+} ions selectively (Wang et al., 2018). The functionalization of CDs improved the mechanical properties of hydrogels, and upon interaction with Fe^{3+} ions, quenching of blue fluorescence emission intensity was observed. The fluorescence response arises from efficient mass transfer through a porous 3D hydrogel network structure.

Generally, organic fluorescent dye molecules suffer from ACQ, reducing fluorescent emission intensity. Chen, Xiao, and coworkers demonstrated anti-ACQ of coumarin by preventing π-π stacking with nanocellulose spacer using hydrogen bonding interaction (Li et al., 2019). Sulfurated moiety in hydroxy coumarin-induced photo-induced electron transfer and weak fluorescence. When weakly emitting cellulose membrane comprising of sulfurated hydroxy coumarin (CAM) interacts with Hg^{2+}, fluorescence intensity was dramatically increased, facilitating the turn-on mode detection of Hg^{2+}. Upon exposure to Hg^{2+}, the chemical reaction between Hg^{2+} and CAM resulted in the desulfuration of sulfurated coumarin into coumarin. This diminishes the photo-induced electron transfer and enhances the intramolecular charge transfer efficiency, increasing the intensity of emission.

Fu and coworkers have developed a disposable cellulose-based turn-on fluorescence sensor by incorporating amino-containing spirolactam rhodamine derivative (Xu et al., 2013). The fluorescent dye was integrated onto cellulose via atom transfer radical polymerization (ATRP), surface-initiated ATRP, and finally with ester-amine reaction. The spirolactam rhodamine-grafted cellulose (Cellulose-g-RD), when in contact with Hg^{2+} metal ion solution, exhibits an increase in fluorescence intensity along with a distinct color change from colorless to pink, which facilitates the solid-state sensing of Hg^{2+} with a detection limit of $5 \times 10^{-5} M$ (Figure 16.3d). The spectral change can be attributed to the formation of highly emissive rhodamine dye from the non-luminescent rhodamine spirocycle upon interaction with Hg^{2+} metal ions.

16.2.4 BIOSENSING

In addition to chemosensing, fluorogenic dye-attached cellulose substrates can also be employed in biosensing applications. But the bottleneck in developing biosensors based on cellulose is the poor reactivity of cellulose fiber surface. Brumer and coworkers have overcome this issue by employing a chemoenzymatic approach and developed 5(6)-carboxyfluorescein-tetraethylene glycol (TEG)-azide appended cellulose paper-based fluorogenic biosensor for the detection of esterase (Figure 16.4a and b) (Derikvand et al., 2016). The non-fluorescent diester of fluorescein was attached to the cellulose paper surface by Cu-catalyzed azide-alkyne click chemistry. The presence of esterase causes the hydrolysis of fluorescein diester, thereby liberating highly fluorescent fluorescein dye, which is attached to cellulose paper. The advantage of the developed sensor lies in the fact that the dye is covalently attached to cellulose and prevents loss of signal and increase in toxicity arising due to chromophore diffusion. Ma and coworkers developed carboxymethyl chitosan (CMC) appended CdTe quantum dots (CMCS-QDs) as turn-off fluorescent sensors for lysosome detection (Song et al., 2014). The lysosome-sensitive probe was developed by modifying CdTe quantum dots with CMC by utilizing the interaction between the amino group and the carboxyl group of CdTe quantum dots. By exploiting the affinity of Zn^{2+} toward chitosan, the fluorescence intensity of quantum dots was enhanced by adding Zn^{2+} to CMCS-QDs. When Zn^{2+}-CMCS-QDs interact with the lysosome, the lysosome brings about the hydrolysis of CMC chains and forms the lysosome-QDS complex by releasing Zn^{2+} from QDs. Since Zn^{2+} facilitates the reduction of non-radiative transition, the formation of the lysosome-QDS complex by releasing Zn^{2+} quenches the emission intensity. Rhee and coworkers developed a ratiometric fast responsive fluorescent urea biosensor based on the urease oxazine 170 perchlorate-ethyl cellulose membrane immobilized with urease (Dinh Duong & Il Rhee, 2015). The urea sensing membrane comprised two layers where ethyl cellulose (EC) was employed as supporting material for the top urease membrane layer and bottom oxazine 170 perchlorate layer. The response time of the probe depends on the protein capture ability of the enzyme layer, and in addition to the role of supporting material, EC also facilitates the quick response of the urease enzyme. The ratiometric probe works because when urea solution comes in contact with the probe membrane, the urease enzyme catalyzes urea hydrolysis into NH_3 and CO_2. Since Oxazine 170 perchlorate is an NH_3-sensitive fluorescent molecule, the OC17-EC membrane in the probe interacts with NH_3 released from urea hydrolysis and delivers a quick response in a ratiometric manner.

Lateral flow assays have been used widely to diagnose various diseases rapidly because of the biodegradable, affordable, and easy-to-use nature of these assays (Mansfield, 2009). A recent example of a lateral flow assay is the rapid antigen test strip for SARS CoV-2. Porous nitrocellulose membranes and paper strips have been widely used as analytical strips for lateral flow assays (Hu et al., 2014; Yetisen et al., 2013). Prazeres and coworkers have utilized nitrocellulose membrane to develop a sandwich-type assay to detect Cardiac Troponin I (cTnI) (Natarajan et al., 2021). The assay required an incubation time of 15 minutes. Though cellulose-based lateral flow assay exhibited less fluorescence, modification of the cellulose with cellulose nanofibers resulted in an amplified fluorescence. A schematic representation of the analytical test strip is shown in Figure 16.4c.

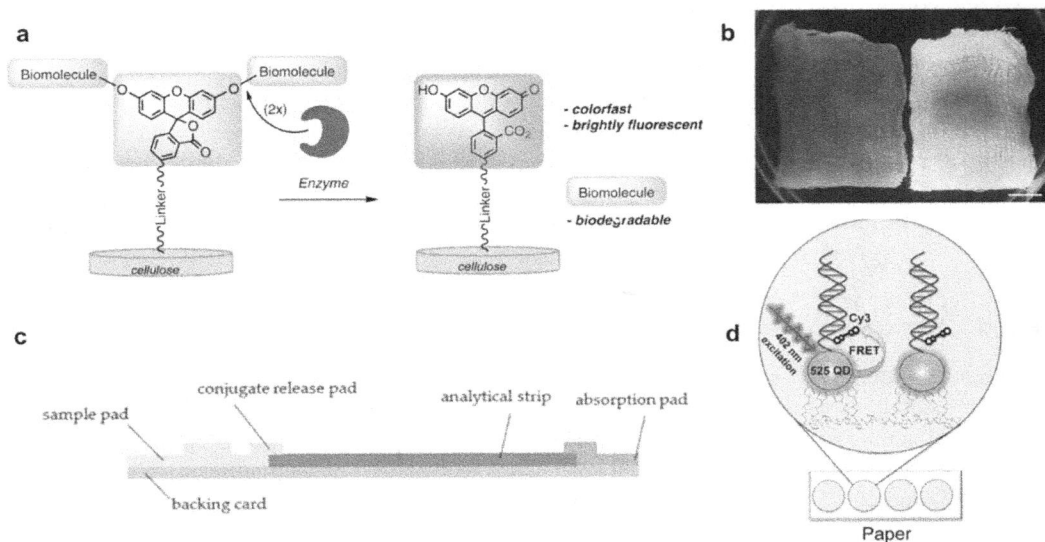

FIGURE 16.4 (a) Surface tethering of cellulose with deacetylated fluorescein, (b) a photograph depicting fluorescence emission change upon interaction with urea, mechanism of pH sensing. Reproduced with permission from Derikvand et al. (2016) © 2016 American Chemical Society. (c) Schematic representation of a lateral flow assay. Reproduced with permission from Natarajan et al. (2021). (d) Schematic representation of the FRET-based nucleic acid hybridization detecting paper strip. Reproduced with permission from Noor and Krull (2013) © 2012 American Chemical Society.

Chang and coworkers have developed a lateral flow assay using nitrocellulose membrane embedded with CdO QDs to detect α-fetoprotein, which acts as a tumor marker for primary hepatic carcinoma (Yang et al., 2011). This biosensor detected α-fetoprotein as low as $1\,\mathrm{ng\,mL^{-1}}$ with a false positive rate of 3.92% and a false negative rate of 4.38%.

Li and coworkers developed a DNA circuit to detect nucleic acids—a diagnostic tool for infectious and genetic diseases. The paper-strip-based immunogenic assay can detect specific single-stranded markers for pathogens and has a detection limit of ~300 nM (Allen et al., 2012). The type of analytes detected can be increased by combining the technique with amplification techniques such as loop-mediated isothermal amplification and enzyme-free catalytic hairpin assembly-amplified circuit. Krull and coworkers have developed a solid-phase assay based on paper for the transduction of nucleic acid hybridization (Noor & Krull, 2013). The surface of the cellulose paper was functionalized with QD–probe oligonucleotide conjugates bound to imidazole (Figure 16.4d). Cy3 acceptors were FRET-paired with the QDs, which gave rise to FRET-sensitized emission, measured as the output signal.

16.2.5 pH Sensing and Food Freshness Monitoring

Fluorescent sensors have attracted considerable attention toward sensing and monitoring pH. Zhang and coworkers have developed a cellulose-based bimodal pH-responsive probe (Phen-MDI-CA) that can exhibit colorimetric and fluorescence responses toward pH change (Nawaz et al., 2019). Phen-MDI-CA consists of pH-responsive phenanthroline chromophore moiety anchored on a cellulose chain with a urea linkage. Under acidic solid and basic buffer conditions, Phen-MDI-CA displays a drastic change in luminescence. As the pH changes from 11 to 14, colorless to yellow-orange color change and blue-green to yellow-red emission change were observed (Figure 16.5a). The shift in emission and a decrease in intensity under basic conditions

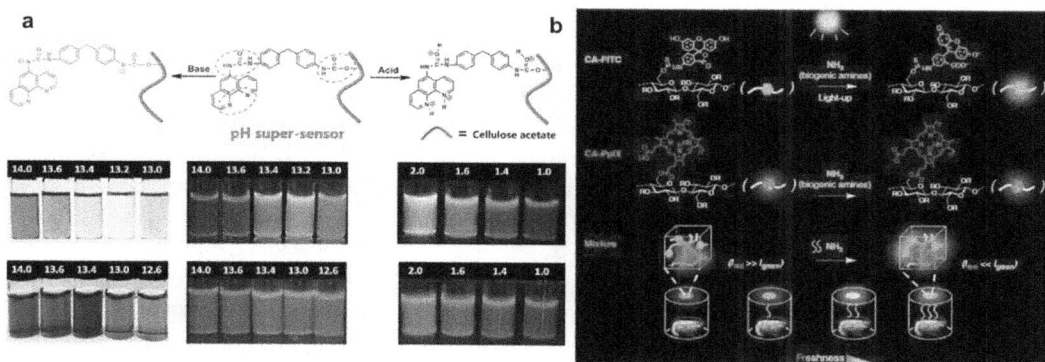

FIGURE 16.5 (a) Mechanism of pH sensing, photographs of fluorescence, and visual color change of Phen-MDI-CA under acidic and basic conditions. Reproduced with permission from Nawaz et al. (2019) © 2019 American Chemical Society. (b) Schematic representation of ratiometric approach for monitoring seafood freshness. Reproduced with permission from Jia et al. (2019b).

is due to the shift in intramolecular charge transfer characteristics with conjugation change upon deprotonation of the amino group of urea linkage. At the same time, under acidic conditions, the protonation of the amino group and carbonyl moiety disrupts the intramolecular charge transfer characteristics, resulting in a spectral shift. As a result, a color change from colorless to yellow in the visible region and blue to colorless under UV light was observed as the pH changed from 2.0 to 1.0. Interestingly, the luminescence shift can be modulated and amplified with the help of pH-irresponsive luminophores. Huang and coworkers developed a pH-responsive cellulose nanocrystal-based fluorescent sensor by covalent modification of cellulose nanocrystals with pH-responsive (5 (and 6)-carboxy-2′,7′-dichlorofluorescein dye using L-leucine amino acid linker (Tang et al., 2016). A turn-on response was obtained on contact with a buffer solution of different pH. On varying the pH from 2.2 to 13, a linear increment was observed in the emission intensity of the probe and pH up to a pH value of 10.8, and the turn-on response can be attributed to the formation of anions of (5 (and 6)-carboxy-2′,7′-dichlorofluorescein dye.

In foodstuffs such as milk, fish and fish products, processed meat, fermented foodstuffs, beverages, and so on, biogenic amines are formed by the degradation of amino acids caused by microbial activity in the food products. Hence biogenic amines can be used as markers for assessing the quality and freshness of foodstuffs during storage and transportation. Zhang and coworkers developed ratiometric cellulose-based amine-responsive fluorescence material for naked-eye detection and real-time monitoring of the freshness of seafood (Figure 16.5b) (Jia et al., 2019b). The design concept was simple and very interesting. Cellulose acetate was covalently linked with a biogenic amine indicator, fluorescein isothiocyanate, and an internal reference, protoporphyrin IX. Fluorescein isothiocyanate is green emissive while protoporphyrin IX is red-emitting, and the cellulose acetate anchoring unit prevents ACQ of both the fluorophores, thereby resulting in highly emissive cellulose acetate linked fluorophores. The ratiometric fluorescent probe was prepared by blending various ratios of cellulose acetate-fluorescein isothiocyanate and cellulose acetate-protoporphyrin IX. A red emissive fluorescent tag prepared by mixing cellulose acetate-protoporphyrin IX and cellulose acetate-fluorescein isothiocyanate in a 5:1 ratio was employed to monitor the freshness of shrimp and crab at room temperature; as time progressed, the red emission of the tag changed to green. The emission change can be attributed to the spoilage of shrimp and crab, as the spoilage releases biogenic amines, which increase the fluorescence intensity of cellulose acetate-fluorescein isothiocyanate. Since the concentration of protoporphyrin IX remains constant, a distinct emission change from red to green was observed.

16.3 POLYSACCHARIDE-BASED FLUORESCENT MATERIALS FOR ANTI-COUNTERFEITING

Counterfeiting has been a severe problem faced by governments and businesses around the globe over the last few decades in terms of scope and magnitude. Counterfeited products, especially currency, documents, food, and drugs, can cause considerable losses to the economy, company reputation, and consumer health. Although several laws have been implemented to prevent counterfeiting, many counterfeited products in terms of currency, food, medicine, apparel, and electronic fields are readily available in markets, especially in developing and underdeveloped countries (Abdollahi et al., 2020).

Numerous anti-counterfeiting strategies such as barcodes, holograms, watermarks, and two-dimensional codes are used to identify and prevent counterfeiting. However, complex identification procedures, high cost of analysis, and easy replication limit the utility of anti-counterfeiting materials in real applications (Arppe & Sørensen, 2017). Hence, anti-counterfeiting materials based on various fluorescent materials, such as organic dyes, CDs, and quantum dots, have been developed (Yu et al., 2021). The polysaccharide-based fluorescent smart materials have gained significant importance in security printing applications because of their cost-effectiveness, biodegradability, easy processability, and fast response time. In this section, we will discuss various polysaccharide-based anti-counterfeiting materials. We shall also discuss polysaccharide-based anti-counterfeiting materials exhibiting dynamic fluorescence under external stimuli such as light, heat, force, electricity, and chemicals to develop fascinating and chameleon-like visual effects resulting in hard-to-duplicate authentication strategies.

Zhang and coworkers developed a simple technique that utilized a cellulose-based bioplastic material for anti-counterfeiting (Wang et al., 2016). They successfully incorporated fluorescent dye molecules such as rhodamine and fluorescein into the cellulose material by using hydrogen bonding between the polar groups of dye molecules and the hydroxyl groups of cellulose material. The fluorescent bioplastic material thus synthesized via the hot-pressing process exhibited excellent mechanical properties with good thermal stability. Accordingly, the composite material could be easily incorporated into papers, making their anti-counterfeiting applicability more facile. Later, the group of Bian reported a similar fluorophore (thiazolopyridinecarboxylic acid) integrated cellulose system with enhanced fluorescence (Chen et al., 2017). They developed a one-pot synthetic methodology of mixing cellulose powder with citric acid and cystine and heating above 80°C, resulting in modified cellulose with fluorescence. The emitting material was thiazolopyridinecarboxylic acid, formed by a series of dehydration reactions between cysteine and citric acid. They printed the concentrated solution of citric acid and cysteine for the counterfeiting application. This paper on temperature treatment showed the written texts under 365 nm UV light illumination and invisible under daylight.

Another class of fluorescent materials with superior photoluminescent behavior is rare-earth ion-based nanomaterials. There are a few reports where cellulose materials are used as a base material to incorporate lanthanide ions and improve their emission properties. In this aspect, Wang, Hou, and coworkers have synthesized a new category of hydrogel materials based on cellulose (carboxy methyl cellulose (CMC)) that finds application in latent-fingerprint detection, decryption, and encryption. Integration of the terbium (III)-complex-functionalized CMC ((Tb (III)–carboxymethylcellulose) with a DPY 2,6-(dimethyl pyridine-4-amine)-functionalized CMC-binding aptamer leads to the formation of hydrogels with green fluorescence and excellent light stability. (Hai et al., 2018). This fluorescence could be quenched by adding ClO^- and reverts with SCN^-. The system retained its behavior in the solution state and the solid-state while incorporated into a filter paper. Highly magnified images of the latent fingerprints with minute details are extracted when the fingerprints are treated with the aforementioned solution (Figure 16.6).

Yu and coworkers have utilized the exceptional photoluminescent behavior of lanthanide nanocrystals by anchoring them to cellulose-based composites (Wang et al., 2018). For this purpose, polyvinylpyrrolidone was incorporated into cellulose fibers, and the resulting composite material

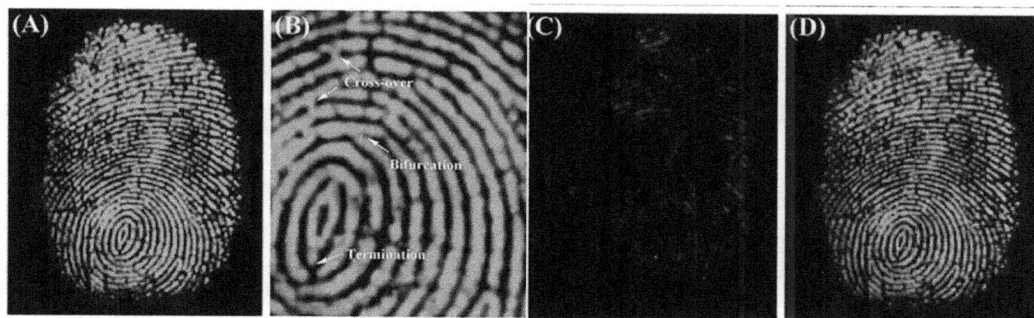

FIGURE 16.6   (a) Luminescence image of a late fingerprint (b) magnified image with details of Tb (III)–carboxymethyl cellulose hydrogel. Photographic depiction of the response of the hydrogel gel in (c) ClO⁻ and (d) SCN⁻. Reproduced with permission from Hai et al. (2018) 2018 Wiley-VCH Verlag GmbH & Co. KGaA, Weinheim.

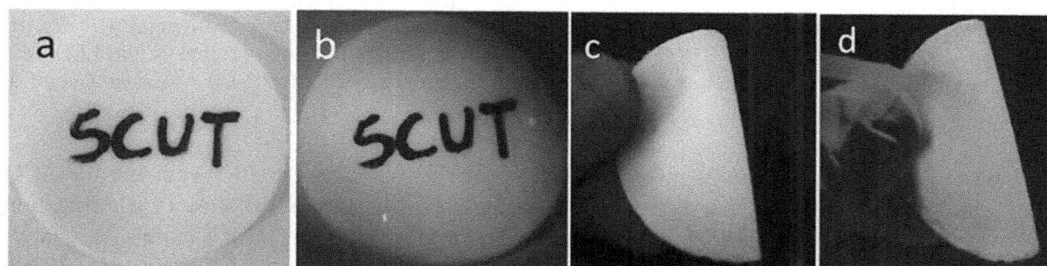

FIGURE 16.7   Photographs of a free-standing paper of bhpfibers-polyvinylpyrrolidone@LaF$_3$:Eu^{3+} complex with written letters (a) before and (b) after 365 nm UV light illumination. (c and d) Optical images demonstrating the flexibility of the papers thus made under normal and 365 nm UV light illumination. Reproduced with permission from Wang et al. (2018) © 2018 American Chemical Society.

was utilized to accommodate the lanthanum ions with enhanced stability. The highly polar −C-N- and C=O bonds in the polyvinylpyrrolidone made the lanthanum ion adhere firmly to the site. The bleached hardwood pulp cellulose fibers (bhp fibers)-polyvinylpyrrolidone@LaF$_3$:Eu^{3+} complex showed stable fluorescence and stability toward acid and alkali. This composite could quickly adsorb into paper and showed better flexibility, printable, and rewritable property (Figure 16.7).

Later, the same group reported the chemical modification of cellulose by 2,2,6,6-tetramethyl-1-piperidinyloxy (TEMPO), which allowed the enhanced adsorption of the LaF$_3$:RE^{3+} via electrostatic interaction (Wang et al., 2020). Incorporation of TEMPO-oxidized cellulose (TOC), LaF$_3$:RE^{3+} composite into TiO$_2$ (TOC@LaF$_3$:Eu^{3+}@TiO$_2$) resulted in a self-standing paper, which upon further modification with HDTMS (hexadecyltrimethoxysilane) gives the additional advantage of superhydrophobicity and self-cleaning ability. The coordination ability of the TEMPO-oxidized cellulose nanofibrils (tCNFs) was used by Zhang and coworkers through an adequately designed self-assembly event (Zhang et al., 2019). Herein, they co-assembled tCNFs with lanthanide complexes of *tris*(2-benzimidazolylmethyl)amine (NTB) ((Tb(NTB)Cl$_3$ and/or Eu(NTB)Cl$_3$). Blending and vacuum filtration of the nanocomposite gave papers with excellent fluorescent properties and unique translucence. This makes them a good candidate for the anti-counterfeiting application.

CDs are a unique set of fluorescent materials with a broad spectrum of applications owing to their low toxicity, resistance to photobleaching, biocompatibility, ease of preparation, and so on. However, the ACQ property of these materials in solid state often restricts their application in the material world. In this aspect, we will discuss a few examples where chemical modification

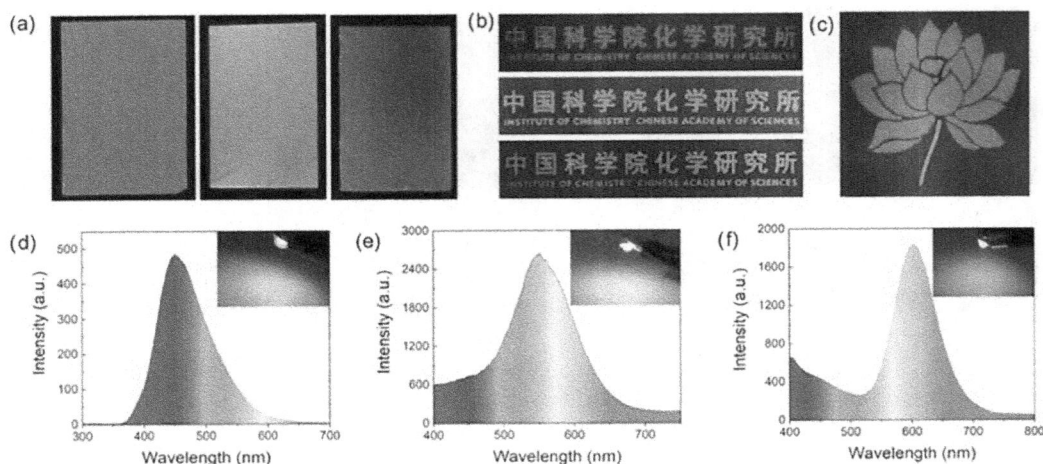

FIGURE 16.8 (a) Optical images of negatively charged CDs@CelN$^+$ (blue), FCDs@CMC(Green), pPD-CDs@CMC (Red) under UV irradiation (365 nm). (b and c) Screen-printed letters and images with the ink prepared with CD@CMCs under 356 nm UV light illumination. (d–f) Emission spectra of the LED with different color emissions. Reproduced with permission from Jin et al. (2020) © 2020 American Chemical Society.

of cellulose with CDs provides a protective layer over each CD and thereby overcome the ACQ effect. Recently, Zhang and coworkers have reported the modification of cellulose by esterification of the acid group with (3-carboxypropyl)trimethylammonium chloride, which leaves the quaternized amine, which can electrostatically interact with the negatively charged CDs (Jin et al., 2020). CDs@cellulose showed enhanced fluorescence and quantum yield compared with the free negatively charged CDs. On a comparable basis, they have extended the work with positively charged carbon dots (FCDs (prepared from 4-fluoro-o-phenylenediamine), pPDs (prepared from p-phenylenediamine)) and negatively charged carboxymethyl cellulose, leading to different fluorescence emissions (Figure 16.8). They also demonstrated the processability of CDs@cellulose systems, which can be readily converted into ink, film, or coated into other substrates.

Later, Jin and coworkers reported a similar system where they synthesized a positively charged cellulose derivative by attaching the imidazole salt group (Im$^+$) and spiropyran (SP) (C-Im$^+$-SP) (Jin et al., 2021). Negatively charged carbon dots (NCD$_S$) were introduced into the above system via electrostatic interaction, with bright fluorescence in the solid-state, otherwise unattainable for NCDs due to ACQ behavior. The proximity of NCDs and SP in the system made an excellent FRET pair, bringing the system additional dynamic behavior. A few CDs get excited on UV light irradiation and emit blue light. However, a few CDs transfer their excited state energy to the nearby SP, which leads to red emission. The ring-opening process of SP to merocyanine occurs in a finite time compared to the relatively faster FRET process, which provides the system with fluorescence depending on the irradiation time. This process could be reversed by visible light irradiation, causing the merocyanine ring-closing phenomenon (Figure 16.9). In addition, UV irradiation causes fluorescence to shift from blue to red, and their intensity ratio could be modulated by varying the composition of NCDs in the system. All these peculiarities make them an excellent candidate for anti-counterfeiting pattering. These composite NCD$_S$@C-Im$^+$-SP can be readily coated into the substrate or act as inks in photo-rewritable papers and anti-counterfeit patterns.

Guttena and coworkers have reported a facile conversion of cigarette butts into CDs with appreciably high quantum yield (Bandi et al., 2018). Extraction of cellulose acetate from cigarette butts followed by one-step hydrothermal methods is observed to synthesize nitrogen and sulfur co-doped CDs. The appreciably high quantum yield (26%) obtained for these CDs made them a

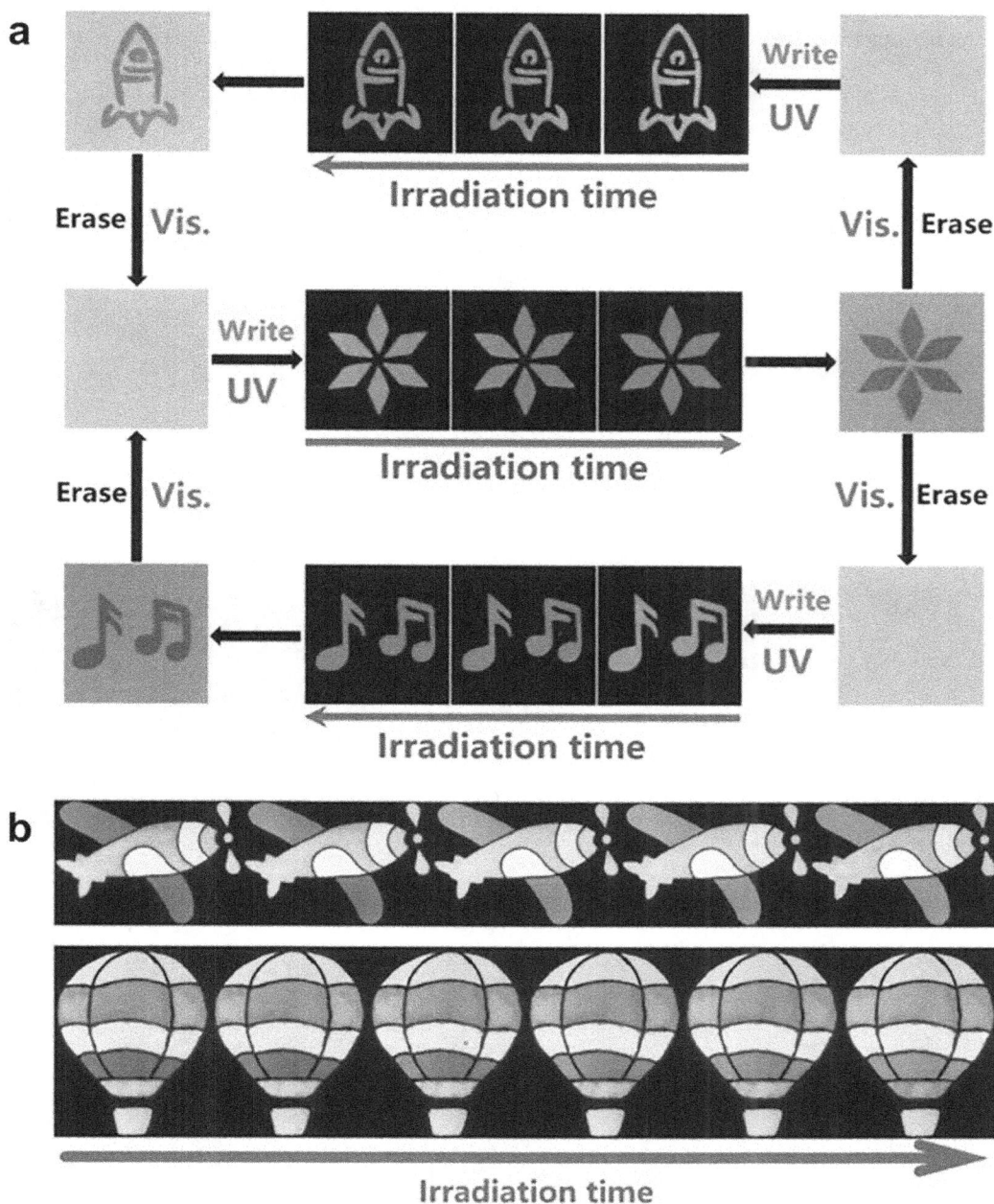

FIGURE 16.9 (a) Photo-rewritable paper and (b) dynamic multi-color fluorescence patterns developed from NCD$_S$@C-Im$^+$-SP. Reproduced with permission from Jin et al. (2021) © 2020 Elsevier BV.

good candidate for security ink-related applications. The inherent behavior of cellulose to undergo photo-oxidation and photodegradation upon UV light irradiation leads to fluorescence. Su, Zheng, and coworkers cleverly explored this property for cellulose-based photolithography for anti-coun-terfeiting applications (Cheng et al., 2021). Cellulose nanofibers extracted from the garlic husk were made into papers with a thickness of 20 mm and excellent mechanical stability. UV irradiation with a pre-defined photomask made the required impression on the paper. Hence, this relatively simple technique developed a renewable cellulose nanofiber that can act as both inks and substrates. Moreover, the information encrypted via UV irradiation was stable under acidic or alkaline conditions, which increased their applicability in security materials.

Dynamic fluorescent materials are considered next-generation materials for anti-counterfeiting applications compared to static fluorescent materials as they are hard to duplicate. Zhang and coworkers extended the possibility of FRET for developing dynamically controlled fluorescent materials (Tian et al., 2018). They have used three chromophores, spiropyran, fluorescein, and pyrene, coupled to cellulose. A simple blending of these three materials resulted in FRET between the chromophore, and the resulting blend showed excellent processability. In addition, they achieved almost full-color fluorescence by controlling the ratio of chromophores. They also prepared a fluorescent QR code that could show the gradual fluorescence change accompanied by visible light changes (Figure 16.10).

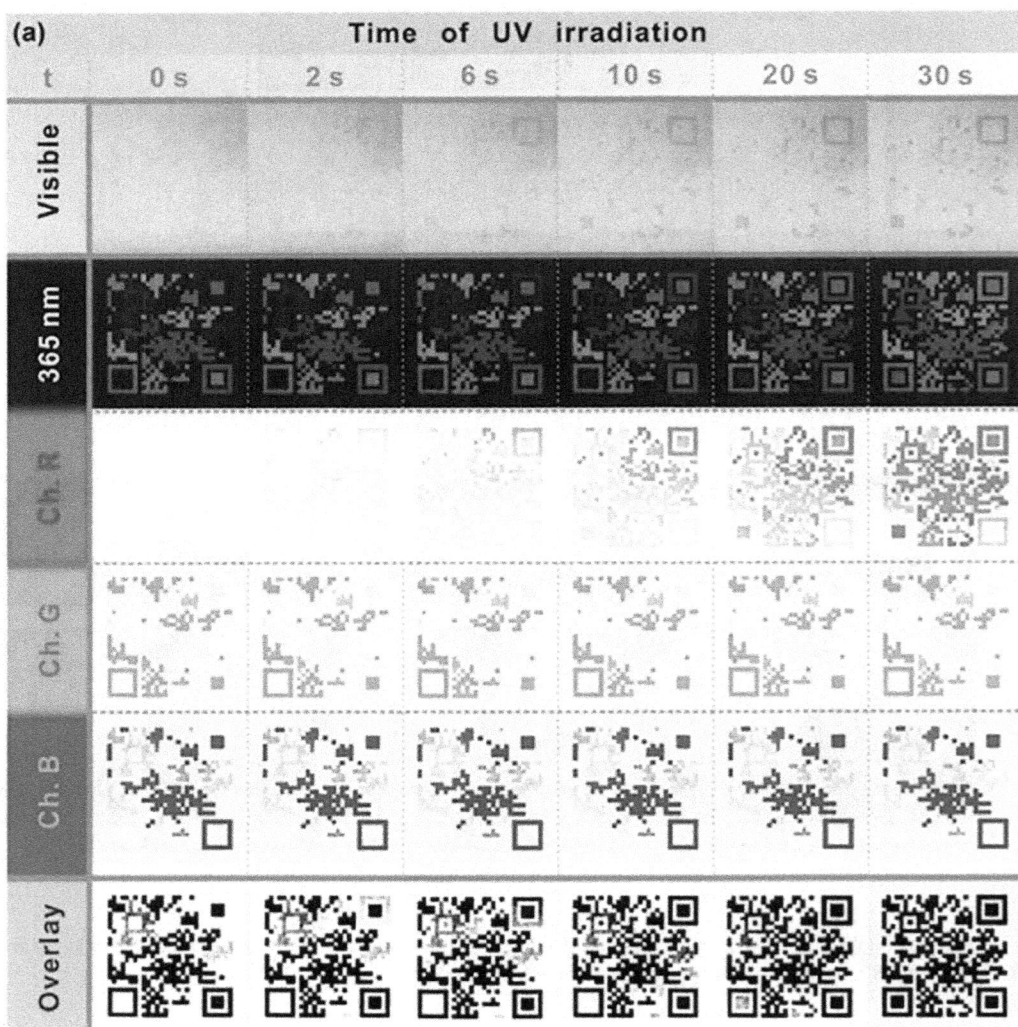

FIGURE 16.10 Dynamic changes observed in the QR code generated using cellulose-based trichromatic (spiropyran-red, fluorescein-green, and pyrene-blue) materials on UV irradiation (365 nm, 200 µW cm^{-2}) time: the images shown are those under visible light, fluorescent images under red channel (R), green channel (G), and blue channel (B) of the fluorescent images, the overlay pattern of the RGB channel, respectively, are shown from top to bottom. Reproduced with permission from Tian et al. (2018) © 2017 WILEY-VCH Verlag GmbH & Co. KGaA, Weinheim.

16.4   CONCLUSIONS AND FUTURE PERSPECTIVES

This chapter summarizes various polysaccharide-based fluorescent materials for sensing and security applications in this chapter. Such bio-based materials are advantageous over instrumental and conventional sensing techniques due to their low toxicity, economical, easy processability, biocompatibility, and fast response time. Enhanced interest in developing such polysaccharide-based fluorescent materials can result in better and greener synthetic strategies for developing such materials. The demonstrated properties of these materials can be improved to obtain materials with better selectivity, sensitivity, photostability, efficiency, reproducibility, and zero toxicity with minimal cost of production. The development of color-tunable stimuli-responsive multi-emissive (fluorescence/phosphorescence) polysaccharide-based anti-counterfeiting solutions as the next-generation anti-counterfeiting materials will be a shot in the arm for the fight against counterfeiting, as duplicating such materials will be near impossible.

ACKNOWLEDGMENTS

AP acknowledges the University Grants Commission (UGC), Government of India for a research fellowship. DE is grateful to the Council of Scientific and Industrial Research (CSIR), Government of India for a research fellowship. VKP is grateful to the CSIR, Government of India and Department of Science and Technology (DST), Nano Mission, Government of India (DST/NM/TUE/EE-02/2019), for the financial support.

REFERENCES

Abdollahi, A., Roghani-Mamaqani, H., Razavi, B., & Salami-Kalajahi, M. (2020). Photoluminescent and chromic nanomaterials for anticounterfeiting technologies: recent advances and future challenges. *ACS Nano, 14*, 14417–14492. https://doi.org/10.1021/acsnano.0c07289

Allen, P. B., Arshad, S. A., Li, B., Chen, X., & Ellington, A. D. (2012). DNA circuits as amplifiers for the detection of nucleic acids on a paperfluidic platform. *Lab on a Chip, 12*, 2951–2958. https://doi.org/10.1039/c2lc40373k

Arppe, R., & Sørensen, T. J. (2017). Physical unclonable functions generated through chemical methods for anti-counterfeiting. *Nature Reviews Chemistry, 1*, 0031. https://doi.org/10.1038/s41570-017-0031

Bandi, R., Devulapalli, N. P., Dadigala, R., Gangapuram, B. R., & Guttena, V. (2018). Facile conversion of toxic cigarette butts to N,S-codoped carbon dots and their application in fluorescent film, security ink, bioimaging, sensing and logic gate operation. *ACS Omega, 3*, 13454–13466. https://doi.org/10.1021/acsomega.8b01743

Basabe-Desmonts, L., Reinhoudt, D. N., & Crego-Calama, M. (2007). Design of fluorescent materials for chemical sensing. *Chemical Society Reviews, 36*, 993–1017. https://doi.org/10.1039/b609548h

Chen, H., Yan, X., Feng, Q., Zhao, P., Xu, X., Ng, D. H. L., & Bian, L. (2017). Citric acid/cysteine-modified cellulose-based materials: green preparation and their applications in anticounterfeiting, chemical sensing, and UV shielding. *ACS Sustainable Chemistry & Engineering, 5*, 11387–11394. https://doi.org/10.1021/acssuschemeng.7b02473

Cheng, H., Wei, X., Qiu, H., Wang, W., Su, W., & Zheng, Y. (2021). Chemically stable fluorescent anti-counterfeiting labels achieved by UV-induced photolysis of nanocellulose. *RSC Advances, 11*, 18381–18386. https://doi.org/10.1039/D1RA02089G

Derikvand, F., Yin, D. T., Barrett, R., & Brumer, H. (2016). Cellulose-based biosensors for esterase detection. *Analytical Chemistry, 88*, 2989–2993. https://doi.org/10.1021/acs.analchem.5b04661

Dinh Duong, H., & Il Rhee, J. (2015). Development of a ratiometric fluorescent urea biosensor based on the urease immobilized onto the oxazine 170 perchlorate-ethyl cellulose membrane. *Talanta, 134*, 333–339. https://doi.org/10.1016/j.talanta.2014.10.064

Fu, L.-H., Qi, C., Ma, M.-G., & Wan, P. (2019). Multifunctional cellulose-based hydrogels for biomedical applications. *Journal of Materials Chemistry B, 7*, 1541–1562. https://doi.org/10.1039/C8TB02331J

Hai, J., Li, T., Su, J., Liu, W., Ju, Y., Wang, B., & Hou, Y. (2018). Reversible response of luminescent terbium(III)-nanocellulose hydrogels to anions for latent fingerprint detection and encryption. *Angewandte Chemie International Edition, 57*, 6786–6790. https://doi.org/10.1002/anie.201800119

Hu, H., Wang, F., Yu, L., Sugimura, K., Zhou, J., & Nishio, Y. (2018). Synthesis of novel fluorescent cellulose derivatives and their applications in detection of nitroaromatic compounds. *ACS Sustainable Chemistry & Engineering*, *6*, 1436–1445. https://doi.org/10.1021/acssuschemeng.7b03855

Hu, J., Wang, S. Q., Wang, L., Li, F., Pingguan-Murphy, B., Lu, T. J., & Xu, F. (2014). Advances in paper-based point-of-care diagnostics. *Biosensors and Bioelectronics*, *54*, 585–597. https://doi.org/10.1016/j.bios.2013.10.075

Jia, R., Tian, W., Bai, H., Zhang, J., Wang, S., & Zhang, J. (2019a). Sunlight-driven wearable and robust antibacterial coatings with water-soluble cellulose-based photosensitizers. *Advanced Healthcare Materials*, *8*, 1801591. https://doi.org/10.1002/adhm.201801591

Jia, R., Tian, W., Bai, H., Zhang, J., Wang, S., & Zhang, J. (2019b). Amine-responsive cellulose-based ratiometric fluorescent materials for real-time and visual detection of shrimp and crab freshness. *Nature Communications*, *10*, 795. https://doi.org/10.1038/s41467-019-08675-3

Jin, K., Ji, X., Yang, T., Zhang, J., Tian, W., Yu, J., Zhang, X., Chen, Z., & Zhang, J. (2021). Facile access to photo-switchable, dynamic-optical, multi-colored and solid-state materials from carbon dots and cellulose for photo-rewritable paper and advanced anti-counterfeiting. *Chemical Engineering Journal*, *406*, 126794. https://doi.org/10.1016/j.cej.2020.126794

Jin, K., Zhang, J., Tian, W., Ji, X., Yu, J., & Zhang, J. (2020). facile access to solid-state carbon dots with high luminescence efficiency and excellent formability via cellulose derivative coatings. *ACS Sustainable Chemistry & Engineering*, *8*, 5937–5945. https://doi.org/10.1021/acssuschemeng.0c00237

Kacmaz, S., Ertekin, K., Mercan, D., Oter, O., Cetinkaya, E., & Celik, E. (2015). An ultra sensitive fluorescent nanosensor for detection of ionic copper. *Spectrochimica Acta Part A: Molecular and Biomolecular Spectroscopy*, *135*, 551–559. https://doi.org/10.1016/j.saa.2014.07.056

Li, M., An, X., Jiang, M., Li, S., Liu, S., Chen, Z., & Xiao, H. (2019). "Cellulose spacer" strategy: anti-aggregation-caused quenching membrane for mercury ion detection and removal. *ACS Sustainable Chemistry & Engineering*, *7*, 15182–15189. https://doi.org/10.1021/acssuschemeng.9b01928

Li, M., Li, X., Xiao, H.-N., & James, T. D. (2017). Fluorescence sensing with cellulose-based materials. *ChemistryOpen*, *6*, 685–696. https://doi.org/10.1002/open.201700133

Li, Y., Xiao, H., Chen, M., Song, Z., & Zhao, Y. (2014). Absorbents based on maleic anhydride-modified cellulose fibers/diatomite for dye removal. *Journal of Materials Science*, *49*, 6696–6704. https://doi.org/10.1007/s10853-014-8270-8

Lu, W., Zhang, J., Huang, Y., Théato, P., Huang, Q., & Chen, T. (2017). Self-diffusion driven ultrafast detection of ppm-level nitroaromatic pollutants in aqueous media using a hydrophilic fluorescent paper sensor. *ACS Applied Materials & Interfaces*, *9*, 23884–23893. https://doi.org/10.1021/acsami.7b08826

Mansfield, M. A. (2009). Lateral Flow Immunoassay. In R. Wong & H. Tse (Eds.), *Medical Mycology Journal* (Vol. 57). Humana Press. https://doi.org/10.1007/978-1-59745-240-3

Mobarak, N. N., Jumaah, F. N., Ghani, M. A., Abdullah, M. P., & Ahmad, A. (2015). Carboxymethyl carrageenan based biopolymer electrolytes. *Electrochimica Acta*, *175*, 224–231. https://doi.org/10.1016/j.electacta.2015.02.200

Mozammil Hasnain, S. M., Hasnain, M. S., & Nayak, A. K. (2019). Natural Polysaccharides. In M. S. Hasnain & A. K. Nayak (Eds.), *Natural Polysaccharides in Drug Delivery and Biomedical Applications* (pp. 1–14). Elsevier. https://doi.org/10.1016/B978-0-12-817055-7.00001-7

Natarajan, S., Jayaraj, J., & Prazeres, D. M. F. (2021). A cellulose paper-based fluorescent lateral flow immunoassay for the quantitative detection of cardiac troponin I. *Biosensors*, *11*, 1–12. https://doi.org/10.3390/bios11020049

Nawaz, H., Tian, W., Zhang, J., Jia, R., Chen, Z., & Zhang, J. (2018). Cellulose-based sensor containing phenanthroline for the highly selective and rapid detection of Fe 2+ ions with naked eye and fluorescent dual modes. *ACS Applied Materials & Interfaces*, *10*, 2114–2121. https://doi.org/10.1021/acsami.7b17342

Nawaz, H., Tian, W., Zhang, J., Jia, R., Yang, T., Yu, J., & Zhang, J. (2019). Visual and precise detection of pH values under extreme acidic and strong basic environments by cellulose-based superior sensor. *Analytical Chemistry*, *91*, 3085–3092. https://doi.org/10.1021/acs.analchem.8b05554

Nawaz, H., Zhang, X., Chen, S., You, T., & Xu, F. (2021). Recent studies on cellulose-based fluorescent smart materials and their applications: a comprehensive review. *Carbohydrate Polymers*, *267*, 118135. https://doi.org/10.1016/j.carbpol.2021.118135

Nešić, A., Cabrera-Barjas, G., Dimitrijević-Branković, S., Davidović, S., Radovanović, N., & Delattre, C. (2019). Prospect of polysaccharide-based materials as advanced food packaging. *Molecules*, *25*, 135. https://doi.org/10.3390/molecules25010135

Noor, M. O., & Krull, U. J. (2013). Paper-based solid-phase multiplexed nucleic acid hybridization assay with tunable dynamic range using immobilized quantum dots as donors in fluorescence resonance energy transfer. *Analytical Chemistry*, *85*, 7502–7511. https://doi.org/10.1021/ac401471n

Pal, K., Sarkar, P., Anis, A., Wiszumirska, K., & Jarzębski, M. (2021). Polysaccharide-based nanocomposites for food packaging applications. *Materials*, *14*, 5549. https://doi.org/10.3390/ma14195549

Peng, N., Wang, Y., Ye, Q., Liang, L., An, Y., Li, Q., & Chang, C. (2016). Biocompatible cellulose-based superabsorbent hydrogels with antimicrobial activity. *Carbohydrate Polymers*, *137*, 59–64. https://doi.org/10.1016/j.carbpol.2015.10.057

Shariatinia, Z. (2019). Pharmaceutical Applications of Natural Polysaccharides. In M. S. Hasnain & A. K. Nayak (Eds.), *Natural Polysaccharides in Drug Delivery and Biomedical Applications* (pp. 15–57). Elsevier. https://doi.org/10.1016/B978-0-12-817055-7.00002-9

Smith, C. J., Wagle, D. V., O'Neill, H. M., Evans, B. R., Baker, S. N., & Baker, G. A. (2017). Bacterial cellulose ionogels as chemosensory supports. *ACS. Applied Materials & Interfaces*, *9*, 38042–38051. https://doi.org/10.1021/acsami.7b12543

Song, Y., Li, Y., Liu, Z., Liu, L., Wang, X., Su, X., & Ma, Q. (2014). A novel ultrasensitive carboxymethyl chitosan-quantum dot-based fluorescence "turn on-off" nanosensor for lysozyme detection. *Biosensors and Bioelectronics*, *61*, 9–13. https://doi.org/10.1016/j.bios.2014.04.036

Tang, L., Li, T., Zhuang, S., Lu, Q., Li, P., & Huang, B. (2016). Synthesis of pH-sensitive fluorescein grafted cellulose nanocrystals with an amino acid spacer. *ACS Sustainable Chemistry & Engineering*, *4*, 4842–4849. https://doi.org/10.1021/acssuschemeng.6b01124

Tian, W., Zhang, J., Yu, J., Wu, J., Nawaz, H., Zhang, J., He, J., & Wang, F. (2016). Cellulose-based solid fluorescent materials. *Advanced Optical Materials*, *4*, 2044–2050. https://doi.org/10.1002/adom.201600500

Tian, W., Zhang, J., Yu, J., Wu, J., Zhang, J., He, J., & Wang, F. (2018). Phototunable full-color emission of cellulose-based dynamic fluorescent materials. *Advanced Functional Materials*, *28*, 1703548. https://doi.org/10.1002/adfm.201703548

Wang, H., Vagin, S. I., Rieger, B., & Meldrum, A. (2020). An ultrasensitive fluorescent paper-based CO2 sensor. *ACS Applied Materials & Interfaces*, *12*, 20507–20513. https://doi.org/10.1021/acsami.0c03405

Wang, M., Li, R., Feng, X., Dang, C., Dai, F., Yin, X., He, M., Liu, D., & Qi, H. (2020). Cellulose nanofiber-reinforced ionic conductors for multifunctional sensors and devices. *ACS Applied Materials & Interfaces*, *12*, 27545–27554. https://doi.org/10.1021/acsami.0c04907

Wang, Qing, Chen, G., Yu, Z., Ouyang, X., Tian, J., & Yu, M. (2018). Photoluminescent composites of lanthanide-based nanocrystal-functionalized cellulose fibers for anticounterfeiting applications. *ACS Sustainable Chemistry & Engineering*, *6*, 13960–13967. https://doi.org/10.1021/acssuschemeng.8b02307

Wang, Qing, Xie, D., Chen, J., Liu, G., & Yu, M. (2020). Facile fabrication of superhydrophobic and photoluminescent TEMPO-oxidized cellulose-based paper for anticounterfeiting application. *ACS Sustainable Chemistry & Engineering*, *8*, 13176–13184. https://doi.org/10.1021/acssuschemeng.0c01559

Wang, Qiyang, Cai, J., Chen, K., Liu, X., & Zhang, L. (2016). Construction of fluorescent cellulose biobased plastics and their potential application in anti-counterfeiting banknotes. *Macromolecular Materials and Engineering*, *301*, 377–382. https://doi.org/10.1002/mame.201500364

Wang, Y., Liang, Z., Su, Z., Zhang, K., Ren, J., Sun, R., & Wang, X. (2018). All-biomass fluorescent hydrogels based on biomass carbon dots and alginate/nanocellulose for biosensing. *ACS Applied Bio Materials*, *1*, 1398–1407. https://doi.org/10.1021/acsabm.8b00348

Wu, S., Wang, W., Yan, K., Ding, F., Shi, X., Deng, H., & Du, Y. (2018). Electrochemical writing on edible polysaccharide films for intelligent food packaging. *Carbohydrate Polymers*, *186*, 236–242. https://doi.org/10.1016/j.carbpol.2018.01.058

Xu, L. Q., Neoh, K.-G., Kang, E.-T., & Fu, G. D. (2013). Rhodamine derivative-modified filter papers for colorimetric and fluorescent detection of Hg2+ in aqueous media. *Journal of Materials Chemistry A*, *1*, 2526. https://doi.org/10.1039/c2ta01072k

Yang, Q., Gong, X., Song, T., Yang, J., Zhu, S., Li, Y., Cui, Y., Li, Y., Zhang, B., & Chang, J. (2011). Quantum dot-based immunochromatography test strip for rapid, quantitative and sensitive detection of alpha fetoprotein. *Biosensors and Bioelectronics*, *30*, 145–150. https://doi.org/10.1016/j.bios.2011.09.002

Yetisen, A. K., Akram, M. S., & Lowe, C. R. (2013). Paper-based microfluidic point-of-care diagnostic devices. *Lab on a Chip*, *13*, 2210–2251. https://doi.org/10.1039/c3lc50169h

Yu, X., Zhang, H., & Yu, J. (2021). Luminescence anti-counterfeiting: from elementary to advanced. *Aggregate*, *2*, 20–34. https://doi.org/10.1002/agt2.15

Zhang, S., Liu, G., Chang, H., Li, X., & Zhang, Z. (2019). Optical haze nanopaper enhanced ultraviolet harvesting for direct soft-fluorescent emission based on lanthanide complex assembly and oxidized cellulose nanofibrils. *ACS Sustainable Chemistry & Engineering*, *7*, 9966–9975. https://doi.org/10.1021/acssuschemeng.9b00970

Zhao, D., Zhu, Y., Cheng, W., Chen, W., Wu, Y., & Yu, H. (2021). Cellulose-based flexible functional materials for emerging intelligent electronics. *Advanced Materials*, *33*, 2000619. https://doi.org/10.1002/adma.202000619

Zheng, Y., Monty, J., & Linhardt, R. J. (2015). Polysaccharide-based nanocomposites and their applications. *Carbohydrate Research*, *405*, 23–32. https://doi.org/10.1016/j.carres.2014.07.016

Zong, A., Cao, H., & Wang, F. (2012). Anticancer polysaccharides from natural resources: a review of recent research. *Carbohydrate Polymers*, *90*, 1395–1410. https://doi.org/10.1016/j.carbpol.2012.07.026

Index

Note: **Bold** page numbers refer to tables and *italic* page numbers refer to figures.

For Product Safety Concerns and Information please contact our EU
representative GPSR@taylorandfrancis.com
Taylor & Francis Verlag GmbH, Kaufingerstraße 24, 80331 München, Germany

www.ingramcontent.com/pod-product-compliance
Lightning Source LLC
Chambersburg PA
CBHW080919220326
41598CB00034B/5622